Advances in Biochemistry in Health and Disease

Volume 36

Advances in Biochemistry in Health and Disease focus on the latest developments in biochemical research with implications for health and disease. This book series consists of original edited volumes and monographs, presented by leading experts in the field and provides an up to date and unique source of information for all those interested in the fundamental, biochemical processes of the latest and emerging topics and techniques.

Covering a wide variety of topics, this book series is a valuable source of information from those at the lab bench through to the Health Care workers.

Naranjan S. Dhalla • Paramjit S. Tappia
Anureet K. Shah
Editors

Functional Biochemistry of Metallic Elements

Springer

Editors
Naranjan S. Dhalla
Institute of Cardiovascular Sciences
St. Boniface Hospital Albrecthsen
Research Centre
Winnipeg, MB, Canada

Paramjit S. Tappia
Asper Clinical Research Institute
and St. Boniface Hospital Albrechtsen
Research Centre
Winnipeg, MB, Canada

Anureet K. Shah
Department of Nutrition and Food Science
California State University
Los Angeles, CA, USA

ISSN 2512-2142 ISSN 2512-2150 (electronic)
Advances in Biochemistry in Health and Disease
ISBN 978-3-032-14444-7 ISBN 978-3-032-14445-4 (eBook)
https://doi.org/10.1007/978-3-032-14445-4

This Springer imprint is published by the registered company Springer Nature Switzerland AG
The registered company address is: Gewerbestrasse 11, 6330 Cham, Switzerland

Preface

Metallic elements can be broadly categorized as trace minerals, toxic trace elements, and major minerals. Essential trace elements are those that are required in small amounts for normal physiological and cellular biochemical processes; however, their deficiency and toxicity can cause negative health effects. In this book, we have put together the views of several experts, from around the world, in the field of human health and disease with specialty in understanding the effects of different metallic elements on the biochemical processes that impact cellular function. Multidisciplinary approaches have been described to cover the biochemical, physiological, and nutritional aspects of metallic elements. This book highlights the multifaceted functionality of the metallic elements to provide a compendium of information from basic fundamental knowledge to recommendations for their use as adjunct therapy in clinical practice.

There are 21 chapters, which have been put together in the form of two sections, namely: (i) Trace Elements in Health and Disease; (ii) Heavy Metals and Major Minerals. Each chapter in the book advances the understanding of the effects of these minerals on cellular, biochemical, and molecular processes in relation to health and disease, which hopefully will stimulate and motivate biomedical researchers and scientists to explore the relationship between cations and micronutrients and cell function. This book is considered to serve as a highly useful resource for health professionals, nutritional scientists, medical students, fellows, residents, and graduate students.

This book has been assembled to focus on the functional biochemistry of specific metallic elements in different cell systems in relation to human health and disease. The intent of this volume is to provide current and basic understanding of the cellular functionality regulated by different metals. It is hoped that readers will gain knowledge and understanding of the importance of metallic minerals and trace elements. The new and novel insights provided by each of the contributing authors will facilitate the understanding as well as the advancements on the role of trace minerals in the prevention of different diseases and pathophysiological conditions that are globally relevant.

Each state-of-the-art review on the biochemistry of metallic elements illustrates their impact on cellular function as well as the outcomes in health and disease. The dynamic interplay between micronutrients and human health is a burgeoning field of study in health and disease and is an active area of research that will gain further momentum. This view is substantiated by the fact that micronutrients are intimately involved in the normal cell function as well as pathogenesis of different diseases that can arise as a consequence of their deficiencies or toxicity. Thus, this volume underscores the significance of the trace elements in preventive medicine and therapeutic interventions.

We are appreciative of the members of the Advisory Board for this series on "Advances in Biochemistry in Health and Disease" for their guidance and important suggestions. We would also like to express our gratitude for the outstanding contributors as well as to Dr. Alexis Rivas for his enthusiastic and passionate efforts and his time and commitment in evaluating and approving this project. Furthermore, we are highly thankful to Bibhuti Bhusan Sharma and his team for the excellent assistance and professionalism in the production of this book. We would also like to thank the St. Boniface Hospital Albrecthsen Research Centre for the infrastructural support for this project.

Winnipeg, MB, Canada — Naranjan S. Dhalla
Winnipeg, MB, Canada — Paramjit S. Tappia
Los Angeles, CA, USA — Anureet K. Shah

Contents

Contributors

Ernest A. Adeghate Department of Anatomy, College of Medicine and Health Sciences, United Arab Emirates University, Al Ain, United Arab Emirates

Zayed Bin Sultan Foundation for Health Sciences, United Arab Emirates University, Al Ain, United Arab Emirates

Mahra H. Alderei Department of Physiology, College of Medicine and Health Sciences, United Arab Emirates University, Al Ain, United Arab Emirates

Dana H. Aldhaheri Department of Physiology, College of Medicine and Health Sciences, United Arab Emirates University, Al Ain, United Arab Emirates

Robby Alkhusairy Department of Integrative Physiology, Tottori University Graduate School and Faculty of Medicine, Tottori University, Yonago, Japan

Siddhi A. Awate Shobhaben Pratapbhai Patel School of Pharmacy & Technology Management, SVKM's Narsee Monjee Institute of Management Studies (NMIMS) Deemed-to-be University, Mumbai, Maharashtra, India

Shamsa M. Baniyas Department of Physiology, College of Medicine & Health Sciences, United Arab Emirates University, Al Ain, UAE

Ambika Binesh Department of Basic Sciences, Institute of Fisheries Post Graduate Studies, Tamil Nadu Dr. J. Jayalalithaa Fisheries University (TNJFU), Chennai, Tamil Nadu, India

Mayara Bocchi Department of Medicine, Federal University of Jataí, Jataí, Goiás, Brazil

Harpal S. Buttar Department of Pathology and Laboratory Medicine, Faculty of Medicine, University of Ottawa, Ottawa, ON, Canada

Mohini Chaurasia Era College of Pharmacy, Era University, Lucknow, India

David Michel de Oliveira Department of Physical Education, Federal University of Jataí, Jataí, Goiás, Brazil

Postgraduate Program in Bioscience and One Health, Federal University of Jataí, Jataí, Goiás, Brazil

Fábio Morato de Oliveira Department of Medicine, Federal University of Jataí, Jataí, Goiás, Brazil

Allison Shaw Devine Division of Cardiology, Department of Internal Medicine, University of Arkansas for Medical Sciences, Little Rock, AR, USA

Crystal M. D'Souza Department of Anatomy, College of Medicine and Health Sciences, United Arab Emirates University, Al Ain, United Arab Emirates

Axel Eickhoff Department of Internal Medicine II, Division of Gastroenterology and Hepatology, Klinikum Hanau, Klinikum Hanau, Hanau, Germany

Academic Teaching Hospital of the Medical Faculty, Goethe University Frankfurt/Main, Frankfurt/Main, Germany

Ghizal Fatima Department of Biotechnology, Chronobiology Laboratory, Era University, Lucknow, India

Jan Fedacko Department of Gediatrics, PJ Safaric University, Kosice, Slovakia

Eduardo Vignoto Fernandes Postgraduate Program in Bioscience and One Health, Federal University of Jataí, Jataí, Goiás, Brazil

Human and Comparative Anatomy Laboratory, Federal University of Jataí, Jataí, Goiás, Brazil

S. J. S. Flora Era College of Pharmacy, Era University, Lucknow, India

Keshav Garg Department of Internal Medicine, University of Arkansas for Medical Sciences, Little Rock, AR, USA

Elle Garriock Asper Clinical Research Institute and St. Boniface Hospital Albrechtsen Research Centre, Winnipeg, MB, Canada

Nitesh Gautam Division of Cardiology, Department of Internal Medicine, University of Arkansas for Medical Sciences, Little Rock, AR, USA

Shawn D. George Asper Clinical Research Institute and St. Boniface Hospital Albrechtsen Research Centre, Winnipeg, MB, Canada

Ramesh K. Goyal Delhi Pharmaceutical Sciences and Research University, New Delhi, India

Nandini Gupta National Forensic Sciences University, Gandhinagar, India

Mana Hatakeyama Department of Integrative Physiology, Tottori University Graduate School and Faculty of Medicine, Tottori University, Yonago, Japan

Takeshi Y. Hiyama Department of Integrative Physiology, Tottori University Graduate School and Faculty of Medicine, Tottori University, Yonago, Japan

International Platform for Dryland Research and Education, Tottori University, Tottori, Japan

Frank Christopher Howarth Department of Physiology, College of Medicine & Health Sciences, United Arab Emirates University, Al Ain, United Arab Emirates

Sisira Joy Department of Basic Sciences, Institute of Fisheries Post Graduate Studies, Tamil Nadu Dr. J. Jayalalithaa Fisheries University (TNJFU), Chennai, Tamil Nadu, India

Pandi Kalaiselvan Department of Basic Sciences, Institute of Fisheries Post Graduate Studies, Tamil Nadu Dr. J. Jayalalithaa Fisheries University, Chennai, India

Ginpreet Kaur Shobhaben Pratapbhai Patel School of Pharmacy and Technology Management, SVKM's Narsee Monjee Institute of Management Studies (NMIMS) Deemed-to-be University, Mumbai, Maharashtra, India

Khushmanjot Kaur Section of Cardiology, Department of Internal Medicine, Max Rady College of Medicine, Rady Faculty of Health Sciences, University of Manitoba, Winnipeg, MB, Canada

Cardiovascular Real-World Evidence (CARDIO-RWE) Research Group, Section of Cardiology, I.H. Asper Institute, St. Boniface Hospital, Winnipeg, MB, Canada

Tarnjot Kaur Department of Pharmaceutical Sciences, School of Health Sciences and Technology, UPES, Dehradun, Uttarakhand, India

Sadaf Khan Department of Biotechnology, Chronobiology Laboratory, Era University, Lucknow, India

Mota X. Kholopo Department of Environmental Health, Faculty of Health Sciences, Doornfontein Campus, University of Johannesburg, Johannesburg, South Africa

Abhiram Kumar Laboratoy of Molecular Medicine, Birla Institute of Technology and Sciences, Pilani, Hyderabad Campus, Hyderabad, India

Swati Kundu Indian Council of Medical Research Headquarters, Ministry of Health and Family Welfare, New Delhi, India

Jayant Kumar Department of Pharmaceutical Sciences, HNB Garhwal University (A Central University), Chauras Campus, Tehri Garhwal, Uttarakhand, India

Setor K. Kunutsor Section of Cardiology, Department of Internal Medicine, Max Rady College of Medicine, Rady Faculty of Health Sciences, University of Manitoba, Winnipeg, MB, Canada

Cardiovascular Real-World Evidence (CARDIO-RWE) Research Group, Section of Cardiology, I.H. Asper Institute, St. Boniface Hospital, Winnipeg, MB, Canada

Department of Physiology and Pathophysiology, Rady Faculty of Health Sciences, University of Manitoba, Winnipeg, MB, Canada

Jawahar L. Mehta Division of Cardiology, Department of Internal Medicine, University of Arkansas for Medical Sciences, Little Rock, AR, USA

Mukesh Nandave Department of Pharmacology, Delhi Pharmaceutical Sciences and Research University (DPSRU), New Delhi, India

Bhoomika M. Patel National Forensic Sciences University, Gandhinagar, India

Mayur M. Patel Institute of Pharmacy, Nirma University, Ahmedabad, India

Fajar Shodiq Permata Department of Anatomy, College of Medicine and Health Sciences, United Arab Emirates University, Al Ain, United Arab Emirates

Faculty of Veterinary Medicine, Universitas Brawijaya, Malang, Indonesia

Ferial Pestano-Gupta Division of Natural Sciences, Berbice Campus, University of Guyana, Tain, Guyana

Anusha Prasannan Department of Basic Sciences, Institute of Fisheries Post Graduate Studies, Tamil Nadu Dr. J. Jayalalithaa Fisheries University (TNJFU), Chennai, Tamil Nadu, India

Bram Ramjiawan Department of Pharmacology and Therapeutics, Max Rady College of Medicine, University of Manitoba, Winnipeg, MB, Canada

Amit Ranjan Department of Basic Sciences, Institute of Fisheries Post Graduate Studies, Tamil Nadu Dr. J. Jayalalithaa Fisheries University (TNJFU), Chennai, Tamil Nadu, India

Phoka C. Rathebe Department of Environmental Health, Faculty of Health Sciences, Doornfontein Campus, University of Johannesburg, Johannesburg, South Africa

Ammar Mehdi Raza Department of Pedodontics, Career Institute of Dental Sciences and Hospital, Lucknow, India

Nicholas Sagadaya Division of Natural Sciences, Berbice Campus, University of Guyana, Tain, Guyana

Prachi Saluja Division of Pulmonary, Critical Care and Sleep Medicine, University of Cincinnati, Cincinnati, OH, USA

Dhanvi Sheth Shobhaben Pratapbhai Patel School of Pharmacy & Technology Management, SVKM's Narsee Monjee Institute of Management Studies (NMIMS) Deemed-to-be University, Mumbai, Maharashtra, India

Jaipaul Singh School of Pharmacy and Biomedical Sciences, University of Central Lancashire, Preston, UK

Ram B. Singh Halberg Hospital and Research Institute, Moradabad, UP, India

Romi Singh Era College of Pharmacy, Era University, Lucknow, India

Paramjit S. Tappia Asper Clinical Research Institute and St. Boniface Hospital Albrechtsen Research Centre, Winnipeg, MB, Canada

Rolf Teschke Department of Internal Medicine II, Division of Gastroenterology and Hepatology, Klinikum Hanau, Hanau, Germany

Academic Teaching Hospital of the Medical Faculty, Goethe University Frankfurt/Main, Frankfurt/Main, Germany

Sofia Morais Tornis Department of Medicine, Federal University of Jataí, Jataí, Goiás, Brazil

Jyoti Upadhyay Department of Pharmaceutical Sciences, School of Health Sciences and Technology, UPES, Dehradun, Uttarakhand, India

Kaliyamurthi Venkatachalam Department of Basic Sciences, Institute of Fisheries Post Graduate Studies, Tamil Nadu Dr. J. Jayalalithaa Fisheries University (TNJFU), Chennai, Tamil Nadu, India

Tran Dang Xuan Center for the Planetary Health and Innovation Science, the IDEC Institute, Smart Agriculture Faculty, Laboratory of Plant Physiology and Biochemistry, Graduate School of Advanced Science and Engineering, Hiroshima University, Higashi-Hiroshima City, Hiroshima, Japan

Kotaro Yamamoto Department of Integrative Physiology, Tottori University Graduate School and Faculty of Medicine, Tottori University, Yonago, Japan

Part I
Trace Elements in Health and Disease

Chapter 1
Modulation of Trace Elements in Pregnant Women and Its Clinical Importance for Fetal Development

Sofia Morais Tornis, Mayara Bocchi, Fábio Morato de Oliveira, David Michel de Oliveira, and Eduardo Vignoto Fernandes

Abstract Pregnancy triggers a series of metabolic, physiological, and hormonal changes in the maternal organism, aiming to sustain fetal development and prepare the body for childbirth and lactation. Among these changes, a heightened demand for micronutrients becomes evident, especially trace elements (TEs), minerals required in small amounts but essential for various biological functions such as enzymatic activity, gene regulation, immune function, and antioxidant defense. Ensuring sufficient intake of these TEs is vital for maintaining maternal health and ensuring healthy fetal development. In this context, the aim of the present narrative review was to understand the modulation and clinical relevance of the main TEs—iron, zinc, iodine, copper, selenium, magnesium, and calcium—during the gestational period. The bibliographic research was conducted in the following databases: National Library of Medicine (PubMed), Cochrane, Regional Health Portal (BVS), Embase, Scopus, Web of Science, Latin American and Caribbean Literature in Health Sciences (LILACS), Scientific Electronic Library Online (SciELO), and Cumulative Index to Nursing and Allied Health Literature (CINAHL-EBSCO), covering the period from 2009 to 2025, with no language restrictions. The results showed that imbalances or deficiencies in these TEs may compromise maternal

S. M. Tornis · M. Bocchi · F. M. de Oliveira
Department of Medicine, Federal University of Jataí, Jataí, Goiás, Brazil

D. M. de Oliveira
Department of Physical Education, Federal University of Jataí, Jataí, Goiás, Brazil

Postgraduate Program in Bioscience and One Health, Federal University of Jataí, Jataí, Goiás, Brazil

E. V. Fernandes (✉)
Postgraduate Program in Bioscience and One Health, Federal University of Jataí, Jataí, Goiás, Brazil

Human and Comparative Anatomy Laboratory, Federal University of Jataí, Jataí, Goiás, Brazil
e-mail: eduardovignoto@ufj.edu.br

N. S. Dhalla et al. (eds.), *Functional Biochemistry of Metallic Elements*, Advances in Biochemistry in Health and Disease 36,
https://doi.org/10.1007/978-3-032-14445-4_1

physiological processes and fetal growth, emphasizing the need for supplementation strategies and an appropriate nutritional approach during prenatal care to promote better outcomes for both mother and baby.

Keywords Pregnancy · Fetal development · Micronutrients · Trace elements · Supplementation · Maternity · Gestation

Introduction

The gestational period results in a series of physiological changes that occur in the woman's body, such as increased body mass, changes in the gastrointestinal, hormonal, cardiac, and hematological systems, in order to ensure the normal development and health of the fetus, as well as to prepare the mother and baby for childbirth and breastfeeding [1]. Moreover, there is also an increase in nutritional demands due to the growth of the fetus, the placenta, and maternal tissues [2].

During pregnancy, dietary intake must meet the needs of the baby, as well as those of the mother, whose health is closely related to the adequate supply of essential elements such as macronutrients (carbohydrates, lipids, and proteins) and micronutrients (vitamins and minerals) [3]. Mainly provided by a balanced diet, micronutrients such as vitamins (e.g., A, B, C, and D) and minerals (e.g., iron, zinc, selenium, iodine, copper, magnesium, and calcium) help to limit the risk of nutritional deficiencies [4].

Among the minerals, trace elements (TEs), also known as oligoelements, stand out. TEs are inorganic chemical elements present in low concentrations in the body, yet they serve critical biological functions in many biological processes, such as the regulation of physiological functioning, enhancement of the immune system, and increase of cellular antioxidant defense. In this regard, TEs influence enzymatic activity, regulate gene expression, and participate in protein synthesis, playing an indispensable role in the balance of life alongside organic elements [5]. Iron (Fe), for instance, has a significant impact on maternal health and fetal development, as it participates in various metabolic processes in the body, such as compound synthesis and support in the oxygen transport mechanism [3].

Zinc (Zn) is an essential TE not only for its involvement in the synthesis of hundreds of proteins [6] but also because its importance begins as early as the oocyte maturation process, which will later be fertilized [7]. Also, among the TEs involved in metabolic reactions, iodine is crucial for thyroid hormone biosynthesis [8], while copper (Cu) acts as a cofactor for various enzymes. However, it must not accumulate in the body, as its excess stimulates the synthesis of reactive oxygen species [9]. Selenium (Se), which will also be addressed in this chapter, assists in combating free radicals due to its antioxidant and anti-inflammatory properties [7]. Magnesium (Mg), in turn, presents important cationic and regulatory characteristics [10], while calcium (Ca) primarily contributes to bone maintenance [11]. Few pieces of evidence have been found regarding other TEs and their relationship with pregnancy, such as fluoride, chromium, and manganese.

Therefore, the search for scientific studies in the literature concerning the relationship between pregnancy and the TEs may be valuable for the implementation of supplementation strategies and improvements in maternal-fetal health. In this context, this narrative review aims to explore the modulation of TEs during pregnancy, highlighting their clinical relevance for fetal development.

Methodology

The present narrative review searched for articles published in the following databases: National Library of Medicine (PubMed), Cochrane, Regional Health Portal (BVS), Embase, Scopus, Web of Science, Latin American and Caribbean Literature in Health Sciences (LILACS), Scientific Electronic Library Online (SciELO), and Cumulative Index to Nursing and Allied Health Literature (CINAHL-EBSCO). The selection of studies involved a search strategy combining terms related to the modulation of TEs during pregnancy, such as importance, effects, fetal development, fetus, iron, iodine, zinc, copper, magnesium, selenium, deficiency, supplementation, pregnancy, and maternity. The search covered the period from 2009 to 2025, with no language restrictions applied to the articles.

Importance of Fe in Maternal-Fetal Supply

During pregnancy, there is an increased demand for Fe due to the rise in plasma volume (40–60%) and the number of red blood cells (25%), making the monitoring of its serum levels essential to avoid deficiency and, consequently, the development of gestational anemia [12].

Fe deficiency is one of the most common conditions during pregnancy, affecting approximately 25% of pregnant women in the third trimester [13]. This condition may lead to iron deficiency anemia, which is characterized by limited hemoglobin (Hb) synthesis due to Fe scarcity, and is associated with risks such as preterm birth, postpartum hemorrhage, and low birth weight [14]. In this process, the body, in an attempt to prioritize Fe supply for hemoglobin production—favored over all other tissues, including the brain—neglects its distribution to other organs and systems, which can impair, for instance, the neural development of the child [13].

Fe is a key component in DNA synthesis, energy production, neurotransmitter synthesis and glial cell development. The fetal brain consumes approximately 60% of the available oxygen, emphasizing the need for Fe in the development of the nervous system [12]. Infants born with Fe deficiency may experience impaired formation of gray matter, resulting in a less complex dendritic architecture and, consequently, compromised memory, slower neural processing, reduced motor function, and socialization difficulties—consequences that may persist into adulthood, even after Fe repletion [15]. Moreover, Fe deficiency is associated with significant long-term neuromorbidities, such as depression, anxiety, autism, and schizophrenia [13].

Regarding the maternal health impacts of low Fe levels (with a recommended daily intake of 30–60 mg according to the World Health Organization), symptoms include appetite disorders, cerebral and myocardial hypoxia, preeclampsia, eclampsia, premature rupture of fetal membranes, abnormal thyroid function, postpartum depression, hemorrhage, poor wound healing, and even maternal death. These phenomena are explained by the body's compensatory mechanism, in which, under Fe deficiency, this trace element is preferentially directed to the fetus via the placenta, even at the expense of the mother's own Fe levels [13].

Given this scenario, Fe supplementation has proven to be an effective strategy for reducing such risks, with a recommended intake of 40 mg/day by the Brazilian Ministry of Health. The study by Froessler et al. [16] demonstrated that Fe supplementation, combined with intravenous infusion, results in better hematologic responses compared to oral therapy alone. Additionally, women who used high Fe supplements (above 51 mg/day) had a 70% lower risk of ovulatory infertility [17].

On the other hand, caution must be taken with excessive Fe intake, as it can interfere with the absorption of other essential TE, such as zinc, copper, and magnesium, due to competition for binding sites. It may also increase the synthesis of reactive oxygen species, which damages placental tissue [3]. Furthermore, Fe oversaturation has been associated with increased risk of miscarriage, prematurity, and low birth weight [9].

Therefore, ensuring adequate nutritional status during pregnancy is essential not only for maternal health but also for fetal and infant development. As breast milk contains low levels of Fe, the Fe stores acquired during fetal life are crucial for development in the first months of life. Thus, Fe intake and supplementation as recommended by the World Health Organization (WHO) during pregnancy should be a priority, ensuring better outcomes for future generations [12].

Zn and Its Role During Pregnancy

Zn is a trace element (TE) that plays fundamental roles in various stages of pregnancy, influencing the organism even before fertilization, through its involvement in oocyte maturation, and extending to the completion of embryonic development [9]. It is the second most abundant TE in the body [18], and its activity spans three major categories: structural, regulatory, and catalytic functions [19]. Among its various functions, it acts as a cofactor for hundreds of enzymes involved in nucleic acid synthesis, replication, and intra- and intercellular signaling [7].

Present in all body tissues, Zn is primarily concentrated in bones, muscles, skin, and liver [20]. It participates in the maturation of the female gamete by increasing its concentration to enable the progression of meiosis I, thereby ensuring the formation of a viable ovum. It also supports symmetric division and proliferation of the embryo before and after fertilization [7]. Therefore, a deficiency in this TE can compromise embryonic maturation, resulting in inadequate organ development, anatomical anomalies, preeclampsia, or even spontaneous abortion [9].

A cohort study involving 3187 pregnant women conducted in China concluded that maternal dietary Zn deficiency increases the risk of reduced gestational size and low birth weight [21]. Furthermore, it can compromise fetal integrity by inducing uterine spasms and altering the composition of the amniotic fluid [22], as well as by reducing the synthesis and degradation of collagen, which is essential for preventing premature rupture of fetal membranes [23].

Zn's greatest impact lies in regulating chromatin structure and function, influencing gene expression and cellular differentiation. For instance, to ensure its role in neural tube development, Garner et al. [24] describe the transplacental transfer of this TE from mother to fetus [25]. However, Zn requires specific transporters to cross the cell membrane. Among them, ZNT (zinc transporter) proteins, responsible for Zn efflux, and ZIP (zinc importer) proteins, responsible for Zn influx, are the most important for facilitating its cellular entry [20].

Although reduced Zn levels in the maternal and fetal organisms can lead to significant complications, excessive levels can also be harmful to both. High doses may disrupt essential biochemical reactions and manifest through symptoms such as cramps, diarrhea, or nausea [3, 26]. Additionally, imbalanced Zn levels may interfere with the absorption of other TEs, such as co Cu, and increase oxidative stress—an effect detrimental to both maternal and fetal health [20].

It is therefore essential to maintain balanced Zn concentrations to minimize risks during pregnancy. According to the *Dietary Reference Intakes* (DRIs) established by the US Institute of Medicine (2001), the recommended daily intake of Zn for pregnant women is 11 mg. If dietary intake is insufficient, supplementation may be a viable alternative, which has already demonstrated positive outcomes, as reported by Grzeszczak et al. [3]. According to the authors, babies of mothers who received Zn supplementation exhibited more efficient heart rates and demonstrated a faster growth curve compared to those whose mothers did not receive nutritional monitoring [3].

Therefore, considering the wide range of functions Zn performs, the importance of this trace element in maintaining maternal-fetal health becomes evident. Its structural, regulatory, and catalytic roles influence essential biological processes. However, its serum levels should be closely monitored, since both excess and deficiency can result in adverse outcomes throughout pregnancy. Accordingly, adequate nutritional monitoring, combined with a balanced diet, is essential to ensure optimal Zn intake and to prevent potential complications during gestation.

Role of Cu in Maternal and Fetal Metabolic Reactions

Cu is the third most abundant trace element (TE) in the human body and plays essential roles throughout the reproductive process [9]. It functions as a cofactor for enzymes known as cuproenzymes, which are involved in Fe metabolism, oxidative processes, and the catalysis of oxidative phosphorylation—a key component of cellular respiration [25].

Like other TEs, Cu is primarily obtained through diet, with sources including seafood, seeds, organ meats, and nuts [7]. In the stomach, copper ions (Cu^{2+}) are reduced to Cu^{+} and absorbed in the intestinal tract [27]. Specific proteins, such as the high-affinity copper uptake protein 1 (CTR1), are responsible for transporting this element to body tissues [3].

During pregnancy, the demand for Cu increases significantly to meet both maternal and fetal needs. In this context, it is important to note that the recommended daily intake of Cu ranges from 0.9 to 1.3 mg/day, increasing to 1 mg/day during pregnancy, according to the *Dietary Reference Intakes* (DRIs) established by the US Institute of Medicine [28]. Studies indicate that maternal blood concentrations of Cu are considerably higher than those in the fetus, since the placenta acts as a barrier, hindering Cu transfer between the mother and the fetus [25]. Maintaining Cu balance is therefore crucial, as elevated plasma levels have been associated with an increased risk of complications during the first trimester of pregnancy [9].

Additionally, Cu plays a role in the functioning of specific enzymes, such as ceruloplasmin, catalase, and peroxidases, as well as other antioxidant compounds, including glutathione and metallothionein, helping to prevent oxidative stress [9]. Nevertheless, it is important to emphasize that imbalances in this TE can impair antioxidant mechanisms, being associated with complications such as spontaneous abortion, congenital malformations, preterm birth, and fetal growth restriction [25].

With regard to supplementation, studies have shown that pregnant women who received 1000 mg of Cu daily starting at the 17th week of gestation experienced fewer depressive symptoms than the control group during the second and third trimesters of pregnancy. Moreover, reduced Cu levels early in gestation may lead to hypertension [3]. Therefore, regulating Cu homeostasis is indispensable for favorable pregnancy outcomes, as evidenced by a study conducted by Wilson et al. [29], which showed that pregnant women with low serum Cu concentrations (average of 10.3 μmol/L) had fewer pregnancy complications compared to those with high concentrations (approximately 52.9 μmol/L) of this TE [29].

The Role of Ca During Pregnancy

Ca is an essential micronutrient in the human body, orchestrating key physiological processes, such as bone mineralization, muscle contraction, nerve impulse transmission, and blood coagulation. During pregnancy, Ca requirements increase to meet both maternal and fetal demands. Deficiency in Ca during this period has been associated with adverse outcomes, including preterm birth, low birth weight, insulin resistance in the offspring, and preeclampsia [11]. The WHO recommends Ca supplementation for pregnant women with low dietary intake [30], with a suggested intake ranging between 1.5 g and 2 g per day.

Such supplementation plays a vital role in maintaining blood pressure by reducing parathyroid hormone levels, thereby ensuring the release of renin by the kidneys and promoting vasodilation [31]. However, it is essential that this intake be monitored, as excessive supplementation may increase the risk of urinary tract infections, urinary calculi, constipation, reduced absorption of other essential elements such as Fe, and even cardiovascular damage due to increased arterial calcification [32].

Randomized controlled trials conducted in India and Tanzania, involving approximately 11,000 nulliparous pregnant women in each study, confirmed that Ca supplementation reduced the risk of preeclampsia. In India, the incidence of this condition was 3% in the 500 mg group and 3.6% in the 1500 mg group, whereas in Tanzania, the rates were 3% and 2.7%, respectively [31]. A randomized clinical trial conducted in Brazil produced similar results. Participants were divided into three groups: placebo, 500 mg/day, and 1500 mg/day from the 20th week of gestation until delivery. The findings confirmed that even at lower doses, biochemical markers of preeclampsia—such as proteinuria, urea, and creatinine—decreased [32].

Ca deficiency compromises not only maternal health but also fetal development, beginning with its fundamental role in skeletal growth. Ca is transported across the placenta by syncytiotrophoblast proteins, including isoforms of the plasma membrane Ca^{2+}-ATPase, and it is estimated that up to 30 grams of Ca are transferred from mother to fetus throughout pregnancy [11]. Furthermore, a study involving supplemented and non-supplemented pregnant adolescents demonstrated that those who received Ca supplementation had lower chances of delivering preterm infants [31].

Thus, it is evident that Ca plays a multifaceted role in ensuring maternal and fetal well-being. Accordingly, nutritional strategies aimed at encouraging the intake of calcium-rich foods (such as milk, fish, and dark green leafy vegetables) are crucial, especially in countries with low to moderate Ca consumption, such as Brazil and Mexico [33].

The Importance of Mg During Pregnancy

Mg is a vital mineral, essential for numerous physiological and cellular processes. As a multivalent cation, it is involved in over 600 enzymatic reactions, including protein and fatty acid synthesis, nerve impulse transmission, and nucleotide metabolism. Additionally, Mg activates coenzymes of important anions—such as ATP—and contributes to bone formation by enhancing Ca absorption and activating vitamin D in the kidneys [34].

Mg deficiency during pregnancy poses several risks to both mother and fetus, including gestational diabetes, preeclampsia, seizures, abnormal neural development, and fetal growth restriction [10, 34]. A study described by Fanni et al. [10] indicates that in women with Mg deficiency during gestation (recommended intake: 310–320 mg/day according to Dietary Reference Intakes), the likelihood of the fetus developing metabolic syndrome and insulin resistance in adulthood increases, especially when newborns present low serum Mg levels at birth [10].

Thus, the administration of magnesium sulfate ($MgSO_4$) is widely recommended in obstetrics, exhibiting neuroprotective effects for fetuses in preterm pregnancies. These benefits are mediated by the blocking of neurotransmitters responsible for neuronal cell death, and modulation of free radicals and pro-inflammatory cytokines [10]. A meta-analysis conducted by Conde-Agudelo and Romero [35] revealed that $MgSO_4$ supplementation before the 34th week of gestation reduced the prevalence of cerebral palsy in premature newborns from 5.6% to 3.9%. Moreover, the meta-analysis by Wolf et al. [36], which included a total of 5917 women, indicated that $MgSO_4$ treatment may serve as an intermediate intervention for fetal neuroprotection [37].

However, despite global clinical practice guidelines recommending the use of $MgSO_4$ for fetal neural protection, results from Magenta [38] indicated that intravenous administration of this trace element between the 30th and 34th weeks of gestation did not reduce the risk of infant mortality or cerebral palsy in children up to 2 years old. Nevertheless, this study had limited power to detect significant differences between the groups [38].

Magnesium sulfate has also been studied for its potential in preventing and treating both eclampsia and preeclampsia. Research has shown that therapy with this compound reduces the risk of seizures in women with preeclampsia by more than half [39, 40]. Supplementation has also been associated with a significant reduction in the incidence of low-birth-weight infants [10].

Finally, magnesium's role in pregnancy also extends to glucose metabolism. As described by Lee et al. [34], the expression of Mg transporters may impact maternal glycemic homeostasis, potentially contributing to the development of gestational diabetes. This factor, combined with the evidence, underscores the importance of adequate Mg intake throughout pregnancy, as maintaining appropriate levels of this trace element is crucial for maternal and fetal health during and after gestation.

Se and Its Mechanism of Action During Pregnancy

Se is an essential trace element that plays important roles during pregnancy due to its integration into selenoproteins, acting in hormonal regulation, as well as in antioxidant and anti-inflammatory defense mechanisms [41]. During gestation, maternal serum concentrations of Se tend to decrease due to its transfer to breast milk and increased placental transport, which may result in various maternal-fetal disorders [9].

Studies indicate the essentiality of Se even before fertilization, given its role in fertility. In women, its importance begins at the embryonic stage, ensuring embryo quality during the cleavage phase and, consequently, increasing the chances of healthy blastocyst formation [42]. In vitro fertilization treatments, Se supplementation has shown positive effects on embryonic development [7].

Epidemiological studies have demonstrated that Se deficiency is associated with various gestational complications, including preeclampsia, gestational diabetes, spontaneous abortion, fetal growth restriction, and preterm birth. This can be explained by the relationship between low Se levels and oxidative stress, since serum concentrations below 100 μg/L impair the activity of glutathione peroxidase (GPX) and thioredoxin reductases, which are antioxidant enzymes [9].

Se deficiency is also related to congenital disorders such as orofacial clefts and neural tube defects and is associated with a higher incidence of neurodevelopmental disorders, including autism spectrum disorder (ASD) and attention deficit hyperactivity disorder (ADHD) [41]. In this context, Se supplementation has been associated with a reduced risk of congenital anomalies, as demonstrated by a case-control study conducted in six hospitals in the city of Xi'an (China), which showed a significant decrease in the risk of congenital heart disease [43]. The positive effects of this supplementation extend further, showing potential to increase cell proliferation, reduce DNA damage, and lower the risks of gestational hypertension, preeclampsia, postpartum depression, fetal growth restriction, and thyroid dysfunctions [41].

Therefore, ensuring an adequate intake of Se (8 mg/day for females, according to the 2006 Dietary Reference Intakes), as well as its supplementation, when necessary, is a crucial measure for the protection of both fetal and maternal health.

Iodine as a Hormonal Modulator During Gestation

Iodine is an essential trace element for the biosynthesis of thyroid hormones, which are fundamental throughout fetal development [7, 8, 15, 44]. During the first half of pregnancy, the fetus depends exclusively on maternal thyroid hormones, as its own thyroid gland is still developing [8]. Given the importance of these hormones, adequate iodine intake is critical for pregnant women, with a recommended daily consumption of 250 μg/day, according to WHO [45].

Insufficient intake of this trace element is associated with several adverse outcomes, including increased risk of spontaneous abortion, preterm birth, fetal growth restriction, congenital anomalies, preeclampsia, and subfertility [7]. Severe iodine deficiency is considered the most significant preventable cause of brain damage in newborns, according to Santos et al. [46]. The syndrome of congenital iodine deficiency, known as cretinism, is characterized by profound intellectual disability, deaf-mutism, strabismus, and dwarfism, representing the most severe impact of iodine insufficiency [47].

During pregnancy, iodine demand and thyroid hormone production increase due to elevated levels of estrogen and human chorionic gonadotropin (hCG), as well as increased maternal renal excretion of iodine [8]. For this reason, supplementation becomes a preventive strategy against potential fetal neurodevelopmental damage. For instance, one study demonstrated that in a Chinese region with severe iodine deficiency, supplementation in pregnant women reduced the fetal microcephaly rate from 27% to 11% [44]. Additionally, iodized oil supplementation in Algeria led to a 6.7% increase in newborn birth weight [44].

Iodine, mainly found in water, fish, and marine shellfish, is absorbed in the small intestine in its ionic form, then incorporated into the thyroid and used for the synthesis of thyroglobulin and tyrosine, compounds involved in the production of thyroid hormones. The concentration of this trace element in the urine is measured in pregnant women to assess the adequacy of its intake [44]. Since more than 90% of ingested iodine is excreted via the urinary system, the median urinary iodine concentration serves as an effective biomarker for estimating recent iodine intake [44].

Given the relevance of this trace element for fetal development and maternal health, measures such as fortifying table salt with iodine and appropriate supplementation during pregnancy are essential to prevent disorders associated with iodine deficiency [44, 48].

TEs in Pregnancy

In order to deepen the understanding of how different TEs play an essential role in overall maternal and fetal health during pregnancy, the TEs and their specific behaviors in the body, as explored above, have been summarized in Fig. 1.1.

Interactions Between TEs During Pregnancy

The homeostasis of TEs is a complex process involving various mechanisms of absorption, metabolic regulation, and transport, in addition to the participation of numerous substances. Throughout pregnancy, these interactions become even more crucial, as they directly affect maternal health and fetal development. Fe and Cu, for instance, interact in such a way that Cu is essential for the incorporation of Fe into the heme group, as it is part of the cytochrome oxidase enzyme, which is required for oxidative phosphorylation— a process that supports this incorporation in the fetal organism [3]. In this context, Fig. 1.2 illustrates this interaction, highlighting the role of Cu in oxidative phosphorylation, which takes place inside the mitochondria of fetal cells.

On the other hand, an increased supply of Cu may impair Fe absorption, as Cu has a high affinity for transferring the iron-transporting enzyme in the blood. The reverse is also true: with high maternal Fe supplementation, both the serum levels and activity of Cu are reduced [3].

The competition between Zn and Fe is also noteworthy during pregnancy. The absorption of both TEs occurs via the divalent metal transporter 1 (DMT-1), located in the membrane of enterocytes in the intestine, which creates direct competition between these elements [49]. This was observed in a study conducted by O'Brien et al., where supplementation with 60 mg/day of Fe during pregnancy significantly

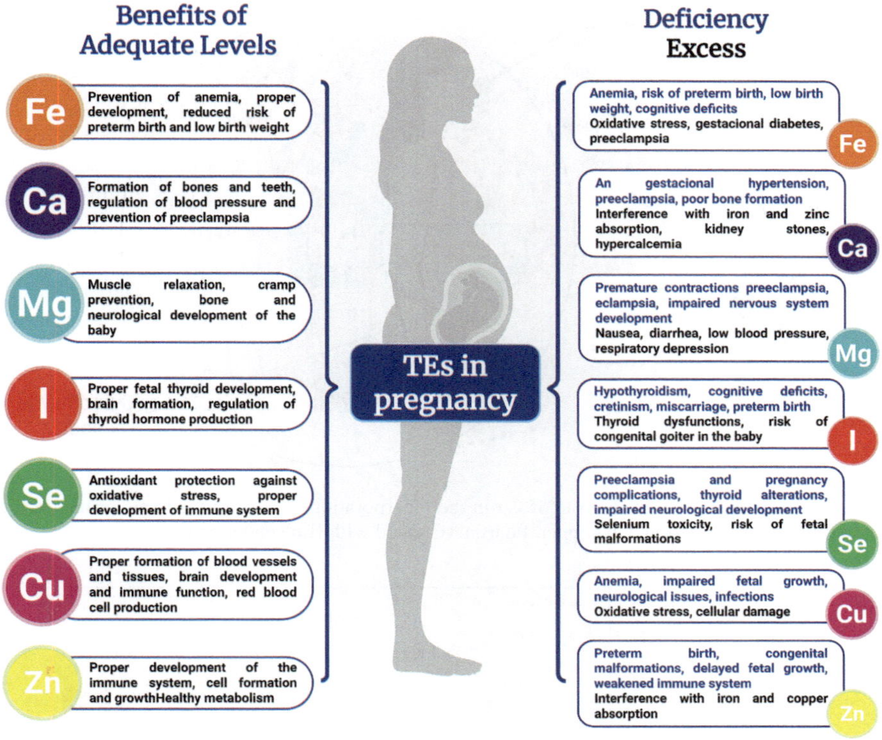

Fig. 1.1 Schematic representation of the importance of TE for pregnant women, as well as the effects of concentrations below or above the normal range. TE trace elements, Fe iron, Ca calcium, Mg magnesium, I iodine, Se selenium, Cu copper, Zn zinc. (Created with BioRender.com)

reduced serum Zn levels. The opposite was also reported, with daily supplementation of 15 mg of Zn resulting in reduced Fe concentration [3].

Figure 1.3 illustrates how this competition between Zn and Fe takes place during the absorption of these TEs in the small intestine. It can be seen that while the ferrous ion is transported from the extracellular to the intracellular environment by a DMT-1 protein, the zinc ion is unable to enter the cell through the same transporter, and vice versa.

In addition to Fe, Zn also interacts antagonistically with Cu. Zn supplementation stimulates the production of metallothionein, which has a higher affinity for Cu. This leads to Cu accumulation in the enterocytes and its subsequent loss through cellular exfoliation, ultimately resulting in Cu deficiency [3]. Fe absorption can also be affected not only by Cu and Zn but by Ca as well. The inhibitory effect of Ca on Fe absorption has been confirmed by studies described in Benkhedda et al. (2010), demonstrating a reduction in Fe absorption ranging from 5% to 60% when Ca is present in meals [50, 51].

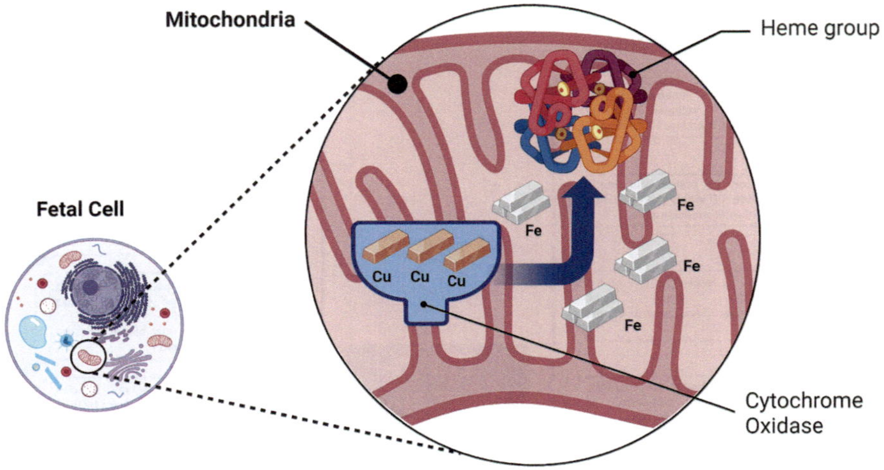

Fig. 1.2 Representation of the role of Cu in the incorporation of Fe into the heme group during oxidative phosphorylation. Cu copper, Fe iron. (Created with BioRender.com)

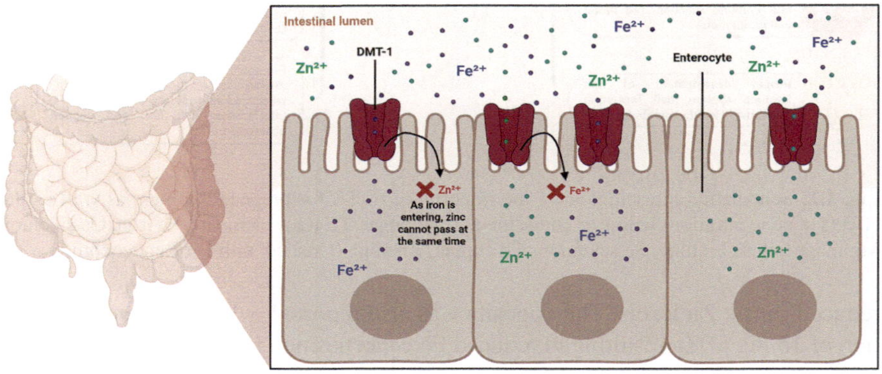

Fig. 1.3 Illustration of the competition between Fe and Zn for DMT-1 during the intestinal absorption process. Fe^{2+} ferrous ion, Zn^{2+} zinc ion, DMT-1 divalent metal transporter 1. (Created with BioRender.com)

Conclusion

Pregnancy is a period of profound physiological and metabolic changes in the maternal body, requiring nutritional adaptations that ensure an adequate intake of trace elements (TEs). This narrative review highlighted the importance of the main TEs—iron, zinc, iodine, copper, selenium, calcium, and magnesium—for maintaining maternal well-being and optimal fetal growth. Each of these elements plays essential roles, from hormonal regulation, energy metabolism, and antioxidant protection to the immunological, neurological, and skeletal development of the fetus.

The findings indicate that both deficiency and excess of these TEs are associated with a range of gestational and neonatal complications, including preterm birth, low birth weight, intrauterine growth restriction, congenital malformations, neurodevelopmental disorders, and increased maternal morbidity risks such as preeclampsia, gestational diabetes, and postpartum depression. In this regard, the clinical implications are substantial and underscore the need for an individualized approach in the nutritional management of pregnant women, with an emphasis on early laboratory screening and well-calibrated supplementation strategies.

However, the current literature still presents significant limitations. Despite the large number of studies, much of the evidence is based on observational data, lacking randomized clinical trials to confirm the efficacy and safety of specific supplementations, especially across different population contexts. Furthermore, the complex interactions among TEs remain poorly understood, contributing to their frequent neglect in clinical practice. There is also no clear consensus on optimal dosages for different scenarios and nutritional profiles, representing a major gap in the formulation of universal guidelines.

Given these challenges, promising perspectives for research and clinical practice involve deepening our understanding of the bioavailability, metabolism, and interaction of TEs during pregnancy. Future investigations should explore personalized medicine approaches and the long-term impact of maternal supplementation on the offspring's metabolic programming. In clinical practice, the integration of individualized nutritional monitoring protocols, combined with strengthened public policies on food fortification, may help prevent deficiencies and promote healthier pregnancies.

In summary, the proper management of TEs during prenatal care represents a strategic axis in promoting maternal-fetal health, and its recognition in the fields of obstetrics and clinical nutrition is essential for building a solid foundation for prevention and care for future generations.

References

1. Jouanne M, Oddoux S, Noël A, Voisin-Chiret AS (2021) Nutrient requirements during pregnancy and lactation. Nutrients 13(2):692. https://doi.org/10.3390/nu13020692
2. Haider BA, Bhutta ZA (2017) Multiple-micronutrient supplementation for women during pregnancy. Cochrane Database Syst Rev 4(4):CD004905. https://doi.org/10.1002/14651858.CD004905.pub5
3. Grzeszczak K, Kwiatkowski S, Kosik-Bogacka D (2020) The role of Fe, Zn, and Cu in pregnancy. Biomolecules 10(8):1176. https://doi.org/10.3390/biom10081176
4. Domínguez L, Fernández-Ruiz V, Cámara M (2023) Micronutrients in food supplements for pregnant women: European health claims assessment. Nutrients 15(21):4592. https://doi.org/10.3390/nu15214592
5. Berger MM, Shenkin A, Schweinlin A et al (2022) ESPEN micronutrient guideline. Clin Nutr 41(6):1357–1424. https://doi.org/10.1016/j.clnu.2022.02.015
6. Zhang H, Pan J, Jiang H et al (2023) A study on the correlation between MTHFR and folic acid combined with trace elements for the prevention of fetal malformations in the first

trimester of pregnancy. Medicine (Baltimore) 102(44):e35330. https://doi.org/10.1097/MD.0000000000035330
7. Dring JC, Forma A, Chilimoniuk Z et al (2021) Essentiality of trace elements in pregnancy, fertility, and gynecologic cancers: a state-of-the-art review. Nutrients 14(1):185. https://doi.org/10.3390/nu14010185
8. Lee SY (2021) Editorial: consequences of iodine deficiency in pregnancy. Front Endocrinol (Lausanne) 12:740239. https://doi.org/10.3389/fendo.2021.740239
9. Grzeszczak K, Łanocha-Arendarczyk N, Malinowski W et al (2023) Oxidative stress in pregnancy. Biomolecules 13(12):1768. https://doi.org/10.3390/biom13121768
10. Fanni D, Gerosa C, Nurchi VM et al (2021) The role of magnesium in pregnancy and in fetal programming of adult diseases. Biol Trace Elem Res 199(10):3647–3657. https://doi.org/10.1007/s12011-020-02513-0
11. Takaya J (2021) Calcium-deficiency during pregnancy affects insulin resistance in offspring. Int J Mol Sci 22(13):7008. https://doi.org/10.3390/ijms22137008
12. Georgieff MK (2020) Iron deficiency in pregnancy. Am J Obstet Gynecol 223(4):516–524. https://doi.org/10.1016/j.ajog.2020.03.006
13. Benson AE, Shatzel JJ, Ryan KS et al (2022) The incidence, complications, and treatment of iron deficiency in pregnancy. Eur J Haematol 109(6):633–642. https://doi.org/10.1111/ejh.13870
14. Zhang Y, Lu Y, Jin L (2022) Iron metabolism and ferroptosis in physiological and pathological pregnancy. Int J Mol Sci 23(16):9395. https://doi.org/10.3390/ijms23169395
15. Cortés-Albornoz MC, García-Guáqueta DP, Velez-van-Meerbeke A, Talero-Gutiérrez C (2021) Maternal nutrition and neurodevelopment: a scoping review. Nutrients 13(10):3530. https://doi.org/10.3390/nu13103530
16. Froessler B, Gajic T, Dekker G, Hodyl NA (2018) Treatment of iron deficiency and iron deficiency anemia with intravenous ferric carboxymaltose in pregnancy. Arch Gynecol Obstet 298(1):75–82. https://doi.org/10.1007/s00404-018-4782-9
17. Georgsen M, Krog MC, Korsholm AS et al (2021) Serum ferritin level is inversely related to the number of previous pregnancy losses in women with recurrent miscarriage. Fertil Steril 115:389–396. https://doi.org/10.1016/j.fertnstert.2020.08.1410
18. Classen HG, Grober U, Low D, Schmidt J, Stracke H (2011) Zinc deficiency: symptoms, causes, diagnosis, and therapy. Med Monatsschr Pharm 34:87–95
19. Chasapis CT, Loutsidou AC, Spiliopoulou CA, Stefanidou M (2012) Zinc and human health: an update. Arch Toxicol 86:521–534. https://doi.org/10.1007/s00204-011-0775-1
20. Michalczyk K, Cymbaluk-Płoska A (2020) The role of zinc and copper in gynecological malignancies. Nutrients 12(12):3732. https://doi.org/10.3390/nu12123732
21. Wang H, Hu YF, Hao JH, Chen YH, Su PY, Wang Y, Yu Z, Fu L, Xu YY, Zhang C, Tao FB, Xu DX (2015) Maternal zinc deficiency during pregnancy elevates the risks of fetal growth restriction: a population-based birth cohort study. Sci Rep 5:11262. https://doi.org/10.1038/srep11262
22. Sultana M, Jahan N, Sultana N, Quraishi SB, Chowdhury TR (2010) Preterm birth: the role of zinc. J Bangladesh Soc Physiol 5:27–33. https://doi.org/10.3329/jbsp.v5i1.5415
23. Siva S, Rubin DT, Gulotta G, Wroblewski K, Pekow J (2017) Zinc deficiency is associated with poor clinical outcomes in patients with inflammatory bowel disease. Inflamm Dis 23:152. https://doi.org/10.1097/MIB.0000000000000989
24. Garner TB, Hester JM, Carothers A, Diaz FJ (2021) Role of zinc in female reproduction. Biol Reprod 104(5):976–994. https://doi.org/10.1093/biolre/ioab023
25. Osredkar J, Geršak ŽM, Karas Kuželički N, Snoj Tratnik J, Mazej D, Falnoga I, Horvat M, Geršak K (2022) Association of Zn and Cu levels in cord blood and maternal milk with pregnancy outcomes among the Slovenian population. Nutrients 14(21):4667. https://doi.org/10.3390/nu14214667
26. Plum LM, Rink L, Haase H (2010) The essential toxin: impact of zinc on human health. Int J Environ Res Public Health 7:1342–1365. https://doi.org/10.3390/ijerph7041342

27. Wang F, Jiao P, Qi M, Frezza M, Dou QP, Yan B (2010) Turning tumor-promoting copper into an anti-cancer weapon via high-throughput chemistry. Curr Med Chem 17:2685–2698. https://doi.org/10.2174/092986710791859315
28. Khayat S, Fanaei H, Ghanbarzehi A (2017) Minerals in pregnancy and lactation: a review article. J Clin Diagn Res 11(9):QE01–QE05. https://doi.org/10.7860/JCDR/2017/28485.10626
29. Wilson RL, Bianco-Miotto T, Leemaqz SY, Grzeskowiak LE, Dekker GA, Roberts CT (2018) Early pregnancy maternal trace mineral status and the association with adverse pregnancy outcome in a cohort of Australian women. J Trace Elem Med Biol 46:103–109. https://doi.org/10.1016/j.jtemb.2017.11.016
30. World Health Organization (WHO) (2013) Guideline: calcium supplementation in pregnant women. WHO, Geneva
31. Dwarkanath P, Muhihi A, Sudfeld CR, Wylie BJ, Wang M, Perumal N, Thomas T, Kinyogoli SM, Bakari M, Fernandez R, Raj JM, Swai NO, Buggi N, Shobha R, Sando MM, Duggan CP, Masanja HM, Kurpad AV, Pembe AB, Fawzi WW (2024) Two randomized trials of low-dose calcium supplementation in pregnancy. N Engl J Med 390(2):143–153. https://doi.org/10.1056/NEJMoa2307212
32. Pitilin EB, Bagatini MD, Gasparin VA, Oliveira PP, Lentsck MH, Baratieri T, Falavina LP et al (2024) Effects of calcium supplementation on markers of preeclampsia: randomized clinical trial. Acta Paul Enferm 37:eAPE01622
33. Shkembi B, Huppertz T (2021) Calcium absorption from food products: food matrix effects. Nutrients 14(1):180. https://doi.org/10.3390/nu14010180
34. Lee CC, Yang PK, Chen LC, Cheong ML, Tsai YL, Tsai MS (2022) Associations between gene expression of magnesium transporters and glucose metabolism in pregnancy. J Formos Med Assoc 121(7):1231–1237. https://doi.org/10.1016/j.jfma.2021.11.006
35. Conde-Agudelo A, Romero R (2009) Antenatal magnesium sulfate for the prevention of cerebral palsy in preterm infants less than 34 weeks' gestation: a systematic review and meta-analysis. Am J Obstet Gynecol 200(6):595–609. https://doi.org/10.1016/j.ajog.2009.04.005
36. Wolf HT, Huusom LD, Henriksen TB, Hegaard HK, Brok J, Pinborg A (2020) Magnesium sulphate for fetal neuroprotection at imminent risk for preterm delivery: a systematic review with meta-analysis and trial sequential analysis. BJOG 127(10):1180–1188. https://doi.org/10.1111/1471-0528.16238
37. Jafarabady K, Shafiee A, Eshraghi N, Salehi SA, Mohammadi I, Rajai S, Zareian Z, Movahed F, Bakhtiyari M (2024) Magnesium sulfate for fetal neuroprotection in preterm pregnancy: a meta-analysis of randomized controlled trials. BMC Pregnancy Childbirth 24(1):519. https://doi.org/10.1186/s12884-024-06703-9
38. Crowther CA, Ashwood P, Middleton PF, McPhee A, Tran T, Harding JE, MAGENTA Study Group (2023) Prenatal intravenous magnesium at 30-34 weeks' gestation and neurodevelopmental outcomes in offspring: the MAGENTA randomized clinical trial. JAMA 330(7):603–614. https://doi.org/10.1001/jama.2023.12357
39. Duley L, Henderson-Smart DJ, Walker GJ, Chou D (2010) Magnesium sulphate versus diazepam for eclampsia. Cochrane Database Syst Rev:CD000127. https://doi.org/10.1002/14651858.CD000127.pub2
40. Duley L, Gülmezoglu AM, Henderson-Smart DJ, Chou D (2010) Magnesium sulphate and other anticonvulsants for women with pre-eclampsia. Cochrane Database Syst Rev:CD000025. https://doi.org/10.1002/14651858.CD000025.pub2
41. Demircan K, Chillon TS, Jensen RC, Jensen TK, Sun Q, Bonnema SJ, Glintborg D, Bilenberg N, Andersen MS, Schomburg L (2024) Maternal selenium deficiency during pregnancy in association with autism and ADHD traits in children: the Odense Child Cohort. Free Radic Biol Med 220:324–332. https://doi.org/10.1016/j.freeradbiomed.2024.05.001
42. Wu S, Wang M, Deng Y, Qiu J, Zhang X, Tan J (2020) Associations of toxic and essential trace elements in serum, follicular fluid, and seminal plasma with in vitro fertilization outcomes. Ecotoxicol Environ Saf 204:110965. https://doi.org/10.1016/j.ecoenv.2020.110965

43. Yang J, Kang Y, Chang Q, Zhang B, Liu X, Zeng L, Yan H, Dang S (2022) Maternal zinc, copper, and selenium intakes during pregnancy and congenital heart defects. Nutrients 14(5):1055. https://doi.org/10.3390/nu14051055
44. Toloza FJK, Motahari H, Maraka S (2020) Consequences of severe iodine deficiency in pregnancy: evidence in humans. Front Endocrinol (Lausanne) 11:409. https://doi.org/10.3389/fendo.2020.00409
45. World Health Organization (WHO) (2014) Guideline: fortification of food-grade salt with iodine for the prevention and control of iodine deficiency disorders. WHO, Geneva. Available from: https://www.ncbi.nlm.nih.gov/books/NBK254243/
46. Santos LKX, Lima LL, Lima SML, Navarro AM (2012) Iodine deficiency and associated factors in infants and preschool children from a municipality in the semi-arid region of Minas Gerais, Brazil, 2008. Cad Saude Publica 28(2):346–356. https://doi.org/10.1590/S0102-311X2012000200013
47. Skeaff SA (2023) An adequate intake of iodine in pregnancy: better safe than sorry. Am J Clin Nutr 118(1):8–9. https://doi.org/10.1016/j.ajcnut.2023.05.013
48. Pontes AAN, Rocha ADM, Leite DFB, Lessa ADF, Adan LFF (2009) Salt iodization in Brazil: a controversial issue. Arq Bras Endocrinol Metabol 53(1):124–126. https://doi.org/10.1590/S0004-27302009000100017
49. Espinoza A, Le Blanc S, Olivares M, Pizarro F, Ruz M, Arredondo M (2012) Iron, copper, and zinc transport: inhibition of divalent metal transporter 1 (DMT1) and human copper transporter 1 (hCTR1) by shRNA. Biol Trace Elem Res 146:281–286. https://doi.org/10.1007/s12011-011-9243-2
50. Benkhedda K, L'Abbé MR, Cockell KA (2010) Effect of calcium on iron absorption in women with marginal iron status. Br J Nutr 103(5):742–748. https://doi.org/10.1017/S0007114509993546
51. Lönnerdal B (2010) Calcium and iron absorption--mechanisms and public health relevance. Int J Vitam Nutr Res 80(4–5):293–299. https://doi.org/10.1024/0300-9831/a000036

Chapter 2
Trace Elements as Molecular Orchestrators of Health and Disease: A Biochemical and Pathophysiological Perspectives

Ghizal Fatima, Sadaf Khan, Jan Fedacko, Ammar Mehdi Raza, Ram B. Singh, and Jaipaul Singh

Abstract Trace elements, though required in minute quantities, play fundamental roles in maintaining physiological homeostasis, enzymatic functions, and cellular integrity. Essential trace elements, such as zinc, copper, selenium, iron, and manganese, act as cofactors for key enzymatic reactions, regulate redox balance, and contribute to immune function, neuroprotection, and metabolic processes. Dysregulation of these elements, either through deficiency or excess, has been implicated in a wide spectrum of diseases, including cardiovascular disorders, neurodegenerative conditions, diabetes, cancer, and others. Zinc and copper homeostasis are critical for immune competence and oxidative stress modulation, while selenium exerts profound effects on redox signalling and thyroid function. Iron imbalance is closely associated with anemia, oxidative stress, and neurodegeneration, whereas manganese plays a pivotal role in mitochondrial function and antioxidant defense. Emerging research highlights the role of trace element interactions, metal ion transporters, and metalloproteins in disease pathophysiology. This chapter explores the intricate biochemical and molecular mechanisms through which trace elements can influence health and diseases, emphasizing their diagnostic and therapeutic potential in precision medicine and clinical nutrition. Understanding the homeostatic

G. Fatima (✉) · S. Khan
Department of Biotechnology, Chronobiology Laboratory, Era University, Lucknow, India

J. Fedacko
Department of Gediatrics, PJ Safaric University, Kosice, Slovakia

A. M. Raza
Department of Pedodontics, Career Institute of Dental Sciences and Hospital, Lucknow, India

R. B. Singh
Halberg Hospital and Research Institute, Moradabad, UP, India
e-mail: rbs@tsimtsoum.net

J. Singh
School of Pharmacy and Biomedical Sciences, University of Central Lancashire, Preston, UK

N. S. Dhalla et al. (eds.), *Functional Biochemistry of Metallic Elements*,
Advances in Biochemistry in Health and Disease 36,
https://doi.org/10.1007/978-3-032-14445-4_2

regulation and pathophysiological implications of trace elements may provide novel insights into disease prevention and targeted interventions.

Keywords Trace elements · Micronutrients · Oxidative stress · Metabolic disorders · Neurodegeneration · Precision medicine

Introduction

Trace elements, also known as microminerals, are essential inorganic nutrients required by the human body in minute quantities and they play pivotal roles in numerous physiological and biochemical processes. These elements, including zinc, copper, selenium, iron, manganese, and chromium, serve as cofactors for enzymes, participate in redox homeostasis, and regulate immune function, cellular signalling, and metabolic pathways. The delicate balance of trace elements is critical for maintaining health, as both deficiencies and excesses can contribute to pathophysiological diseases and dysfunctions [1]. Table 2.1 illustrates the roles of trace Elements in health and disease. The significance of trace elements in human health extends to nearly all organ systems. Zinc, for example, is indispensable for immune function, wound healing, and enzymatic activity, while copper plays a role in mitochondrial respiration, connective tissue formation, and neurotransmitter synthesis [2]. Selenium, an essential component of selenoproteins, is crucial for antioxidant defense and thyroid hormone metabolism [3]. Iron, a key element in hemoglobin and myoglobin, ensures efficient oxygen transport, whereas manganese is required for bone development and enzymatic reactions involving metabolism and antioxidant defense [4]. Chromium, though less studied, has been implicated in insulin signalling and glucose homeostasis [5].

Dysregulation of trace elements has been linked to various chronic diseases, including cardiovascular disorders (CVDs), neurodegenerative diseases (NDD), diabetes, and cancer [6]. Iron overload, as observed in hereditary hemochromatosis, leads to oxidative damage and organ dysfunction, while iron deficiency is a major cause of anemia and impaired cognitive function [7]. Zinc deficiency has been associated with immune dysfunction, delayed wound healing, and increased susceptibility to infections [8]. Selenium imbalance is implicated in thyroid disorders and increased oxidative stress, contributing to chronic inflammatory conditions [9]. The role of copper dysregulation in neurodegenerative disorders such as Alzheimer's and Parkinson's disease has gained increasing attention, since either copper accumulation or deficiency can alter neuronal function and contribute to disease progression [10]. The main toxic effects attributable to manganese are extrapyramidal side effects that closely resemble symptoms of Parkinson's disease. These adverse effects are secondary to their deposition in specific components of the basal ganglia and alteration of dopaminergic neuronal enzyme activity. Other prominent effects include cardiotoxicity, hepatotoxicity, and increased mortality in infants [11]. The toxicity associated with chronic chromate (CrO_4^{2-}) exposure has been known for

Table 2.1 Role of trace elements in health and disease

Trace elements	Biological functions	Deficiency effects	Excess effects	Associated diseases
Zinc (Zn)	Immune function, enzyme activity, wound healing, DNA synthesis, antioxidant defense	Impaired immunity, delayed wound healing, growth retardation, increased infections	Nausea, vomiting, copper deficiency, immune suppression [8]	Immune disorders, growth failure, neurodegenerative diseases
Copper (Cu)	Mitochondrial function, iron metabolism, neurotransmitter synthesis, antioxidant defense	Anemia, neutropenia, neurological disorders, osteoporosis	Wilson's disease, liver damage, neurotoxicity [10]	Alzheimer's disease, Parkinson's disease, Menkes disease, Wilson's disease
Selenium (Se)	Antioxidant defense, thyroid hormone metabolism, DNA synthesis, immune function	Oxidative stress, hypothyroidism, weakened immunity, cardiomyopathy (Keshan disease)	Selenosis (nausea, hair loss, neurological issues) [9].	Thyroid disorders, cardiovascular diseases, cancer risk
Iron (Fe)	Oxygen transport (hemoglobin, myoglobin), energy production, cognitive function	Anemia, fatigue, impaired cognitive function, immune dysfunction	Iron overload (hemochromatosis), liver damage, oxidative stress [7].	Anemia, neurodegenerative disorders, heart disease, infections
Manganese (Mn)	Bone formation, metabolism, antioxidant enzyme function (SOD), neurotransmitter regulation	Bone deformities, impaired metabolism, neurological issues	Manganism (neurotoxicity), tremors, cognitive impairment [11].	Neurological disorders, metabolic disorders, osteoporosis
Chromium (Cr)	Glucose metabolism, insulin signaling, lipid metabolism	Impaired glucose tolerance, insulin resistance, increased risk of diabetes	Potential DNA damage, liver and kidney toxicity [12].	Type 2 diabetes, metabolic syndrome, cardiovascular disease

more than 200 years. During World War II exposure to hexavalent chromium was linked to an increased risk of lung cancer [12].

The mechanisms underlying the impact of trace elements on health and disease involve their participation in enzymatic catalysis, redox balance, gene regulation, and cellular metabolism [13]. Metalloproteins, metal ion transporters, and metal-responsive transcription factors play key roles in maintaining trace element homeostasis [14]. Disruptions in these regulatory systems can lead to oxidative stress, mitochondrial dysfunction, and inflammatory responses, exacerbating disease pathology. Furthermore, interactions between trace elements, such as the antagonistic relationship between zinc and copper or the synergistic effects of selenium and iodine in thyroid function, highlight the complexity of their biological roles. Given their crucial functions, trace elements have garnered interest in the field of precision medicine and clinical nutrition [14]. Advances in analytical techniques allow for precise measurement of trace element levels in biological fluids, providing potential biomarkers for disease diagnosis and prognosis. Moreover, therapeutic strategies involving trace element supplementation, chelation therapy, and dietary interventions are being explored to either prevent or mitigate disease progression.

This chapter aims to provide a comprehensive understanding of the role of trace elements in health and diseases, highlighting their biochemical functions, regulatory mechanisms, and clinical implications. By elucidating the molecular pathways through which these elements influence physiological processes, it is possible to develop targeted strategies for disease prevention and therapeutic interventions. Understanding the interplay of trace elements in health and disease offers novel insights into personalized medicine and nutritional approaches for improving human health.

Biochemical Role of Trace Elements

Trace elements are vital minerals that are necessary for many biological processes but that organisms need in trace amounts. Notwithstanding their modest levels in the body, they are essential for preserving homeostasis, promoting the activity of enzymes, and guaranteeing regular metabolic processes. Iron, zinc, copper, manganese, selenium, iodine, and molybdenum are examples of common trace elements. Numerous illnesses or metabolic abnormalities can result from their imbalance or lack, as they frequently function as cofactors or structural elements of enzymes [15].

Trace elements play a variety of roles in biochemistry, including controlling protein synthesis, redox reactions, and metabolic activities. They support the correct operation of biochemical pathways by stabilizing macromolecules such as proteins and nucleic acids. Research on nutrition, health, and disease prevention has made understanding the biochemical functions of trace elements a major focus [15].

The Major Trace Elements and Their Roles

Trace elements are essential micronutrients that play critical roles in various biological processes, including immune function, enzyme activity, antioxidant defense, and metabolism. Figure 2.1 illustrates the role of essential trace elements in human health and biological functions.

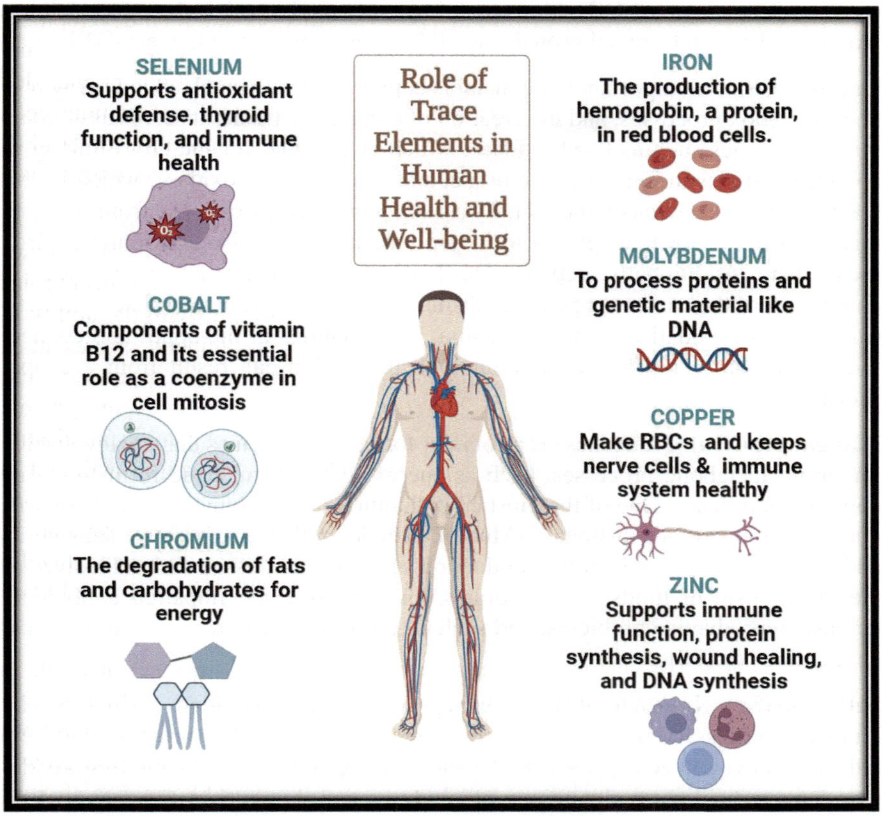

Fig. 2.1 Role of essential trace elements in human health and biological functions

Iron (Fe) One of the most important trace elements, iron is mainly engaged in electron transport, oxygen transport, and cellular respiration. Iron is a fundamental part of the heme groups in hemoglobin and myoglobin in its ferrous (Fe^{2+}) form, which allows oxygen to bind and be released in muscles and blood. A variety of enzymes, especially those engaged in redox processes, depend on iron as a cofactor. Among these are cytochromes, which are essential for oxidative phosphorylation in the mitochondria [16]. Anemia from iron shortage can affect oxygen transport, resulting in weakness, exhaustion, and immune system failure [17].

Zinc (Zn) One of the most functionally varied trace elements in biochemistry, zinc is involved in more than 300 enzymatic processes. It is necessary for the action of enzymes that are involved in wound healing, cellular division, protein synthesis, and DNA synthesis. By attaching to histidine and cysteine residues, zinc stabilizes protein structures and helps to produce zinc fingers, which are motifs that enable proteins to attach to DNA. Additionally, zinc is essential for immune function maintenance, signaling pathway regulation, and as a cofactor for the antioxidant enzyme superoxide dismutase (SOD). Skin conditions, immunological dysfunction, delayed wound healing, and stunted growth can all be caused by zinc deficiency [18].

Copper (Cu) An important trace metal is copper, which is involved in energy production, redox processes, and the creation of connective tissue. It serves as a cofactor for enzymes such as lysyl oxidase, which is essential for the cross-linking of collagen and elastin in connective tissue, and cytochrome c oxidase, which is crucial for the electron transport chain. By assisting in the release of iron from cells and making it easier for hemoglobin to incorporate it, copper also contributes to iron metabolism. Additionally, copper contributes to antioxidant defense by being a component of the enzyme superoxide dismutase (SOD), which neutralizes free radicals to reduce oxidative damage. Anemia, abnormalities of the bones and connective tissue, and compromised immunological function can result from a copper shortage [19].

Manganese (Mn) Manganese is a cofactor for several enzymes that are involved in important metabolic processes, such as energy metabolism, bone formation, and antioxidant defense. One of the most crucial antioxidant enzymes in mitochondria, manganese superoxide dismutase (MnSOD) shields cells from oxidative damage. In addition, manganese aids in the production of connective tissue and is involved in the metabolism of lipids, carbohydrates, and amino acids. Impaired antioxidant defense, reproductive problems, and skeletal deformities can result from a manganese deficiency [20].

Selenium (Se) Known for its function in protecting against oxidative damage, selenium is a necessary trace element. It is integrated into selenoproteins that help reduce reactive oxygen species (ROS) and shield cellular constituents from oxidative damage, including glutathione peroxidases and thioredoxin reductases. As a cofactor for deiodinases, which activate or deactivate thyroid hormones, selenium also contributes to thyroid hormone metabolism. Increased oxidative stress, compromised antioxidant defenses, and diseases like Kashin-Beck disease (an osteoarthritis) and Keshan disease (a cardiomyopathy) are all linked to selenium deficiency [21].

Iodine (I) Iodine is a trace element that is mostly linked to the production of thyroid hormones, specifically triiodothyronine and thyroxine, which control growth, development, and metabolism. These hormones, which are essential for healthy development during pregnancy and infancy, especially for brain development, are produced by the thyroid gland utilizing iodine. Thyroid conditions including goiter (enlargement of the thyroid gland) and hypothyroidism can be brought on by an iodine deficit. These conditions can cause mental retardation in infants and cognitive impairment in adults [22].

Molybdenum (Mo) Sulfite oxidase and xanthine oxidase are two crucial enzymes that depend on molybdenum, a vital trace metal. Purine metabolism and the metabolism of amino acids containing sulfur are both mediated by these enzymes. Although very uncommon, molybdenum deficiency can cause brain damage and other problems linked to sulfur metabolism [23].

Mechanisms of Action in Enzyme Catalysis and Metabolism

Enzyme catalysis is one of the most extensively studied applications of trace elements. Proteins called enzymes catalyze biological reactions, and many of them need metal ions as cofactors to function [24]. These active sites of enzymes, which take part in a variety of catalytic processes, are essential for trace elements like magnesium, manganese, copper, iron, and zinc. For instance, zinc aids in the binding of substrate molecules or stabilizes the structure of the enzyme, both of which are important in enzyme catalysis. An important enzyme in the respiratory chain of the mitochondria, cytochrome c oxidase, is one example of an electron transport reaction that involves copper [25]. Heme groups, which contain iron, are necessary for the catalytic activity of enzymes such as hemoglobin and cytochrome P450 [26]. Magnesium promotes enzyme activity by maintaining ATP, while manganese is a cofactor for enzymes involved in the creation of compounds like carbohydrates [27].

These trace elements also play a role in more general metabolic processes such as the citric acid cycle, oxidative phosphorylation, and glycolysis. Iron deficiency and zinc insufficiency are two examples of metabolic illnesses that can result from deficiencies or excesses of these metals that affect enzyme performance, disrupt metabolic pathways, and decrease immunological function.

Role in Redox Reactions and Antioxidant Defense

Redox processes, which entail the transfer of electrons and are essential for preserving cellular energy generation and homeostasis, involve trace elements as major participants. By serving as electron donors or acceptors, several trace elements, including iron, copper, manganese, and selenium, directly contribute to the facilitation of redox processes [28]. The function of selenium in the antioxidant defense system is very well-known. It is a crucial part of selenoproteins such as glutathione peroxidases, which prevent oxidative damage to cells by catalyzing the reduction of lipid hydroperoxides and hydrogen peroxide. Through enzymes like superoxide dismutase (SOD), which catalyze the conversion of superoxide radicals to hydrogen peroxide and oxygen, copper and zinc also contribute to antioxidant defense by lowering oxidative stress [29].

These trace elements assist in controlling the equilibrium between the generation and neutralization of free radicals by promoting the transfer of electrons in redox processes. Protecting cellular constituents (such as lipids, proteins, and DNA) against oxidative damage which is connected to aging, cancer, cardiovascular disorders, and other long-term medical conditions requires this equilibrium [30].

Mechanisms of Trace Element Imbalance and Pathophysiology

Essential micronutrients known as trace elements are vital for immunological response, hormone synthesis, enzyme activity, and general metabolic equilibrium. However, serious pathophysiological disorders can arise from their imbalance, either in excess or in shortage. Trace element control includes cellular uptake, use, excretion, gastrointestinal absorption, and dietary intake. Disease states can result from disturbances at any of these levels brought on by nutritional, environmental, or hereditary causes. Creating successful therapeutic and preventative interventions requires an understanding of the mechanisms behind these imbalances [31].

Deficiency States and Their Impact on Health

Deficiencies in trace elements occur due to inadequate dietary intake, malabsorption, increased physiological demand, or chronic diseases that affect nutrient metabolism. These deficiencies can impair biological functions and lead to severe health consequences [32]. Below are some key trace element deficiencies and their mechanisms and clinical implications. Table 2.2 *shows the mechanisms and health impacts of trace element deficiencies, while* Fig. 2.2 provides a comprehensive overview of the relationship between trace element deficiencies and their associated diseases, emphasizing the crucial role of micronutrient balance in maintaining health.

Toxicity of Trace Elements: Mechanisms and Clinical Manifestations

Excessive supplementing, industrial contamination, environmental exposure, or compromised excretion processes can all lead to an excessive buildup of trace elements. Oxidative stress, enzyme inhibition, and systemic toxicity are frequently caused by trace element poisoning. Table 2.3 outlines the mechanisms and clinical manifestations of trace element toxicity, emphasizing the consequences of excessive accumulation in the body.

Trace Elements in Disease Mechanisms

A variety of physiological and biochemical processes depend heavily on trace elements, which are vital micronutrients. Because of their roles in cellular signaling, redox balance, and enzyme function, they are essential for preserving homeostasis.

Table 2.2 Mechanisms and health impacts of trace element deficiencies

Trace elements	Mechanisms	Impacts
Iron deficiency	Iron is essential for hemoglobin, myoglobin, and enzymatic systems. Deficiency results from poor intake, chronic blood loss (e.g., GI bleeding, heavy menstruation), or malabsorption (e.g., celiac disease) [33]	Fatigue, pallor, weakness, cognitive decline, reduced immunity, and severe cases may lead to heart failure due to inadequate oxygen transport
Zinc deficiency	Zinc is crucial for immune function, wound healing, and enzymatic activity. Causes include poor diet, chronic illness, genetic disorders (e.g., acrodermatitis enteropathica), and excessive loss through diarrhea or kidney disease [34]	Growth retardation, delayed wound healing, impaired taste and smell, immune dysfunction, and neurological disorders such as depression and memory loss
Iodine deficiency	Iodine is necessary for thyroid hormone synthesis. Deficiency occurs due to inadequate dietary intake, particularly in regions with iodine-deficient soils [35]	Goiter, hypothyroidism, developmental delays in children, cognitive impairment, and severe conditions like cretinism, which leads to mental retardation and stunted growth
Selenium deficiency	Selenium is a cofactor for antioxidant enzymes such as glutathione peroxidase. Deficiency arises from poor dietary intake or malabsorption syndromes [36]	Increased oxidative stress, weakened immune system, Keshan disease (cardiomyopathy), and predisposition to thyroid dysfunction
Copper deficiency	Copper is essential for iron metabolism, connective tissue integrity, and neurological function. Causes include genetic conditions (e.g., Menkes disease), excessive zinc intake (which inhibits copper absorption), or gastrointestinal disorders affecting absorption [37]	Anemia, neutropenia, connective tissue disorders, osteoporosis, neurological impairments (ataxia, neuropathy), and cardiovascular abnormalities

The pathophysiology of many diseases, including cancer, metabolic syndromes, autoimmune disorders, cardiovascular diseases, and neurodegenerative disorders, can be attributed to imbalances, either toxicity or deficiency. Knowing how trace elements function in certain disease processes helps to clarify possible treatment and preventative strategies [43]. Figure 2.3 illustrates the impact of trace element imbalances on various human disease mechanisms, emphasizing their role in neurodegenerative diseases, cardiovascular diseases, metabolic disorders, and autoimmune conditions. It highlights how excess or deficiency of essential elements, such as iron, copper, zinc, and selenium, contribute to pathological processes, including oxidative stress, inflammation, insulin resistance, and immune dysfunction.

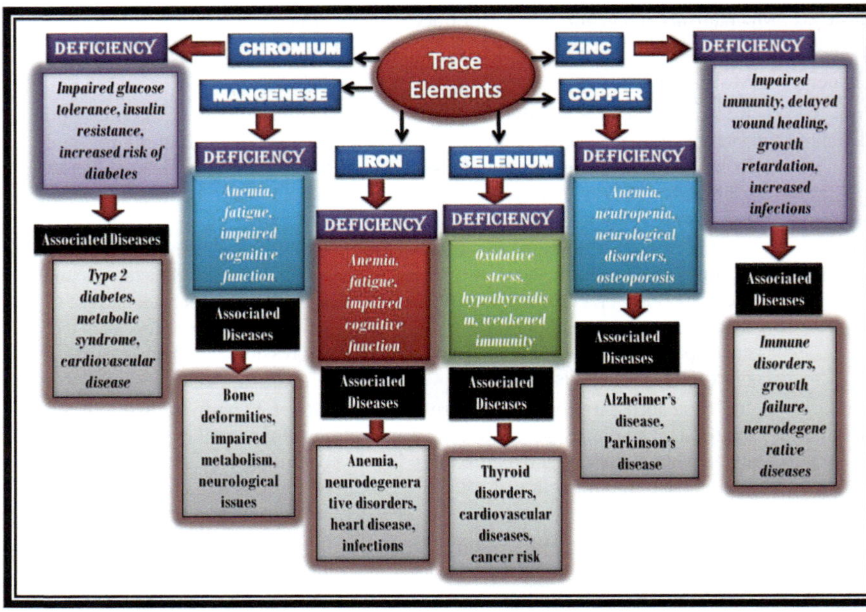

Fig. 2.2 Trace elements: deficiency and associated diseases—the figure illustrates the relationship between trace element deficiencies and their associated diseases. Each trace element (Zn, Cu, Se, Fe, Mn, Cr) is mapped to its deficiency consequences, highlighting their impact on various physiological functions and disease pathogenesis, emphasizing the critical role of micronutrient balance in health

Table 2.3 Trace element toxicity: mechanisms and clinical manifestations

Trace elements	Mechanism	Clinical manifestations
Iron toxicity	Excess iron leads to oxidative stress via Fenton reactions, causing tissue damage	Liver cirrhosis, diabetes, cardiomyopathy, and neurodegeneration (hemochromatosis) [38]
Zinc toxicity	Disruption of copper metabolism and impaired immune function	Nausea, vomiting, immune suppression, and neurological disorders [39]
Copper toxicity	Excess copper induces oxidative stress, leading to cellular damage	Liver damage (Wilson's disease), neurological impairment, and gastrointestinal distress [40]
Selenium toxicity	Disruption of sulfur-containing enzymes, leading to oxidative damage	Hair loss, nail brittleness, gastrointestinal symptoms, and neurological abnormalities [41]
Iodine toxicity	Overstimulation or inhibition of thyroid function	Hyperthyroidism, hypothyroidism, thyroiditis, and goitre [42]

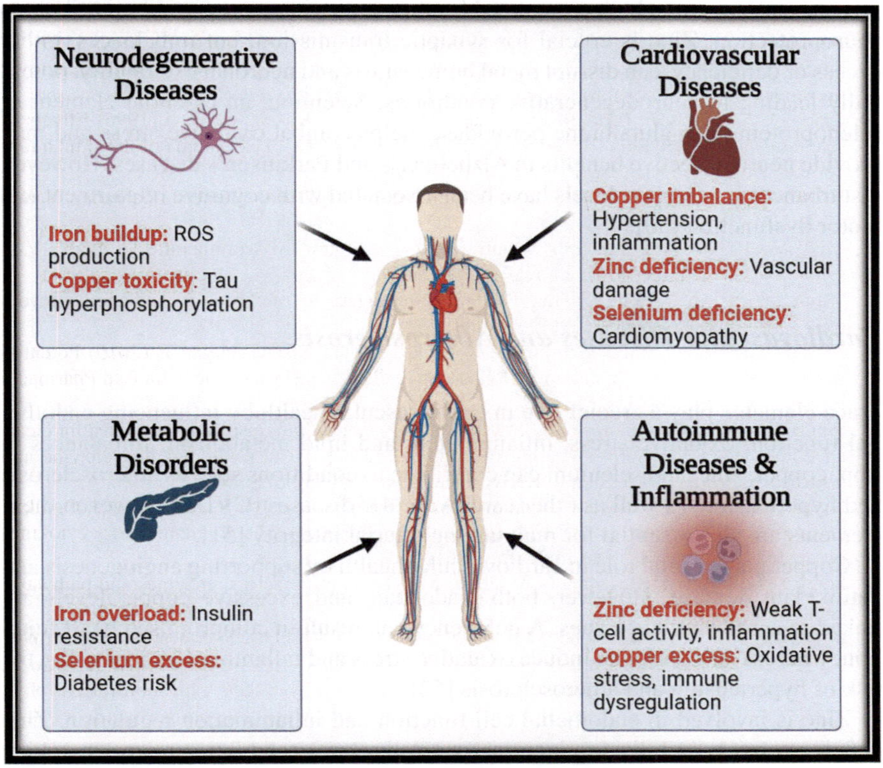

Fig. 2.3 Role of trace element imbalances in human disease mechanisms

Neurodegenerative Diseases

Neurodegenerative diseases are characterized by progressive neuronal loss, protein misfolding, oxidative stress, and neuroinflammation [44, 45]. Trace elements such as iron, copper, zinc, and selenium play critical roles in maintaining neuronal function, but their dysregulation can exacerbate neurodegenerative processes.

One of the main characteristics of many neurodegenerative diseases is iron buildup. In Alzheimer's disease (AD), excessive iron catalyzes the formation of reactive oxygen species (ROS) via the Fenton reaction, leading to oxidative damage and neuronal death. Amyloid-beta plaque, a defining feature of AD, has also been linked to iron overload [46]. In Parkinson's disease (PD), increased iron levels in the substantia nigra contribute to dopaminergic neuron degeneration and α-synuclein aggregation [47].

Copper imbalance also plays a role in neurodegeneration. While copper is vital for various enzymatic functions, an excess of it can trigger oxidative stress and enhance tau hyperphosphorylation in Alzheimer's disease [48]. In Wilson's disease, excessive copper buildup in the brain results in significant neurological damage [49].

Zinc and selenium, both recognized for their antioxidant properties, contribute to neuroprotection. Zinc is crucial for synaptic transmission, but imbalances, either excess or deficiency, can disrupt metal homeostasis and neuronal excitability, potentially leading to neurodegenerative conditions. Selenium, an essential element of selenoproteins like glutathione peroxidase, helps combat oxidative stress and may provide neuroprotective benefits in Alzheimer's and Parkinson's diseases. However, disturbances in selenium levels have been associated with cognitive impairment and motor dysfunction [50].

Cardiovascular Diseases and Atherosclerosis

Trace elements play a crucial role in cardiovascular health by influencing endothelial function, oxidative stress, inflammation, and lipid metabolism. Imbalances in iron, copper, zinc, and selenium can contribute to conditions such as atherosclerosis and hypertension, as well as other cardiovascular diseases (CVDs). However, these elements are also essential for maintaining arterial integrity [51].

Copper plays a vital role in cardiovascular health by supporting angiogenesis and antioxidant defense. However, both inadequate and excessive copper levels are linked to cardiovascular issues. A deficiency can result in compromised heart function, whereas an excess can induce oxidative stress and inflammation, increasing the risk of hypertension and atherosclerosis [52].

Zinc is involved in endothelial cell function and inflammation regulation. Zinc deficiency has been linked to increased oxidative stress and inflammatory cytokine production, leading to vascular damage and plaque instability in atherosclerosis. On the other hand, excess zinc can interfere with copper metabolism and disrupt antioxidant defenses [53].

Selenium plays a protective role in cardiovascular health through its involvement in selenoproteins such as glutathione peroxidase, which reduce oxidative stress. Selenium deficiency has been linked to Keshan disease, a cardiomyopathy prevalent in selenium-deficient populations. However, excessive selenium intake may have prooxidant effects and contribute to metabolic disturbances [54].

Metabolic Disorders

Trace elements play a fundamental role in regulating various metabolic pathways, influencing key processes, such as insulin function, glucose homeostasis, lipid metabolism, and overall energy balance. The proper balance of essential elements, including zinc, chromium, iron, and selenium, is crucial for maintaining metabolic health. However, imbalances in these elements have been associated with the development of metabolic disorders, including diabetes and obesity [55].

Zinc is particularly important in the regulation of insulin production, storage, and secretion within pancreatic beta cells. It plays a crucial role in insulin crystallization, ensuring its proper function, while also acting as a protective agent against oxidative stress. Studies have shown that zinc deficiency is linked to insulin resistance, leading to an increased risk of developing type 2 diabetes. On the other hand, excessive zinc levels can have negative consequences, potentially impairing glucose metabolism and disrupting insulin signaling pathways, which may contribute to metabolic dysfunction [56].

Chromium is another essential trace element known for its role in enhancing insulin sensitivity. It serves as a key component of chromodulin, a protein that facilitates the activation of insulin receptors, thereby improving glucose uptake by cells. A deficiency in chromium has been associated with impaired glucose tolerance and a heightened risk of developing diabetes. Due to its beneficial effects on insulin function, chromium supplementation has been explored as a potential therapeutic strategy for managing blood sugar levels in individuals with diabetes or insulin resistance [57].

Iron also plays a significant role in metabolic health, but an overload of this element has been recognized as a major risk factor for diabetes. Excess iron contributes to insulin resistance and pancreatic beta-cell dysfunction, primarily by promoting oxidative stress and inflammation within metabolic tissues. Conditions such as hemochromatosis, which lead to excessive iron accumulation in the body, are strongly associated with an increased risk of diabetes and metabolic syndrome, highlighting the importance of maintaining iron homeostasis [58].

Selenium, another critical trace element, has a complex and dual relationship with metabolic health. While it is essential for antioxidant defense and protects cells from oxidative damage, excessive selenium intake has been linked to an increased risk of type 2 diabetes. This may be due to its influence on insulin signaling pathways and mitochondrial function, which, when disrupted, can contribute to metabolic disturbances. Therefore, while selenium is necessary for maintaining cellular health, its levels must be carefully regulated to prevent adverse metabolic outcomes [59].

Autoimmune Diseases and Inflammation

The immune system depends on trace elements to maintain proper inflammatory responses, regulate cytokine production, and support immune cell activity. Disruptions in the balance of zinc, copper, selenium, and iodine have been linked to autoimmune conditions, such as rheumatoid arthritis, multiple sclerosis, and thyroid disorders [60].

Zinc is vital for immune regulation and controlling inflammation. A zinc deficiency weakens T cell activity, heightens the production of pro-inflammatory cytokines, and worsens autoimmune diseases like rheumatoid arthritis and lupus. Previous research has explored zinc supplementation as a potential strategy to help balance immune responses [61]. Copper plays an essential role in activating immune cells and regulating inflammatory processes. Both a shortage and an excess of copper have been associated with autoimmune diseases. While excessive copper levels can contribute to inflammation and oxidative stress, insufficient copper compromises immune function [62]. Selenium is critical for controlling inflammation and managing autoimmune reactions. A deficiency in selenium has been linked to a greater risk of developing Hashimoto's thyroiditis and other diseases. However, consuming too much selenium can also lead to autoimmune thyroid dysfunction [63]. Iodine is necessary to produce thyroid hormones, yet an imbalance in iodine levels can lead to autoimmune thyroid disorders. Excess iodine intake has been shown to trigger Hashimoto's thyroiditis or worsen hyperthyroidism, whereas insufficient iodine levels are connected to goiter and hypothyroidism [64].

Trace Elements and Their Role in Immune Function

Trace elements are vital for sustaining a well-functioning immune system, as they contribute to the regulation of inflammatory responses and the activation of immune cells. Essential elements, such as zinc, copper, selenium, iron, and manganese, play a pivotal role in immune defense by serving as cofactors for various enzymes, influencing the production of cytokines, and helping to control oxidative stress within the body. These elements work in tandem to support both innate and adaptive immune responses, ensuring the body can effectively fight off infections and maintain immune balance [65].

An imbalance in these trace elements, whether due to deficiency or excessive accumulation, can lead to significant immune dysfunction. A lack of essential trace elements may weaken immune defenses, making individuals more susceptible to infections, while an overabundance may trigger chronic inflammation and contribute to the development of autoimmune disorders [66]. For instance, inadequate zinc levels can impair T cell function and hinder proper immune signaling, whereas excessive copper can promote oxidative stress and inflammation. Selenium, iron, and manganese are also critical in maintaining immune equilibrium, and disruptions in their levels can negatively impact overall health [67].

Maintaining a proper balance of these trace elements is crucial for preventing immune-related disorders, reducing inflammation, and ensuring the body's ability to respond effectively to pathogens. A well-regulated intake of these nutrients through diet or supplementation, when necessary, can help support immune resilience and lower the risk of chronic diseases associated with immune dysfunction.

Role in Immune Response and Inflammatory Pathways

Trace elements are fundamental to the proper functioning of both the innate and adaptive immune systems. They play a crucial role in activating immune cells, regulating cytokine production, and ensuring the resolution of inflammation. These elements serve as essential cofactors for various enzymes and proteins that contribute to immune defense and inflammatory balance. The immune system relies on an optimal supply of trace elements to mount an effective response against pathogens while preventing excessive or prolonged inflammation that can lead to tissue damage.

Deficiencies or imbalances in trace elements can disrupt immune homeostasis, leading to increased susceptibility to infections, impaired wound healing, and a heightened risk of chronic inflammatory and autoimmune diseases. Each trace element, such as zinc, copper, selenium, iron, and manganese, has distinct yet interconnected functions in immune regulation. Zinc is essential for T cell and B cell development, while copper facilitates pathogen elimination through reactive oxygen species (ROS). Selenium, through its role in selenoproteins, protects immune cells from oxidative damage and modulates inflammation. Iron is necessary for immune cell proliferation but must be carefully regulated to avoid promoting pathogen growth, and manganese contributes to antioxidant defense and T cell activation [31]. Table 2.4 outlines the specific immune functions of these key trace elements, their effects on immune regulation, and the consequences of their deficiency or excess.

Interaction of Trace Elements with Immune Cells and Signaling Pathways

Trace elements play a critical role in modulating immune cell activity and influencing key signaling pathways that regulate immune responses. They are essential for the proper functioning of both innate and adaptive immunity, ensuring that immune cells can effectively recognize and eliminate pathogens while maintaining immune balance. Through their involvement in various biochemical processes, these elements help coordinate immune signaling, inflammatory responses, and oxidative stress regulation.

- *Role of Trace Elements in T Cells and B Cells*

T cells and B cells are fundamental components of the adaptive immune system, and trace elements are vital for their development, activation, and function. Zinc is particularly crucial for the maturation of T cells in the thymus and plays a key role in determining T-helper (Th) cell differentiation [8]. A deficiency in zinc disrupts both Th1 and Th2 responses, leading to weakened adaptive immunity and increased susceptibility to infections. Selenium contributes to B cell function by supporting the production of antibodies, which are essential for targeting specific pathogens

Table 2.4 Immune functions of key trace elements: roles, deficiency effects, and excess consequences

Trace elements	Role in immunity	Effects of deficiency	Effects of excess
Zinc	Regulates immune responses, supports T cell, B cell, and NK cell function, and modulates inflammatory cytokines (IL-6, TNF-α) [68]	Increased infection susceptibility, delayed wound healing, and excessive inflammation	Can disrupt immune balance and impair cellular function [69]
Copper	Facilitates antimicrobial activity via ROS production in macrophages; involved in immune cell activation [70]	Weakened immune defenses, increased infection risk	Promotes oxidative stress and chronic inflammation
Selenium	Essential for immune regulation through selenoproteins; protects immune cells from oxidative stress and balances inflammatory responses [59]	Reduced immune function, higher risk of severe viral infections	May trigger autoimmune dysfunction
Iron	Supports immune cell proliferation (macrophages and lymphocytes); ferritin regulates immune responses [71]	Weakens immune defenses and hampers immune cell function	Promotes pathogen growth by serving as a nutrient source
Manganese	Produces antioxidant enzymes (SOD), protects immune cells from oxidative stress, and aids in T cell differentiation [72]	Impaired immune function, increased inflammation	Potential disruption of immune homeostasis

[59]. Additionally, copper is involved in T cell activation, ensuring an appropriate immune response when the body encounters harmful invaders [73].

- *Impact on Macrophages and Neutrophils*

Macrophages and neutrophils serve as the first line of defense in the immune system by identifying, engulfing, and destroying pathogens. Iron is necessary for macrophage function, but its levels must be carefully regulated, as excessive iron can create an environment that promotes pathogen growth, increasing susceptibility to infection [74]. Zinc and selenium contribute to the effectiveness of macrophages by enhancing their ability to carry out phagocytosis and produce reactive oxygen species (ROS), which help neutralize invading microbes [75]. Copper also plays an essential role in neutrophil function by facilitating oxidative burst activity, a mechanism that enables neutrophils to generate ROS and effectively eliminate pathogens.

- *Oxidative Stress Management and Inflammation Control*

Balancing oxidative stress is crucial for preventing chronic inflammation and immune dysfunction. Selenium and manganese contribute to the body's antioxidant defenses by supporting the activity of key enzymes, such as glutathione peroxidase and superoxide dismutase (SOD), which help neutralize ROS and protect immune cells from oxidative damage [76]. However, an imbalance in iron or copper levels can lead to excessive ROS production, increasing the risk of chronic inflammation and tissue damage. When iron and copper accumulate beyond necessary levels, they can contribute to prolonged oxidative stress, which has been linked to various inflammatory and autoimmune diseases.

Therapeutic Implications and Clinical Perspectives on Trace Elements

Trace elements are fundamental to maintaining physiological balance and overall health. Their precise regulation is essential, as both deficiencies and excesses are linked to a wide range of diseases, including neurodegenerative conditions, cardiovascular disorders, cancer, metabolic syndromes, and immune system dysfunctions. Given their integral role in numerous biochemical and physiological processes, ensuring adequate levels of these micronutrients is critical for disease prevention and management [31].

Understanding the therapeutic potential of trace elements has opened new avenues in clinical medicine. Supplementation strategies are being explored to correct deficiencies and enhance disease outcomes, particularly in conditions where trace element imbalances contribute to disease progression. For example, zinc and selenium supplementation have been investigated for their potential to modulate immune function and reduce oxidative stress in chronic illnesses [75], while iron management plays a key role in treating anemia and metabolic disorders [77]. The growing field of personalized medicine highlights the importance of tailoring trace element intake to individual needs based on genetic, metabolic, and environmental factors. Since different individuals metabolize and utilize trace elements differently, precision medicine approaches aim to optimize supplementation or dietary intake based on a person's specific biochemical profile. This personalized approach enhances treatment efficacy and reduces the risk of toxicity from excessive intake.

Furthermore, the identification and validation of reliable biomarkers for assessing trace element status are crucial for accurate diagnosis and targeted therapy. Biomarkers, such as serum zinc, copper levels, selenium-dependent enzyme activity, and iron-binding capacity, help in detecting deficiencies or toxicities early, allowing for timely intervention. Research continues to explore novel diagnostic tools, such as metabolomic and proteomic analyses, to provide deeper insights into trace element imbalances and their impact on human health.

By integrating trace element research into clinical practice, healthcare professionals can develop more effective prevention and treatment strategies, improving patient outcomes across various diseases. Ensuring the appropriate intake and regulation of trace elements through diet, supplementation, or medical intervention remains a key component of maintaining long-term health and preventing chronic disease progression. Figure 2.4 illustrates the therapeutic implications and clinical

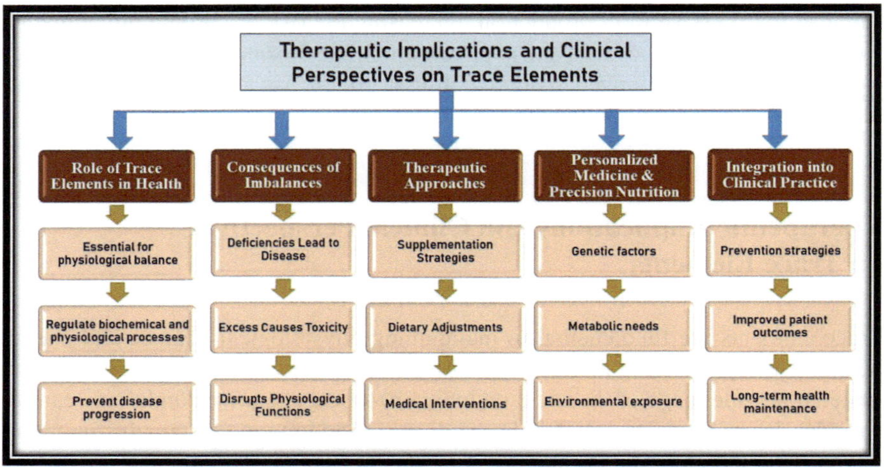

Fig. 2.4 Trace elements in health and disease: clinical implications and management Strategies

perspectives on trace elements, highlighting their essential role in maintaining physiological balance and preventing disease progression.

Research Challenges and Future Directions

Trace elements play a vital role in numerous physiological processes, and their dysregulation is associated with a wide range of diseases. Despite significant advancements in trace element research, several challenges remain, including the need for more precise analytical techniques, a deeper understanding of trace element metabolism in health and disease, and the development of targeted therapeutic approaches. Addressing these challenges will help improve disease prevention, diagnosis, and treatment strategies [51].

Advancements in Trace Element Analysis: Emerging Techniques and Their Applications

The precise measurement of trace element concentrations in biological samples is essential for gaining insights into their physiological roles, interactions, and implications in health and disease. Recent advancements in analytical technologies have significantly enhanced the sensitivity, specificity, and accuracy of trace element detection. These improvements facilitate more comprehensive studies on trace element metabolism, enabling better clinical diagnostics and environmental assessments. Several innovative techniques have emerged, offering distinct advantages for

Table 2.5 Emerging techniques for trace element analysis

Analytical technique	Description	Key applications
Inductively coupled plasma mass spectrometry (ICP-MS)	A highly sensitive technique capable of detecting multiple trace elements simultaneously at ultra-trace levels [79]	Used in clinical diagnostics, environmental monitoring, and toxicology studies to assess trace element levels in blood, urine, and tissues
X-ray fluorescence spectroscopy (XRF)	A non-destructive method that enables in situ analysis of trace elements in solid biological samples [80]	Ideal for evaluating long-term trace element exposure in hair, nails, and bones
Laser ablation inductively coupled plasma mass spectrometry (LA-ICP-MS)	Provides spatially resolved mapping of trace elements within tissues and cells [81]	Applied in neurodegenerative disease and cancer research to study metal distribution at the microscopic level
Neutron activation analysis (NAA)	A precise technique that does not require sample digestion, minimizing contamination risks [82]	Utilized in forensic investigations, environmental assessments, and biomedical research

the quantitative and spatial analysis of trace elements in biological matrices [78]. Table 2.5 illustrates the emerging techniques for the analysis of trace elements. The figure further presents various analytical techniques used for detecting and quantifying trace elements in biological and environmental samples. It also highlights the principles, sensitivity, and key applications of each method.

Despite these technological advancements, several challenges persist, including the need for standardization of analytical methodologies, improved quality control in sample collection and preparation, and enhanced assessment techniques for determining trace element bioavailability in various biological systems. Continued research and innovation in analytical chemistry are essential for refining these methods and expanding their applicability in clinical and research settings.

Conclusion

In conclusion, it is important to emphasise that trace elements play a pivotal role as molecular orchestrators in maintaining human health and in preventing several diseases. These micronutrients, though required in small amounts, are indispensable for a wide array of biological functions, including enzymatic catalysis, redox balance, immune modulation, and genetic regulation. Their intricate involvement in metabolic pathways underscores their significance in cellular homeostasis, where even minor imbalances, either deficiencies or toxic accumulations, can lead to severe physiological consequences. This review highlights the essential contributions of elements such as zinc, copper, selenium, iron, cobalt, chromium, and molybdenum in various biochemical processes, emphasizing their roles in enzymatic activity, protein synthesis, and immune defense. Research has consistently demonstrated that inadequate levels of these elements are linked to numerous diseases, including neurodegenerative disorders, cardiovascular conditions, and metabolic syndromes. Moreover, advancements in molecular biology have revealed their interactions at the genetic level, influencing gene expression, DNA repair mechanisms, and cell signaling pathways.

A growing body of evidence suggests that targeted nutritional interventions, fortified diets, and supplementation can mitigate the risks associated with trace element imbalances. However, despite their therapeutic potential, a deeper understanding of their bioavailability, synergistic effects, and toxicity thresholds is crucial to developing effective clinical applications. Future research should focus on precision medicine approaches, leveraging genomics and metabolomics to tailor interventions that optimize trace element levels based on individual physiological and genetic profiles.

Conflict of Interest None

References

1. Mertz W (1981) The essential trace elements. Science 213(4514):1332–1338. https://doi.org/10.1126/science.7022654
2. Muhamed PK, Vadstrup S (2014) Zinc is the most important trace element. Ugeskr Laeger 176(5):V11120654. Danish PMID: 25096007.
3. Barceloux DG (1999) Selenium. J Toxicol Clin Toxicol 37(2):145–1472. https://doi.org/10.1081/clt-100102417
4. Bárány E, Bergdahl IA, Bratteby LE, Lundh T, Samuelson G, Skerfving S, Oskarsson A (2005) Iron status influences trace element levels in human blood and serum. Environ Res 98(2):215–223. https://doi.org/10.1016/j.envres.2004.09.010
5. Chen WY, Chen CJ, Liu CH, Mao FC (2009) Chromium supplementation enhances insulin signalling in skeletal muscle of obese KK/HlJ diabetic mice. Diabetes Obes Metab 11(4):293–303. https://doi.org/10.1111/j.1463-1326.2008.00936.x
6. Fatima G, Dzupina A, Alhmadi HB, Magomedova A, Siddiqui Z, Mehdi A, Hadi N (2024) Magnesium matters: a comprehensive review of its vital role in health and diseases. Cureus 16(10):e71392. https://doi.org/10.7759/cureus.71392
7. Kawabata T (2022) Iron-induced oxidative stress in human diseases. Cells 11(14):2152. Published 2022 Jul 8. https://doi.org/10.3390/cells11142152
8. Wessels I, Maywald M, Rink L (2017) Zinc as a gatekeeper of immune function. Nutrients 9(12):1286. Published 2017 Nov 25. https://doi.org/10.3390/nu9121286
9. Mojadadi A, Au A, Salah W, Witting P, Ahmad G (2021) Role for selenium in metabolic homeostasis and human reproduction. Nutrients 13(9):3256. Published 2021 Sep 18. https://doi.org/10.3390/nu13093256
10. Zhong G, Wang X, Li J et al (2024) Insights into the role of copper in neurodegenerative diseases and the therapeutic potential of natural compounds. Curr Neuropharmacol 22(10):1650–1671. https://doi.org/10.2174/1570159X22666231103085859
11. O’Neal SL, Zheng W (2015) Manganese toxicity upon overexposure: a decade in review. Curr Environ Health Rep 2(3):315–328. https://doi.org/10.1007/s40572-015-0056-x

12. Gad SC (1989) Acute and chronic systemic chromium toxicity. Sci Total Environ 86(1–2):149–157. https://doi.org/10.1016/0048-9697(89)90201-5
13. Cabral M, Kuxhaus O, Eichelmann F et al (2021) Trace element profile and incidence of type 2 diabetes, cardiovascular disease and colorectal cancer: results from the EPIC-Potsdam cohort study. Eur J Nutr 60(6):3267–3278. https://doi.org/10.1007/s00394-021-02494-3
14. Chen L, Shen Q, Liu Y et al (2025) Homeostasis and metabolism of iron and other metal ions in neurodegenerative diseases. Signal Transduct Target Ther 10:31. https://doi.org/10.1038/s41392-024-02071-0
15. National Research Council (US) Committee on Diet and Health (1989) Diet and health: implications for reducing chronic disease risk. National Academies Press (US), Washington (DC). 14, Trace Elements. Available from: https://www.ncbi.nlm.nih.gov/books/NBK218751/
16. Ems T, St Lucia K, Huecker MR (2025) Biochemistry, iron absorption. [Updated 2023 Apr 17]. In: StatPearls [Internet]. Stat-Pearls Publishing, Treasure Island. Available from: https://www.ncbi.nlm.nih.gov/books/NBK448204/
17. Shukla V, Parvez S, Fatima G, Singh S, Magomedova A, Batiha GE-S, Alexiou A, Papadakis M, Welson NN, Hadi N (2024) Micronutrient interactions: magnesium and its synergies in maternal–fetal health. Food Sci Nutr 12(10):6913–6928. https://doi.org/10.1002/fsn3.4316
18. Costa MI, Sarmento-Ribeiro AB, Gonçalves AC (2023) Zinc: from biological functions to therapeutic potential. Int J Mol Sci 24(5):4822. PMID: 36902254; PMCID: PMC10003636. https://doi.org/10.3390/ijms24054822
19. Chen L, Min J, Wang F (2022) Copper homeostasis and cuproptosis in health and disease. Signal Transduct Target Ther 7(1):378. PMID: 36414625; PMCID: PMC9681860. https://doi.org/10.1038/s41392-022-01229-y
20. Li L, Yang X (2018) The essential element manganese, oxidative stress, and metabolic diseases: links and interactions. Oxid Med Cell Longev 2018:7580707. PMID: 29849912; PMCID: PMC5907490. https://doi.org/10.1155/2018/7580707
21. Shahidin, Wang Y, Wu Y, Chen T, Wu X, Yuan W, Zhu Q, Wang X, Zi C (2025) Selenium and selenoproteins: mechanisms, health functions, and emerging applications. Molecules 30(3):437. https://doi.org/10.3390/molecules30030437
22. Andersson M, Braegger CP (2022) The role of iodine for thyroid function in lactating women and infants. Endocr Rev 43(3):469–506. https://doi.org/10.1210/endrev/bnab029
23. Adamus JP, Ruszczyńska A, Wyczałkowska-Tomasik A (2024) Molybdenum's role as an essential element in enzymes catabolizing redox reactions: a review. Biomolecules 14(7):869. Published 2024 Jul 19. https://doi.org/10.3390/biom14070869
24. Cooper GM (2000) The cell: a molecular approach. In: The central role of enzymes as biological catalysts, 2nd edn. Sinauer Associates, Sunderland. Available from: https://www.ncbi.nlm.nih.gov/books/NBK9921/
25. Lee SR (2018) Critical role of zinc as either an antioxidant or a prooxidant in cellular systems. Oxid Med Cell Longev 2018:9156285. Published 2018 Mar 20. https://doi.org/10.1155/2018/9156285
26. Ogun AS, Joy NV, Valentine M (2025) Biochemistry, heme synthesis [Updated 2023 May 1]. In: Stat-Pearls [Internet]. Stat-Pearls Publishing, Treasure Island. Available from: https://www.ncbi.nlm.nih.gov/books/NBK537329/
27. Fiorentini D, Cappadone C, Farruggia G, Prata C (2021) Magnesium: biochemistry, nutrition, detection, and social impact of diseases linked to its deficiency. Nutrients 13(4):1136. Published 2021 Mar 30. https://doi.org/10.3390/nu13041136
28. Wróblewska J, Nuszkiewicz J, Wróblewski M, Wróblewska W, Woźniak A (2024) Selected trace elements and their impact on redox homeostasis in eye health. Biomolecules 14(11):1356. Published 2024 Oct 24. https://doi.org/10.3390/biom14111356
29. Hariharan S, Dharmaraj S (2020) Selenium and selenoproteins: it's role in regulation of inflammation. Inflammopharmacology 28(3):667–695. https://doi.org/10.1007/s10787-020-00690-x

30. Chandimali N, Bak SG, Park EH et al (2025) Free radicals and their impact on health and antioxidant defenses: a review. Cell Death Discov 11:19. https://doi.org/10.1038/s41420-024-02278-8
31. Mehri A (2020) Trace elements in human nutrition (II) - an update. Int J Prev Med 11:2. Published 2020 Jan 3. https://doi.org/10.4103/ijpvm.IJPVM_48_19
32. Kiani AK, Dhuli K, Donato K et al (2022) Main nutritional deficiencies. J Prev Med Hyg 63(Suppl 3):E93–E101. Published 2022 Oct 17. https://doi.org/10.15167/2421-4248/jpmh2022.63.2S3.2752
33. Kumar A, Sharma E, Marley A, Samaan MA, Brookes MJ (2022) Iron deficiency anaemia: pathophysiology, assessment, practical management. BMJ Open Gastroenterol 9(1):e000759. https://doi.org/10.1136/bmjgast-2021-000759
34. Maxfield L, Shukla S, Crane JS (2025) Zinc deficiency. [Updated 2023 Jun 28]. In: StatPearls [Internet]. Stat-Pearls Publishing, Treasure Island. Available from: https://www.ncbi.nlm.nih.gov/books/NBK493231/
35. Eastman CJ, Zimmermann MB (2000) The iodine deficiency disorders. [Updated 2018 Feb 6]. In: Feingold KR, Anawalt B, Blackman MR et al (eds) Endotext [Internet]. MDText.com, Inc., South Dartmouth. Available from: https://www.ncbi.nlm.nih.gov/books/NBK285556/
36. Barchielli G, Capperucci A, Tanini D (2022) The role of selenium in pathologies: an updated review. Antioxidants (Basel) 11(2):251. Published 2022 Jan 27. https://doi.org/10.3390/antiox11020251
37. Burkhead JL, Collins JF (2022) Nutrition information brief-copper. Adv Nutr 13(2):681–683. https://doi.org/10.1093/advances/nmab157
38. Kumfu S, Chattipakorn SC, Chattipakorn N (2022) Iron overload cardiomyopathy: using the latest evidence to inform future applications. Exp Biol Med (Maywood) 247(7):574–583. https://doi.org/10.1177/15353702221076397
39. Schoofs H, Schmit J, Rink L (2024) Zinc toxicity: understanding the limits. Molecules 29(13):3130. Published 2024 Jul 1. https://doi.org/10.3390/molecules29133130
40. Immergluck J, Anilkumar AC (2025) Wilson disease. [Updated 2023 Aug 7]. In: StatPearls [Internet]. Stat-Pearls Publishing, Treasure Island. Available from: https://www.ncbi.nlm.nih.gov/books/NBK441990/
41. MacFarquhar JK, Broussard DL, Melstrom P et al (2010) Acute selenium toxicity associated with a dietary supplement. Arch Intern Med 170(3):256–261. https://doi.org/10.1001/archinternmed.2009.495
42. Southern AP, Anastasopoulou C, Jwayyed S (2025) Iodine toxicity. [Updated 2024 May 2]. In: StatPearls [Internet]. Stat-Pearls Publishing, Treasure Island. Available from: https://www.ncbi.nlm.nih.gov/books/NBK560770/
43. Jahankhani K, Taghipour N, Mashhadi Rafiee M, Nikoonezhad M, Mehdizadeh M, Mosaffa N (2023) Therapeutic effect of trace elements on multiple myeloma and mechanisms of cancer process. Food Chem Toxicol 179:113983. https://doi.org/10.1016/j.fct.2023.113983
44. Michalska P, León R (2020) When it comes to an end: oxidative stress crosstalk with protein aggregation and neuroinflammation induce neurodegeneration. Antioxidants (Basel) 9(8):740. Published 2020 Aug 12. https://doi.org/10.3390/antiox9080740
45. Khan MA, Khan ZA, Shoeb F, Fatima G, Khan RH, Khan MM (2023) Role of de novo lipogenesis in inflammation and insulin resistance in Alzheimer's disease. Int J Biol Macromol 242(Pt 2):124859. https://doi.org/10.1016/j.ijbiomac.2023.124859
46. Liu JL, Fan YG, Yang ZS, Wang ZY, Guo C (2018) Iron and Alzheimer's disease: from pathogenesis to therapeutic implications. Front Neurosci 12:632. Published 2018 Sep 10. https://doi.org/10.3389/fnins.2018.00632
47. Wise RM, Wagener A, Fietzek UM, Klopstock T, Mosharov EV, Zucca FA, Sulzer D, Zecca L, Burbulla LF (2022) Interactions of dopamine, iron, and alpha-synuclein linked to dopaminergic neuron vulnerability in Parkinson's disease and Neurodegeneration with Brain Iron Accumulation disorders. Neurobiol Dis 175:105920. https://doi.org/10.1016/j.nbd.2022.105920

48. Zubčić K, Hof PR, Šimić G, Jazvinšćak Jembrek M (2020) The role of copper in Tau-related pathology in Alzheimer's disease. Front Mol Neurosci 13:572308. Published 2020 Sep 10. https://doi.org/10.3389/fnmol.2020.572308
49. Immergluck J, Anilkumar AC (2025) Wilson disease. [Updated 2023 Aug 7]. In: Stat-Pearls [Internet]. Stat-Pearls Publishing, Treasure Island. Available from: https://www.ncbi.nlm.nih.gov/books/NBK441990/
50. Duţă C, Muscurel C, Dogaru CB, Stoian I (2025) Selenoproteins: zoom-in to their metal-binding properties in neurodegenerative diseases. Int J Mol Sci 26(3):1305. https://doi.org/10.3390/ijms26031305
51. Wechselberger C, Messner B, Bernhard D (2023) The role of trace elements in cardiovascular diseases. Toxics 11(12):956. Published 2023 Nov 23. https://doi.org/10.3390/toxics11120956
52. Wang YM, Feng LS, Xu A, Ma XH, Zhang MT, Zhang J (2024) Copper ions: the invisible killer of cardiovascular disease (review). Mol Med Rep 30(5):210. https://doi.org/10.3892/mmr.2024.13334
53. Shen T, Zhao Q, Luo Y, Wang T (2022) Investigating the role of Zinc in atherosclerosis: a review. Biomolecules 12(10):1358. Published 2022 Sep 23. https://doi.org/10.3390/biom12101358
54. Shimada BK, Alfulaij N, Seale LA (2021) The impact of selenium deficiency on cardiovascular function. Int J Mol Sci 22(19):10713. Published 2021 Oct 2. https://doi.org/10.3390/ijms221910713
55. Dubey P, Thakur V, Chattopadhyay M (2020) Role of minerals and trace elements in diabetes and insulin resistance. Nutrients 12(6):1864. Published 2020 Jun 23. https://doi.org/10.3390/nu12061864
56. Lemaire K, Chimienti F, Schuit F (2012) Zinc transporters and their role in the pancreatic β-cell. J Diabetes Investig 3(3):202–211. https://doi.org/10.1111/j.2040-1124.2012.00199.x
57. Hua Y, Clark S, Ren J, Sreejayan N (2012) Molecular mechanisms of chromium in alleviating insulin resistance. J Nutr Biochem 23(4):313–319. https://doi.org/10.1016/j.jnutbio.2011.11.001
58. Marku A, Galli A, Marciani P, Dule N, Perego C, Castagna M (2021) Iron metabolism in pancreatic beta-cell function and dysfunction. Cells 10(11):2841. Published 2021 Oct 22. https://doi.org/10.3390/cells10112841
59. Zhang F, Li X, Wei Y (2023) Selenium and selenoproteins in health. Biomolecules 13(5):799. Published 2023 May 8. https://doi.org/10.3390/biom13050799
60. Jia MJ, Yin HF, Liang YC, Jiang F, Li HL (2025) Effect of trace elements and nutrients on 21 autoimmune diseases: a Mendelian randomization study. Front Immunol 15:1462815. Published 2025 Jan 20. https://doi.org/10.3389/fimmu.2024.1462815
61. Gammoh NZ, Rink L (2017) Zinc in infection and inflammation. Nutrients 9(6):624. https://doi.org/10.3390/nu9060624
62. Liu Y, Zhu J, Xu L, Wang B, Lin W, Luo Y (2022) Copper regulation of immune response and potential implications for treating orthopedic disorders. Front Mol Biosci 9:1065265. Published 2022 Dec 5. https://doi.org/10.3389/fmolb.2022.1065265
63. Huwiler VV, Maissen-Abgottspon S, Stanga Z et al (2024) Selenium supplementation in patients with Hashimoto thyroiditis: a systematic review and meta-analysis of randomized clinical trials. Thyroid 34(3):295–313. https://doi.org/10.1089/thy.2023.0556
64. Farebrother J, Zimmermann MB, Andersson M (2019) Excess iodine intake: sources, assessment, and effects on thyroid function. Ann N Y Acad Sci 1446(1):44–65. https://doi.org/10.1111/nyas.14041. Epub 2019 Mar 20.
65. Weyh C, Krüger K, Peeling P, Castell L (2022) The role of minerals in the optimal functioning of the immune system. Nutrients 14(3):644. Published 2022 Feb 2. https://doi.org/10.3390/nu14030644
66. Cannas D, Loi E, Serra M, Firinu D, Valera P, Zavattari P (2020) Relevance of essential trace elements in nutrition and drinking water for human health and autoimmune disease risk. Nutrients 12(7):2074. Published 2020 Jul 13. https://doi.org/10.3390/nu12072074

67. Stefanache A, Lungu II, Butnariu IA et al (2023) Understanding how minerals contribute to optimal immune function. J Immunol Res 2023:3355733. Published 2023 Nov 1. https://doi.org/10.1155/2023/3355733
68. Maggini S, Wintergerst ES, Beveridge S, Hornig DH (2007) Selected vitamins and trace elements support immune function by strengthening epithelial barriers and cellular and humoral immune responses. Br J Nutr 98(S1):S29–S35. https://doi.org/10.1017/S0007114507832971
69. Karunakaran V, Harding K, Sarnowski A, Walter E (2025) Trace elements: clinical perspectives in the critically ill. J Intensive Care Soc. Published online February 5, 2025. https://doi.org/10.1177/17511437241305269
70. Cheng F, Peng G, Lu Y et al (2022) Relationship between copper and immunity: the potential role of copper in tumor immunity. Front Oncol 12:1019153. Published 2022 Nov 7. https://doi.org/10.3389/fonc.2022.1019153
71. Ni S, Yuan Y, Kuang Y, Li X (2022) Iron metabolism and immune regulation. Front Immunol 13:816282. Published 2022 Mar 23. https://doi.org/10.3389/fimmu.2022.816282
72. Liu M, Sun X, Chen B, Dai R, Xi Z, Xu H (2022) Insights into manganese superoxide dismutase and human diseases. Int J Mol Sci 23(24):15893. Published 2022 Dec 14. https://doi.org/10.3390/ijms232415893
73. Li J, Liang Y, Mao H, Deng W, Zhang J (2014) Effects of B-lymphocyte dysfunction on the serum copper, selenium and zinc levels of rheumatoid arthritis patients. Pak J Med Sci 30(5):1064–1067. https://doi.org/10.12669/pjms.305.5214
74. Buttar H, Singh R, Juneja L, De Meester F, Wilczynska A, Takahashi T, Fatima G, Wilson D, Isaza A (2024) Diet and lifestyle guidelines for immunomodulation with reference to coronavirus pandemic: a scientific statement of the International College of Nutrition. Food Sci Nutr 12(10):6913–6928. https://doi.org/10.1002/fsn3.4316
75. Elmadfa I, Meyer AL (2019) The role of the status of selected micronutrients in shaping the immune function. Endocr Metab Immune Disord Drug Targets 19(8):1100–1115. https://doi.org/10.2174/1871530319666190529101816
76. Lee J-G, Jang J-Y, Baik S-M (2025) Selenium as an antioxidant: roles and clinical applications in critically ill and trauma patients: a narrative review. Antioxidants 14(3):294. https://doi.org/10.3390/antiox14030294
77. Nguyen M, Tadi P (2025) Iron supplementation. [Updated 2023 Jul 3]. In: StatPearls [Internet]. Sta-tPearls Publishing, Treasure Island. Available from: https://www.ncbi.nlm.nih.gov/books/NBK557376/
78. Oviedo MN, Luján CE, Lemos AA, Botella MB, Llaver M, Wuilloud RG (2024) An overview of preconcentration techniques combined with inductively coupled plasma mass spectrometry for trace element determination in biological studies. Anal Bioanal Chem 416(11):2641–2656. Epub 2024 Jan 20. https://doi.org/10.1007/s00216-024-05124-z
79. Wilschefski SC, Baxter MR (2019) Inductively coupled plasma mass spectrometry: introduction to analytical aspects. Clin Biochem Rev 40(3):115–133. https://doi.org/10.33176/AACB-19-00024
80. Bali V, Khajuria Y, Maniyar V et al (2023) Quantitative analysis of human hairs and nails. Biophys Rev 15(3):401–417. Published 2023 Jun 20. https://doi.org/10.1007/s12551-023-01069-2
81. Becker JS, Matusch A, Wu B (2014) Bioimaging mass spectrometry of trace elements—recent advance and applications of LA-ICP-MS: a review. Anal Chim Acta 835:1–18. https://doi.org/10.1016/j.aca.2014.04.048
82. Das DD, Sharma N, Chawla PA (2024) Neutron activation analysis: an excellent nondestructive analytical technique for trace metal analysis. Crit Rev Anal Chem 54(7):2450–2466. https://doi.org/10.1080/10408347.2023

Chapter 3
Trace Minerals (Fe, Zn, Cr, Cu, Se, V, Mn, Mo, Co) in Human Health and Disease

Kaliyamurthi Venkatachalam, Ambika Binesh, Amit Ranjan, Anusha Prasannan, and Sisira Joy

Abstract Trace minerals, including iron, zinc, copper, selenium, chromium, manganese, vanadium, cobalt, and molybdenum, are the essential micronutrients that have established their vital role in cellular homeostasis, metabolic pathways, immune regulation, and enzymatic action. Both deficiency and toxicity have a significant impact on human health, which is discussed in this chapter. These minerals have complex synergistic or antagonistic interactions, as well as competitive absorption. Distinct trace mineral requirements are essential for individuals in different age groups. For monitoring the levels of these trace minerals, advanced analytical techniques are used, which involve mass spectrometry and metallomics. In spite of progressive research in this area, research gaps still remain in defining the optimal levels recommended for intake, understanding the disease-related changes, and also in the evaluation of the effects of long-term supplementation. Strategies should be taken that focus on the levels of balanced dietary intake for preventing deficiency and toxicity, and also omics-based approaches are recommended to improve optimization of trace minerals to minimize associated risks.

Keywords Trace mineral interactions · Deficiency and toxicity · Biomarkers · Precision nutrition · Enzymatic catalysis · Oxidative stress

Introduction

Trace minerals, or trace elements, are inorganic micronutrients essential in lesser amounts for human health. They have crucial functions in a host of physiological processes, such as cellular homeostasis, enzymatic catalysis, and regulation of the

K. Venkatachalam (✉) · A. Binesh · A. Ranjan · A. Prasannan · S. Joy
Department of Basic Sciences, Institute of Fisheries Post Graduate Studies, Tamil Nadu Dr. J. Jayalalithaa Fisheries University (TNJFU), Chennai, Tamil Nadu, India
e-mail: kaliyamurthi@tnfu.ac.in

N. S. Dhalla et al. (eds.), *Functional Biochemistry of Metallic Elements*, Advances in Biochemistry in Health and Disease 36,
https://doi.org/10.1007/978-3-032-14445-4_3

immune system. Trace minerals can be placed into two broad categories: essential and possibly essential elements. Essential trace minerals, including iron (Fe), zinc (Zn), copper (Cu), and selenium (Se), are unequivocally necessary for normal physiological function, while possibly essential elements, like chromium (Cr), manganese (Mn), molybdenum (Mo), and cobalt (Co), are not yet generally accepted as being essential to humans, but the evidence indicates they are involved in the regulation of metabolism and health ([7, 14]). Global trace mineral dietary intake patterns differ significantly among populations, depending on geographical, socioeconomic, and dietary considerations. In most low- and middle-income countries, micronutrient deficiencies are still widespread due to inadequate dietary variation and limited access to nutrient-dense food. On the other hand, in some areas or certain occupational and environmental settings, excessive exposure or intake of these nutrients may cause toxicity, thus complicating nutritional guidelines and public health advice [42, 53, 76]. Physiologically, trace minerals play a crucial role in preserving metabolic homeostasis. They often function as cofactors in enzymatic processes, supporting biochemical activities like antioxidant defense, DNA synthesis, and immune response regulation. Iron, for exemplify, is required for oxygen transport and energy metabolism [7, 76]. These elements when in imbalance, such as deficiency or excessive, can severely compromise health. Deficiencies are connected to compromised immunity, anemia, growth retardation, and susceptibility to infectious diseases, while toxicities lead to neurological damage, organ damage, and metabolic dysfunction [42, 52]. From a public perspective, the double challenge of preventing both deficiency and toxicity highlights the need for balanced dietary consumption. Attaining and sustaining maximum trace mineral status is important for the prevention of disease, proper development, and regulation of general well-being. For designing effective nutritional interventions and evidence-based public health policies at both individual and population levels, it is essential to understand the physiological functions, dietary sources, and tolerable intake levels of trace minerals.

Iron (Fe)

Iron stands as one of the most vital metals in the living system among the transition metals. Regulatory proteins like hemoglobin, transferrin, and ferritin balance iron absorption, transport, storage, and its utilization [74, 57].

Absorption, Transport, and Storage (Role of Transferrin, Ferritin, Hepcidin)

Iron absorption occurs in the small intestine, especially in the duodenum and upper jejunum. Based on the iron's dietary form (heme iron or nonheme iron), the process of absorption differs. The heme iron, which is derived from animal sources, is absorbed by the heme carrier protein-1 (HCP-1), and the nonheme iron found in plant-based foods, as ferric iron (Fe^{3+}), which to get absorbed is converted to ferrous form (Fe^{2+}) by vitamin C. This ferrous form is then absorbed through divalent metal-ion transporter-1(DMT 1) and transported into the bloodstream via ferroportin 1 (FPN 1). The ferric iron form (Fe^{3+}), in circulation, is bound by transferrin and plays the role of delivering iron to various target tissues in the body by interacting with transferrin receptors (TfR) on the surface of cells. Ferritin, which is an iron-storage protein, makes sure that free iron is not available in the system to catalyze unwanted, harmful reactions and also ensures that it is immediately available when in demand [34]. The iron regulatory proteins (IRP1 and IRP2) modulate the iron concentrations, and the iron balance in the whole human system is controlled by hepcidin, which is a liver-derived peptide [28]. Hepcidin synthesis is also said to be influenced by inflammatory signals, where interleukin 6 (IL-6), a pro-inflammatory cytokine, is reported to induce hepcidin production. Hepcidin degrades ferroportin and its receptors, which reduces the availability of functional ferroportin, thus effectively limiting the export of iron from enterocytes, macrophages [33, 65], and hepatocytes and finally leads to reduced iron availability in the plasma [24].

Functions and Toxicity

The well-known function of iron is in the transport of oxygen via hemoglobin, followed by serving its role in cellular energy metabolism, cofactor for enzymatic reactions like DNA synthesis, component of the catalase enzyme, which is involved in the antioxidant defense mechanism and also in ATP production. If iron is present in excess, it leads to toxicity by the Fenton reaction, releasing hydroxyl radicals, which are highly damaging reactive oxygen species. It leads to oxidative damage and a condition called hemochromatosis, in which the body absorbs and stores an excessive amount of iron in various organs [58]. Iron imbalance also leads to neurological disorders, restless leg syndrome, pediatric stroke, and cranial nerve paralysis. Iron imbalances also contribute to neurological disorders like Alzheimer's and Parkinson's disease [57].

Deficiency and Role in Infection, Cardiovascular Disease, and Neurodegeneration

Iron deficiency, which is a primary cause of anemia, is a common nutritional deficiency globally, affecting almost everyone in different age groups, from children aged 6 to 59 months, pregnant women, and women aged between 15 and 49 years, according to a survey by the World Health Organization [21, 90].

Role in Infection

Iron plays a dual role in infection by influencing both the host defense mechanism as well as the pathogen multiplication. Even though iron supplementation is essential for the proper functioning of the host immune system, excessive supplementation can lead to morbidity and mortality from infections. This happens because the pathogen takes up excess iron for its growth and expansion. People with hemochromatosis are at a higher risk of infection as excess iron is available for the utilization of the microorganism. Within hours of infection, our innate immune system counteracts infection by reducing the availability of iron by a process called hypoferremia, where inflammatory cytokines cascade increases the hepcidin level, which in turn degrades ferroportin, which traps iron within hepatocytes, macrophages, and enterocytes, thereby limiting the availability of iron in blood [57].

Cardiovascular Disease

Iron plays a crucial role in cardiac diseases, as its deficiency significantly impacts the dynamics of blood circulation and oxygen delivery to various target tissues. In the case of patients with iron deficiency, iron therapy has been shown to improve the condition. There is a clear connection between inflammation and iron metabolism concerning cardiovascular disease, as the inflammatory stimuli trigger the release of hepcidin, leading to hypoferremia and iron trapping inside tissues, which eventually leads to anemia due to inflammation and affects cardiac functioning [11, 12, 45].

Neurodegeneration

Anemia involves a reduced oxygen-carrying capacity, which causes ischemia and chronic cerebral hypoxia that can lead to cognitive disorders, from mild to severe dementia like conditions. This hypoxia-like condition also leads to neuroinflammation due to the production of reactive oxygen species (ROS), which proves harmful to endothelial cells and affects the integrity of the blood-brain barrier (BBB) [83].

Zinc (Zn)

Zinc is an important trace element essential to human health, being involved in many biological functions, such as enzymatic catalysis, gene expression, immune functions, and cellular metabolism. The physiological significance of zinc arises due to its vital function in absorption, transport, storage, and actions as a structural and catalytic cofactor. Zinc absorption takes place mainly in the small intestine (duodenum), facilitated by specific transporters like ZIP4 (enterocyte uptake) and ZnT1 (efflux to circulation). Dietary factors affect the bioavailability of zinc such as phytates reduce absorption, whereas amino acids increase it [69]. Zinc in blood is highly connected with albumin and red blood cells. Most of the zinc is sequestered in skeletal muscle, bone, and the liver, where intracellular buffering a detoxification of the metal are controlled by metallothionein proteins. Zinc is an essential structural unit of zinc finger proteins, which control DNA binding, gene expression, and often cellular processes [54]. Additionally, zinc is a cofactor for more than 300 enzymes involved in essential biochemical pathways like the synthesis of nucleic acids, antioxidant defense (e.g., superoxide dismutase), wound healing, and immune modulation [54]. Zinc is vital in immune protection, supporting mucosal barrier function, and modulating cytokine activity. It has important roles in wound healing, antioxidant defense, cell growth, and apoptosis regulation. Nonetheless, overconsumption of zinc might result in toxicity, producing gastrointestinal upset, copper deficiency via competitive absorption, and suppression of the immune system [69]. Therefore, zinc content must be tightly controlled to avoid negative consequences. Deficiency of zinc occurs through growth retardation and taste disturbance (ageusia) [30, 54]. Supplementation has been clearly effective in the treatment of pediatric diarrhea and is applied in the treatment of Wilson's disease. The role of zinc as an immunomodulator was also researched during the COVID-19 pandemic, especially in the relief of complications [60]. However, zinc supplementation is not definitely established as an effective treatment for severe cases of COVID-19 [15].

The diverse roles of zinc in human physiology speak to its importance as an essential micronutrient. Although its deficiency has severe health consequences, over-supplementation is risky in terms of toxicity and an imbalance of nutrients. Therapy with zinc supplementation thus needs to be undertaken cautiously, considering individual requirements and interactions. Continuing research is necessary to develop zinc-based interventions and maximize their clinical efficiency.

Chromium (Cr)

Chromium functions in maintaining the normal metabolism of glucose, lipid, and protein in the body and the dynamic balance of blood glucose and improving levels of blood sugar and lipid to a certain extent. Chromium is an important active component of the glucose tolerance factor (GTF), which increases the sensitivity of tissue receptors toward insulin. GTF can regulate the metabolism of biological macromolecules, deposition of muscle tissue, and utilization of cholesterol by increasing the affinity between the tissue receptors and insulin. Chromium is also considered a promising anti-heat stress element that can reduce the need for antibiotics due to its strong antioxidant effect, which prevents reactive oxygen species (ROS) from degrading lipid film structures, thereby preventing lipid peroxidation and cell damage. In clinical applications, intravenous chromium has shown effectiveness in managing severe insulin resistance in critically ill patients [70, 81].

Absorption, Transport, and Storage

The absorption of chromium from dietary sources is low, with organic chromium showing higher absorption than the inorganic forms. The presence of vitamin C is considered to increase chromium absorption. The bioavailability of chromium picolinate, which is a common supplement form, is higher when compared to chromium nicotinate [51]. Once chromium is absorbed, it is transported into the bloodstream with the help of transferrin. Chromium is stored in different tissues (liver, spleen, kidneys, heart, muscle, pancreas, brain) of the human body in different concentrations, out of which the liver serves as the primary storage site. The absorbed chromium is then excreted from the body in the urine, and the remaining is excreted through feces, small amounts of milk, bile, and sweat [5, 36].

Role in Glucose Metabolism and Insulin Signaling (Controversial)

The role of chromium in glucose metabolism and insulin signaling is widely investigated, but due to its intricate mechanism, there are still facts that need clarifications. Its role in glucose metabolism was first recognized in the 1950s, further establishing it as an essential trace metal. The tyrosine kinase activity of the insulin receptor is enhanced when chromium binds with chromodulin, thus downregulating the activity of phosphotyrosine phosphatase and amplifying the insulin signaling cascade. It also promotes GLUT4 membrane translocation and IRS-1 phosphorylation [31, 80]. The deficiency of chromium is said to demonstrate insulin resistance in malnutrition patients, and recovery is observed after the administration of chromium [56]. Individuals with type 2 diabetes have reported 20–40% lower blood chromium levels when compared to healthy individuals. Bioavailable compounds like chromium picolinate have shown more effect in type 2 diabetes patients than using higher doses of chromium itself. Even though there are an ample number of findings yet controversy does exist with respect to chromium supplementation and its clinical effectiveness on grounds of the form of chromium used, dosage, exposure, and patient age group [4, 80].

Functions and Toxicity

Trivalent chromium (Cr^{3+}) is a pharmacologically beneficial substance that is involved in carbohydrate, lipid, and protein metabolism, thus enhancing the action of insulin and reducing inflammation. The biological effects of chromium are dependent upon its oxidative state. Cr^{3+} reduces cholesterol metabolism by inhibiting 3-hydroxy-3-methyl-glutaryl-CoA reductase and enhances insulin activity by roughly three times, paving the elimination of glucose from the circulation.

Hexavalent chromium (Cr^{6+}) is a strong human carcinogen associated with occupational lung cancer, leading to ROS generation, DNA damage, and angiogenesis via multiple pathways because it has higher mobility and easier cell penetration than trivalent chromium, which demonstrates low toxicity [16, 46].

Chromium Deficiency and Supplementation in Diabetes Management

Even though chromium deficiency is not common, it can result in symptoms with characteristics similar to those associated with diabetes and cardiovascular disease, namely, exhaustion, raised cholesterol levels, elevated blood pressure, reduced resistance to infection, disturbances in hormones, and neurological problems. Deficient patients exhibit substantially lower blood chromium levels (0.55 ng/ml compared with normal 4.9–9.5 ng/ml) and hair chromium levels (154–175 ng/g compared to >500 ng/g), based on laboratory tests. Severe deficiency, which happens frequently in patients who are undernourished or receiving parenteral nutrition without chromium supplement, can result not only in normal insulin levels but also in insulin resistance, diabetes, hypertriglyceridemia, nerve damage, decreased body weight, metabolic syndrome, compromised nitrogen metabolism, and glucose intolerance. Many of these implications can be reversed with dietary supplements, which will enhance neurological function, metabolic equilibrium, and overall wellness [80, 88].

Supplementation with chromium picolinate for type 2 diabetes, insulin resistance, and metabolic syndrome has demonstrated considerably improved glucose metabolism in clinical trials. The effectiveness of the supplementation is dose-dependent; weakly glucose-intolerant individuals benefit from consuming ≥200 μg/day, while people with clear diabetes benefit more by consuming ≥500 μg twice daily. Brewer's yeast chromium slightly reduces fasting glucose without changing HbA1c. The factors that affect the outcome are based on the patient population, baseline chromium status, dosage, duration of time, and supplement type. The American Diabetes Association and other major health organizations do not advise routine chromium supplementation for diabetes due to insufficient and inconsistent evidence, even though intravenous chromium has rapidly boosted glucose control in critically ill individuals with severe insulin resistance ([80, 85]).

Copper (Cu)

After iron and zinc, copper is the third most abundant trace metal, which plays multiple roles in biological processes, but still, based on its precise concentration, it can show both its beneficial as well as detrimental effects [14].

Absorption, Transport, and Storage

The dietary form of copper, cupric (Cu^{2+}), is primarily absorbed in the stomach and duodenum and reduced into its cuprous (Cu^{+}) ion state by DCYTB (duodenal cytochrome b), which is then transported into enterocytes, which are the intestinal epithelial cells, by a high-affinity copper transporter 1(Ctr1). Once they are absorbed in the enterocyte, copper is transported to various compartments by metallochaperons like Atox1. Again, from the enterocytes, copper is transported into the portal blood circulation by a copper-transporting P-type ATPase (ATP7A) where they are found complexed with albumin, alpha-2-macroglobulin, and also amino acids like histidine. Copper serves as one of the main components of ceruloplasmin, secreted by hepatocytes, which is vital for iron absorption and transport. Hepatocytes play a major role in copper uptake, storage, and biliary excretion, thus preventing copper overload [25, 29].

Functions and Toxicity

Copper serves as an essential cofactor for enzymes playing significant roles in energy production, like cytochrome c oxidase (COX), which is essential in ATP synthesis. Its role in the antioxidant defense mechanism is by being part of the superoxide dismutase (SOD), which helps protect against the negative effects of free radicals and reactive oxygen species (ROS). It also establishes its role in connective tissue formation as it is essential for the synthesis of collagen and elastin, which provide strength and stability to bones and blood vessels. Copper catalyzes heme synthesis and iron transport, the synthesis of brain neurotransmitters like catecholamines, and also plays a role in immune function, particularly neutrophil activities. As a whole, it is essential for proper growth involving bone strength, immunity, brain development, skin pigmentation, and taste sensation. The toxic effect of copper is when an excess amount is accumulated in the body, which leads to an imbalance in the normal physiological functions of the body and cellular damage. It leads to dysfunction of enzymes involved in energy production, antioxidant defense, and neurotransmitter synthesis. Copper can interact with various proteins and cause alterations in its structure, causing conditions like Alzheimer's disease. Copper toxicity can be due to its excessive intake and dysfunction of copper metabolism because of genetic disorders or can be due to environmental exposure. Our body has evolved to balance this copper overload via transporters like ATP7A and ATP7B, but any mutation in genes can lead to dysfunction and accumulation of copper, further leading to toxic effects [19, 55].

Role in Mitochondrial Respiration

Copper plays a significant role in mitochondrial respiration by serving as a part of COX, also known as Complex IV, which catalyzes the last step of the electron transport chain. CuA and CuB are the copper ion binding sites of COX. Copper enters the cell via transporters (Ctr1) and is transported into the mitochondria with the help of a chaperon, COX17, which then transfers copper to the synthesis of cytochrome C oxidase 1 (SCO1), synthesis of cytochrome C oxidase 2 (SCO2), and COX11, involved in the CuA site of MT-CO2/COX2 and CuB site of MT-CO1/COX1. A new terminology, known as cuproptosis, has recently been studied, which involves copper-dependent cell death caused by mitochondrial damage and stress [25].

Diseases: Deficiency, Emerging Links to Cancer and Cardiovascular Risk

Copper deficiency is linked with various health concerns like anemia, connective tissue disorders, reduced immunity, and neurological impairments. The accumulation of copper with increased tissue and serum copper levels is observed in various cancers, like breast, stomach, prostate, lung, and thyroid. It is also linked with metastatic cancer and cancer angiogenesis, which is blood vessel formation. The therapeutic potential of copper also shows its ability to be used as an anticancer alternative [27].

Copper is essential for the proper functioning of the cardiac system, but excess copper promotes oxidative stress, causing DNA damage and lipid metabolism disruption, leading to myocardial injury and atherosclerosis. Mitochondrial dysfunction due to deficiency and excess of copper serves as a key contributor to cardiovascular diseases. It is also a main causative agent of arrhythmia, promoting modification of low-density lipoprotein (LDL) and plaque formation [13, 84, 86].

Selenium (Se)

Selenium (Se) is another trace element that is required in very minute amounts but still holds a crucial role in health. It is a nonmetal and is known to be part of the amino acid selenocysteine. From the human proteome, a total of 25 selenoproteins have been identified so far.

Absorption, Transport, and Storage

Food is the main source of selenium, and based on the content, its concentration in blood and tissues is being determined. Selenium sulfide and selenium oxide are the poorly utilized forms of selenium, and selenates, selenites, selenium-containing amino acid derivatives are the easily absorbable forms. Different transport mechanisms (Na^+ independent, OH^- antiporter or $Na^+/K^+/Cl^-$ cotransporter) are involved in the absorption of the selenium forms in the duodenum and small intestine. The liver serves as the main organ for selenium regulation. The absorbed selenium is excreted in the urine in the form of methylated selenosugars, and people who have higher intake of selenium, exhale it as volatile dimethyldiselenide [44].

Functions and Toxicity

Selenium serves its role in combination with other minerals, like copper, zinc, manganese, and magnesium, as part of the antioxidant defense mechanism, DNA replication and repair, cellular homeostasis, and various metabolic processes. Its level is reported to decline with age, and its deficiency is noted with an increase in mortality of aged people, especially those above 60 years. Selenium exerts its role by incorporating it into the 25 selenoproteins available in humans. It plays a major role in being part of the enzymes involved in the antioxidant system, glutathione peroxidase (GPXs) and thioredoxin reductase (TrxRs), and part of immune cells by protecting the thymus and enhancing antibody production. It has shown anti-inflammatory activity, reproductive health improvement, cardiovascular protection, anticancer effects, and a significant role in thyroid function regulation. Excess daily selenium intake leads to selenosis, serving as the toxic effect of selenium [8].

Role in Redox Regulation and Thyroid Hormone Metabolism

Selenoproteins carry out selenium's role in redox regulation and the thyroid hormone mechanism. Concerning redox regulation, selenium is incorporated into selenoproteins that possess antioxidant functions, such as GPXs and TrxRs. Glutathione is reduced by GPXs, thereby converting toxic peroxides into nontoxic hydroxyl compounds and the breakdown of hydrogen peroxide into water. Similarly, TrxRs reduce thioredoxin, thereby enhancing the cell's overall antioxidant capacity. Selenium also plays a vital role in thyroid hormone metabolism in the regulation of interconversion between triiodothyronine (T3) and thyroxine (T4) with the help of selenoproteins, like iodothyronine deiodinases (DIO1, DIO2, and DIO3). This regulation is essential for maintaining the balance of thyroid hormones in the body, which further controls various metabolic processes, like the metabolism of carbohydrates, fats, minerals, proteins, and vitamins [8].

Diseases: Deficiency (Keshan Disease) and Toxicity (Selenosis)

Selenium deficiency occurs due to its insufficient daily intake, and its occurrence is relatively not common in healthy populations. Keshan disease is a classic example of selenium deficiency disease, which is characterized by a type of heart muscle disease known as cardiomyopathy with manifestations like cardiogenic shock, heart failure, cardiomegaly, and cardiac arrhythmia. As a prevention and treatment measure, supplementation with selenium has proved to be effective. There is one more selenium deficiency disease known as Kashin-Beck disease, which is characterized by osteochondropathy, a condition that affects bone and cartilage [34].

Selenosis, also known as selenium poisoning, is the toxic effect of an excessive amount of selenium. It had been noted in areas that have higher levels of selenium in soil that the plants grown in these areas will have high selenium content. It is associated with headache, fever, ataxia, nausea, brittle nails, hair loss, fatigue, and vomiting. It was first reported by Marco Polo in the thirteenth century in his horses [8, 34].

Vanadium (V)

Vanadium (V) is an ultra-trace element that displays a paradoxical function in biological systems, being both a functional micronutrient and a toxicant. Vanadium is poorly absorbed in the intestine and is greatly influenced by its oxidation state and chemical form. In biological systems, vanadium is found mainly in prevalent (V^{5+}) and tetravalent (V^{4+}) states. After absorption, vanadium competes with iron for binding to transferrin and lactoferrin, allowing it to reach systemic distribution [75]. Despite limited bioavailability, vanadium is stored in certain tissues, with significant accumulation observed in bone, liver, kidney, and spleen. The storage of the element is regulated by dietary intake and availability of binding proteins, although effective renal excretion processes keep its trace levels within physiological ranges. Vanadium acts biologically through the inhibition of protein tyrosine (PTPs) to increase insulin receptor phosphorylation and the uptake of glucose. The phosphatase inhibitory activity underpins the mechanistic action of vanadium as an insulin mimetic [75]. Experimental vanadium complexes, including vanadyl sulfate and bis(matolato)oxovanadium (BMOV), are effective in enhancing glucose homeostasis, lipid profiles, and insulin sensitivity in models of diabetes. Nonetheless, the therapeutic potential of vanadium is hampered by its dose-dependent, chemical speciation-dependent toxicity profile. Oxidative stress, peroxidation, and mitochondrial damage can be caused by elevated doses of vanadium, leading to cellular and systemic toxicity [23]. Structural analogy with phosphate of vanadium enables it to interfere with key enzyme systems as well as processes of metabolism, and thus, there are issues about the long-term safety of vanadium. Vanadium deficiency, although poorly documented in humans, is implicated in the deregulation of carbohydrate and lipid metabolism, which could lead to metabolic syndromes [71]. There is experimental evidence suggesting therapeutic application of vanadium compounds in type 2 diabetes mellitus, wherein their insulin-mimetic activities can reduce hyperglycemia and enhance insulin sensitivity. In addition, vanadium in cancer therapy has been investigated based on its ability to cause cell cycle arrest, apoptosis, and tumor cell proliferation suppression by inhibiting phosphatase [64]. However, translation of vanadium therapy to clinical settings is hindered by limited therapeutic windows and off-target effects. Vanadium is a biologically active trace element that holds great therapeutic value for metabolic and oncological diseases. Its ability to modulate essential enzymatic pathways, notably by inhibition of phosphatase and sensitization of insulin receptors, highlights its importance in biomedical research. Nevertheless, the duality of its therapeutic and toxic effects necessitates a thorough evaluation of its pharmacokinetics, safety profile, and therapeutic indices. The future direction for research should be to characterize the specific molecular mechanisms that control biological action and create target-specific delivery systems of vanadium to limit toxicity and enhance therapeutic efficacy.

Manganese (Mn)

Manganese is the fifth most abundant metal on the earth and an essential trace element that is needed by the body in very minute amounts, found almost naturally everywhere in the surroundings. It plays a role as an enzyme cofactor for various metabolic processes, antioxidant defense, DNA replication and transcription, and also in mitochondrial respiration. In certain metabolic conditions, serum manganese levels are elevated and are reported in children with obesity [26].

Absorption, Transport, and Storage

In the human diet, the main source of manganese is through the consumption of legumes, whole grains, nuts, and rice. Absorption mainly occurs in the duodenum through active transport involving divalent metal transporter 1 (DMT1), which serves as the main transporter of manganese into most cells. Its absorption is also influenced by the presence of tannins, oxalic acid, and phytic acid, which can inhibit it. Infants are reported to absorb and retain elevated levels of manganese when compared with adults due to the higher demand. Alternations in iron concentrations also influence their absorption because of the shared transport mechanisms. Once manganese is absorbed, it is distributed to the tissues via plasma, where it exists in different oxidation states, of which Mn^{2+} is predominant. Manganese is found to be bound with several blood components, including citrate, transferrin (Tf), albumin, and bicarbonate for transport. The storage sites of manganese are preferably the liver and pancreas, as these tissues are rich in mitochondria; the kidney and the bones are also other sites [59].

Functions and Toxicity

Manganese has a diverse physiological function, playing roles in maintaining homeostatic balance, a cofactor for enzymes involved in metabolic processes, and antioxidant defense by serving as a key element of mitochondrial superoxide dismutase (MnSOD), healthy cartilage and bone formation, normal functioning of the mammary gland, and maintaining neural signaling by being vital part of neurotransmitter synthesis (dopamine, gamma-aminobutyric acid (GABA), and glutamate). The toxic effect of manganese due to its overexposure is known as manganism, which shares similarities with Parkinson's disease, and its accumulation in the brain also causes manganese-induced damage. Manganism is found in individuals whose occupation is welding, mining, battery manufacturing, and fungicide usage. Severe toxicity leads to oxidative stress, mitochondrial dysfunction, inflammation [65, 66], and neurotransmitter dysfunction [41].

Role in Bone Formation, Reproduction, and Metabolism

Manganese is essential for bone formation and its healthy maintenance by adding to the synthesis of bone collagen and cartilage. It is involved in the process of bone mineralization, where macro minerals, like calcium and phosphorus, are deposited for bone density and strength. Studies have reported that women with osteoporosis have significantly lower levels of manganese in serum when compared with normal women without the condition. In vitro studies have proved that manganese helps in the mineralization process by inducing osteogenic differentiation of human mesenchymal stem cells. Also, supplementation studies have proved that a combination of manganese with calcium, zinc, and copper has been shown to reduce bone loss in the lumbar region of postmenopausal women. Deficiency or very low levels of manganese can contribute to infertility if prolonged, which may lead to sterility. It also serves as a part of numerous enzymes involved in carbohydrate metabolism (pyruvate carboxylase, phosphoenolpyruvate), amino acid metabolism (arginase), ATP production, and antioxidant defense (MnSOD, which detoxifies free radicals in mitochondria) [41].

Diseases: Deficiency and Excess (E.g., Manganism, a Neurotoxic Condition)

Deficiency of manganese is rare, but if it occurs, it will hinder the normal functions that it does in the body. As already discussed, manganism is a neurotoxic condition due to excess manganese, which arises mainly due to occupational exposure, as dietary sources barely lead to toxicity due to their poor absorption and efficient excretion mechanism. The symptoms involve motor dysfunction, cognitive, and psychological disorders. If manganese is inhaled, it can be transported directly to the brain, bypassing the liver metabolic pathway [22].

Molybdenum (Mo)

Molybdenum (Mo) is a critical trace element involved in human metabolism by being incorporated into the molybdenum cofactor (Moco). The cofactor is crucial for the catalytic function of various molybdoenzymes, such as xanthine oxidase, aldehyde oxidase, sulfite oxidase, and mitochondrial amidoxime reducing component (mARC), which are responsible for purine catabolism, aldehyde detoxification, and sulfur amino acid metabolism. The absence of molybdenum or its cofactor causes serious metabolic impairment, highlighting its biological significance. Molybdenum is mainly absorbed as molybdate (MoO_4^{2-}) in the proximal gastrointestinal tract. Its absorption is efficient and regulated by food constituents. After being absorbed, molybdenum travels in plasma in a mostly albumin-bound state and is eliminated by excretion via the kidneys, with urinary excretion being a good measure of recent intake [49]. Cereals, legumes, and milk are major food sources, contributing to a typical intake range of 100–170 μg/day in different populations. Physiological roles of molybdenum are all mediated by molybdenum-dependent enzymes, such as the following: xanthine oxidase/dehydrogenase (XO/XDH), hydroxylates and oxidizes purines to uric acid; aldehyde oxidase (AO), involvement in detoxification of aldehydes and N-heterocyclic compounds; sulfite oxidase (SO), oxidizes sulfite to sulfate, which is crucial in cysteine metabolism; and mARC enzymes, plays a role in the reduction of N-hydroxylated substrates, being part of detoxification processes. These enzymes catalyze essential redox reactions that are at the core of purine metabolism, sulfur metabolism, and xenobiotic detoxification. Molybdenum deficiency is extremely uncommon in human beings owing to its widespread availability cofactor deficiency (MoCD), which appears neonatally with intractable seizures, severe neurodegeneration, ocular lens dislocation, and premature death. Biochemically, it results in a combined deficiency of all molybdoenzymes with accumulation of sulfite, xanthine, and S-sulfocysteine, with resultant severe neurological injury [35, 47]. Although a primary molybdenum deficiency due to poor intake is basically nonexistent in healthy people, secondary deficiency can be seen in patients on long-term total parenteral nutrition (TPN) without supplementation of molybdenum, characterized by tachycardia, headache, confusion, and sulfite toxicity [61]. Human molybdenum toxicity is uncommon, with documented cases of adverse effects being scarce. Yet, long-term exposure to elevated environmental levels could result in gouty symptoms because of the resultant overproduction of uric acid through hyperactivity of xanthine oxidase. In ruminants, copper metabolism is disturbed due to excessive intake of molybdenum, and symptoms of copper deficiency result [6]. At present, treatment for MoCD is poor. Recent development involves cyclic pyraoprotein monophosphate (cPMP) replacement therapy, aiming at upstream Moco biosynthesis defects, with promising results in prolonging survival and decreasing neurological impairment in type A MoCD patients. There are no approved treatments for types B and C MoCD, making continued investigation into targeted therapeutics of value [62]. Molybdenum is a vital micronutrient that is a cofactor for essential metabolic pathways. Although it is uncommon to find a deficiency, genetic molybdenum cofactor deficiencies constitute a serious clinical condition with disastrous consequences that are undiagnosed. Molybdenum homeostasis regulation is strong, which reduces the risk of toxicity under normal diet intakes. Technical progress in the field of genetic diagnostics and new targeted therapies gives better control of MoCD, which emphasizes the demand for early diagnosis and treatment.

Cobalt (Co)

Biologically significant, cobalt has a dual role: while necessary in trace amounts for health, its excess exposure leads to toxicity, underscoring the necessity of strict regulation of its body content. Cobalt from foods, primarily as vitamin B_{12} or inorganic compounds, is taken by gastrointestinal tract with bioavailability between 5% and 45%, varying with solubility [38, 39]. After absorption, it is carried through the blood to the liver, kidneys, pancreas, heart, and skeletal muscle. Excretion mainly takes place through urine [78]. Inhalation is another major pathway of exposure in occupational environments, especially in the hard metal industry [39]. The main biological function of cobalt is the central atom in vitamin B_{12}, which facilitates hematopoiesis and ensure neurological integrity [10]. Nonetheless, high levels of serum cobalt (>300 μg/L) can result from chronic exposure [39]. Cobalt fumes cause occupational asthma, lung fibrosis, and cancer. Cobalt, through vitamin B_{12}, is necessary for erythrocyte maturation and the synthesis of neurotransmitters. Deficiency affects DNA synthesis, which results in megaloblastic anemia and neurodegenerative integrity [10]. Excess cobalt, on the other hand, may lead to neurotoxicity in the form of optic atrophy and peripheral neuropathy [17]. Cobalt deficiency is exceedingly uncommon in human beings and typically represents vitamin B_{12} deficiency, presenting as anemia, neuropathy, and dementia [78]. Many individuals obtain cobalt needs from an adequate diet with animal foods, whereas deficiency is more common in strict vegans or those with compromised vitamin B_{12} absorption. Cobalt is vital for human health via its function in vitamin B_{12}-dependent biochemical mechanisms, notably in hematopoiesis and neurological processes. Nevertheless, its tight margin between optimal and toxic concentrations requires stringent surveillance of exposure, particularly in occupational and medical settings. More research is needed to optimize safe exposure limits and determine long-term health effects of subclinical cobalt poisoning.

The physiological roles, deficiency, toxicity, and recommended daily intake (RDI) values for adults of the trace minerals are given in Table 3.1.

Table 3.1 Classification of trace minerals

Mineral	Physiological role	Defeciency	Toxicity	Recommended Daily intake (adult)	Reference
Iron (Fe)	Enzyme cofactor, oxygen transport, energy metabolism	Anemia, cognitive impairment, fatigue	Hemochromatosis, oxidative stress, liver damage, cardiovascular issues	8–11 mg	[48, 74]
Zinc (Zn)	Enzyme cofactor, DNA synthesis, cell growth	Immune dysfunction, skin lesions, growth retardation	Immune suppression, vomiting, nausea, headaches	8–11 mg	[48, 69]
Chromium (Cr)	Insulin signaling and glucose metabolism	Impaired glucose tolerance, insulin resistance, and weight loss	Liver and kidney damage, Carcinogenic effect, skin irritation	25–35 μg	[48, 70]
Copper (Cu)	Mitochondrial respiration, antioxidant defense, bone formation	Dysregulated immune function, anemia, neurological disorder, bone damage	Oxidative stress, Liver, kidney, and neurological damage (Wilson's disease)	0.8–2.4 mg	[48, 25]
Selenium (Se)	Redox regulation, immune function and thyroid hormone metabolism	Keshan disease, depleted immune function, skeletal muscle dysfunction	Selenosis, neurological disorder, gastrointestinal issues	26–55 μg	[44, 48]
Vanadium (V)	Phosphatase inhibition, insulin like activity	Not enough documentation	GI irritation, reduced red blood cells, respiratory irritation	Not established	[48, 75]
Manganese (Mn)	Bone formation, enzyme cofactor, reproduction	Bone demineralization, fertility issues, poor growth	Manganism, lung damage	2–5 mg	[48, 59]
Molybdenum (Mo)	Enzyme cofactor, purine metabolism, detoxification	Arrhythmia, brain damage, seizures (MoCD)	Gout-like symptoms, achy joints, high uric acid content	45 μg	[48, 49]
Cobalt (Co)	Component of vit. B12, hematopoiesis	Megaloblastic anemia, neurodegenerative integrity	Neurotoxicity, polycythemia, cardiomyopathy	Not established as part of vit B12	[48, 38]

Interactions Among Trace Minerals

Competitive Absorption and Antagonistic/Synergistic Effects

Trace minerals, which include chromium, iron, manganese, zinc, cobalt, molybdenum, and copper, interact closely, where the imbalance in one can affect others, influencing cellular homeostasis and contributing to oxidative stress and inflammation [50]. They are required in less than 100 mg per day yet play a significant role. According to Chung et al. [20], there's a strong antagonistic interaction between iron and copper. While iron deficiencies promote copper absorption, excess iron inhibits it, and too much maternal iron intake diminishes copper in the mother and offspring. Zinc and iron compete for absorption as well, and it is less apparent when zinc is consumed in food; however, supplements with more than 25 mg of iron decrease zinc uptake [32]. Interactions among zinc and copper are of clinical significance as a high zinc level causes metallothionein, which traps copper in enterocytes, whereas a low zinc level enhances copper absorption. To avoid oxidative stress, the ideal zinc-to-copper ratio is 8:1 to 15:1 [32]. Figure 3.1 shows the antagonistic effects of Fe, Zn, and Cu. Magnesium influences the equilibrium of calcium, neurotransmitter synthesis, vitamin D activation, and vitamin B6 metabolism in in addition to interacting with calcium and iron [67].

Importance of Balance in Supplementation and Dietary Intake

Management of trace mineral deficiency is difficult because of the need to avoid toxicity while avoiding deficiencies, which is made worse by complex nutrient interactions that are frequently overlooked in the risk assessment models used

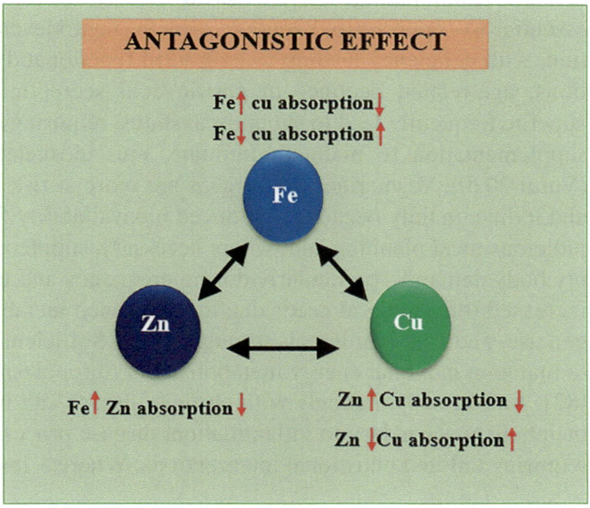

Fig. 3.1 Antagonistic interaction between Fe, Zn, and Cu

currently [82]. Variations regarding dietary intake, metabolic processes, health issues, supplementary use, and diagnostic thresholds confuse clinical assessment, but deficiencies in any trace element can result in pathological diseases that may be curable with customized supplementation. According to Adetunji and Obeng-Gyasi [3], micronutrients often have competitive or synergistic effects, as evidenced by the copper, zinc, and iron interaction. As a result, their uptake and bioavailability are extremely interconnected. According to population studies, targeted supplementation and food fortification with minerals such as copper, selenium, iodine, and zinc have been shown to reduce the prevalence of deficiencies. Toxic metal exposure and excessive supplementation can impact the equilibrium of vital components, leading to oxidative stress and inflammation, both of which are associated with cardiovascular disease. The lack of thorough studies on combined supplementation with micronutrients and their mechanistic interactions highlights the need for customized approaches that take baseline status, drug-nutrient interactions, and the cumulative impact of altering one mineral's metabolism on others into account.

Trace Minerals in Special Populations

Trace elements, such as iron, zinc, copper, selenium, manganese, and iodine, are essential micronutrients for enzymatic function, antioxidant protection, growth, and immune function. Their physiological importance is increased in special groups when nutritional needs or absorption rates differ from those of the general population. During pregnancy and lactation, trace mineral needs significantly rise to sustain fetal development, placental operation, and maternal metabolic adjustment. Deficiencies in iron, zinc, copper, and selenium have been linked with negative outcomes, such as preeclampsia, preterm delivery, and low birth weight ([68]). The evidence suggests that specific supplementation can reduce these risks and enhance neonatal health parameters [2]. During infancy, maximum trace mineral nutrition is essential to support rapid somatic growth, neurodevelopment, and immune function, with deficiency leading to long-term functional dysfunction. In older populations, age-related declines in gastric acid secretion, intestinal absorption, and appetite frequently lead to suboptimal status, requiring careful dietary evaluation or supplementation to maintain immune, musculoskeletal, and cognitive function (Vural 2020). Vegetarians and vegans are more at risk of inadequacy in iron, zinc, and iodine, mainly because of reduced bioavailability from plant foods. Therefore, judicious meal planning and, where necessary, supplementation are advised to supply body demands, particularly during pregnancy and lactation [63]. Athletes show increased trace mineral needs due to heightened metabolic activity, enhanced oxygen use, and loss of minerals through sweat. Sufficient zinc and iron are necessary to maintain maximal energy metabolism, erythropoiesis, and recovery mechanisms [82]. Analogously, patients with chronic disease can have deranged trace mineral metabolism secondary to inflammation, disease process, or pharmacotherapy, thus requiring tailored nutritional interventions. Whereas insufficiency must be avoided,

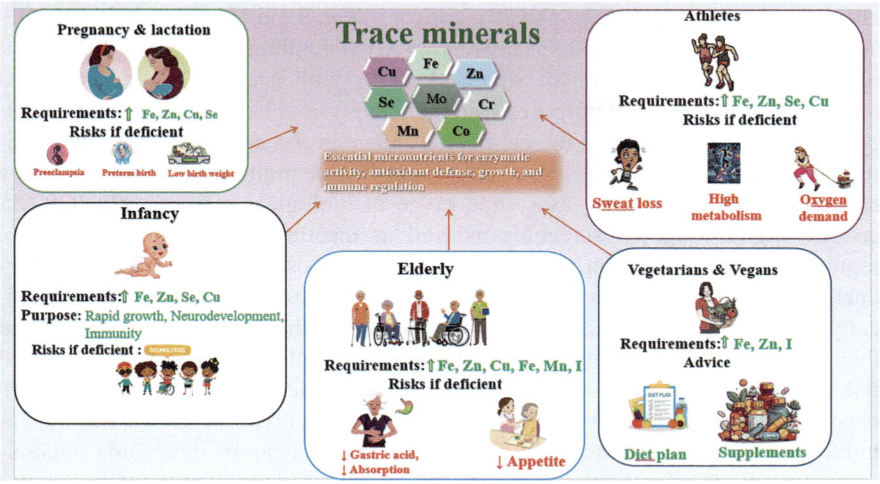

Fig. 3.2 Role of trace minerals in special populations

excess leads to toxicity and adverse effects on health. Therefore, accurate measurements of trace mineral status and evidence-based advice by health professionals are crucial to achieving an optimal balance between deficiency and excess. Figure 3.2 illustrates how trace elements affect special population such as pregnant ladies, athletes, infants, elderly people, vegetarians, and vegans.

Advances in Assessment and Biomarkers

Trace minerals, such as iron, zinc, copper, selenium, and manganese, are crucial for many biochemical functions, and dysregulation has been associated with a variety of disease conditions. Therefore, sensitive and specific biomarkers are essential for accurately identifying deficiencies or toxicities, guiding clinical management, and elucidating underlying disease mechanisms. Biomarkers from biological samples like blood, urine, and hair give useful information about trace mineral status. Plasma and serum analyses by blood-based assays are still the gold standard for clinical applications because they are convenient and comparatively highly reliable [76]. Urinary excretion rates are very helpful for the assessment of short-term variations or acute exposure to trace elements. Hair mineral content analysis provides a more long-term view of mineral depletion or accumulation but is limited in its clinical usefulness by sensitivity to environmental contaminations, potentially leading to misleading results [43]. New methods, including plasma metallomics, have shown promise for the detection of condition-specific biomarkers. For instance, element-pair ratios are good markers for acute myocardial infarction [37]. Analytical chemistry developments have played a critical role in improving trace mineral measurement. Atomic absorption spectroscopy (AAS) continues to be a popular technique for the determination of single elements with acceptable sensitivity and accuracy for most clinical purposes. Yet, inductively coupled plasma mass spectroscopy (ICP-MS) has become the method of choice in research and regulatory environments because it possesses excellent sensitivity, multielement determination capability, and amenability to a wide range of biological matrices [9]. ICP-MS enables high-throughput screening as well as meeting rigorous quality control requirements in application areas from clinical diagnostics to pharmaceutical production and food safety analysis. The convergence of omics technologies specifically metallomics and proteomics has revolutionized the field of trace mineral and their interactions within biology and offers insights into disease pathogenesis and risk stratification. In cardiovascular science, metallomic profiling has detected both beneficial and harmful element associations with acute myocardial infarction, highlighting its clinical relevance [37]. Proteomics fills this gap by describing mineral-dependent protein pathways, which allow for the identification of molecular signatures potentially guiding treatment regimens for an individual [76]. These system-level, high-throughput strategies have the potential to drive precision nutrition and personalized medicine. The synergistic application of standard analytical tools and personalized medicine. The synergetic application of standard tools and innovative omics-based platforms presents a solid framework for evaluating trace mineral status. Although challenges persist, such as reducing environmental contamination and standardizing biomarker interpretation, technological improvements continue to broaden the sensitivity, specificity, and clinical utility of such tools. Future research will be anticipated to optimize biomarker panels, enhance diagnostic precision, and improve integration of trace mineral monitoring into preventive and therapeutic interventions. Table 3.2 consists of the biomarkers for trace minerals along with their advantages and limitations.

Clinical and Public Health Implications

The maintenance of balance between trace minerals is important in public health as both excess and deficiency have paved the way for several health consequences, like cancers, neurodegenerative disorders, and cardiovascular conditions. Deficiency of iron and zinc is the main cause of anemia, affecting 25% of the population globally. This situation can be mitigated by ensuring adequate intake of essential minerals, reducing exposure to toxic elements, and also by giving importance to analyzing trace minerals in clinical practice for early detection of any kind of imbalance, which can be soon rectified by dietary modifications and nutritional supplementations. Proper awareness about dietary sources of minerals is also essential for minimizing health risks and ensuring well-being. There is also a need for strict regulations regarding the presence of trace elements in food and drinking water, for which the legislative body and health workers should acquire updated knowledge of both the

Table 3.2 Common biomarkers for trace mineral status

Trace mineral	Biomarker	Advantage	Limitations	Reference
Iron (Fe)	Serum ferritin	Reflects Fe stores widely used in clinical practice	Acute-phase reactant during inflammation	[40]
	Serum Fe	Indicates circulatory Fe levels	Fluctuate due to dietary intake and diurnal rhythms	
	Transferrin saturation	Reflects Fe transport capacity	Affected by malnutrition and inflammation	
Zinc (Zn)	Serum/plasma Zn	Simple and widely available	Sensitive to fasting Diurnal variations and inflammation Low specificity	[40]
	Urinary Zn	Reflects recent Zn intake Useful in supplementation studies	Poor correlation with body stores Affected by renal function	
Chromium (Cr)	Serum/plasma Cr	Useful in exposure and supplementation studies	Reflect recent intake, not long-term status	[40]
	Urinary Cr	Effective for monitoring Occupational exposure	Insensitive to dietary chromium intake	
Copper (Cu)	Serum Cu	Reflects overall Cu status	Elevated in inflammatory states and pregnancy	[40]
	Ceruloplasmin	Indicates functional Cu pool	Acute-phase reactant Influenced by external factors	
Selenium (Se)	Serum Se	Reflects recent dietary intake	Not specific to functional Se pool Affected by dietary forms	[40]
	Urinary Se	Reflects recent Se intake Useful in population studies	High variability Not reliable for assessing body stores or long-term status	
Vanadium (V)	Serum V	Detects decent vanadium exposure Correlates with urinary V levels in exposed individuals	No established quantitative relationship with exposure level Influenced by recent intake	[72]
	Urinary V	Useful for detecting occupational exposure	Reflects only recent exposure	
	Hair V	Can indicate past or chronic exposure	No established relationship between exposure and hair level	
Manganese (Mn)	Serum/plasma Mn	Indicates Mn exposure	Rapidly excreted Prone to contamination during sample handling	[40]

(continued)

Table 3.2 (continued)

Trace mineral	Biomarker	Advantage	Limitations	Reference
Molybdenum (Mo)	Serum/plasma Mo	Reflects recent short-term intake	No validated biomarkers for deficiency	[40]
Cobalt (Co)	Blood Co	Detects recent occupational or environmental exposure Correlates with air and dermal exposure levels in some studies	Influenced by compound solubility Weaker association for less-soluble cobalt	[79]
	Urine Co	Sensitive to inhalation exposure Good for assessing soluble cobalt compounds	Reflects mainly recent exposure Levels affected by smoking and oral co-exposure from hand-to-mouth transfer	
	Hair Co	Potential indicator or chronic or environmental exposure	Contamination from dust is common Unclear relationship to acute internal dose	

toxic and beneficial effects of the trace elements [16, 40, 57]. Addressing disparities in health resulting from mineral imbalances requires the incorporation of geochemical data into environmental management policy and public health campaigns. Both locally and globally, the medical community has to devote close attention to the assessment and monitoring of low-dose trace element exposure in the general public, especially among vulnerable people [18, 77, 89].

Future Directions and Research Gaps

Significant gaps remain in understanding how plasma trace mineral levels change during critical illness, making it difficult to know whether these changes reflect acute-phase reactions, true deficiencies, reduced bioavailability, or increased metabolism, which complicates clinical management in infections, cancers, inflammatory diseases, and heart, liver, and kidney failure [18]. Even though there are techniques that allow the measurement of essential and nonessential elements yet more research is essential to innovate effective nutritional breakthroughs to prevent and reduce the risk of disease in adults who are more vulnerable [1]. To determine the optimal dietary amounts, further investigation is advised to examine how micronutrients influence immune cell metabolism during infections, as the metabolic processes by which trace minerals enable their protective effects remain poorly understood. Addressing iron, zinc, and selenium deficiencies while controlling elevated exposure to hazardous elements resulting from urbanization necessitates collaborative international research. Selenium's potential cardioprotective and anticancer benefits are a key field of investigation recently [87]. Advanced X-ray-based techniques like

XAS (X-ray absorption spectroscopy), XANES (X-ray absorption near edge structure) imaging, and nanoparticle XRF (X-ray fluorescence tomography), along with emerging omics approaches like metallomics and metalloproteomics, in combination with ICP-MS (inductively coupled plasma mass spectrometry), synchrotron-based XAS, and ESI-MS (electrospray ionization mass spectrometry), will serve as powerful tools for investigating the element distribution, protein-metal interactions, and also their role in homeostasis [16, 73].

Conclusion

Trace minerals, such as iron, zinc, chromium, copper, selenium, vanadium, manganese, molybdenum, and cobalt, play indispensable roles in human health through their involvement in enzymatic catalysis, cellular homeostasis, immune regulation, and metabolic pathways. Their physiological importance is underscored by the fact that even marginal deficiencies can impair growth, cognitive development, and immune defense, while excessive intake can cause toxicity and contribute to chronic disease pathogenesis. The delicate balance of these micronutrients is further complicated by competitive absorption, synergistic and antagonistic interactions, and the influence of physiological states, such as pregnancy, aging, and chronic illness. Advances in analytical techniques and biomarkers have improved the capacity to monitor trace mineral status, yet significant gaps remain in understanding optimal intake ranges, long-term supplementation effects, and their molecular mechanisms in disease modulation. From a public health perspective, ensuring adequate but safe levels of trace mineral intake remains a global challenge, requiring targeted nutritional interventions, improved dietary assessment tools, and integration of emerging omics-based approaches. Future research should prioritize elucidating the nuanced interplay between these minerals and chronic disease risk while developing precision nutrition strategies to optimize their benefits across diverse populations.

Funding Declaration This study was supported by the grant from the Indian Council of Medical Research, Govt of India, ICMR—IIRP—0508/2023.

Competing Interests The authors declare that they have no competing interests.

References

1. Aaseth JO (2022) Toxic and essential metals in human health and disease 2021. Biomolecules 12(10):1375
2. Adams JB, Sorenson JC, Pollard EL, Kirby JK, Audhya T (2021) Evidence-based recommendations for an optimal prenatal supplement for women in the US, part two: minerals. Nutrients 13(6):1849

3. Adetunji AG, Obeng-Gyasi E (2025) Investigating the interplay of toxic metals and essential elements in cardiovascular disease. J Xenobiotics 15(3):68
4. Alkhalidi F (2023) A comparative study to assess the use of chromium in type 2 diabetes mellitus. J Med Life 16(8):1178
5. Alkhatib AJ (2021) Chromium in controlling diabetes and metabolic aspects. Adv Obes Weight Manag Control 11(3):86–88
6. Anke M, Seifert M, Arnhold W, Anke S, Schäfer U (2010) The biological and toxicological importance of molybdenum in the environment and in the nutrition of plants, animals and man: part V: essentiality and toxicity of molybdenum. Acta Aliment 39(1):12–26
7. Attar T (2020) A mini-review on importance and role of trace elements in the human organism. Chem Rev Lett 3(3):117–130
8. Bai S, Zhang M, Tang S, Li M, Wu R, Wan S et al (2024) Effects and impact of selenium on human health, a review. Molecules 30(1):50
9. Barik P, Mehta A, Makhija R, Saha M, Asati V (2025) Recent advancements in inductively coupled plasma mass spectrometry in trace element analysis. Curr Anal Chem 21:1129
10. Batyrova G (2024) The role of cobalt in human health: a brief overview. West Kazakhstan Med J 66(1):6–15
11. Binesh A, Devaraj Sivasitambaram N, Halagowder D (2020) Monocytes treated with ciprofloxacin and oxyLDL express myristate, priming atherosclerosis. J Biochem Mol Toxicol 34(3):e22442
12. Binesh A, Devaraj SN, Halagowder D (2018) Atherogenic diet induced lipid accumulation induced NFκB level in heart, liver and brain of Wistar rat and diosgenin as an anti-inflammatory agent. Life Sci 196:28–37
13. Binesh A, Venkatachalam K, Saravanan R (2025) Investigating the therapeutic potential of Gracilaria edulis extract in copper-exposed Zebrafish embryos: an anti-inflammatory perspective. Nat Prod J 15(5):e22103155294834
14. Binesh A, Venkatachalam K (2024) Copper in human health and disease: a comprehensive review. J Biochem Mol Toxicol 38(11):e70052
15. Briassoulis G, Briassoulis P, Ilia S, Miliaraki M, Briassouli E (2023) The anti-oxidative, anti-inflammatory, anti-apoptotic, and anti-necroptotic role of zinc in COVID-19 and sepsis. Antioxidants 12(11):1942
16. Cannas D, Loi E, Serra M, Firinu D, Valera P, Zavattari P (2020) Relevance of essential trace elements in nutrition and drinking water for human health and autoimmune disease risk. Nutrients 12(7):2074
17. Catalani S, Rizzetti MC, Padovani A, Apostoli P (2012) Neurotoxicity of cobalt. Hum Exp Toxicol 31(5):421–437
18. Cedeño Y, Miranda M, Orjales I, Herrero-Latorre C, Suárez M, Luna D, López-Alonso M (2020) Trace element levels in serum are potentially valuable diagnostic markers in dogs. Animals 10(12):2316
19. Charkiewicz AE (2024) Is copper still safe for us? What do we know and what are the latest literature statements? Curr Issues Mol Biol 46(8):8441–8463
20. Chung HY, Lee MK, Kim W, Choi MK, Kim SH, Kim E et al (2022) Issues pertaining to Mg, Zn and Cu in the 2020 dietary reference intakes for Koreans. Nutr Res Pract 16(Suppl 1):S113–S125
21. Coutinho GGPL, Goloni-Bertollo EM, Bertelli ÉCP (2005) Iron deficiency anemia in children: a challenge for public health and for society. Sao Paulo Med J 123:88–92
22. Dobson AW, Erikson KM, Aschner M (2004) Manganese neurotoxicity. Ann N Y Acad Sci 1012(1):115–128
23. Gad SC (2001) Vanadium. In: Patty's toxicology. Wiley, pp 1–36
24. Ganz T, Nemeth E (2012) Hepcidin and iron homeostasis. Biochimica et Biophysica Acta (BBA) 1823(9):1434–1443
25. Garza NM, Swaminathan AB, Maremanda KP, Zulkifli M, Gohil VM (2023) Mitochondrial copper in human genetic disorders. Trends Endocrinol Metab 34(1):21–33

26. González-Domínguez Á, Jurado-Sumariva L, González-Domínguez R (2024) Association between childhood obesity, trace elements, and heavy metals: recent discoveries and future perspectives. Obes Rev 25(8):e13764
27. Górska A, Markiewicz-Gospodarek A, Trubalski M, Żerebiec M, Poleszak J, Markiewicz R (2024) Assessment of the impact of trace essential metals on cancer development. Int J Mol Sci 25(13):6842
28. Gozzelino R, Arosio P (2016) Iron homeostasis in health and disease. Int J Mol Sci 17(1):130
29. Hamza I, Gitlin JD (2013) Hepatic copper transport. In: Madame Curie bioscience database [Internet]. Landes Bioscience
30. Hassan A, Sada KK, Ketheeswaran S, Dubey AK, Bhat MS (2020) Role of zinc in mucosal health and disease: a review of physiological, biochemical, and molecular processes. Cureus 12(5)
31. Jalali M, Bahadoran Z, Mirmiran P, Azizi F (2025) Dietary mineral patterns are associated with the pre-diabetes regression and progression: the Tehran lipid and glucose study (TLGS). BMC Nutr 11(1):67
32. Jin D, Wei X, He Y, Zhong L, Lu H, Lan J et al (2024) The nutritional roles of zinc for immune system and COVID-19 patients. Front Nutr 11:1385591
33. Joy S, Prasannan A, Venkatachalam K, Binesh A (2025) Molecular mechanism of notch signaling and macrophages in deep vein thrombosis: a comprehensive review. Cell Biochem Biophys 83:2831
34. Lahhob QR, Al-sanaf AE, Mohammed NY, Abbas HJ, Ibrahim ZJ, Najim MK, Malik ZA, Budaiwi ZK, Abdul-Jabbar ZH, Kadhamk MJ (2023) Mineral and trace elements, dietary sources, biological effects, deficiency, and toxicity: a review. Eurasian Chem. Commun 5(6):536–555
35. Lee EJ, Dandamudi R, Granadillo JL, Grange DK, Kakajiwala A (2021) Rare cause of xanthinuria: a pediatric case of molybdenum cofactor deficiency B. CEN Case Rep 10(3):378–382
36. Lewicki S, Zdanowski R, Krzyzowska M, Lewicka A, Debski B, Niemcewicz M, Goniewicz M (2014) The role of chromium III in the organism and its possible use in diabetes and obesity treatment. Ann Agric Environ Med 21(2):331
37. Lim SY, Dayal H, Seah SJ, Tan RPW, Low ZE, Laserna AKC et al (2023) Plasma metallomics reveals potential biomarkers and insights into the ambivalent associations of elements with acute myocardial infarction. J Trace Elem Med Biol 77:127148
38. Lison D (2015) Cobalt. In: Handbook on the toxicology of metals. Academic Press, pp 743–763
39. Lison D (2022) Cobalt. In: Handbook on the toxicology of metals. Academic Press, pp 221–242
40. López-Alonso M, Rivas I, Miranda M (2025) Trace mineral imbalances in global health: challenges, biomarkers, and the role of serum analysis. Nutrients 17(13):2241
41. Magro G, Laterza V, Tosto F, Torrente A (2025) Manganese neurotoxicity: a comprehensive review of pathophysiology and inherited and acquired disorders. J Xenobiotics 15(2):54
42. Mehri A (2020) Trace elements in human nutrition (II)–an update. Int J Prev Med 11(1):2
43. Milne DB (2000) Laboratory assessment of trace element and mineral status. In: Clinical nutrition of the essential trace elements and minerals: the guide for health professionals. Humana Press, Totowa, pp 69–90
44. Minich WB (2022) Selenium metabolism and biosynthesis of selenoproteins in the human body. Biochem Mosc 87(Suppl 1):S168–S177
45. Mohammadifard N, Humphries KH, Gotay C, Mena-Sánchez G, Salas-Salvadó J, Esmaillzadeh A et al (2019) Trace minerals intake: risks and benefits for cardiovascular health. Crit Rev Food Sci Nutr 59(8):1334–1346
46. Moradi F, Maleki V, Saleh-Ghadimi S, Kooshki F, Pourghassem Gargari B (2019) Potential roles of chromium on inflammatory biomarkers in diabetes: a systematic. Clin Exp Pharmacol Physiol 46(11):975–983
47. Nagappa M, Bindu PS, Taly AB, Sinha S, Bharath RD (2015) Child neurology: molybdenum cofactor deficiency. Neurology 85(23):e175–e178

48. National Research Council, Division on Earth, Commission on Life Sciences, & Committee on Diet (1989) Diet and health: implications for reducing chronic disease risk. National Academic Press
49. Oskarsson A, Kippler M (2023) Molybdenum–a scoping review for Nordic nutrition recommendations 2023. Food Nutr Res 67:10–29219
50. Pacheco D, Cotas J, Domingues A, Ressurreição S, Bahcevandziev K, Pereira L (2021) Chondracanthus teedei var. lusitanicus: the nutraceutical potential of an unexploited marine resource. Mar Drugs 19(10):570
51. Pechova A, Pavlata L (2007) Chromium as an essential nutrient: a review. Vet Med 52(1):1–18
52. Piyushbhai MK, Binesh A, Shanmugam SA, Venkatachalam K (2024) Effects of mercury chloride on antioxidant and inflammatory cytokines in zebrafish embryos. J Biochem Mol Toxicol 38(1):e23589
53. Piyushbhai MK, Binesh A, Shanmugam SA, Venkatachalam K (2023) Exposure to low-dose arsenic caused teratogenicity and upregulation of proinflammatory cytokines in zebrafish embryos. Biol Trace Elem Res 201(7):3487–3496
54. Prasad AS (1996) Zinc: the biology and therapeutics of an ion. Ann Intern Med 125(2):142–143
55. Prohaska JR (2008) Role of copper transporters in copper homeostasis. Am J Clin Nutr 88(3):826S–829S
56. Rajpathak S, Rimm EB, Li T, Morris JS, Stampfer MJ, Willett WC, Hu FB (2004) Lower toenail chromium in men with diabetes and cardiovascular disease compared with healthy men. Diabetes Care 27(9):2211–2216
57. Razzaque MS, Wimalawansa SJ (2025) Minerals and human health: from deficiency to toxicity. Nutrients 17(3):454
58. Rosenblum SL (2023) Inflammation, dysregulated iron metabolism, and cardiovascular disease. Front Aging 4:1124178
59. Roth J, Ponzoni S, Aschner M (2012) Manganese homeostasis and transport. In: Metallomics and the cell. Springer Netherlands, Dordrecht, pp 169–201
60. Santos HO (2022) Therapeutic supplementation with zinc in the management of COVID-19–related diarrhea and ageusia/dysgeusia: mechanisms and clues for a personalized dosage regimen. Nutr Rev 80(5):1086–1093
61. Sardesai VM (1993) Molybdenum: an essential trace element. Nutr Clin Pract 8(6):277–281
62. Schwarz G, Veldman A (2013) Molybdenum cofactor disorders. In: Physician's guide to the diagnosis, treatment, and follow-up of inherited metabolic diseases. Springer Berlin Heidelberg, Berlin, Heidelberg, pp 191–203
63. Sebastiani G, Herranz Barbero A, Borrás-Novell C, Alsina Casanova M, Aldecoa-Bilbao V, Andreu-Fernández V et al (2019) The effects of vegetarian and vegan diet during pregnancy on the health of mothers and offspring. Nutrients 11(3):557
64. Sharfalddin AA, Al-Younis IM, Mohammed HA, Dhahri M, Mouffouk F, Abu Ali H et al (2022) Therapeutic properties of vanadium complexes. Inorganics 10(12):244
65. Sharma P, Piyushbhai MK, Venkatachalam K, Binesh A (2024a) Exploring the role of secondary metabolites from plants and microbes as modulators of macrophage differentiation. Cardiovasc Hematol Disord Drug Targets 23(4):134–150
66. Sharma P, Venkatachalam K, Binesh A (2024b) Decades long involvement of THP-1 cells as a model for macrophage research: a comprehensive review. Antiinflamm Antiallergy Agents Med Chem 23(2):85–104
67. Shukla V, Parvez S, Fatima G, Singh S, Magomedova A, Batiha GES et al (2024) Micronutrient interactions: magnesium and its synergies in maternal–fetal health. Food Sci Nutr 12(10):6913–6928
68. Spencer B, Vanderlelie J, Perkins A (2015) Essentiality of trace element micronutrition in human pregnancy: a systematic review. J Preg Child Health 2(3):1000157–1000151
69. Stiles LI, Ferrao K, Mehta KJ (2024) Role of zinc in health and disease. Clin Exp Med 24(1):38
70. Surani SR, Ratnani I, Guntupalli B, Bopparaju S (2012) Severe insulin resistance treatment with intravenous chromium in septic shock patient. World J Diabetes 3(9):170

71. Tarantin AV, Zemlyanova MA (2015) Vanadium essential role and toxic effects. Ekologiya cheloveka (Human Ecology) 22(12):59–64
72. Taylor J (2012) Toxicological profile for vanadium. US Department of Health and Human Services, Public Health Service, Agency for Toxic Substances and Disease Registry
73. Terzano R, Denecke MA, Falkenberg G, Miller B, Paterson D, Janssens K (2019) Recent advances in analysis of trace elements in environmental samples by X-ray based techniques (IUPAC technical report). Pure Appl Chem 91(6):1029–1063
74. Timoshnikov VA, Selyutina OY, Polyakov NE, Didichenko V, Kontoghiorghes GJ (2022) Mechanistic insights of chelator complexes with essential transition metals: antioxidant/pro-oxidant activity and applications in medicine. Int J Mol Sci 23(3):1247
75. Tripathi D, Mani V, Pal RP (2018) Vanadium in biosphere and its role in biological processes. Biol Trace Elem Res 186(1):52–67
76. Tsuji PA, Canter JA, Rosso LE (2016) Trace minerals and trace elements. In: Encyclopedia of food and health, pp 331–338
77. Tu K, Liu K, Wang Y, Jiang Y, Zhang C (2023) Association of dietary intake of zinc and selenium with breast cancer risk: a case-control study in Chinese women. Nutrients 15(14):3253
78. Uddin MH, Rumman M (2020) Cobalt toxicity and human health. In: Metal toxicology handbook. CRC Press, pp 273–285
79. Valerio LG Jr (2024) Predictive analytics for toxicology: applications in discovery science. CRC Press
80. Via M (2012) The malnutrition of obesity: micronutrient deficiencies that promote diabetes. Int Sch Res Notices 2012(1):103472
81. Via M, Scurlock C, Raikhelkar J, Luozzo GD, Mechanick JI (2008) Chromium infusion reverses extreme insulin resistance in a cardiothoracic ICU patient. Nutr Clin Pract 23(3):325–328
82. Vural Z, Avery A, Kalogiros DI, Coneyworth LJ, Welham SJ (2020) Trace mineral intake and deficiencies in older adults living in the community and institutions: a systematic review. Nutrients 12(4):1072
83. Wang X, Bakulski KM, Karvonen-Gutierrez CA, Park SK, Morgan D, Jackson BP et al (2025) Blood essential trace elements and Alzheimer's disease biomarkers in midlife. Front Aging Neurosci 17:1539749
84. Wang YM, Feng LS, Xu A, Ma XH, Zhang MT, Zhang J (2024) Copper ions: the invisible killer of cardiovascular disease. Mol Med Rep 30(5):210
85. Wang ZQ, Yu Y, Zhang XH, Komorowski J (2014) Chromium–insulin reduces insulin clearance and enhances insulin signaling by suppressing hepatic insulin-degrading enzyme and proteasome protein expression in KKAy mice. Front Endocrinol 5:99
86. Yadav S, Prasannan A, Venkatachalam K, Binesh A (2025) Exploring the mechanism and crosstalk between IL-6 and IL-1β on M2 macrophages under metabolic stress conditions. Cytokine 186:156852
87. Yao JH, Ortega EF, Panda A (2025) Impact of zinc on immunometabolism and its putative role on respiratory diseases. Immunometabolism 7(1):e00057
88. Yin RV, Phung OJ (2015) Effect of chromium supplementation on glycated hemoglobin and fasting plasma glucose in patients with diabetes mellitus. Nutr J 14(1):14
89. Zhao S, Wang S, Yang X, Shen L (2023) Dose–response relationship between multiple trace elements and risk of all-cause mortality: a prospective cohort study. Front Nutr 10:1205537
90. Zheng J, Wu F, Wang F, Cheng J, Zou H, Li Y et al (2023) Biomarkers of micronutrients and phytonutrients and their application in epidemiological studies. Nutrients 15(4):970

Chapter 4
Sources and Role of Trace Elements in Nutrition

Amit Ranjan, Pandi Kalaiselvan, Ambika Binesh, and Kaliyamurthi Venkatachalam

Abstract Trace elements, despite being needed in extremely small amounts, play a crucial role in maintaining physiological functions, growth, and health of organisms. Trace elements, such as iron, zinc, and selenium, contribute to various physiological functions, from immune support to enzyme activity. They have a very important role in the growth and development of the human body. The trace elements are divided into three groups, namely, essential elements, elements which are probably essential, and potentially toxic elements. These components are necessary for a number of physiological functions, such as hormone production, immune system control, enzyme activity, and tissue integrity maintenance. A balanced intake of trace elements through a diverse and nutrient-dense diet is essential to supporting general well-being, even if excesses or deficits of these elements can cause major health issues. This chapter discusses sources and the role of these essential trace elements.

Keywords Trace elements · Bioavailability · Diet · Physiology · Metabolism

Introduction

Trace elements play a vital role in many physiological processes of the organism. Although they are present in extremely small quantities of less than 0.01% in the human body; however, they are important for growth, development, maintenance, and the recovery of health [1]. Trace elements are usually defined as the elements that are required in amounts between 1 and 100 mg/day by adults or make up less than 0.01% of total body weight [2]. Based on the nutritional significance, the trace

A. Ranjan (✉) · P. Kalaiselvan · A. Binesh · K. Venkatachalam
Department of Basic Sciences, Institute of Fisheries Post Graduate Studies, Tamil Nadu Dr. J. Jayalalithaa Fisheries University, Chennai, India
e-mail: amitranjan@tnfu.ac.in

N. S. Dhalla et al. (eds.), *Functional Biochemistry of Metallic Elements*, Advances in Biochemistry in Health and Disease 36,
https://doi.org/10.1007/978-3-032-14445-4_4

elements have been divided into three groups: (a) essential elements (chromium (Cr), copper (Cu), zinc (Zn), selenium (Se), molybdenum (Mb), cobalt (Co), and iodine (I)), (b) elements which are probably essential (manganese (Mn), silicon (Si), nickel (Ni), boron (B), and vanadium (V)), and (c) potentially toxic elements (lead (Pb), cadmium (Cd), fluorine (F), mercury (Hg), arsenic (As), aluminum (Al), barium (Ba), lithium (Li), and tin (Sn)) [3].

Sources of Different Trace Minerals

Iron

Iron is a mineral found naturally in numerous foods, added to certain products through fortification, and available as a dietary supplement. There are two primary forms of dietary iron: heme and nonheme [4]. The iron content in various plant and animal foods is documented in [5]. Plant-based foods and those fortified with iron contain only nonheme iron, while meat, poultry, and seafood provide both heme and nonheme iron [6]. Lean meats and seafood are the most abundant sources of heme iron, whereas nuts, beans, vegetables, and fortified grains are sources of nonheme iron [7]. Among plant foods, spinach is a notable source of iron, though its bioavailability is lower than that of animal sources due to the presence of iron absorption inhibitors like polyphenols [8]. The liver from animals, such as calves, pigs, sheep, horses, and ducks, is the most well-known animal source of iron.

Other significant animal sources of iron include kidneys, brewer's yeast, meat, chicken egg yolk, and fish [9]. Table 4.1 lists the various dietary sources of Fe. During food processing, products like cereals, wheat, and other flours are fortified with artificial iron compounds to meet iron requirements in many countries [10]. Commercially, several iron compounds are available, including ferrous and ferric

Table 4.1 Dietary iron food sources [9]

Source	Average iron content (mg/100 g)
Raw yolk of chicken eggs	5
Raw fish	5
Raw meat (beef and veal)	4
Raw animal liver	20
Cow milk	0.2
Dried oregano	18
Cocoa powder	14.3
Dried soy flour and wheat bran	8
Raw edible mushroom	1–2
Arabica coffee powder	12
Red wine	0.9–1.1
Dried pulses (beans, lentils)	9

iron salts like ferric citrate, ferric sulfate, ferrous gluconate, and ferrous sulfate [11, 12]. Ferrous iron is more beneficial than ferric iron due to its higher bioavailability. Other forms, such as carbonyl iron, polysaccharide-iron complexes, iron amino acid chelates, and heme iron polypeptides, are available but less commonly used commercially because they cause fewer gastrointestinal side effects than ferric and ferrous salts [12]. These iron salts are also included in multimineral/multivitamin supplements and used as dietary supplements. Generally, different forms of iron are combined in supplements with varying levels of elemental iron. Nowadays, iron nanoparticles are produced and utilized in hyperthermia, biomedical, catalytic, magnetic, and electrical applications [13].

Zinc

Zinc can be found in a diverse range of foods, such as shellfish and red meat [14]. Oysters and beef, in particular, offer more zinc than any other food sources [15]. According to Samman [16], liver and cheese are also abundant in Zn. Eggs and dairy products provide zinc in amounts ranging from 0.6 to 1.0 mg per serving [17]. Animal-based sources of zinc have relatively high bioavailability due to the lack of compounds that hinder absorption and the presence of certain amino acids that facilitate it [14]. Although nuts, whole grains, beans, and fortified foods are also good zinc sources, the bioavailability of zinc from plant-based sources is lower because of anti-nutritional factors that bind zinc and prevent its absorption [14]. For instance, phytates, which are the storage form of phosphorus in plants, bind to minerals like zinc in the intestine, forming an insoluble complex that blocks zinc absorption [18]. Fruits and vegetables of plant origin have minimal zinc content. Breakfast cereals are frequently fortified with synthetic zinc compounds, such as zinc sulfate and zinc oxide, which are considered significant artificial zinc sources [17]. The zinc content of various food sources has been documented in [19].

Copper

Foods with the highest levels of zinc often also have the most copper. Beef liver is an excellent source of both these minerals. Foods with more than 0.2 mg of copper per 100 gm are regarded as good sources [20]. According to Ensoy et al. [21], crustaceans and shellfish are the primary seafood sources of copper, while whole grains, beans, seeds, nuts, dried fruits, and dark leafy greens are the main plant-based sources. To combat micronutrient deficiencies, synthetic copper is added to cereal flour. Copper supplementation is advised alongside zinc supplementation [22]. Copper gluconate is preferred over copper sulfate as a fortificant due to its stability and minimal physicochemical interactions in staple food matrices. The recommended copper addition to staple foods is between 1.2 and 3.0 mg/kg of flour [22].

Iodine

Iodine is a crucial trace mineral, primarily due to its role in thyroid hormones, which are essential for growth and brain development both before and during pregnancy. In numerous countries, animal-derived sources, like fish, eggs, and milk, account for over 60% of iodine consumption, with the iodine content in both animal and plant foods documented by Nicol et al. [23]. Milk and dairy products are the main contributors to iodine intake, providing about 12–53% of the daily requirement [24]. Although cow's milk typically has a lower iodine content, it has become a significant iodine source through the use of iodine-enriched cattle feed and iodophors, iodine-based disinfectants, during milking [25]. Additionally, dairy products, such as cheese and yogurt, are also rich in iodine. In contrast, plant-based alternatives, like yogurt and cheese, are often not fortified with iodine [26], and substituting these for dairy products may lead to reduced iodine intake. Chicken eggs are another good iodine source, containing 50 μg/100 g, or about 25 μg per egg [23].

While fish and seafood naturally contain high levels of iodine, the exact amounts vary by species, offering a wide range of options (e.g., different fish species, fresh or processed products) and iodine content per serving [23]. Following fish and seafood, seaweed is becoming popular as a functional ingredient in plant-based products to enhance flavor, improve the food matrix, or enrich products with bioactive compounds [27, 28]. Seaweed is a rich plant-based iodine source, with kelp being particularly high in iodine [29, 30]. However, not all seaweed species have high iodine levels, and those with suitable iodine concentrations can be used to address iodine deficiency through plant-based ingredients. Besides food sources, iodine-fortified food-grade salts are used to prevent and manage iodine deficiency [31]. Countries like the Netherlands, Belgium, and Denmark have introduced non-animal iodine sources by fortifying bread with iodized salt [32].

Manganese

The primary way the general public is exposed to manganese is through their diet [33, 34]. Plants are a rich source of manganese, with foods like grain cereals, various nuts, tea, legumes, beans, almonds, chickpeas, and coffee being particularly high in this mineral [35]. In contrast, animal-based foods contain minimal amounts of manganese. Similar to zinc, the presence of compounds, such as phytic acid, oxalic acid, and tannins, which are prevalent in plant-based foods, can inhibit manganese absorption [36]. According to ATSDR [34], the average manganese levels in different foods show that nuts and nut products have the highest concentrations, ranging from 18.2 to 46.8 mg/kg, followed by grains and grain products, which range from 0.42 to 40.7 mg/kg. Milk and dairy products have the lowest manganese content, with levels ranging from 0.02 to 0.49 mg/kg.

Cobalt

Cobalt is an essential trace mineral in the human diet, primarily because it is a component of vitamin B12, which is vital for nerve and blood cell health and DNA synthesis [37]. Foods high in vitamin B12 contain cobalt. In humans, animal products are the sole natural sources of vitamin B12. These include animal liver and kidney; fish and shellfish; meats, like beef, lamb, and pork; and dairy items, such as milk, cheese, yogurt, and eggs [38]. Nonetheless, for individuals who do not consume animal products, breakfast cereals and nutritional yeast can supply the necessary vitamin B12 as sources of cobalt. The cobalt levels in different food categories showed considerable variation, as illustrated in Table 4.2.

Molybdenum

Molybdenum is commonly found in the human diet. In plant-based foods, the amount of molybdenum is largely influenced by the soil's mineral content in the area, while in meat, it is determined by the diet of the animals. The primary sources of molybdenum in our diet include legumes, grain products, and nuts [40, 41]. Tsongas et al. [42] conducted measurements of molybdenum levels across various foods, reporting the following concentrations per 100 g of wet weight: vegetables contain 0.1 to 0.8 mg; legumes range from 2 to 87 mg; eggs, milk, and cheese have 0.5 to 1.0 mg; chicken, beef, and pork contain 0.2 to 0.5 mg; bread, pasta, and rice have 3.0 to 4.0 mg; and fruit contains 0.1 to 0.9 mg. Legumes are the most abundant in molybdenum, as illustrated in Table 4.3, with lima beans offering 87 mg of molybdenum per 100 g.

Table 4.2 Range of cobalt content in various food sources [39]

Source	Cobalt content (μg/kg)
Cereals and cereal products	4.03–13.11
Milk and dairy products	3.60–12.24
Meat and meat products	2.01 and 5.52
Eggs	0.94 to 5.48
Fish and seafoods	2.98 and 7.45
Vegetables	3.71–13.92
Legumes	39.28–104.29
Potatoes	5.07–12.80
Fresh fruits	1.97–6.43

Table 4.3 Molybdenum content of legumes [42]

Legume	Molybdenum content (mg/100 g)
Kidney beans	8
Butter lima beans	17
Pinto beans	10
Lima beans	87
Small white beans	45
Black-eyed beans	12
Red beans	20
Yellow split beans	22
Green split beans	15
Whole green peas	13
Sweet peas	2

Selenium

Products derived from plants and animals are the primary sources of this element. In animals, selenium is found to be selenocysteine and selenomethionine, with selenocysteine being the predominant form [43]. Humans consume selenium in this form, and it undergoes further transformation, binding to amino acids and proteins [44]. Plants provide selenium in the form of inorganic selenate compounds, which are subsequently converted into organic forms. The selenium content in diets varies based on the geographical location of plants and animals [45]. The diversity in dietary selenium levels is influenced by the region where plants grow and animals reside [46, 47]. This is determined by the selenium concentration in the soil, soil type, agroclimatic conditions, and crop type [48]. Key dietary sources of selenium include cereals, meat, dairy products, fish, seafood, milk, and nuts [49]. Rich selenium sources are found in sea salt, eggs (when feed is supplemented with Se-yeast), giblets, selenium-enriched yeast, bread, mushrooms, garlic, asparagus, and kohlrabi [50]. Fruits and vegetables generally have low selenium content, as it is mainly present in the protein fraction, making low-protein plants and vegetables poor selenium sources [51]. High selenium levels are found in hyperaccumulator plants like *Astragalus bisulcatus* and some *Brassicaceae* species [45].

Animal-derived food products exhibit varying selenium levels, depending on the geographical area and dietary selenium supply (Table 4.4). Besides food sources, selenium supplementation has become important due to widespread human deficiencies. Direct supplementation involves using food supplements with vitamins and micronutrients [49]. Indirectly, selenium can be added to fertilizers or animal feed [52]. Generally, food enriched with inorganic selenium or supplements based on selenium yeast biomass is considered the most direct supplementation method [53]. However, selenomethionine supplementation is preferred for enriching food with selenium to prevent deficiencies, as it is a protein-bound selenium source [54]. Thus, selenium is supplemented or enriched in both organic forms (selenocysteine, selenocysteine, and selenoglutathione) and inorganic forms (Na2SeO3 and Na2SeO4).

Table 4.4 Selenium content in food products

Source	Selenium content (μg/g)	References
Brazil nuts	0.2–512	[55]
Yeast	500–4000	[46, 50]
Bread	0.09–0.20	[51, 56]
Fish	0.06–0.63	[46, 56, 57]
Eggs	0.09–0.25	[46, 58]
Chicken	0.15	[56]
Beef	0.01–0.73	[46, 58]
Pork	0.27–0.35	[51, 56]
Broccoli	0.012	[58]
Milk	0.01–0.06	[56, 57]
Chocolate	0.04	[46, 58]
Liver	0.3–0.4	[59]

Chromium

Chromium (Cr), often called as glucose tolerance factor, is an essential nutrient that potentiates insulin, and thus influences carbohydrate, lipid, and protein metabolism [3]. The chromium levels in various foods differ significantly [60]. Mussels have the highest amount, with 128 μg per 100 g, and Brazil nuts are close behind with 100 μg. Oysters contain about 57 μg, while dried dates and pears have 29 μg and 27 μg, respectively. Brown shrimp and whole-meal flour offer 26 μg and 21 μg, respectively. Tomatoes have around 20 μg of chromium, and mushrooms contain 17 μg. Broccoli provides 16 μg, whereas whole-grain barley has 13 μg. Hazelnuts contain about 12 μg, and pork chops have 10 μg. Whole-grain maize has 9 μg, while egg yolks contain 6 μg. Beef and herring have lower chromium levels, with 3 μg and 2 μg, respectively, per 100 g [60].

Role of Different Trace Elements

Iron (Fe) Iron is a plentiful and vital element on earth, which is crucial for the survival of all living beings [61]. Despite its abundance, it often acts as a limiting factor for growth in various environments [62]. The human body needs iron to produce oxygen-carrying proteins, like hemoglobin and myoglobin, as well as to form heme enzymes and other iron-dependent enzymes that participate in electron transfer and redox reactions [63]. Additionally, iron is necessary for developing antioxidants that neutralize reactive oxygen species. For instance, catalase, an iron-based enzyme found in cell peroxisomes, facilitates the breakdown of hydrogen peroxide into water and oxygen [43]. Iron is also crucial for producing microbiocidal hypochlorite acids, and a deficiency of iron has been linked to a decrease in the average concentration of blood T cells [64].

Zinc (Zn) Zinc is an essential trace element often deficient in human nutrition. Zinc is omnipresent within cells, predominantly found in specific cellular components (about 95% of zinc), and plays well-defined physiological roles [65]. Zinc is closely associated with metalloproteins as a structural element or with metal enzymes as cofactors. Its biological functions can be categorized into three main areas: catalytic, structural, and regulatory [66]. Zinc is vital for the metabolic activity of over 300 enzymes in the body and is essential for protein synthesis, DNA creation, and cell division [67]. It is involved in numerous critical bodily functions, including the growth and maintenance of connective tissue, wound healing, prostaglandin production, proper thyroid function, maintaining normal serum testosterone levels, cognitive function, blood clotting, sperm production, and fetal development [67]. Zinc also plays a crucial role in the development of bones, cartilage, and muscles; the regulation of oxidative stress (reducing the production of compounds due to lipid and DNA oxidation); and the proliferation and maturation of immune cells [68]. Zinc reduces the generation of ROS, which leads to activation of NF-kB, which in turn modulate the immune response to infections whose dysfunction can cause autoimmune diseases and tumors [69].

Copper (Cu) Copper is vital for all living beings and forms a key part of the cytochrome oxidase enzyme complex. Copper-containing proteins play significant roles in biological processes, such as electron and oxygen transport, with the ability to easily switch between Cu (I) and Cu (II) states [70]. This mineral is present in superoxide dismutase, an enzyme that facilitates the conversion of superoxide into oxygen peroxide and hydrogen [71]. Copper is also crucial in systems responsible for hemoglobin production and the creation of cross-links among collagen, elastin, and keratin fibers in hair [71]. Copper has a regulatory role in cell growth and in maintaining homeostasis in the immune system [1]. Copper contributes to various physiological functions, including bone mineralization, temperature regulation, connective tissue synthesis, melanin production, cholesterol metabolism, myelin formation, heart function, immune response, catecholamine metabolism, glucose regulation, and oxidative phosphorylation [72].

Iodine (I) Iodine is a vital non-metallic trace element which plays a critical role in normal growth and development in vertebrates by supporting adequate thyroid hormone production [73]. Its sole known metabolic role is its integration into thyroid hormones (T3 and T4), which are essential for metabolic regulation and growth [74]. Iodine regulates all metabolic pathways related to thyroid hormones, thyroxine, and triiodothyronine and exerts a decisive role in fetal development. It is one of the most powerful antioxidants required for protective functions in inflammatory processes and cancer development [75].

Manganese (Mn) Manganese is an essential element for living organisms, which is required for growth, development, and health maintenance [76]. Manganese is involved in the synthesis and metabolism of carbohydrates, fats, and cholesterol [77]. It is crucial for various metabolic functions, including energy metabolism, nervous system operation, metalloenzyme activation, reproductive hormone function, antioxidant system function, and immune system function [34, 78]. Manganese also plays a significant role in blood clotting, connective tissue and bone growth, and cellular energy regulation. It serves as an important cofactor for enzymes involved in neurotransmitter synthesis and metabolism [79]. Manganese is also essential for brain development and maintaining cellular homeostasis [80].

Cobalt (Co) Cobalt is essential for maintaining healthy nerve and blood cells and for DNA synthesis [37]. It serves as a cofactor for vital enzymes, like methylmalonyl-CoA mutase and methionine synthase, which are involved in the synthesis and breakdown of amino acids and fatty acids [81]. During pregnancy, it plays a key role in regulating the immune system [82]. Research by Yahn et al. (2021) confirmed that cobalt present in vit. B12 is effective in treating viral infections and ischemic stroke. The importance of molybdenum in metabolism was recognized after its role as a cofactor for the enzyme xanthine oxidase was discovered in 1953. Molybdenum acts as a cofactor for three additional enzymes: sulfite oxidase, aldehyde oxidase, and mitochondrial amidoxime [83]. Sulfite oxidase facilitates the conversion of sulfite to sulfate, which is crucial for metabolizing sulfur-containing amino acids. A deficiency or absence of sulfite oxidase can lead to neurological issues and early mortality. Xanthine oxidase is responsible for the oxidative hydroxylation of purines and pyridines, including the transformation of hypoxanthine into xanthine and xanthine into uric acid. Aldehyde oxidase oxidizes purines, pyrimidines, and pteridines and is involved in the metabolism of nicotinic acid [84]. In each of these enzymes, molybdenum is attached to a complex, multi-ring organic component known as molybdopterin, which facilitates redox reactions essential for metabolism and detoxification [83].

Selenium (Se) Selenium is an essential trace mineral and plays an important role in the human body. It is a natural component of glutathione peroxidase and other selenoproteins [1]. Selenium was initially recognized for its toxic effects on the body until the 1940s. By 1976, numerous studies had shown that selenium is vital for humans, particularly in reducing cardiovascular diseases and tumors [84]. Selenium has chemopreventive properties, such as boosting the immune system, protecting against oxygen radicals, modifying body proteins, safeguarding metabolic enzymes, reducing the oxidative effects of carcinogens, and producing selenium metabolites [85]. Organic selenium compounds, including seleno-amino acids, seleno derivatives of nucleic acids, peptides containing selenium, and other compounds, play a significant role in biological processes [86]. Selenium is renowned for its potent antioxidant properties through the production of selenoproteins, which capture or neutralize free oxygen radicals, thereby preventing oxidative stress, short-term tissue damage, and inflammation [87]. Additionally, selenium is a component of the glutathione peroxidase enzyme. Glutathione peroxidase breaks down lipoperoxides into water and alcohol, which are produced from the oxidation of unsaturated fatty acids by free radicals. This enzyme consists of four identical subunits, each containing one selenium atom in the form of selenocysteine. Selenium is crucial for the production of thyroid hormones, which are responsible for maintaining healthy hair, skin, and eyes. Selenium works synergistically with vitamin E to perform antioxidant functions by overlapping with glutathione peroxidase [88].

Chromium (Cr) Chromium (Cr) is an essential element involved in protein, lipid, and carbohydrate metabolism [1]. Cr (III) also plays a crucial role in nutrient metabolism, aiding in the regulation of insulin activity and improving lipid and carbohydrate metabolism [43]. Numerous studies have indicated that exposure to chromium can lead to cancer by generating Cr (III) and an excess of reactive oxygen species, which may cause DNA damage and oxidative stress [89]. The immune response to chromium exposure can vary based on the compound type, dosage, and method of exposure [60]. A deficiency in Cr intake is linked to cardiovascular diseases and diabetes [90]. Research has shown that individuals with type 2 diabetes tend to have lower chromium levels compared to those without the condition. Insulin resistance, a common precursor to cardiovascular disease, can be mitigated by chromium picolinate, which has been found to decrease insulin resistance and lower the risk of developing type 2 diabetes [91]. Chromium can function as an immune modulator, either inhibiting or enhancing immune responses by influencing immune cells and cytokine production [92].

Molybdenum (Mo) Molybdenum is a metal and an essential nutrient that is a component of nearly seven enzymes that support the functioning of many systems in the human body. It is a component of enzymes such as aldehyde oxidase, xanthine oxidase, and sulfite oxidase, which are involved in aldehyde oxidation, nucleotide metabolism, and sulfite detoxification [93]. Molybdenum is crucial in the synthesis and dissociation of epinephrine, norepinephrine, serotonin, and melatonin neurotransmitters [94].

References

1. Cannas D, Loi E, Serra M, Firinu D, Valera P, Zavattari P (2020) Relevance of essential trace elements in nutrition and drinking water for human health and autoimmune disease risk. Nutrients 12(7):2074
2. Aliasgharpour M, Rahnamaye Farzami M (2013) Trace elements in human nutrition: a review. J Emerg Health Care 2(3):115–128
3. World Health Organization (1996) Trace elements in human nutrition and health. WHO Library Cataloguing in Publication Data, pp 105–122
4. Ross AC, Caballero B, Cousins RJ, Tucker KL (2020) Modern nutrition in health and disease. Jones & Bartlett Learning
5. U.S. Department of Agriculture, Agricultural Research Service. (2019). FoodData Central. https://fdc.nal.usda.gov/
6. Erdman JW Jr, Macdonald IA, Zeisel SH (eds) (2012) Present knowledge in nutrition. John Wiley & Sons
7. U.S. Department of Agriculture and U.S. Department of Health and Human Services. (2020). Dietary Guidelines for Americans, 2020–2025 (9th ed.). https://www.dietaryguidelines.gov/
8. Rutzke CJ, Glahn RP, Rutzke MA, Welch RM, Langhans RW, Albright LD et al (2004) Bioavailability of iron from spinach using an in vitro/human Caco-2 cell bioassay model. Habitation 10(1):7–14
9. Briguglio M, Hrelia S, Malaguti M, Lombardi G, Riso P, Porrini M et al (2020) The central role of iron in human nutrition: from folk to contemporary medicine. Nutrients 12(6):1761

10. Whittaker P, Tufaro PR, Rader JI (2001) Iron and folate in fortified cereals. J Am Coll Nutr 20(3):247–254
11. Manoguerra AS, Erdman AR, Booze LL, Christianson G, Wax PM, Scharman EJ et al (2005) Iron ingestion: an evidence-based consensus guideline for out-of-hospital management. Clin Toxicol 43(6):553–570
12. Murray-Kolbe LE, Beard J (2010) Iron. In: Coates PM, Betz JM, Blackman MR et al (eds) Encyclopedia of dietary supplements, 2nd edn. Informa Healthcare, pp 432–438
13. Huber DL (2005) Synthesis, properties, and applications of iron nanoparticles. Small 1(5):482–501
14. Huang L, Wong CP, Ho E (2025) Zinc. Adv Nutr 16:100408
15. Huth PJ, Fulgoni VL, Keast DR, Park K, Auestad N (2013) Major food sources of calories, added sugars, and saturated fat and their contribution to essential nutrient intakes in the U.S. diet: Data from the National Health and Nutrition Examination Survey (2003–2006). Nutr J 12:116
16. Samman S (2012) Zinc. In: Mann JI, Truswell AS (eds) Essentials of human nutrition, 4th edn. Oxford University Press, pp 171–175
17. King JC, Cousins RJ (2014) Zinc. In: Ross AC, Caballero B, Cousins RJ, Tucker KL, Ziegler TR (eds) Modern nutrition in health and disease, 11th edn. Lippincott Williams & Wilkins, Baltimore, pp 189–205
18. King JC, Brown KH, Gibson RS, Krebs NF, Lowe NM, Siekmann JH et al (2016) Biomarkers of nutrition for development (BOND)—zinc review. J Nutr 146(Suppl):858S–885S
19. U.S. Department of Agriculture. (2022). FoodData Central. https://fdc.nal.usda.gov/
20. Lawler MR, Klevay LM (1984) Copper and zinc in selected foods. J Am Diet Assoc 84(9):1028–1030
21. Ensoy C, Rakhmawati TW, Faes C, Aerts M (2015) Final evidence report as part of preparatory work for the setting of dietary reference values for magnesium, copper and phosphorus. EFSA Supporting Publications 12(10):691E
22. Rosado JL (2003) Zinc and copper: proposed fortification levels and recommended zinc compounds. J Nutr 133(9):2985S–2989S
23. Nicol K, Nugent AP, Woodside JV, Hart KH, Bath SC (2024) Iodine and plant-based diets: a narrative review and calculation of iodine content. Br J Nutr 131(2):265–275
24. Bath SC, Verkaik-Kloosterman J, Sabatier M, Ter Borg S, Eilander A, Hora K et al (2022) A systematic review of iodine intake in children, adults, and pregnant women in Europe—comparison against dietary recommendations and evaluation of dietary iodine sources. Nutr Rev 80(11):2154–2177
25. van der Reijden OL, Zimmermann MB, Galetti V (2017) Iodine in dairy milk: sources, concentrations and importance to human health. Best Pract Res Clin Endocrinol Metab 31(4):385–395
26. Nicol K, Thomas EL, Nugent AP, Woodside JV, Hart KH, Bath SC (2023) Iodine fortification of plant-based dairy and fish alternatives: the effect of substitution on iodine intake based on a market survey in the UK. Br J Nutr 129(5):832–842
27. Kyriakopoulou K, Keppler JK, van Der Goot AJ (2021) Functionality of ingredients and additives in plant-based meat analogues. Foods 10(3):600
28. Garciarena IN, Poulsen M, Lindholm Bøgh K (2022) Risk–benefit assessment of seaweed Allergenicity risk assessment of novel protein. EFSA J 20:e200414
29. Katagiri R, Asakura K, Sasaki S, Hirota N, Notsu A, Miura A et al (2015) Estimation of habitual iodine intake in Japanese adults using 16 d diet records over four seasons with a newly developed food composition database for iodine. Br J Nutr 114(4):624–634
30. Jeon MJ, Kim WG, Kwon H, Kim M, Park S, Oh HS et al (2017) Excessive iodine intake and thyrotropin reference interval: data from the Korean National Health and Nutrition Examination Survey. Thyroid 27(7):967–972
31. World Health Organization, UNICEF & ICCIDD (2007) Assessment of iodine deficiency disorders and monitoring their elimination. WHO, Geneva

32. Jakobsen LS, Nielsen JO, Paulsen SE, Outzen M, Linneberg A, Møllehave LT et al (2022) Risk–benefit assessment of an increase in the iodine fortification level of foods in Denmark—a pilot study. Foods 11(9):1281
33. US EPA. Health effects support document for manganese; US Environmental Protection Agency, Office of Water: Washington, DC, USA, 2002
34. ATSDR. Toxicological Profile for Manganese; U.S. Department of Health and Human Services. Public Health Service. Agency for Toxic Substances and Disease Registry: Atlanta, 2012; 556
35. Filippini T, Cilloni S, Malavolti M, Violi F, Malagoli C, Tesauro M et al (2018) Dietary intake of cadmium, chromium, copper, manganese, selenium and zinc in a Northern Italy community. J Trace Elem Med Biol 50:508–517
36. Choi MK, Bae YJ (2020) Dietary intake and urinary excretion of manganese in Korean healthy adults. Biol Trace Elem Res 196(2):384–392
37. Osman D, Cooke A, Young TR, Deery E, Robinson NJ, Warren MJ (2021) The requirement for cobalt in vitamin B12: a paradigm for protein metalation. Biochim Biophys Acta Mol Cell Res 1868(1):118896
38. Genchi G, Lauria G, Catalano A, Carocci A, Sinicropi MS (2023) Prevalence of cobalt in the environment and its role in biological processes. Biology 12(10):1335
39. Filippini T, Tancredi S, Malagoli C, Malavolti M, Bargellini A, Vescovi L et al (2020) Dietary estimated intake of trace elements: risk assessment in an Italian population. Expo Health 12(4):641–655
40. Pennington JA, Jones JW (1987) Molybdenum, nickel, cobalt, vanadium, and strontium in total diets. J Am Diet Assoc 87(12):1644–1650
41. Hattori H, Ashida A, ITÔ C, Yoshida M (2004) Determination of molybdenum in foods and human milk, and an estimate of average molybdenum intake in the Japanese population. J Nutr Sci Vitaminol 50(6):404–409
42. Tsongas TA, Meglen RR, Walravens PA, Chappell WR (1980) Molybdenum in the diet: an estimate of average daily intake in the United States. Am J Clin Nutr 33(5):1103–1107
43. Lahhob QR, Mohammed NY, Abbas HJ, Ibrahim ZJ, Najim MK, Malik ZA et al (2023) Mineral and trace elements, dietary sources, biological effects, deficiency, and toxicity: a review. Eurasian Chem Commun 5(2023):536–555
44. Pezzarossa B, Petruzzelli G, Petacco F, Malorgio F, Ferri T (2007) Absorption of selenium by Lactuca sativa as affected by carboxymethylcellulose. Chemosphere 67(2):322–329
45. Kieliszek M (2019) Selenium–fascinating microelement, properties and sources in food. Molecules 24(7):1298
46. Navarro-Alarcon M, Cabrera-Vique C (2008) Selenium in food and the human body: a review. Sci Total Environ 400(1–3):115–141
47. Mehdi Y, Hornick JL, Istasse L, Dufrasne I (2013) Selenium in the environment, metabolism and involvement in body functions. Molecules 18(3):3292–3311
48. Qin HB, Zhu JM, Liang L, Wang MS, Su H (2013) The bioavailability of selenium and risk assessment for human selenium poisoning in high-se areas, China. Environ Int 52:66–74
49. Frączek A, Pasternak K (2013) Selenium in medicine and treatment. J Elem 18(1)
50. Kieliszek M, Błażejak S (2016) Current knowledge on the importance of selenium in food for living organisms: a review. Molecules 21(5):609
51. dos Santos M, da Silva Júnior FMR, Muccillo-Baisch AL (2017) Selenium content of Brazilian foods: a review of the literature values. J Food Compos Anal 58:10–15
52. Pavlovic Z, Miletic I, Zekovic M, Nikolic M, Glibetic M (2018) Impact of selenium addition to animal feeds on human selenium status in Serbia. Nutrients 10(2):225
53. Roman M, Jitaru P, Barbante C (2014) Selenium biochemistry and its role for human health. Metallomics 6(1):25–54
54. Pillai R, Uyehara-Lock JH, Bellinger FP (2014) Selenium and selenoprotein function in brain disorders. IUBMB Life 66(4):229–239

55. Junior ES, Wadt LDO, Silva KE, Lima RMB, Batista KD, Guedes MC et al (2017) Natural variation of selenium in Brazil nuts and soils from the Amazon region. Chemosphere 188:650–658
56. Skalny AV, Burtseva TI, Salnikova EV, Ajsuvakova OP, Skalnaya MG, Kirichuk AA, Tinkov AA (2019) Geographic variation of environmental, food, and human hair selenium content in an industrial region of Russia. Environ Res 171:293–301
57. Klapec T, Mandić ML, Grgić J, Primorac L, Perl A, Krstanović V (2004) Selenium in selected foods grown or purchased in eastern Croatia. Food Chem 85(3):445–452
58. Pappa EC, Pappas AC, Surai PF (2006) Selenium content in selected foods from the Greek market and estimation of the daily intake. Sci Total Environ 372(1):100–108
59. Pilarczyk B, Tomza-Marciniak A, Pilarczyk R, Kuba J, Hendzel D, Udała J, Tarasewicz Z (2019) Eggs as a source of selenium in the human diet. J Food Compos Anal 78:19–23
60. Milanković V, Tasić T, Leskovac A, Petrović S, Mitić M, Lazarević-Pašti T et al (2024) Metals on the menu—analyzing the presence, importance, and consequences. Foods 13(12):1890
61. Lieu PT, Heiskala M, Peterson PA, Yang Y (2001) The roles of iron in health and disease. Mol Asp Med 22(1–2):1–87
62. Quintero-Gutiérrez AG, González-Rosendo G, Sánchez-Muñoz J, Polo-Pozo J, Rodríguez-Jerez JJ (2008) Bioavailability of heme iron in biscuit filling using piglets as an animal model for humans. Int J Biol Sci 4(1):58
63. Abbaspour N, Hurrell R, Kelishadi R (2014) Review on iron and its importance for human health. J Res Med Sci 19(2):164
64. Shah KK, Verma R, Oleske JM, Scolpino A, Bogden JD (2019) Essential trace elements and progression and management of HIV infection. Nutr Res 71:21–29
65. Roohani N, Hurrell R, Kelishadi R, Schulin R (2013) Zinc and its importance for human health: an integrative review. J Res Med Sci 18(2):144
66. Cousins RJ (1996) Zinc. In: Filer LJ, Ziegler EE (eds) Present knowledge in nutrition, 7th edn. International Life Sciences Institute Nutrition Foundation, Washington, DC, pp 293–306
67. Deshpande JD, Joshi MM, Giri PA (2013) Zinc: the trace element of major importance in human nutrition and health. Int J Med Sci Public Health 2(1):1–6
68. Baltaci AK, Yuce K, Mogulkoc R (2018) Zinc metabolism and metallothioneins. Biol Trace Elem Res 183(1):22–31
69. Mezzaroba L, Alfieri DF, Simão ANC, Reiche EMV (2019) The role of zinc, copper, manganese and iron in neurodegenerative diseases. Neurotoxicology 74:230–241
70. Lippard SJ, Berg JM (1994) Principles of bioinorganic chemistry. University Science Books, Mill Valley. ISBN 0-935702-73 3
71. Avram OR, Caragea G, Varzaru CA (2021) Copper and its role in the human body–the importance of establishing copper concentrations in the body. Rom J Mil Med 124(2):254
72. Solomons NW (1985) Biochemical, metabolic, and clinical role of copper in human nutrition. J Am Coll Nutr 4(1):83–105
73. Lazarus JH (2015) The importance of iodine in public health. Environ Geochem Health 37(4):605–618
74. Zimmermann MB (2011) The role of iodine in human growth and development. Semin Cell Dev Biol 22(6):645–652. Academic Press
75. Zimmermann MB, Boelaert K (2015) Iodine deficiency and thyroid disorders. Lancet Diabetes Endocrinol 3(4):286–295
76. Avila DS, Puntel RL, Aschner M (2013) Manganese in health and disease. In: Interrelations between essential metal ions and human diseases, pp 199–227
77. Konikowska K, Mandecka A (2018) Trace elements in human nutrition. In: Recent Advances in Trace Elements, pp 339–372
78. IOM (Institute of Medicine) (2011) Dietary reference intakes: vitamin A, vitamin K, arsenic, boron, chromium, copper, iodine, iron, manganese, molybdenum, nickel, silicon, vanadium, and zinc. National Academies Press, Washington, DC
79. Erikson KM, Aschner M (2003) Manganese neurotoxicity and glutamate-GABA interaction. Neurochem Int 43(4–5):475–480

80. Culotta VC, Yang M, O'Halloran TV (2006) Activation of superoxide dismutases: putting the metal to the pedal. Biochim Biophys Acta Mol Cell Res 1763(7):747–758
81. Sokolovskaya OM, Plessl T, Bailey H, Mackinnon S, Baumgartner MR, Yue WW et al (2021) Naturally occurring cobalamin (B12) analogs can function as cofactors for human methylmalonyl-CoA mutase. Biochimie 183:35–43
82. Wang HX, Wahlin A, Basun H, Fastbom J, Winblad B, Fratiglioni L (2001) Vitamin B12 and folate in relation to the development of Alzheimer's disease. Neurology 56(9):1188–1194
83. Novotny JA (2011) Molybdenum nutriture in humans. J Evid Based Complementary Altern Med 16(3):164–168
84. Hegedűsová A, Hegedűs O, Jakabová S, Andrejiová A, Šlosár M, Mezeyová I, Golian M (2021) The role of selenium in human nutrition. In: Selenium supplementation in horticultural crops. Springer International Publishing, Cham, pp 3–45. Turnlund JR, Keyes WR, Peiffer GL, Chiang G (1995) Molybdenum absorption, excretion, and retention studied with stable isotopes in young men during depletion and repletion. Am J Clin Nutr, 61(5), 1102–1109
85. Ginter E (1995) The role of antioxidants in the prevention of cancer. (Úloha antioxidantov v prevencii nádorových ochorení). Bratisl Med J (Bratisl Lek Listy) 96(4):192–194
86. Clark LC, Combs GF, Turnbull BW, Slate EH, Chalker DK, Chow J et al (1996) Effects of selenium supplementation for cancer prevention in patients with carcinoma of the skin: a randomized controlled trial. JAMA 276(24):1957–1963
87. Hrušovský Š (2000) Again more complete protection (Opäť kompletnejšia ochrana). Nature 42(2):10
88. Miko M (1994a) Microminerals in nutrition V.—selénium (Mikrominerálie vo výžive V.—selén). Nutr Health (Výživa a zdravie) 39(4):63–64
89. Wang Y, Su H, Gu Y, Song X, Zhao J (2017) Carcinogenicity of chromium and chemoprevention: a brief update. Onco Target Ther 10:4065–4079
90. Moradi F, Maleki V, Saleh-Ghadimi S, Kooshki F, Pourghassem Gargari B (2019) Potential roles of chromium on inflammatory biomarkers in diabetes: a systematic. Clin Exp Pharmacol Physiol 46(11):975–983
91. Wang ZQ, Yu Y, Zhang XH, Komorowski J (2014) Chromium–insulin reduces insulin clearance and enhances insulin signaling by suppressing hepatic insulin-degrading enzyme and proteasome protein expression in KKAy mice. Front Endocrinol 5:99
92. Dworzański W, Sembratowicz I, Cholewińska E, Tutaj K, Fotschki B, Juśkiewicz J, Ognik K (2021) Effects of different chromium compounds on hematology and inflammatory cytokines in rats fed high-fat diet. Front Immunol 12:614000
93. Schwarz G, Mendel RR, Ribbe MW (2009) Molybdenum cofactors, enzymes and pathways. Nature 460(7257):839–847
94. Havemeyer A, Lang J, Clement B (2011) The fourth mammalian molybdenum enzyme mARC: current state of research. Drug Metab Rev 43(4):524–539

Chapter 5
The Neuroprotective Role of Micronutrients: A Systematic Review of Preclinical and Clinical Studies on Antioxidant Vitamins, Essential Minerals, and Trace Elements in Managing Neurodegenerative Diseases

Dhanvi Sheth, Siddhi A. Awate, Harpal S. Buttar, and Ginpreet Kaur

Abstract The progressive loss of neurons and neurotransmitters characterizes neurodegenerative diseases. Various antioxidant vitamins, essential minerals, and trace elements are critical for maintaining neuronal integrity of the central nervous system and reducing the risk of chronic neurodegenerative diseases (e.g., Parkinson's and Alzheimer's disease). These micronutrients exert multifaceted metabolic effects at both cellular and molecular levels, including reducing oxidative stress, enhancing mitochondrial function, and preserving synaptic plasticity. Antioxidant vitamins (A, B_1, B_2, B_6, B_9, B_{12}, D, E, and K) and minerals/trace elements (calcium, phosphorus, potassium, sodium, chloride, magnesium, iron, zinc, iodine, sulfur, chromium, copper, fluoride, manganese, and selenium) have specific neurobiological functions, which reduce oxidative stress and slow down the progression of mental ailments. They help to maintain neuronal integrity, reduce reactive oxygen species, and provide neuroprotective actions. Additionally, essential minerals and trace elements, such as calcium, sodium, potassium, and magnesium, are necessary for neuronal excitability and synaptic transmission, providing neuroprotective benefits. In contrast, zinc, iron, copper, manganese, and selenium serve as enzymatic cofactors that maintain redox homeostasis. The blood-brain barrier strictly regulates the bioavailability and brain uptake of these metallic nutrients, utilizing selective transport

D. Sheth · S. A. Awate · G. Kaur (✉)
Shobhaben Pratapbhai Patel School of Pharmacy and Technology Management, SVKM's Narsee Monjee Institute of Management Studies (NMIMS) Deemed-to-be University, Mumbai, Maharashtra, India
e-mail: ginpreet.Kaur@nmims.edu

H. S. Buttar
Department of Pathology and Laboratory Medicine, Faculty of Medicine, University of Ottawa, Ottawa, ON, Canada

N. S. Dhalla et al. (eds.), *Functional Biochemistry of Metallic Elements*, Advances in Biochemistry in Health and Disease 36,
https://doi.org/10.1007/978-3-032-14445-4_5

mechanisms to preserve brain homeostasis. Disruptions in the supply of micronutrients, whether due to dietary deficiencies or environmental toxicants, are linked to increased susceptibility to neurodegenerative processes, resulting in cognitive decline, dementia, and depression. This chapter synthesizes current knowledge regarding the potential role of antioxidant micronutrients in maintaining mental health, as well as novel therapeutic approaches for preventing cognitive decline, dementia, and managing various neurodegenerative disorders. The results of preclinical and clinical studies done with micronutrients, as well as the underlying mechanisms, are also discussed.

Keywords Antioxidant micronutrients · Vitamins · Trace elements · Mental health · Cognitive function · Dementia · Neurodegenerative diseases · Preclinical and clinical neurodegenerative studies

Abbreviations

Akt	Protein kinase B
ApoER2	Apolipoprotein E receptor 2
APP/PS1	Amyloid precursor protein/presenilin 1
ARE	Antioxidant response element
ATP	Adenosine triphosphate
ATRA	All-trans retinoic acid
BBB	Blood-Brain Barrier
BDNF	Brain-derived neurotrophic factor
CAT	Catalase
CRBP2	Cellular retinol-binding protein 2
CREB	cAMP response element-binding protein
CVD	Cardiovascular disease
DASH	Dietary approaches to stop hypertension
DHA	Docosahexaenoic acid
DMT1	Divalent metal transporter 1
FAD	Flavin adenine dinucleotide
FMN	Flavin mononucleotide
Fpn	Ferroportin
GPX	Glutathione peroxidase
GPX4	Glutathione peroxidase 4
GSH	Glutathione (reduced form)
HDL	High-density lipoprotein
HMGB1	High mobility group box 1
IL-10	Interleukin 10
IL-6	Interleukin 6
ISX	Intestine-specific homeobox *(not explicitly expanded in text)*
LDL	Low-density lipoprotein

LTD	Long-term depression
LTP	Long-term potentiation
LVDCCs	L-type voltage-dependent calcium channels
MIND	Mediterranean-DASH intervention for neurodegenerative delay
MnSOD	Manganese superoxide dismutase
MRI	Magnetic resonance imaging
MTHF	5-methyltetrahydrofolate
MTHFR	5-methyltetrahydrofolate reductase
NAD	Nicotinamide adenine dinucleotide
NADP	Nicotinamide adenine dinucleotide phosphate
NFAT	Nuclear factor of activated T cells
NF-κB	Nuclear factor kappa-light-chain-enhancer of activated B cells
NGF	Nerve growth factor
Nrf2	Nuclear factor erythroid 2–related factor 2
NRF2	Nuclear factor erythroid 2–related factor 2
PI3K	Phosphatidylinositol 3-kinase
PLP	Pyridoxal phosphate
RBP4	Retinol binding protein 4
ROS	Reactive oxygen species
SeNPs	Selenium nanoparticles
SeP	Selenoprotein P
SLC30A10	Solute carrier family 30 member 10
SOD	Superoxide dismutase
TK	Transketolase
TMP	Thiamine monophosphate
TNF-α	Tumor necrosis factor-alpha
TPP	Thiamine pyrophosphate
TRPV1	Transient receptor potential vanilloid 1
TrxR	Thioredoxin reductase
TTP	Thiamine triphosphate
VDR	Vitamin D receptor
ZIP8/ZIP14	ZRT/IRT-like protein 8 and 14

Introduction

Neurodegenerative diseases, such as cognitive decline, dementia, Alzheimer's disease (AD), and Parkinson's disease (PD), are believed to develop due to oxidative stress, the generation of reactive oxygen species (ROS), and the accumulation of toxic proteins in the brain, ultimately leading to neuronal loss. Neurotransmission and synaptic plasticity are disrupted by an increase in ROS [1]. There are nearly 20 essential minerals and trace elements required for the healthy functioning of the brain and the maintenance of homeostasis in the human body, of which 50% fall under the category of minerals and the remaining fall under the category of non-minerals [2]. Mineral ions are essential mediators for many intricate signaling pathways in the central nervous system (CNS) through the development of metalloregulatory mechanisms within the cells [3]. Physiological functions in the human body are susceptible to oxidative stress when highly reactive free radicals overwhelm the antioxidant defenses, leading to pathological injury and tissue necrosis. Oxidative stress has a particularly adverse effect on the brain, primarily due to its high oxygen and glucose consumption, as well as its high lipid content [4]. Protection from oxidative stress is provided by endogenous antioxidant substances, such as glutathione, superoxide dismutase, and catalase, as well as exogenous antioxidant vitamins, including vitamins A, B, C, D, K, and E, as shown in Table 5.1. The amount of oxidative stress in the body depends on lifestyle choices (smoking, excessive drinking), including unhealthy dietary habits and lack of physical activity. Consuming a well-balanced diet rich in antioxidants helps protect against neurodegeneration and neuronal loss. Oxidative stress generates deleterious effects in the two major brain regions, the hippocampus and amygdala, thereby affecting memory and cognitive processes [5]. The ROS-induced dysregulation of proteins, lipids, and DNA leads to neurodegeneration. Maintaining adequate antioxidant control establishes the necessary mental health conditions and protects against oxidative stress-related disorders [6]. Recent research has demonstrated that dietary patterns and food choices have a substantial impact on brain function and mental health. For instance, consumption of the Mediterranean diet, Intervention for Neurodegenerative Delay (MIND), and Dietary Approaches to Stop Hypertension (DASH) diets have been repeatedly proven to reduce the risk of cognitive decline and Alzheimer's Disease [7]. The intake of antioxidant-rich fruits, vegetables, whole grains, nuts, olive oil, poultry, fish, and low-fat dairy products is prioritized in the Mediterranean diet for its cardiovascular health and anti-inflammatory effects. The DASH diet aims to reduce sodium intake and increase consumption of nutrient-dense foods, such as lean proteins and vegetables, to improve blood flow to the brain [7]. The MIND diet, a combination of components from the Mediterranean and DASH diets, targets brain health by prioritizing green leafy vegetables, berries, nuts, and fish while limiting red meat and sugar-rich pastries and sweets [8]. Additionally, these nutrient-dense diets contain components, such as omega-3 fatty acids, including docosahexaenoic acid (DHA), vitamin B, antioxidants, and polyphenols. These supplements can reduce oxidative stress, promote neurotransmitter synthesis, and preserve neuronal integrity [9]. Thus, adopting these food habits could reduce neuroinflammation, protect cardiovascular health, and prevent age-related cognitive decline. Several studies have suggested that adherence to these diets is associated with a significantly lower risk of developing Alzheimer's and Parkinson's disease. Overall, this chapter will explore the potential of integrating these dietary strategies, supplementing the diet with wholesome foods and micronutrients, to prevent neurodegenerative disease pathologies.

Table 5.1 Summary of the potential role of vitamins in neuroprotection: mechanisms of preclinical and clinical outcomes

Metals	Daily requirements in humans	Significance	Mechanism of action	Clinical trials	Preclinical studies	Ref
Vitamin A	700–900 μg/day	Supports cognitive function, Reduces oxidative stress	Regulates gene expression, antioxidant properties	Shown to reduce oxidative stress and inflammation in AD patients— Efficacy in isolation unclear	Supports cognitive function, reduces aβ accumulation in AD models	[15]
Vitamin B_1 (thiamine)	1.2 mg/day	Energy metabolism Cognitive function	Coenzyme for pyruvate dehydrogenase and α-ketoglutarate dehydrogenase	Deficiency linked to cognitive impairment; supplementation improves memory in the elderly	Enhances memory and neuronal survival in animal models	[18]
Vitamin B_2 (riboflavin)	1.1 mg/day	Antioxidant properties, neuroprotective agent	Coenzyme for flavin adenine dinucleotide (FAD)	To establish its role in cognition	Reduces oxidative stress; supports neuronal health	[19]
Vitamin B_6 (pyridoxine)	1.3 mg/day	Neurotransmitter synthesis Helps in behavioral regulation	Coenzyme for enzymes involved in neurotransmitter synthesis and neuroprotection	Supplementation with B6, B12, and folate reduces brain atrophy in MCI	Supports neurotransmitter balance, reduces oxidative stress	[20]
Vitamin B_{12} (cobalamin)	2.4 μg/day	Myelin formation, enhances cognitive ability	Coenzyme for methionine synthase and methylmalonyl-CoA mutase	Supplementation with Vitamin B12 slows brain atrophy, improves cognition in the elderly	Prevents demyelination, supports cognition in animal models	[21]
Vitamin D	600–800 IU/day	Neuroinflammation regulator agent	Modulates gene expression, influences immune response	800 IU/day for 12 months improved cognition in MCI and AD patients	Reduces neuroinflammation, Supports synaptic function	[22]
Vitamin E	15 mg/day	Antioxidant effects, neuroprotection	Protects neuronal cells from oxidative damage, as an antioxidant	2000 IU/day shows benefit in AD, but the results are unclear	Reduces oxidative stress, delays neurodegeneration	[23]
Vitamin K	90 μg/day	Brain health and cognitive function	Involved in sphingolipid metabolism and neuronal health, helps in blood clotting	Limited human data on its role in cognition	Improves memory, reduces amyloid in animal models	[24]

Impact of Antioxidant Vitamins in the Prevention of Neurodegenerative Diseases

Vitamin A

Vitamin A, also called retinol, is a fat-soluble vitamin. This is essential for the eye's retina health and is also involved in regulating immune function. Vitamin A deficiency during pregnancy causes blindness in the newborn. The daily dietary requirements of vitamin A range from 300 to 700 μg for children and 700 to 900 μg for adults. It is obtained from both plant- and animal-based products. Plant sources provide precursors of vitamin A in the form of carotenoids such as β-carotene, α-carotene, and β-cryptoxanthin [10]. Fish, meat, and dairy products are rich in vitamin A (retinol and retinyl ester). This vitamin is involved in various signaling pathways that regulate gene expression, specifically in the central nervous system (CNS) [11]. Vitamin A regulates vital processes, including the control of neural cell differentiation and patterning during neural tube formation in the fetus. The active metabolite retinoic acid (RA) plays a significant neuroprotective role in the CNS, particularly in the progression of neurodegenerative diseases [12]. It scavenges free radicals, preventing oxidative damage to neuronal cells, reducing lipid peroxidation, and maintaining cellular integrity. RA modulates gene transcription by binding to nuclear receptors, thereby affecting various neuronal processes, including cell differentiation, inflammation control, and neuronal plasticity [13]. These mechanisms are vital in the survival and function of neurons, especially dopaminergic neurons in the substantia nigra region adversely affected in Parkinson's disease [14]. Different derivatives of vitamin A also help regulate neuroinflammation, synaptic plasticity, and homeostatic mechanisms in the brain. The dysregulation of vitamin A signaling leads to impairment in long-term synaptic plasticity, reduction in neurogenesis, and diminished dendritic arborization, collectively contributing to cognitive decline and increased vulnerability to neurodegenerative processes, as seen in diseases such as Alzheimer's disease and Parkinson's disease [15]. As the aging body's ability to mobilize and deliver vitamin A to the brain declines, it exacerbates the risks of neurodegeneration. External nutritional supplementation with vitamin A has been shown to improve its bioavailability in the brain, thereby enhancing learning and memory and potentially slowing the progression of age-related cognitive decline [9]. Preclinical and clinical studies have demonstrated the potential therapeutic effects of vitamin A across multiple animal models, indicating that vitamin A and its active metabolites can reduce oxidative stress and modulate neurotransmitter systems, such as an increase in dopamine production, which is especially relevant in Parkinson's disease [16]. In models of multiple sclerosis, retinoic acid has been shown to suppress the activity of inflammatory T cells, thereby reducing myelin damage and neuroinflammation [17]. Supplementation studies suggest that vitamin A can prevent abnormal neuronal functioning by attenuating oxidative damage, inhibiting amyloid-beta and tau pathology, and further protecting against α-synuclein toxicity, which are central anatomical features of Alzheimer's and Parkinson's disease [16]. In clinical studies, it was found that lower circulating levels of vitamin A are associated with a higher risk of Alzheimer's disease. The neuroprotective roles of different vitamins are shown in Table 5.1.

Different Types of B Vitamins

Vitamins B_1 (Thiamine), B_2 (riboflavin), B_3 (niacin), B_9 (folate), and B_{12} (cobalamin) are the water-soluble vitamins involved in the neuroprotective action primarily due to their antioxidant properties and central roles in neuronal metabolism, neurotransmitter synthesis, and maintenance of neural structure and function [18]. Thiamine was studied initially in pigeons for its role in brain energy metabolism. It is vital in glucose metabolism to fuel energy production in mitochondria, facilitate muscle contraction, and support nerve signal conduction [25]. Vitamin B2 is a powerful antioxidant with critical roles in neurological health and homocysteine metabolism [26]. Riboflavin plays a crucial role in the glutathione redox cycle (GSH), mitigating oxidative stress associated with chronic diseases, including cancer, cardiovascular disease, diabetes, and neurodegenerative disorders. Riboflavin also prevents neurotoxicity by disrupting a self-perpetuating neurotoxic cycle that includes oxidative stress, energy dysfunction, neurogenic inflammation, glutamate excitotoxicity, and homocysteine toxicity [27]. It reduces species with its reduced form, dihydro-riboflavin, stimulating glutathione (GSH) activity by promoting GSH reductase and increasing antioxidant enzymes, such as superoxide dismutase (SOD) and catalase. Riboflavin can also inhibit inflammatory mediators, such as NF-κB and HMGB1, and supports vitamin D production, both of which will optimize an anti-inflammatory response [28]. Riboflavin also suppresses glutamate release, protects against excitotoxicity, and acts on glutamate receptors, modulating the tryptophan–kynurenine pathway. This underlines its multifaceted roles, particularly in disease conditions such as Parkinson's disease and migraine [19]. Niacin exists in two primary forms, nicotinic acid and niacinamide. This vitamin is primarily obtained from food or supplements. Niacin maintains nervous system function by facilitating the synthesis of neurotransmitters [29]. Niacin deficiency has been associated with cognitive impairments, such as brain fog, psychiatric manifestations, and even neurodegenerative diseases. The extreme niacin deficiency can lead to dementia and dermatitis [30]. Vitamin B_9 (folate) is naturally present in foods and in synthetic form as folic acid in supplements and fortified foods, and a deficiency of this vitamin can cause neurological diseases. Its deficiency can lead to depression, irritability, memory problems, and mood changes. Folate and homocysteine deficiency are associated with an increased risk of dementia and depression [31]. Long nerve fiber peripheral neuropathy may result from severe deficiency and, rarely, subacute combined degeneration of the spinal cord. Hereditary folate malabsorption can also cause severe neurological decline in infants. To prevent folate deficiency, a diet rich in folate or a folate supplement is beneficial [32]. Vitamin B_{12} (cobalamin) plays a critical role in scavenging reactive oxygen species (ROS), particularly superoxide, and preserves GSH, a key intracellular antioxidant, further reducing the oxidative stress in neural tissues [33]. Cobalamin modulates cytokine and growth factor production, protecting neurons from inflammation-induced oxidative stress [34]. Recent preclinical findings suggest that vitamin B_{12} supplementation protects neuronal cells from oxidative injury, such as hydrogen peroxide-induced damage, and promotes cell survival in a dose-dependent manner [35]. Clinically, B_{12} deficiency is linked to increased risk of neurological disorders and cognitive impairment and associated with increased levels of pro-inflammatory cytokines and oxidative stress markers. Furthermore, its role is significant in myelin synthesis, essential for maintaining nerve conduction and overall neuronal health [21]. Thus, maintaining normal levels of B vitamins in the body, as well as their specific types, shows promising potential for neuroprotective effects in various neurological diseases and mental health disorders.

Vitamin D

Vitamin D is a neuroactive steroid critical for brain and cognitive function. The primary effects of this vitamin are on D receptors (VDRs) in brain regions, like the hippocampus and cerebral cortex, which regulate the transcription of genes that facilitate synaptic transmission, cell communication, and G protein signaling [36]. Vitamin D is believed to regulate the expression of neurotrophic factors, such as nerve growth factor (NGF) and brain-derived neurotrophic factor (BDNF), which are crucial for the survival and differentiation of neurons [37]. In neurological health, vitamin D regulates calcium entry through L-type voltage-dependent calcium channels (LVDCCs) and transient receptor potential vanilloid 1 (TRPV1) channels to ensure normal neuronal function. Vitamin D also exhibits antioxidant and anti-inflammatory properties through the regulation of antioxidant and regulatory enzymes, as well as cytokine expression. This effect prevents inflammation and oxidative stress, thereby protecting neurons from damage [38]. Its neuroprotective role includes the preservation of neurons and the clearance of beta-amyloid plaques, a hallmark of Alzheimer's disease. Vitamin D deficiency may lead to compromised cognitive function, affecting memory, learning, and other mental functions. In animal models, adequate levels of vitamin D performed better on maze tests in rats, which measure hippocampal-dependent learning and memory, than vitamin D-deficient rats [39]. Vitamin D deficiency can also affect dopamine neuron specification and function, causing neurodevelopmental disorders. Vitamin D can reduce cognitive impairment in neurodegenerative disorders like Alzheimer's disease through the regulation of oxidative stress and amyloid-beta accumulation in the brain. Recent studies have shown that vitamin D deficiency can disrupt the structural stability of perineuronal nets, which hold neuronal connections, and that such disruption can cause cognitive impairment and has been linked to diseases like schizophrenia [40]. Its supplementation has been seen to be effective at all doses, particularly in ensuring optimal cognitive performance and, possibly, preventing the progression of Alzheimer's disease and other neurodegenerative abnormalities [5]. Preclinical studies in animal models have demonstrated that adequate vitamin D levels are crucial for maintaining cognitive function, enhancing memory, and improving learning ability, whereas deficiency exacerbates cognitive impairment [41]. A limited number of clinical trials suggest that vitamin D supplementation may be beneficial in reducing the risk of neurodegeneration; however, further research is needed to determine the optimal dose for achieving the therapeutic effect [22].

Vitamin E

Vitamin E is a lipid-soluble antioxidant that plays a significant role in protecting neuronal cells against oxidative damage, thereby slowing the aging process and mitigating neurodegeneration. Vitamin E suppresses lipid peroxidation, which damages neuronal membranes, as these membranes are most vulnerable to oxidative damage due to their high lipid content [23]. This effect is exerted by scavenging ROS and integrating into the neuronal membrane, competing with ROS and maintaining cellular integrity. Tocopherol and tocotrienol are the two principal forms of vitamin E, functioning as hydrogen donors to reduce lipid radicals and terminate chain reactions of oxidative damage [42]. Vitamin E suppresses the aggregation of amyloid beta plaques, a hallmark of Alzheimer's disease, by suppressing amyloid beta-induced peroxide-mediated lipid peroxidation, thereby safeguarding neurons from amyloid beta toxicity. Tocotrienols are more effective in suppressing amyloid beta-induced oxidative stress and neuronal cell damage induced by hydrogen peroxide, demonstrating greater protection compared to tocopherols [43]. Under healthy conditions, vitamin E is a potent antioxidant that maintains the integrity of cell membranes and DNA in neurons, which are metabolically active and highly vulnerable to oxidative damage. It is crucial for synaptic function because it suppresses the breakdown of docosahexaenoic acid phosphatidylcholine (DHA-PC), an integral part of neuronal membranes. Vitamin E deficiency reduces the levels of DHA-PC, compromising membrane repair mechanisms and leading to enhanced neuronal death [44]. The neuroprotective functions of vitamin E are particularly significant in neurodegenerative disorders, where oxidative stress is the principal causative factor in disease development. In Alzheimer's disease, vitamin E has been found to decrease amyloid-beta deposition and tau protein hyperphosphorylation, slowing functional impairment in patients [43]. In Parkinson's disease, where oxidative stress contributes to dopaminergic neuron loss, vitamin E may potentially prevent neuronal damage [45]. Vitamin E also reduces brain damage from ischemic stroke through its antioxidant effect, thereby preserving neuronal viability and decreasing excitotoxicity. The protective doses of vitamin E vary by disease and individual. Clinical studies indicate that doses of 200 to 1000 IU daily are beneficial for antioxidant effects, particularly in age-related and neurodegenerative diseases [46]. Clinical studies have also shown that increased dietary intake of vitamin E is associated with slower cognitive decline in older adults, especially in those at risk for Alzheimer's disease [47]. Preclinical evidence supports the neuroprotective effects of vitamin E in animal models, where deficiency results in memory impairment, neuronal damage, and the development of Alzheimer's disease pathology [48]. Moreover, tocotrienols have been found to provide greater neuroprotection than tocopherols, due to their greater capacity to inhibit lipid peroxidation and modulate the inflammatory pathways. Despite these encouraging findings, further clinical studies are needed to determine the role of vitamin E in the prevention and treatment of neurodegenerative disorders.

Vitamin K

Vitamin K has long been recognized for its essential role in blood coagulation; however, recent data suggest that this vitamin also plays a significant role in maintaining cognitive function and mental health [24]. It is involved in numerous brain function processes, including the preservation of neuronal integrity and cognitive function. Sphingolipid metabolism is regulated by vitamin K, specifically menaquinone-4 (MK-4), the primary form of this vitamin in the brain. The interaction between them is essential for the proliferation, differentiation, and survival of brain cells, and changes in sphingolipid profiles are associated with neuroinflammation and neurodegeneration [49]. Vitamin K not only facilitates the synthesis of sphingolipids but also exerts an anti-inflammatory effect through the modulation of cytokine production and the activation of proteins, such as Gas6 and Protein S, which have essential functions in preserving neuronal integrity and function [50]. Recent research suggests that the neuroprotective effects of vitamin K against neurodegeneration may be associated with reducing the density of neurofibrillary tangles and decreasing scores of Alzheimer's disease pathology [51]. Lower MK4 levels have been associated with cognitive impairments and increased risk of developing dementia or mild cognitive impairment. Clinical studies have shown that higher levels of MK4 in the brain are associated with improved cognitive function and a lower susceptibility to dementia, highlighting the significant role of vitamin K in protecting cognitive function as individuals age [52]. Moreover, dietary intake of vitamin K has been linked to a reduced risk of cognitive impairment in older adults, as observed in observational studies. Vitamin K also plays a significant role in maintaining a healthy cerebral structure and function [53]. Several animal studies have shown that vitamin K2 supplementation may have potential benefits for enhancing spatial memory and cognitive function. The anti-inflammatory and antioxidant effects of vitamin K are being explored for their potential neuroprotective benefits in conditions such as stroke and dementia. [54]. The purported mechanisms of action for vitamins are shown in Fig. 5.1.

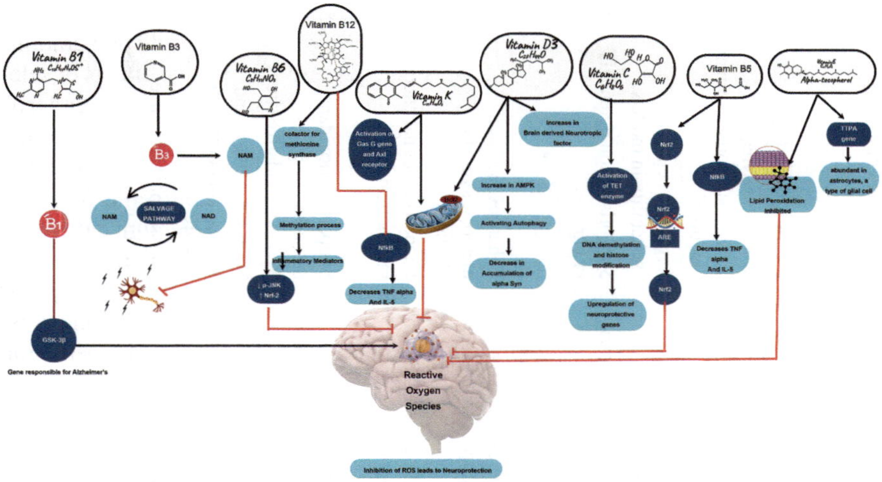

Fig. 5.1 Schematic representation of vitamins in neuroprotection

The Impact of Essential Minerals and Trace Elements on CNS Function

Exogenous minerals are essential substances that our bodies need to function normally. Several minerals are necessary for mental health, including calcium, phosphorus, potassium, sodium, chloride, magnesium, iron, zinc, iodine, sulfur, cobalt, copper, fluoride, manganese, and selenium. Research indicates that minerals found in nuts, fruits, and vegetables hold promise for exhibiting antioxidant and neuroprotective functions (summarized in Table 5.2). Although less understood, these minerals may also enhance cognitive function and neuronal integrity. It appears that various minerals play a significant role in the physiological functions of neurons in the brain. Dysregulation and a decrease in these minerals lead to an increase in oxidative stress, which enhances damage to neuronal structure and function. Neurodegenerative diseases may develop due to disturbances in brain homeostasis, genetic expression, and molecular abnormalities.

Calcium

Calcium (Ca^{++}) is a divalent metal that plays a crucial role in neuroprotection against neurodegenerative abnormalities. It primarily acts through an intracellular signaling pathway and reduces oxidative stress [62]. Calcium does not directly act as an antioxidant, but its homeostasis is crucial for maintaining neuronal survival and function. Imbalance in the calcium levels can lead to the progression of

Table 5.2 Summary of the daily requirements and potential role of essential minerals in neuroprotection: mechanisms of preclinical and clinical outcomes

Mineral	Daily Requirements In Humans	Significance	Mechanism of action	Clinical findings	Preclinical Findings	Ref
Calcium	1000 mg/day	Neurotransmission and signaling	Essential for neuronal signaling, especially in muscles and synaptic plasticity	High serum calcium is linked to cognitive decline	Modulates synaptic activity, supports neuronal function	[20]
Potassium	4700 mg/day	Maintain the resting membrane potential	Maintains stable neuronal membrane potential—repolarization	Linked to reduced cerebral stroke risk	Regulates neuronal excitability in models	[2]
Sodium	1500 mg/day	Action potentials Transmit nerve impulses	Essential for neuronal signaling and conduction Helps in the transmission of nerve impulses	Essential for neuronal function	Supports action potentials in vitro	[2]
Magnesium	400 mg/day	Neuroprotection	Involved in neuronal signaling and synaptic plasticity	Some evidence for cognitive enhancement	Enhances learning, memory, synaptic plasticity	[55]
Phosphorus	1000 mg/day	Energy metabolism Regulation of cellular function	Essential for ATP production and cellular processes involving phosphorylation	Limited clinical data on cognition	Supports neuronal energy metabolism	[56]
Sulfur	800 μg/day	Antioxidant defence systems	Involved in glutathione synthesis and antioxidant defences	No direct clinical neuroprotection data	Supports antioxidant defenses Reduces oxidative stress	[57]
Copper	900 μg/day	Enzyme function promoter	Coenzyme for superoxide dismutase and other enzymes	Low copper linked to cognitive decline	Supports memory and antioxidant defence in models	[58]
Chromium	50–100 μg/day	Glucose metabolism maintain neurological health	Enhances insulin sensitivity and glucose metabolism	Limited clinical evidence for neuroprotection	May support cognitive function in animal models	[59]

Iron	8 mg/day	Oxygen transport	Essential for hemoglobin and oxygen transport	Iron supplementation improves cognition in deficiency	Prevents cognitive deficits in iron-deficient animals	[55]
Zinc	8–11 mg/day	Synaptic plasticity memory formation	Coenzyme for superoxide dismutase and other enzymes. Helps as an immunity modulator	Low zinc linked to cognitive decline in animal models	Enhances memory reduces oxidative stress	[60]
Manganese	2.3 mg/day	Antioxidant enzyme functions	Coenzyme for superoxide dismutase and cofactor for other enzymes like pyruvate carboxylase and arginase	Limited clinical neuroprotection data	Supports antioxidant defence Reduces neurotoxicity	[60]
Selenium	55 µg/day	Antioxidant properties	Coenzyme for glutathione peroxidase	Low selenium linked to cognitive decline	Reduces neurodegeneration	[61]

various neurodegenerative chronic diseases, like Alzheimer's disease [63]. Increased calcium levels are also associated with oxidative stress, which promotes the formation of reactive oxygen species and mitochondrial dysfunction, ultimately leading to neuronal death. Preclinical studies have shown that maintaining a balance in the number of calcium ions can mitigate various forms of oxidative damage and promote neuroprotection. Stabilization of calcium signaling pathways can reduce oxidative stress-induced neuronal injury [64]. Calcium is found to act through an indirect mechanism by modulating cellular processes that affect ROS generation, rather than acting as a classical antioxidant like vitamin E or coenzyme Q_{10} [65]. There exists sufficient clinical evidence to highlight the antioxidant role of calcium in neuroprotection; however, most of these are by the indirect mechanisms, highlighting the potential therapeutic effect that it carries in preserving neuronal homeostasis, which can reduce oxidative stress and neurodegeneration [66].

Potassium

Potassium (K^+), a monovalent mineral, plays a significant role in neuroprotection through its involvement in an antioxidant mechanism [67]. Potassium ions are also vital in maintaining the neuronal membrane structure and impulse transmission, as well as for the proper functioning of the sodium-potassium ATPase pump, which is crucial for maintaining cellular ion homeostasis and neuronal excitability [68]. Any dysfunction in potassium regulation can create oxidative stress, which can lead to severe neuronal damage and neurodegenerative diseases. It has been found in the preclinical studies that antioxidant therapies influence potassium channel activity and ion homeostasis, helping to reduce oxidative damage and improve neuronal survival in models of epilepsy and other neurological disorders. Antioxidant treatment has been shown to restore the sodium-potassium ATPase activity in epileptic animal models, suggesting its neuroprotective effect on the ion pump function and neuron stability [68]. Restoration helps to reduce oxidative stress markers, such as nitric oxide levels, which are crucial in the interplay between the antioxidant defense and potassium regulation in the brain. Antioxidants support mitochondrial function, reduce oxidative stress in the heart, and enhance the neurons' cellular defense mechanisms, such as the Nrf2-ARE pathway, which regulates the expression of antioxidant enzymes and helps maintain redox homeostasis [69]. Clinical evidence has also highlighted that the involvement of potassium as an antioxidant in neurodegenerative diseases is sparse; however, its importance cannot be neglected, as maintaining ionic balance, especially of potassium, helps in employing various neuroprotective strategies [70].

Sodium

Sodium (Na^{+}) is a monovalent mineral found in group 1 of the periodic table. It has similar mechanisms to those of calcium. In contrast, it is not directly recognized for its antioxidant role in neuroprotection, but rather by mediating various sodium-related compounds and enzymes that reduce oxidative stress, such as the activity of the superoxide dismutase (SOD) enzyme, which plays a crucial role in neuroprotection [71]. SOD enzymes have various genetic forms containing manganese, copper, and zinc. These enzymes catalyze the dismutation of superoxide radicals into much less harmful molecules, protecting neurons from oxidative stress. Deficiency of the SOD genes is linked to neurodegenerative diseases like amyotrophic lateral sclerosis (ALS) and Alzheimer's disease where oxidative stress contributes to neuronal damage [72]. Overexpression or supplementation of SOD in preclinical models has been shown to significantly reduce oxidative stress, promote mitochondrial function, and improve neuronal survival, which is essential for the neuroprotective potential [73]. SOD-2 activity reduces synaptic failure and neuronal death in animal models induced by oxidative stress. Extracellular SOD administration inhibits cell death in excitotoxicity models. These results suggest that just enhancing the antioxidant enzyme activity related to sodium-dependent processes can eliminate neurodegeneration [74].

In clinical trials, antioxidants, such as coenzyme Q10, vitamin E, and uric acid, have been investigated for their neuroprotective effects, which are achieved by reducing reactive oxygen species (ROS) and enhancing mitochondrial function [75]. These antioxidants maintain neuronal cellular energy production and reduce oxidative damage, which are key drivers in the progression of neurodegenerative diseases like Parkinson's and Alzheimer's. Sodium arsenate, a sodium-containing compound, has demonstrated neuroprotection in experimental traumatic brain injury by improving neurological function and reducing brain edema, suggesting that some sodium compounds may have therapeutic potential [76].

Magnesium

Magnesium (Mg^{++}), an essential mineral found in the human diet, is vital for supporting metabolism and energy production, as well as the synthesis of various proteins. Another multifaceted role of magnesium is that it acts as an antioxidant and as a neuroprotective agent in the context of neurodegenerative diseases. It acts by modulating neuroinflammation and maintaining neuronal ion homeostasis, which is essential for counteracting and reducing oxidative stress, all of which form the center of the pathogenesis and progression of various neurodegenerative conditions, such as Alzheimer's and Parkinson's disease [77]. Magnesium's antioxidant properties are attributed to its ability to neutralize reactive oxygen species and prevent oxidative damage to neurons. In vitro studies have shown that magnesium compounds protect cerebellar neurons from glutamate-induced toxicity and oxidative stress [78], while in Vivo models have demonstrated reduced oxidative damage and normalized antioxidant systems in the brain with the help of various magnesium supplementation [79]. Preclinical studies have given robust evidence that magnesium has neuroprotective effects in various models of Alzheimer's disease. Magnesium supplementation reduced neuroinflammation by lowering proinflammatory cytokine levels and increasing anti-inflammatory mediators, thereby attenuating cognitive decline [80]. Similarly, in the case of Parkinson's disease models, increased cellular magnesium inhibited ROS production, preserved ATP generation, and protected neurons from toxicity [81]. Magnesium also supports myelin formation and increases resistance to cellular stress in oligodendrocytes, highlighting its neuroprotective role. Clinical trials have also proven that magnesium's neuroprotective potential is supported by evidence linking low magnesium levels to increased neurological deficits and oxidative stress in patients, as well as by studies indicating that magnesium supplementation can improve markers of brain health and reduce neuroinflammatory markers [82]. Magnesium supplements have also been associated with increased serum brain-derived neurotrophic factor (BDNF) levels in patients, which is why they are still used to support neuronal plasticity and cognitive function. Meta-analysis studies have found that magnesium intake is also associated with a reduction in plasma fibrinogen and other inflammatory biomarkers. However, the effects of magnesium's total antioxidant capacity in humans need to be evaluated [81].

Phosphorus

Phosphorus (P), an inorganic element present in the regular human diet, is known to possess neuroprotection properties through indirect mechanisms by its involvement in antioxidant mechanisms that help to mitigate oxidative stress, which is an important causative factor for neurodegenerative diseases [83]. Many antioxidant systems operating in the brain, including enzymatic pathways, are regulated by signaling molecules composed of phosphorus-containing compounds, such as phosphatidylinositol 3-kinase (PI3K) and protein kinase B (Akt), which are part of the TrkB signaling pathway. This pathway promotes activation and nuclear translocation of the Nrf2-ARE antioxidant system, which upregulates genes encoding antioxidant enzymes, thus protecting neurons from oxidative damage and supporting survival and synaptic plasticity. Preclinical studies have demonstrated that the activation of Nrf2 and TrkB pathways reduces the oxidative damage observed in several animal models, improves their cognitive functions, and prevents ischemic brain injury [84]. Nrf2 overexpression in APP/PS1 mouse models of Alzheimer's disease shows neuroprotection against amyloid-beta toxicity, while Nrf2 knockout worsens oxidative damage. Similarly, pharmacological activation of Nrf2 confers neuroprotection in stroke models by rescuing neurons from ischemic injury. In the clinical studies reported so far, there is limited data published to support the hypothesis. However, it has been proven that phosphorus possesses immense antioxidant abilities and can promote neuroprotection by preventing the reduction in scavenging reactive oxygen species and neuronal death, as well as enhancing cognitive abilities [85]. Phosphorus is also vital in the generation of adenosine triphosphate, which is the energy currency of the cell. It has also been shown to be beneficial in improving mitochondrial function and preventing lipid peroxidation, thereby highlighting its importance as an antioxidant [86].

Sulfur

Sulfur (S), a crucial dietary component commonly found in various proteins in the human diet, is a vital antioxidant that plays a significant role in neuroprotection. It helps mitigate various neurodegenerative diseases through the presence of natural sulfur-containing compounds, such as glutathione, taurine, and allyl sulfide. These compounds reduce oxidative stress in the CNS, which is considered a causative agent for neurotoxicity and excitotoxicity, and a key contributor in various new neurodegenerative diseases, such as Alzheimer's (AD) and Parkinson's disease (PD) [87]. Sulfur compounds activate the Nrf2 pathway, leading to the upregulation of antioxidant response element (ARE)-driven genes, and modulate inflammatory pathways such as NF-κB, thereby exerting both antioxidant and anti-inflammatory effects [88]. Preclinical studies in various animal models have demonstrated that sulforaphane, a sulfur-containing isothiocyanate, enhances neurocognitive abilities and reduces neuronal damage in models of multiple sclerosis (MS), AD, and traumatic brain injury. It was also found that SFN treatment reduced memory deficits and cholinergic neuronal loss in AD and MS models, enhanced microglial clearance and amyloid beta aggregation, preserved blood-brain barrier integrity, and elevated anti-inflammatory cytokines, such as IL-10. In vitro studies have also been successful in highlighting the role of sulfur antioxidants, such as N-acetylcysteine (NAC), which protect neuronal cells from oxidative damage by scavenging reactive oxygen species (ROS) and boosting catalase and glutathione levels, as well as activating protective transcription factors like Nrf2 and heme oxygenase 1. In clinical trials, it has also been found that sulfur compounds, such as sulforaphane, are well-tolerated with minor reported side effects, indicating their potential as a safer alternative therapy for neurodegenerative conditions. These studies also highlight the fact that the emerging antioxidant and anti-inflammatory properties of sulfur support its therapeutic promise in promoting neuroprotection in various neurodegenerative diseases, such as AD and MS, by mitigating neuroinflammation and the formation of reactive oxygen species, which are key factors in these disorders [89].

Copper

Copper (Cu), an essential trace mineral that the body cannot synthesize on its own, is vital for physiological functioning, including iron absorption, blood cell formation, and maintaining immunity. Copper plays a crucial role in neuroprotection as an antioxidant by acting as a cofactor for various antioxidant enzymes, such as copper-zinc superoxide dismutase, which catalyzes the dismutation of superoxide radicals, thereby reducing oxidative stress in neuronal cells [58]. Antioxidant activity is crucial for maintaining neuronal integrity and function, as well as for mitigating oxidative stress, a significant contributor to neurodegenerative diseases. In neurodegenerative diseases, such as Alzheimer's disease, Parkinson's disease, Huntington's disease, amyotrophic lateral sclerosis, and prion disease, disrupted copper homeostasis has been observed [90]. In Alzheimer's disease, a copper imbalance affects the metabolism of the amyloid precursor proteins. It promotes the aggregation of amyloid beta peptides through copper binding, which can generate reactive oxygen species, leading to oxidative damage and neuroinflammation. It also regulates the clearance of amyloid beta peptides via low-density lipoprotein receptor-related proteins, indicating a dual role in disease progression [91]. In Parkinson's disease, decreased copper levels in the substantia nigra correlate with the disease pathology, suggesting that copper is involved in the survival of dopaminergic neurons and the synthesis of the dopamine neurotransmitter [92]. Preclinical studies have also demonstrated that it can influence disease outcomes, reducing tau phosphorylation and amyloid beta aggregation in cellular models. At the same time, copper supplementation or restoration in the copper homeostasis improves the neuropathological features and improves cognition in animal models of prion disease copper exhibits a paradoxical role that is it helps the neurons by maintaining regular prion protein conformation and inhibiting excitotoxicity and excessive copper binding to promote prion protein misfolding and aggregation [58]. In clinical studies, natural compounds, such as polyphenols, have shown promise in regulating copper homeostasis and exerting neuroprotective and antioxidant effects. The aim is to restore copper balance, reduce oxidative stress, and inhibit the neurodegenerative pathway by employing various metal-protein attenuating compounds, such as copper chelators and copper supplements [93].

Chromium

Chromium's (Cr) role as an antioxidant in neuroprotection is complex and content-dependent, while the trivalent form of chromium is essentially used in glucose metabolism. Its direct antioxidant and neuroprotective effects are less established than those of the other micronutrients discussed in this chapter. In contrast, hexavalent chromium, a potent neurotoxic agent that induces oxidative stress, damages

DNA and causes apoptosis of neuronal tissue by generating reactive oxygen species, leading to neurodegeneration and impairment [94]. Preclinical studies in various rat models have demonstrated that exposure to CR (V1) leads to increased levels of neurotransmitters, such as dopamine and serotonin, and a decrease in the activity of enzymes like monooxygenase A and acetylcholinesterase. The increased levels of these biomarkers indicate oxidative stress and apoptosis. Chromium picolinate, a commonly used supplemented form of Cr III, has demonstrated neuroprotective effects in various animal models in which neurodegeneration was induced in rotenone-induced rat models of Parkinson's disease. Chromium picolinate administration prevented the development of oxidative and nitrosative stress in the brain, restoring endogenous hydrogen sulfide production, and suggests its role as a neuroprotective agent [95]. Chromium supplements are used to counteract oxidative damage and support neuronal health in similar conditions. The preclinical findings to date have no direct clinical evidence linking chromium's antioxidant activity to neuroprotection. It has been observed that in human neurodegenerative diseases, such as Alzheimer's and Parkinson's, the results are still limited. Further studies are needed in this area to establish the therapeutic and safety potential of chromium compounds in the context of neurodegenerative illnesses [95].

Iron

Iron (Fe), an essential element found in the human diet, is required for the production of hemoglobin in the red blood cell that transports oxygen throughout the body. It is abundantly found in many animal-based foods as well as in some plant-based foods. Iron plays a complex role in the central nervous system, as a factor in numerous physiological processes, and is a potential contributor to oxidative stress if homeostasis is disrupted in any manner under normal conditions. Iron is mainly responsible for oxidative phosphorylation, the production of myelin sheath, neurotransmitter synthesis, and oxygen transport to the brain [96]. It is an essential component of the antioxidant enzymes, such as catalase and peroxidase, which mainly neutralize the reactive oxygen species and protect the neurons from any damage; however, if iron gets accumulated excessively in the brain, it can catalyze the formation of highly reactive hydroxyl radicals via Fenton reactions, causing endogenous antioxidant defenses to fail, and lead to lipid peroxidation, protein modification, and DNA damage, leading to significant neurodegenerative progression [97]. Neurodegenerative diseases like Parkinson's and Alzheimer's are caused by dysregulation in the iron homeostasis, leading to accumulation in regions of the brain, promoting oxidative stress and neuronal loss. Various animal models of Parkinson's disease have shown increased iron in the substantia Niagara associated with dopaminergic neuron degeneration. Reduced iron levels are associated with neuroprotective effects, including reducing oxidative damage and reducing inflammation, as well as preserving mitochondrial functions in both in vitro and in vivo models [98]. Alpha lipoic acid has also been shown to reverse iron-induced neurotoxicity in microglial activation and reduce oxidative stress in zebrafish models and cell culture models. Sinopic acid, a phenolic compound with iron chelator, has antioxidant properties that improve motor symptoms and decrease neuron loss in rotenone-induced Parkinson's disease models, likely by reducing ferroptosis and regulating iron metabolism [99]. Increased brain iron concentration can be detected clinically in patients with neurodegenerative diseases using MRI and iron chelation therapy. Despite limited clinical trials on the use of iron chelators to reduce iron accumulation and oxidative stress associated with its overaccumulation, iron homeostasis holds promise for neuroprotection and inhibiting the progression of neurodegenerative disorders.

Zinc

Zinc (Zn) plays a crucial role in neuroprotection by regulating oxidative stress and maintaining neuronal integrity. Zinc acts as a pro-antioxidant by way of preserving sulfhydryl groups in proteins, competing with redox-active metals, and presenting an antioxidant response, including the upregulation of proteins and enzymes, such as metallothioneine and superoxide dismutase [100]. The actions of zinc can inhibit the creation and accumulation of reactive oxygen species (ROS) and lipid peroxidation, while stabilizing cell membranes, all of which are protective against oxidative injury to neural tissue. Zinc also contributes to synaptic activity and neuronal plasticity, further promoting neuroprotection.

Zinc is a versatile mineral that plays a crucial role in supporting optimal cellular development, health, and well-being. Zinc's diverse roles are evident in many physiological functions, including cell growth and proliferation, immune system modulation, apoptosis, and the catalytic activity of numerous zinc-binding proteins, which are essential for maintaining the structure and function of cells in both adults and children. Zinc also has pivotal functions in the pathophysiology of critical neurodegenerative illnesses like Alzheimer's disease and psychiatric disorders, cancer, brain aging, obesity, diabetes, depression, and Wilson's disease. Zinc deficiency has been linked to serious diseases that are of paramount importance to public health. Preclinical models have demonstrated that zinc supplementation reduces neurological deficits, neuronal apoptosis, and cerebral infarct volume in animal brain injury models, primarily by serving as an antioxidant and anti-inflammatory agent [101]. Clinically, zinc deficiencies have been shown to contribute to elevated oxidative stress, cognitive decline, and increased risk of neurodegenerative diseases, such as Alzheimer's and Parkinson's disease. Zinc homeostasis has been linked to improved mental health and reduced neuronal damage. However, excessive zinc can cause neurotoxic effects, justifying the need to maintain zinc levels within a healthy window [102].

Manganese

Manganese (Mn) is an essential antioxidant trace element in neuroprotection, playing a crucial role as a cofactor of manganese superoxide dismutase (MnSOD), the antioxidant enzyme that catalyzes the detoxification of reactive oxygen species (ROS) in the brain. MnSOD catalyzes the dismutation of superoxide radicals into hydrogen peroxide and oxygen, thus maintaining redox balance and protecting neuronal cells from oxidative stress, a contributing factor for the pathology of many neurodegenerative diseases [103]. Preclinical studies have demonstrated that manganese complexes can catalyze antioxidant activity, thereby reducing levels of reactive oxygen species (ROS), enhancing mitochondrial membrane potential, and even mitigating cell death caused by oxidative stress in neuronal models. Various manganese-based compounds also reduced apoptosis by enhancing the expression of antiapoptotic factors, such as Bcl-2, and inhibiting the activation of caspase-3, thereby supporting their neuroprotective characteristics [104].

Selenium

Selenium (Se) is an essential trace element that humans require in small amounts, mainly through the diet and foods like grains, nuts, meat, and seafood. Selenium is involved in the formation of selenoproteins, which are necessary antioxidant enzymes that help cells avoid oxidative damage and support various immune and thyroid functions. Selenium has been identified as a potent antioxidant that contributes to neuroprotection mainly through its incorporation into selenoproteins, such as glutathione peroxidase 4 (GPX4), which have an essential role in neutralizing reactive oxygen species (ROS) and reducing oxidative stress in neural tissues [105]. In various preclinical studies, selenium supplementation, primarily in the form of selenium nanoparticles (SeNPs), has been demonstrated to exhibit significant neuroprotective effects in animal models of neurodegenerative diseases, seizures, and brain injuries. SeNPs can be loaded into lipid or polymeric nanoparticles, distinguishing them in the market as they can cross the blood-brain barrier, biodistribute, and accumulate in the brain, while exhibiting strong antioxidant capacity and efficacy with lower toxicity compared to other selenium formulations. Selenium appears to increase the expression of antioxidant proteins (e.g., GPX, SOD2, NRF-2) and modulate calcium flux, mitochondrial dynamics, and neuroinflammatory pathways that are well-known to be involved in neurodegeneration [106]. In animal models of ischemia and stroke, selenium treatment decreases neuronal death by preventing ferroptosis (a form of iron-driven cell death from lipid peroxidation) by upregulating GPX4 and decreasing free iron and hydroxyl radical production. SeNPs decrease beta-amyloid aggregation and toxicity in models of Alzheimer's disease, decrease proinflammatory cytokines, and enhance antioxidant ability in brain cells in models of Parkinson's and Huntington's diseases [107]. Preclinical evidence from cell cultures and animal studies consistently demonstrates that selenium, particularly SeNPs, reduces oxidative stress, inhibits apoptosis, and enhances functional recovery after neuronal injury by triggering adaptive stress response pathways and promoting selenoprotein expression.

Iodine

Iodine (I), a trace metal whose primary function is in the treatment of thyroid-related disorders, is found to have considerable antioxidant capabilities that confer neuroprotection, primarily via the neutralization of reactive oxygen species (ROS) and oxidative stress, a hallmark of neurodegenerative pathways [108]. As an electron donor, iodine quenches free radicals like hydroxyl radicals (OH^-) and hydrogen peroxide (H_2O_2) into less reactive compounds like hypoiodous acid and water. In neurodegenerative contexts, iodine's utility extends beyond antioxidant activity. Iodine inhibits proinflammatory pathways by inhibiting nitric oxide production in macrophages and reducing tumor necrosis factor-α (TNF-α) expression to dampen neuroinflammation [109]. Some preclinical studies suggest that iodine may have antioxidant actions that reduce oxidative damage in brain cells. For instance, iodide (I^-) has been shown to protect rat neurons against lipid peroxidation. Furthermore, because iodine is involved in the synthesis of thyroid hormones, iodine can indirectly confer neuroprotection through thyroid hormones, as they regulate oligodendrocyte precursor cell differentiation and myelination, which maintains a healthy white matter, a process disrupted in Alzheimer's disease [110]. While direct evidence of iodine, as a potential candidate in neurodegenerative models, is lacking, its ability to regulate oxidative and inflammatory pathways may provide an attractive potential therapeutic pathway in diseases like Alzheimer's disease, which is driven by mitochondrial dysfunction and ROS accumulation in the brain. The purported mechanism of action for trace minerals is shown in Fig. 5.2.

Conclusions and Future Perspectives

Multivitamins and multiminerals (MVMs) are popular dietary supplements taken by children and adults worldwide. The daily requirements of MVMs are generally met by consuming a variety of wholesome foods, fruits, nuts, and seeds, which are essential for a healthy diet. A deficiency of bare essential minerals, trace elements, and vitamins can pose health risks, including chronic neurodegenerative disorders and mental ailments. However, taking them in high doses or using many different supplements may lead to adverse consequences.

Nutritional strategies play a critical role in neuroprotection and help lower the risk of neurodegenerative diseases such as Parkinson's and Alzheimer's disease. Specifically, antioxidant vitamins and trace minerals play a significant role in maintaining mental health and mitigating the progression of neurodegenerative diseases. The vulnerability of the human brain to oxidative stress, due to its high metabolic

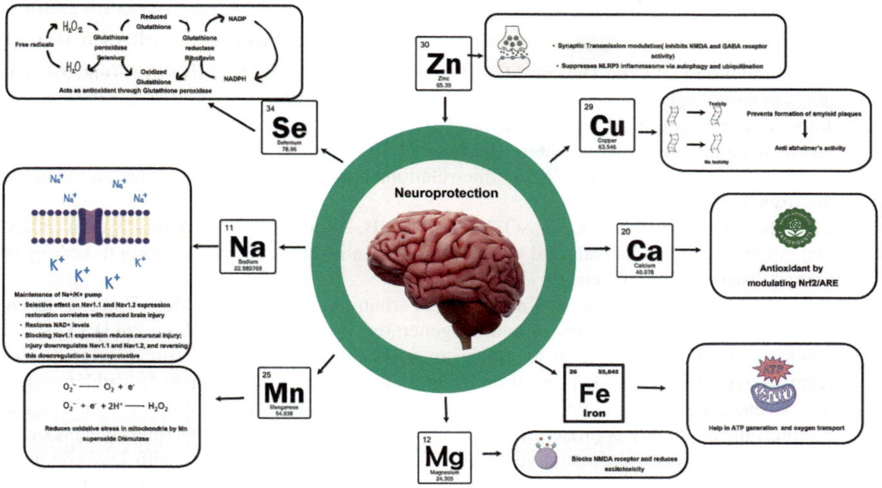

Fig. 5.2 Schematic representation of the role of trace minerals in neuroprotection

demand for oxygen and nutrients, as well as its lipid-rich composition, underscores the importance of antioxidant vitamins A, B, C, D, E, and K in neutralizing free radicals and preserving neuronal integrity. Moreover, various trace minerals (Cu, zinc, iron, manganese, selenium) serve as essential cofactors in enzymatic antioxidant defenses, cell signaling, and neurotransmitter regulation. Dietary patterns rich in micronutrients, as found in the Mediterranean, MIND, and DASH diets, are a significant public health approach to mitigating neurodegenerative disorders. These diets have demonstrated neuroprotective effects through the synergistic interactions of vitamins and trace minerals, provided in adequate amounts, which collectively reduce oxidative damage, neuroinflammation, and support synaptic plasticity. Several clinical studies suggest that adherence to such diets can reduce the risk of Alzheimer's disease by 53%. These diets maintain neuronal integrity, reduce brain inflammation, and enhance memory through nutrients, such as fiber derived from whole grains, antioxidants and vitamins from blueberries, and omega-3 fatty acids from walnuts. A limited number of preclinical and clinical studies have also demonstrated that the homeostasis of these vitamins and trace minerals is crucial for neurological health, and reduced levels should be supplemented with nutritional supplements to prevent the onset and progression of neurodegenerative diseases. Future research and clinical approaches could focus on evidence-based nutritional interventions that consider individual variability and long-term adherence to optimize cognitive resilience and delay brain aging. This collective approach offers promising potential for combating the growing burden of neurodegenerative disorders globally. However, further well-designed, placebo-controlled clinical studies are needed to establish optimal daily nutritional requirements and long-term safety aspects of micronutrients, especially in elderly populations.

Conflict of Interest None declared.

References

1. Olufunmilayo EO, Gerke-Duncan MB, Holsinger RMD (2023) Oxidative stress and antioxidants in neurodegenerative disorders. Antioxidants 12(2):517
2. Jomova K, Makova M, Alomar SY, Alwasel SH, Nepovimova E, Kuca K et al (2022) Essential metals in health and disease. Chem Biol Interact 367:110173
3. Sigel A, Sigel H, Sigel RKO (2013) Interrelations between essential metal ions and human diseases, vol 13. Springer
4. Chaudhary P, Janmeda P, Docea AO, Yeskaliyeva B, Abdull Razis AF, Modu B et al (2023) Oxidative stress, free radicals and antioxidants: potential crosstalk in the pathophysiology of human diseases. Front Chem 11:1158198
5. Orywal K, Socha K, Iwaniuk P, Kaczyński P, Farhan JA, Zoń W et al (2025) Vitamins in the prevention and support therapy of neurodegenerative diseases. Int J Mol Sci 26(3):1333
6. Houldsworth A (2024) Role of oxidative stress in neurodegenerative disorders: a review of reactive oxygen species and prevention by antioxidants. Brain Commun 6(1):fcad356
7. den Brink AC, Brouwer-Brolsma EM, Berendsen AAM, van de Rest O (2019) The Mediterranean, dietary approaches to stop hypertension (DASH), and Mediterranean-DASH intervention for neurodegenerative delay (MIND) diets are associated with less cognitive decline and a lower risk of Alzheimer's disease—a review. Adv Nutr 10(6):1040–1065
8. Aderinto N, Olatunji G, Abdulbasit M, Olajide TN, Kokori E (2023) Examining the efficacy of the Mediterranean-DASH diet intervention for neurodegenerative delay in mitigating cognitive decline. Egypt J Neurol Psychiatry Neurosurg 59(1):145
9. Fekete M, Lehoczki A, Tarantini S, Fazekas-Pongor V, Csípő T, Csizmadia Z et al (2023) Improving cognitive function with nutritional supplements in aging: a comprehensive narrative review of clinical studies investigating the effects of vitamins, minerals, antioxidants, and other dietary supplements. Nutrients 15(24):5116
10. A\csk\in Ö, Uzunçakmak TKÜ, Altunkalem N, Tüzün Y (2021) Vitamin deficiencies/hypervitaminosis and the skin. Clin Dermatol 39(5):847–857
11. Beer K, Singh A, Ravi SC, Gupta AK, Kumar A, Sharma MM (2024) A comprehensive review on the role of vitamin A on human health and nutrition. J Environ Biol 45(6):645–653
12. Pareek A, Singhal R, Pareek A, Ghazi T, Kapoor DU, Ratan Y et al (2024) Retinoic acid in Parkinson's disease: molecular insights, therapeutic advances, and future prospects. Life Sci:123010
13. Ashok A, Andrabi SS, Mansoor S, Kuang Y, Kwon BK, Labhasetwar V (2022) Antioxidant therapy in oxidative stress-induced neurodegenerative diseases: role of nanoparticle-based drug delivery systems in clinical translation. Antioxidants. 11(2):408
14. Guo JD, Zhao X, Li Y, Li GR, Liu XL (2018) Damage to dopaminergic neurons by oxidative stress in Parkinson's disease. Int J Mol Med 41(4):1817–1825
15. Marie A, Darricau M, Touyarot K, Parr-Brownlie LC, Bosch-Bouju C (2021) Role and mechanism of vitamin a metabolism in the pathophysiology of Parkinson's disease. J Parkinsons Dis 11(3):949–970
16. Rai SN, Singh P, Steinbusch HWM, Vamanu E, Ashraf G, Singh MP (2021) The role of vitamins in neurodegenerative disease: an update. Biomedicine 9(10):1284
17. Pouso MR, Cairrao E (2022) Effect of retinoic acid on the neurovascular unit: a review. Brain Res Bull 184:34–45
18. Calderón-Ospina CA, Nava-Mesa MO (2020) B vitamins in the nervous system: current knowledge of the biochemical modes of action and synergies of thiamine, pyridoxine, and cobalamin. CNS Neurosci & Ther 26(1):5–13
19. Wang SJ, Wu WM, Yang FL, Hsu GSW, Huang CY (2008) Vitamin B2 inhibits glutamate release from rat cerebrocortical nerve terminals. Neuroreport 19(13):1335–1338
20. Huang SK, Lu CW, Lin TY, Wang SJ (2022) Neuroprotective role of the B vitamins in the modulation of the central glutamatergic neurotransmission. CNS & Neurological Disorders 21(4):292–301

21. Umekar M, Premchandani T, Tatode A, Qutub M, Raut N, Taksande J et al (2025) Vitamin B12 deficiency and cognitive impairment: a comprehensive review of neurological impact. Brain Disord 18:100220
22. Yang T, Wang H, Xiong Y, Chen C, Duan K, Jia J et al (2020) Vitamin D supplementation improves cognitive function through reducing oxidative stress regulated by telomere length in older adults with mild cognitive impairment: a 12-month randomized controlled trial. J Alzheimer's Dis 78(4):1509–1518
23. Niki E (2014) Role of vitamin E as a lipid-soluble peroxyl radical scavenger: in vitro and in vivo evidence. Free Radic Biol Med 66:3–12
24. Emekli-Alturfan E, Alturfan AA (2023) The emerging relationship between vitamin K and neurodegenerative diseases: a review of current evidence. Mol Biol Rep 50(1):815–828
25. Sambon M, Wins P, Bettendorff L (2021) Neuroprotective effects of thiamine and precursors with higher bioavailability: focus on benfotiamine and dibenzoylthiamine. Int J Mol Sci 22(11):5418
26. Ashoori M, Saedisomeolia A (2014) Riboflavin (vitamin B2) and oxidative stress: a review. Br J Nutr 111(11):1985–1991
27. Marashly ET, Bohlega SA (2017) Riboflavin has neuroprotective potential: focus on Parkinson's disease and migraine. Front Neurol 8:333
28. Suwannasom N, Kao I, Pruß A, Georgieva R, Bäumler H (2020) Riboflavin: the health benefits of a forgotten natural vitamin. Int J Mol Sci 21(3):950
29. Zhang L, Yang S, Liu X, Wang C, Tan G, Wang X et al (2024) Association between dietary niacin intake and risk of Parkinson's disease in US adults: cross-sectional analysis of survey data from NHANES 2005--2018. Front Nutr 11:1387802
30. Gasperi V, Sibilano M, Savini I, Catani MV (2019) Niacin in the central nervous system: an update of biological aspects and clinical applications. Int J Mol Sci 20(4):974
31. Mitchell ES, Conus N, Kaput J (2014) B vitamin polymorphisms and behavior: evidence of associations with neurodevelopment, depression, schizophrenia, bipolar disorder and cognitive decline. Neurosci & Biobehav Rev 47:307–320
32. Balashova OA, Visina O, Borodinsky LN (2018) Folate action in nervous system development and disease. Dev Neurobiol 78(4):391–402
33. van de Lagemaat EE, de Groot LC, van den Heuvel EGHM (2019) Vitamin B12 in relation to oxidative stress: a systematic review. Nutrients 11(2):482
34. Simonenko SY, Bogdanova DA, Kuldyushev NA (2024) Emerging roles of vitamin B12 in aging and inflammation. Int J Mol Sci 25(9):5044
35. Lauer AA, Grimm HS, Apel B, Golobrodska N, Kruse L, Ratanski E et al (2022) Mechanistic link between vitamin B12 and Alzheimer's disease. Biomolecules 12(1):129
36. Cui X, Eyles DW (2022) Vitamin D and the central nervous system: causative and preventative mechanisms in brain disorders. Nutrients 14(20):4353
37. Sailike B, Onzhanova Z, Akbay B, Tokay T, Molnár F (2024) Vitamin D in central nervous system: implications for neurological disorders. Int J Mol Sci 25(14):7809
38. Long W, Fatehi M, Soni S, Panigrahi R, Philippaert K, Yu Y et al (2020) Vitamin D is an endogenous partial agonist of the transient receptor potential vanilloid 1 channel. J Physiol 598(19):4321–4338
39. Banerjee A, Khemka VK, Ganguly A, Roy D, Ganguly U, Chakrabarti S (2015) Vitamin D and Alzheimer's disease: neurocognition to therapeutics. Int J Alzheimers Dis 2015(1):192747
40. Wang W, Li Y, Meng X (2023) Vitamin D and neurodegenerative diseases. Heliyon 9(1)
41. Morello M, Landel V, Lacassagne E, Baranger K, Annweiler C, Féron F et al (2018) Vitamin D improves neurogenesis and cognition in a mouse model of Alzheimer's disease. Mol Neurobiol 55:6463–6479
42. Jiang Q (2014) Natural forms of vitamin E: metabolism, antioxidant, and anti-inflammatory activities and their role in disease prevention and therapy. Free Radic Biol Med 72:76–90

43. González-Reyes RE, Nava-Mesa MO, Vargas-Sánchez K, Ariza-Salamanca D, Mora-Muñoz L (2017) Involvement of astrocytes in Alzheimer's disease from a neuroinflammatory and oxidative stress perspective. Front Mol Neurosci 10:427
44. Wen M, Ding L, Zhang L, Zhou M, Xu J, Wang J et al (2016) DHA-PC and DHA-PS improved Aβ1—40 induced cognitive deficiency uncoupled with an increase in brain DHA in rats. J Funct Foods 22:417–430
45. Schirinzi T, Martella G, Imbriani P, Di Lazzaro G, Franco D, Colona VL et al (2019) Dietary vitamin E as a protective factor for Parkinson's disease: clinical and experimental evidence. Front Neurol 10:148
46. Vatassery GT, Bauer T, Dysken M (1999) High doses of vitamin E in the treatment of disorders of the central nervous system in the aged. Am J Clin Nutr 70(5):793–801
47. Grundman M (2000) Vitamin E and Alzheimer disease: the basis for additional clinical trials. Am J Clin Nutr 71(2):630S–636S
48. La Fata G, Weber P, Mohajeri MH (2014) Effects of vitamin E on cognitive performance during ageing and in Alzheimer's disease. Nutrients 6(12):5453–5472
49. Orticello M, Cavallaro RA, Antinori D, Raia T, Lucarelli M, Fuso A (2023) Amyloidogenic and neuroinflammatory molecular pathways are contrasted using menaquinone 4 (MK4) and reduced menaquinone 7 (MK7R) in association with increased DNA methylation in SK-N-BE neuroblastoma cell line. Cells 13(1):58
50. Popa DS, Bigman G, Rusu ME (2021) The role of vitamin K in humans: Implication in aging and age-associated diseases. Antioxidants. 10:566. Antioxidants Age-Related Dis Anti-Aging Strateg 2021;179
51. O'Keefe JH, Bergman N, Carrera-Bastos P, Fontes-Villalba M, Di Nicolantonio JJ, Cordain L (2016) Nutritional strategies for skeletal and cardiovascular health: hard bones, soft arteries, rather than vice versa. Open Hear 3(1)
52. Diachenko AI, Rodin IA, Krasnova TN, Klychnikov OI, Nefedova LN (2024) The role of vitamin K in the development of neurodegenerative diseases. Biochemist 89(Suppl 1):S57–S70
53. Maresz K (2021) Growing evidence of a proven mechanism shows vitamin K2 can impact health conditions beyond bone and cardiovascular. Integr Med A Clin J 20(4):34
54. Imbrescia K, Moszczynski Z (2023) Vitamin K.[Updated 2023 Feb 13]. StatPearls [Internet] Treasure Isl StatPearls Publ
55. Patel V, Akimbekov NS, Grant WB, Dean C, Fang X, Razzaque MS (2024) Neuroprotective effects of magnesium: implications for neuroinflammation and cognitive decline. Front Endocrinol (Lausanne) 15:1406455
56. Rroji M, Figurek A, Viggiano D, Capasso G, Spasovski G (2022) Phosphate in the context of cognitive impairment and other neurological disorders occurrence in chronic kidney disease. Int J Mol Sci 23(13):7362
57. Rana A, Katiyar A, Arun A, Berrios JN, Kumar G (2025) Natural sulfur compounds in mental health and neurological disorders: insights from observational and intervention studies. Front Nutr 12:1534000
58. Giampietro R, Spinelli F, Contino M, Colabufo NA (2018) The pivotal role of copper in neurodegeneration: a new strategy for the therapy of neurodegenerative disorders. Mol Pharm 15(3):808–820
59. Krikorian R, Eliassen JC, Boespflug EL, Nash TA, Shidler MD (2010) Improved cognitive-cerebral function in older adults with chromium supplementation. Nutr Neurosci 13(3):116–122
60. Szewczyk B (2013) Zinc homeostasis and neurodegenerative disorders. Front Aging Neurosci 5:33
61. Oliveira CS, Piccoli BC, Nogara PA, Pereira ME, de Carvalho KAT, Skalny AV et al (2023) Selenium neuroprotection in neurodegenerative disorders. In: Handbook of neurotoxicity. Springer, pp 2489–2523
62. Mattson MP (2007) Calcium and neurodegeneration. Aging Cell 6(3):337–350

63. Supnet C, Bezprozvanny I (2010) The dysregulation of intracellular calcium in Alzheimer disease. Cell Calcium 47(2):183–189
64. Green KN, LaFerla FM (2008) Linking calcium to Aβ and Alzheimer's disease. Neuron 59(2):190–194
65. Elhefnawei DM, Mahmoud AH, Kadry MO, Al-Mokaddem AK, Badawy MA, El-Desouky MA (2023) Calcium voltage-gated channel subunit alpha 1 C and glial fibrillary acidic protein signaling pathways as a selective biomarker in predicting the efficacy of liposomal loaded co-enzyme Q in the autistic rat model. Toxicol Reports 10:17–26
66. He Y, Zhang H, Wang T, Han Z, Ni Q bin, Wang K et al (2020) Impact of serum calcium levels on Alzheimer's disease: a Mendelian randomization study. J Alzheimers Dis 76(2):713–724
67. Lv J, Xiao X, Bi M, Tang T, Kong D, Diao M et al (2022 Sep) ATP-sensitive potassium channels: a double-edged sword in neurodegenerative diseases. Ageing Res Rev 80:101676
68. Ding F, Sun Q, Long C, Rasmussen RN, Peng S, Xu Q et al (2024) Dysregulation of extracellular potassium distinguishes healthy ageing from neurodegeneration. Brain 147(5):1726–1739
69. Silvestro S, Mazzon E (2022) Nrf2 activation: involvement in central nervous system traumatic injuries. A promising therapeutic target of natural compounds. Int J Mol Sci 24(1):199
70. Udensi UK, Tchounwou PB (2017) Potassium homeostasis, oxidative stress, and human disease. Int J Clin Exp Physiol 4(3):111
71. Rose CR, Verkhratsky A (2016) Principles of sodium homeostasis and sodium signalling in astroglia. Glia 64(10):1611–1627
72. Van Eersel J, Ke YD, Liu X, Delerue F, Kril JJ, Götz J et al (2010) Sodium selenate mitigates tau pathology, neurodegeneration, and functional deficits in Alzheimer's disease models. Proc Natl Acad Sci 107(31):13888–13893
73. Jana A, Modi KK, Roy A, Anderson JA, Van Breemen RB, Pahan K (2013) Up-regulation of neurotrophic factors by cinnamon and its metabolite sodium benzoate: therapeutic implications for neurodegenerative disorders. J Neuroimmune Pharmacol 8:739–755
74. Bayazid AB, Jang YA, Kim YM, Kim JG, Lim BO (2021) Neuroprotective effects of sodium butyrate through suppressing neuroinflammation and modulating antioxidant enzymes. Neurochem Res 46(9):2348–2358
75. Azimi M, Ashour AE, Fuaat AA, Mohamed WMY (2020) Neuroprotective effects of co-administration of coenzyme Q10 and vitamin-E in chronic cerebral hypoperfusion-induced neurodegeneration in rats. Int J Nutr Pharmacol Neurol Dis 10(2):35–42
76. Xu F, Jiang Y, Wang X, Shen L, Yan Y, Guo D et al (2023) Sodium aescinate inhibits microglia activation through NF-κB pathway and exerts neuroprotective effect. Front Pharmacol 14:1086429
77. Maier JAM, Locatelli L, Fedele G, Cazzaniga A, Mazur A (2022) Magnesium and the brain: a focus on neuroinflammation and neurodegeneration. Int J Mol Sci 24(1):223
78. Kozin S, Kravtsov A, Ivashchenko L, Dotsenko V, Vasilyeva L, Vasilyev A et al (2023) Study of the magnesium comenate structure, its neuroprotective and stress-protective activity. Int J Mol Sci 24(9):8046
79. Xiong Y, Ruan YT, Zhao J, Yang YW, Chen LP, Mai YR et al (2022) Magnesium-L-threonate exhibited a neuroprotective effect against oxidative stress damage in HT22 cells and Alzheimer's disease mouse model. World J psychiatry 12(3):410
80. Toffa DH, Magnerou MA, Kassab A, Djibo FH, Sow AD (2019) Can magnesium reduce central neurodegeneration in Alzheimer's disease? Basic evidences and research needs. Neurochem Int 126:195–202
81. Jin X, Liu MY, Zhang DF, Gao H, Wei MJ (2018) Elevated circulating magnesium levels in patients with Parkinson's disease: a meta-analysis. Neuropsychiatr Dis Treat:3159–3168
82. Oyanagi K, Hashimoto T. Magnesium in Parkinson's disease: an update in clinical and basic aspects. 2018;

83. Chen W, Ouyang J, Yi X, Xu Y, Niu C, Zhang W et al (2018) Black phosphorus nanosheets as a neuroprotective nanomedicine for neurodegenerative disorder therapy. Adv Mater 30(3):1703458
84. Saha S, Buttari B, Profumo E, Tucci P, Saso L (2022) A perspective on Nrf2 signaling pathway for neuroinflammation: a potential therapeutic target in Alzheimer's and Parkinson's diseases. Front Cell Neurosci 15:787258
85. Ottaviani MF, Mazzeo R, Cangiotti M, Fiorani L, Majoral JP, Caminade AM et al (2010) Time evolution of the aggregation process of peptides involved in neurodegenerative diseases and preventing aggregation effect of phosphorus dendrimers studied by EPR. Biomacromolecules 11(11):3014–3021
86. Wasiak T, Marcinkowska M, Pieszynski I, Zablocka M, Caminade AM, Majoral JP et al (2015) Cationic phosphorus dendrimers and therapy for Alzheimer's disease. New J Chem 39(6):4852–4859
87. Gong QH, Shi XR, Hong ZY, Pan LL, Liu XH, Zhu YZ (2011) A new hope for neurodegeneration: possible role of hydrogen sulfide. J Alzheimer's Dis 24(s2):173–182
88. Paul BD, Pieper AA (2024) Neuroprotective signaling by hydrogen sulfide and its dysregulation in Alzheimer's disease. Curr Opin Chem Biol 82:102511
89. Schepici G, Bramanti P, Mazzon E (2020) Efficacy of sulforaphane in neurodegenerative diseases. Int J Mol Sci 21(22):8637
90. Waggoner DJ, Bartnikas TB, Gitlin JD (1999) The role of copper in neurodegenerative disease. Neurobiol Dis 6(4):221–230
91. Hung YH, Bush AI, Cherny RA (2010) Copper in the brain and Alzheimer's disease. JBIC J Biol Inorg Chem 15:61–76
92. Montes S, Rivera-Mancia S, Diaz-Ruiz A, Tristan-Lopez L, Rios C (2014) Copper and copper proteins in Parkinson's disease. Oxidative Med Cell Longev 2014(1):147251
93. Gromadzka G, Tarnacka B, Flaga A, Adamczyk A (2020) Copper dyshomeostasis in neurodegenerative diseases—therapeutic implications. Int J Mol Sci 21(23):9259
94. Ajibo DN, Orish CN, Ruggieri F, Bocca B, Battistini B, Frazzoli C et al (2024) An update overview on mechanistic data and biomarker levels in cobalt and chromium-induced neurodegenerative diseases. Biol Trace Elem Res 202(8):3538–3564
95. Akhtar A, Dhaliwal J, Saroj P, Uniyal A, Bishnoi M, Sah SP (2020) Chromium picolinate attenuates cognitive deficit in ICV-STZ rat paradigm of sporadic Alzheimer's-like dementia via targeting neuroinflammatory and IRS-1/PI3K/AKT/GSK-3β pathway. Inflammopharmacology 28:385–400
96. Biasiotto G, Di Lorenzo D, Archetti S, Zanella I (2016) Iron and neurodegeneration: is ferritinophagy the link? Mol Neurobiol 53:5542–5574
97. Batista-Nascimento L, Pimentel C, Andrade Menezes R, Rodrigues-Pousada C (2012) Iron and neurodegeneration: from cellular homeostasis to disease. Oxidative Med Cell Longev 2012(1):128647
98. Kamalinia G, Khodagholi F, Atyabi F, Amini M, Shaerzadeh F, Sharifzadeh M et al (2013) Enhanced brain delivery of deferasirox—lactoferrin conjugates for iron chelation therapy in neurodegenerative disorders: in vitro and in vivo studies. Mol Pharm 10(12):4418–4431
99. Avc\i B, Günayd\in C, Külbay M, Kuruca N, Güvenç T, Bilge SS (2022) Neuroprotective effects of sinapic acid involve the iron regulatory role on the rotenone-induced Parkinson's disease model. Braz J Pharm Sci 58:e20942
100. Mocchegiani E, Bertoni-Freddari C, Marcellini F, Malavolta M (2005) Brain, aging and neurodegeneration: role of zinc ion availability. Prog Neurobiol 75(6):367–390
101. Dineley KE, Votyakova TV, Reynolds IJ (2003) Zinc inhibition of cellular energy production: implications for mitochondria and neurodegeneration. J Neurochem 85(3):563–570
102. Konoha K, Sadakane Y, Kawahara M (2006) Zinc neurotoxicity and its role in neurodegenerative diseases. J Heal Sci 52(1):1–8
103. Bowman AB, Kwakye GF, Hernández EH, Aschner M (2011) Role of manganese in neurodegenerative diseases. J Trace Elem Med Biol 25(4):191–203

104. Martins JAC, Gubert P, Villas Boas GR, Meirelles Paes M, Santamaria A, Lee E et al (2020) Manganese-induced neurodegenerative diseases and possible therapeutic approaches. Expert Rev Neurother 20(11):1109–1121
105. Solovyev ND (2015) Importance of selenium and selenoprotein for brain function: from antioxidant protection to neuronal signalling. J Inorg Biochem 153:1–12
106. Steinbrenner H, Sies H (2013) Selenium homeostasis and antioxidant selenoproteins in brain: implications for disorders in the central nervous system. Arch Biochem Biophys 536(2):152–157
107. Cardoso BR, Roberts BR, Bush AI, Hare DJ (2015) Selenium, selenoproteins and neurodegenerative diseases. Metallomics 7(8):1213–1228
108. Botchway BOA, Liu X, Zhou Y, Fang M (2023) Biometals in Alzheimer disease: emerging therapeutic and diagnostic potential of molybdenum and iodine. J Transl Med 21(1):351
109. Davtyan TK, Mkhitaryan LM, Gabrielyan ES (2009) Design of iodine-lithium-α-dextrin liquid crystal with potent antimicrobial and anti-inflammatory properties. Curr Pharm Des 15(11):1172–1186
110. Ajibawo-Aganbi U, Saleem S, Khan SZA, Veliginti S, Bastidas MVP, Lungba RM et al (2020) Can nutritional adequacy help evade neurodegeneration in older age? A review. Cureus 12(10)

Chapter 6
Chromium in the Prevention of Diabetes-Induced and Cardiovascular and Renal Disorders: Where Do We Stand Now?

Ramesh K. Goyal, Nandini Gupta, Bhoomika M. Patel, and Mayur M. Patel

Abstract Although evidence of antidiabetic affects various trace elements, chromium has existed for more than a century, there has been a sudden increase in therapeutic use of these elements in the prevention and treatment of diabetes-induced complications. This is because of two reasons. First, there has been an increased awareness about health management through the use of micronutrients. Second, effective and safe pharmaceutical formulations of trace elements have been developed and are now available. The molecular mechanisms of action are relatively clear for the trace elements. Chromium was known earlier as the "glucose tolerance factor" (GTF). With alteration in insulin levels, it is transported from the bloodstream to insulin-sensitive cells through the iron-transport protein transferrin. On reaching the site, it is sequestered by apochromodulin, a molecule with a high binding affinity for chromic ions. The resulting holochromodulin (Cr4-chromodulin) then interacts with the insulin-stimulated insulin receptor and amplifies insulin signaling. Insulin influences the recycling of transferrin receptors to the plasma membrane and, through internalization, releases chromium in the acidic environment of newly formed vesicles, thereby leading to chromium transport from the blood to insulin-sensitive cells, involving chromodulin. Among various compounds and formulations for chromium picolinate, it has emerged as a potential micronutrient trace element commonly used for the prevention and treatment of diabetes and associated cardiovascular and renal complications. Today, the chromium picolinate market is projected to reach USD 368.92 billion by 2032, exhibiting a CAGR of 3.41% during the forecast period, and this has been due to rising health consciousness for dietary supplements.

R. K. Goyal (✉)
Delhi Pharmaceutical Sciences and Research University, New Delhi, India

N. Gupta · B. M. Patel
National Forensic Sciences University, Gandhinagar, India

M. M. Patel
Institute of Pharmacy, Nirma University, Ahmedabad, India

N. S. Dhalla et al. (eds.), *Functional Biochemistry of Metallic Elements*,
Advances in Biochemistry in Health and Disease 36,
https://doi.org/10.1007/978-3-032-14445-4_6

Keywords Chromium picolinate · Micronutrient · Diabetes · Cardiovascular · Renal

Introduction

There is increasing evidence that micronutrient status is altered in diabetes mellitus, and micronutrient intake may be important in promoting optimum health of diabetic patients [1]. Apart from zinc, selenium, and vanadium, chromium has emerged as one of the most important trace elements useful as dietary supplements for the prevention of diabetes-induced cardiovascular and renal complications. Both dietary deficiencies and excesses of some trace metals, resulting in impaired insulin release, glucose intolerance, and insulin resistance, have been reported [1, 2]. Few studies have also addressed the possibility of trace element supplementation as an adjunct therapy in human diabetic subjects, with mixed results [3]. Interest in the use of trace elements for the treatment of diabetes was rekindled when, in 1979, compounds of trace elements of vanadium were shown to increase glucose transport and oxidation in adipocytes [4]. Today, the chromium picolinate market is projected to reach USD 368.92 billion by 2032, exhibiting a CAGR of 3.41% during the forecast period [5]. This is due to an increase in health consciousness and demand for dietary supplements for chronic and lifestyle diseases like diabetes. As such, there is a growing prevalence of diabetes and cardiovascular diseases, which has led to increased demand for chromium picolinate as a dietary supplement. In this chapter, we have made an attempt to review the current understanding of the molecular mechanisms of chromium and therapeutic opportunities.

Biochemistry of Chromium

Chromium (Cr), a d-block element, is vital for human health, playing a particularly important role in glucose metabolism [6]. The body's absorption and the subsequent effects of chromium are significantly influenced by its oxidation state, with Cr(VI) being more readily absorbed than Cr(III) [7]. Chromium is ubiquitous in our diet, with almost all food materials containing it in concentrations ranging from 20 to 500 μg/kg. Notably, meats, mollusks, and crustaceans are especially rich sources of this essential element. The recommended daily oral intake of chromium varies with age (Fig. 6.1); infants (1 year) are advised to consume 33–45 μg/day, children (11 years) 123–171 μg/day, and adults 246–343 μg/day. Beyond diet, chromium can enter the body through inhalation and dermal exposure [8, 9].

When considering the different forms of chromium, Cr (VI) stands out due to its efficient absorption, primarily entering cells via nonspecific anion channels. Once inside, it is frequently converted to Cr (III), particularly within red blood cells (RBCs) and lung tissues. Upon reduction to Cr (III), it strongly binds to RBCs,

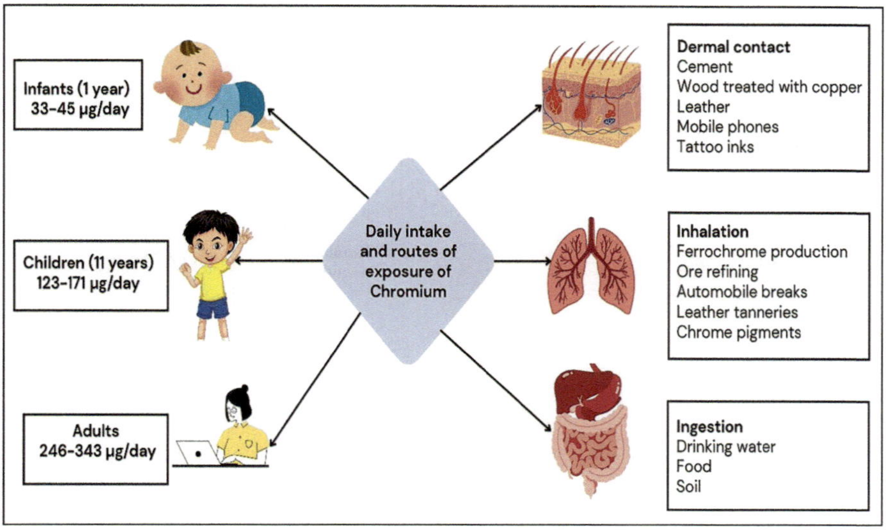

Fig. 6.1 Daily intake and routes of exposure to chromium

making it difficult for the element to exit these cells. Cr (VI) potent oxidative power allows it to traverse biological membranes and interact with crucial biomolecules, like proteins and nucleic acids, potentially leading to molecular alterations. In contrast, Cr (III) enters cells predominantly through passive diffusion. Interestingly, the body tends to accumulate higher concentrations of chromium in the kidneys, liver, and spleen. The elimination of Cr(VI) from the human body occurs in a phased manner, with excretion observed approximately at 7 h, 15–30 days, and a longer period of 3–5 years [10, 11].

Chromium's role in glucose metabolism is crucial, enhancing insulin action even though the precise enzymes involved remain unidentified. One biologically active form, glucose tolerance factor (GTF), found in brewer's yeast and kidney powder, is a complex containing Cr (III), along with nicotinic acid and specific amino acids, like glutamic acid, glycine, and cysteine. GTF is believed to transport chromium to cells that are deficient in it [12, 13]. Another important entity is the low-molecular-weight chromium (LMWCr)-binding substance, an oligopeptide comprising cysteine, glutamate, aspartate, and glycine. This substance is thought to amplify insulin signaling by binding to the inactive form of the insulin receptor tyrosine, thereby stabilizing its active conformation and facilitating insulin's effects [14, 15].

The transport of chromium within the body, particularly from the bloodstream to insulin-sensitive cells, is closely linked to insulin levels and the iron-transport protein transferrin. Transferrin, an 80,000-Da blood serum protein primarily recognized for its role in iron binding, is also considered a major physiological transporter for chromium. Research suggests that as insulin levels rise, such as after glucose intake, there's a corresponding increase in the movement of chromium from the blood into insulin-dependent cells, a process likely mediated by transferrin [16].

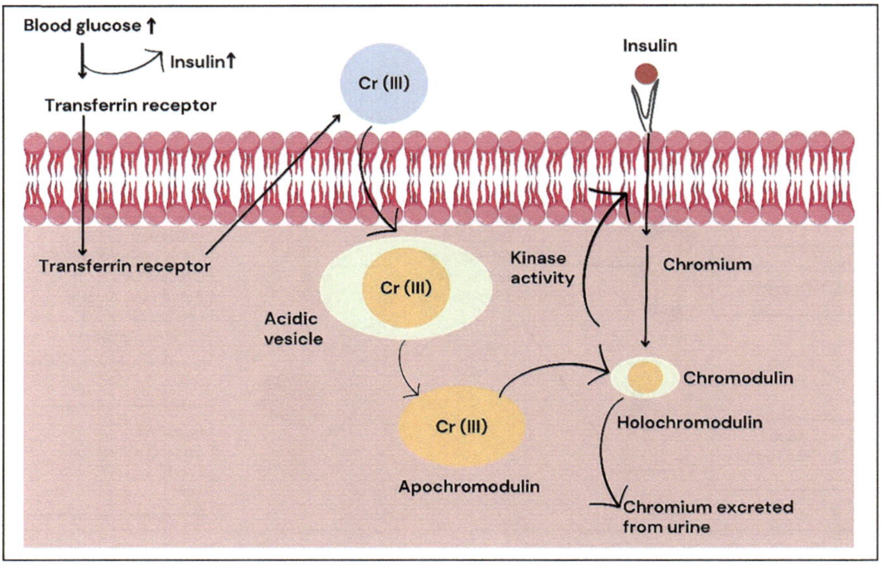

Fig. 6.2 Mechanism of chromium-mediated amplification of insulin signaling via chromodulin

Once chromium reaches insulin-sensitive cells, it is sequestered by apochromodulin, a molecule with a high binding affinity for chromic ions. The resulting holochromodulin (Cr4-chromodulin) then interacts with the insulin-stimulated insulin receptor, helping to sustain its active conformation and amplify insulin signaling. Furthermore, insulin influences the recycling of transferrin receptors to the plasma membrane. These receptors bind metal-saturated transferrin (including that carrying chromium), which is then internalized through endocytosis, releasing the metal in the acidic environment of newly formed vesicles (Fig. 6.2). This mechanism helps explain why elevated insulin levels lead to chromium transport from the blood to insulin-sensitive cells, ultimately involving chromodulin [17].

It has been observed that increases in blood insulin concentrations, often triggered by rising blood sugar, lead to significant decreases in plasma chromium levels and a rise in urinary chromium excretion. This rapid loss of chromium points to a specific transfer mechanism, with chromodulin appearing to be the primary form of chromium excreted in urine under certain conditions [18].

It's important to note that chromium picolinate, a popular dietary supplement, exhibits different absorption and effects compared to dietary chromium. It remains remarkably stable in the digestive system and appears to be absorbed largely intact. However, the picolinate ligands alter the redox potential of the chromium within the complex, making it vulnerable to reduction by biological agents. This can result in the generation of harmful hydroxyl radicals, which, in vitro studies have shown, can cause DNA cleavage. The enhanced absorption of chromium picolinate might be inextricably linked to this potentially detrimental chemistry, underscoring the critical need for long-term studies to fully assess its safety [19, 20].

Evidence of Role of Chromium in Diabetes and Associated Diseases

Suboptimal intake of chromium (III) is strongly linked to type 2 diabetes and cardiovascular diseases (CVD), increasing blood glucose, insulin, cholesterol, and triglycerides, while decreasing lean body mass. These issues are reversible with adequate chromium supplementation. Discovered about five decades ago, chromium is known as the "glucose tolerance factor" (GTF) due to its essential role in maintaining normal glucose tolerance. Thus, chromium supplementation shows promise in preventing metabolic syndrome-related diseases [6, 13].

A 6-week study in rat models of chemically induced type 1 and type 2 diabetes demonstrated that chromium picolinate significantly improved both glucose and lipid metabolism. The treatment notably reduced blood glucose levels and enhanced insulin sensitivity without altering insulin secretion, pointing to a direct insulin-sensitizing effect. Furthermore, elevated cholesterol and triglyceride levels were significantly lowered. In vitro investigations using C2C12 myoblasts and 3 T3-L1 adipocytes corroborated these findings, showing that chromium picolinate amplified insulin-stimulated glucose uptake and triglyceride synthesis. These results suggest that chromium picolinate effectively mitigates metabolic dysregulation in experimental diabetes by potentiating peripheral insulin action within skeletal muscle and adipose tissue [21].

Animal studies, utilizing a rat model of type 2 diabetes (T2DM), indicate that chromium picolinate significantly improves glycemic control and lipid profiles, reducing blood sugar by 63%, cholesterol by 9.7%, and triglycerides by 6.6%. Chromium picolinate also demonstrates protective effects against microvascular complications, improving renal and hepatic histopathology. These findings underscore chromium's essentiality in diabetes management and prevention of its complications. While chromium histidine shows promise due to superior absorption, further human utilization studies are warranted [22].

Another animal study, involving 60 male Wistar rats, investigated the protective effects of chromium histidinate against diabetic renal impairment. Rats were divided into four groups: a control group (12% fat diet), a control group supplemented with chromium histidinate, a high-fat diet followed by streptozotocin to induce diabetes, and an HFD/STZ group additionally supplemented with 110 μg chromium histidinate/kg body weight/day. The study found that chromium histidinate supplementation significantly reduced markers of renal oxidative stress, specifically elevated levels of malondialdehyde (MDA) and 8-isoprostane, and decreased lipid peroxidation. These results indicate that chromium histidine is efficacious in reducing renal risk factors and mitigating impairment caused by diabetes [23].

Chromium is crucial for human nutrition, as evidenced by patients on total parenteral nutrition who developed severe, insulin-resistant diabetic symptoms. These symptoms, including glucose intolerance and neurological issues, resolved upon chromium supplementation, eliminating the need for exogenous insulin. Many healthy individuals on Western diets have suboptimal chromium intake, which can be further compromised by stress or trauma. For short-term TPN patients, daily supplementation of 10–20 micrograms of chromium appears sufficient. However, for long-term TPN patients, chromium levels in fluids should be monitored due to variable contamination [24].

Studies, including one with niacin-bound chromium (NBC), have shown that it can improve insulin sensitivity and significantly lower fasting blood glucose, triglycerides, and HbA1c, a long-term glucose control marker. This underscores chromium's critical impact on metabolic regulation. While some research highlights chromium's metabolic benefits, not all studies demonstrate consistent positive outcomes, especially concerning specific chromium forms and patient populations [25].

A 6-month randomized, double-blind, placebo-controlled clinical trial investigating chromium picolinate in overweight Type 2 Diabetes patients (receiving >50 units of daily insulin) found no improvement in lipid profile, BMI, blood pressure, or insulin requirements. Similarly, the same research group conducted another study using chromium yeast, which also failed to show improved glycemic control in type 2 diabetes patients. However, a subsequent evaluation by the same group, focusing on the effects of high-dose chromium picolinate administration in middle-aged healthy subjects with normal body weight and BMI but diagnosed with type 2 diabetes, yielded different results, indicating improved glucose and insulin metabolism. These conflicting findings suggest that the efficacy of chromium supplementation may depend on factors such as the specific form of chromium used, the dosage, and the baseline characteristics and metabolic status of the patient population [26].

Chromium in Polycystic Ovary Syndrome

Polycystic ovary syndrome (PCOS), a common endocrine disorder affecting women, is often associated with insulin resistance, type 2 diabetes, and cardiovascular disease. Various studies have shown chromium supplements to help maintain glycemic control in PCOS patients and reduce and reduce lipid levels [27]. Meta-analyses of randomized clinical trials reported that although chromium supplementation had no effect on fasting blood glucose, total testosterone, dehydroepiandrosterone, follicle-stimulating hormone, or luteinizing hormone levels, the supplementation significantly reduced body mass index, free testosterone levels, and fasting insulin levels [28, 29]. Thus, there are mixed findings with respect to the beneficial effects of chromium in patients with PCOS. Additional studies with sufficient sample sizes and duration in well-defined populations are needed [27].

Numerous studies have shown an association between poor chromium status and elevated blood cholesterol levels [30], and it is hypothesized that chromium supplementation might improve blood lipid levels. Some have shown that chromium supplementation increases HDL cholesterol levels [30, 31]. Additional research is needed to determine whether chromium supplementation has any clinically significant effects on dyslipidemia.

Current Regulatory Status of Chromium as a Dietary Supplement/Micronutrient

In 2001, the Food and Nutrition Board (FNB) of the National Academies of Sciences, Engineering, and Medicine recommended chromium to be an essential nutrient based on its effects on insulin action [32]. Later, based on research, it was suggested that although chromium might have benefits at pharmacologic amounts (e.g., in the hundreds of mcg), it is not an essential mineral because an absence or deficiency of chromium does not produce abnormalities that can be reversed with the addition of chromium [33–36]. In 2014, the European Food Safety Authority Panel on Dietetic Products, Nutrition, and Allergies could not conclude that there is convincing evidence to show that chromium is an essential nutrient [36]. Currently, there is no direct recommendation for the daily requirement of chromium; however, many multivitamin/mineral supplements containing 35–120 mcg chromium and diet supplements containing only chromium are also available that provide 200 mcg to 1,000 mcg chromium [37]. In spite of inconclusive versions about the deficiency of chromium in diabetes and associated complications [34–36, 38], chromium is also added to TPN solutions to provide 10–15 mcg chromium per day [39]. The American Society for Parenteral and Enteral Nutrition and other experts also recommend a systematic research on parenteral chromium requirements to determine whether chromium levels in TPN solutions should be lowered [34, 39, 40].

Chromium supplements are widely marketed in the United States for people with type 2 diabetes. Many adults use chromium supplements in the hope that they will reduce their risk of diabetes or improve their glycemic control [41–43]. However, the US-FDA has allowed the health claim that chromium picolinate may reduce the risk of insulin resistance and, therefore, possibly may reduce the risk of type 2 diabetes, although the US-FDA concludes that there is uncertainty about the existence of a clear relationship between chromium picolinate with insulin resistance and type 2 diabetes is highly uncertain [44].

Challenges of Formulation Development: The Key to Success for Chromium Marketing

The key to the success of the usage of trace elements for human health depends on the availability of an appropriate formulation with proven efficacy and safety through clinical trials. One of the major requirements for the formulation to be used in humans is bioavailability. Chromium is commercially available in various chemical forms, including Cr picolinate, nicotinate, and chloride. Among these, Cr picolinate has gained particular attention in the nutraceutical field owing to its superior bioavailability. This enhanced absorption is attributed to the chelation of Cr with picolinic acid, forming a molecular complex that facilitates efficient gastrointestinal uptake and improved systemic utilization compared to other Cr salts such as Cr chloride. A recent study evaluated the acute absorption profile of Cr in healthy, non-overweight young adult females administered 200 micrograms of Cr from four different supplement types: Cr picolinate, two forms of Cr nicotinate, and Cr chloride. Utilizing a randomized, crossover design, each intervention was separated by a minimum 1-week washout period and was coadministered with a standardized meal. The Cr absorption was assessed via 24-hour urinary Cr excretion, a validated surrogate marker for bioavailability. The results demonstrated that Cr picolinate elicited significantly greater urinary excretion compared to the other supplement, confirming its superior acute bioavailability. These findings substantiated the rationale for selecting Cr picolinate in supplement design to achieve enhanced therapeutic efficacy [45]. Nonetheless, excessive consumption of Cr and other trace elements may pose toxicological risks. These considerations not only emphasize the need for professional consultation prior to supplementation but also provide opportunities for the advancement of safer, more bioavailable, and precisely dosed Cr formulations [19].

The development of solid oral dosage forms, tablets, and capsules requires a careful balance of formulation science, material properties, and process engineering. Challenges frequently arise due to the physicochemical nature of active pharmaceutical ingredients (APIs) or nutraceutical ingredients and the excipients selected for manufacturing.

For a formulation wherein the quantity of active ingredient is less than 1 mcg, besides routine factors like adhesiveness or "stickiness, and flowability, the environmental sensitivity of certain nutraceutical ingredients is yet another complexity in the formulation development. To address this issue, formulations are protected using desiccant excipients (e.g., microcrystalline cellulose, silicon dioxide), moisture-barrier capsules such as hydroxypropyl methylcellulose (HPMC), and protective film coatings. Recently, one innovative Cr formulation technology was developed, offering a promising solution to many of these challenges. This has been patented (WO2021111407A1) as a novel solid oral dosage formulation of chromium incorporating Cr (Sucrosomial Chromium®) in a sucrosomial delivery system. This platform employs a phospholipid vesicle matrix built on phosphatidylcholine, which encapsulates the active nutrients and is stabilized using sucrose esters (sucresters) and starch-based carriers. The sucrosomial construct enhances gastrointestinal tolerance and protects micronutrients from acidic degradation. Moreover, it enables improved mucosal permeability and controlled intestinal release, significantly enhancing bioavailability over traditional salt forms. This patented system has an advantage for clinical scenarios requiring steady-state delivery of micronutrients, like chromium, and has been recommended for patients with metabolic disorders, including insulin resistance, by minimizing gastrointestinal side effects and improving pharmacokinetics. Moreover, the sucrosomial platform opens new avenues for the development of next-generation Cr supplements that are both efficacious and patient-friendly [46]. To support such advanced formulations, LFA Machines offers services including formulation consultation, pre- and post-mixing evaluations, and small-scale feasibility trials. Complimentary proof-of-concept testing is also available to validate blend performance in tablet form, ensuring manufacturability and quality assurance from concept to final production [47].

The chromium picolinate market is projected to reach USD 368.92 billion by 2032, exhibiting a CAGR of 3.41% during the forecast period. Rising health consciousness and increasing demand for dietary supplements are key factors driving market growth. The growing prevalence of diabetes and cardiovascular diseases has led to increased demand for chromium picolinate as a dietary supplement. Additionally, rising disposable income and changing lifestyles in developing countries are contributing to market expansion. Recently, many new players have been coming in for the formulation's development for the launch of new chromium picolinate products. Thus, besides understanding the molecular mechanisms for chromium, it is essential to have a focus on appropriate formulations to be developed, and the clinical trials are to be conducted for the use in diabetes and associated complications.

Conclusions

The increased health awareness about long-term cardiovascular complications, including those on the kidneys, has generated interest in research on trace elements. As of now, there is substantial evidence available for its efficacy in diabetes, insulin resistance, and associated cardiovascular complications and renal disease. The molecular mechanism of action of chromium is relatively clear for chromium. With alteration in insulin levels, it is transported from the bloodstream to insulin-sensitive cells through the iron-transport protein transferrin. On reaching the site, it is sequestered by apochromodulin, a molecule with a high binding affinity for chromic ions. The resulting holochromodulin (Cr4-chromodulin) then interacts with the insulin-stimulated insulin receptor and amplifies insulin signaling.

While chromium is an essential trace mineral with various health benefits, its usage as a nutraceutical or supplement with multivitamin formulations presents several challenges. One of such challenges is the development of appropriate formulations and their systematic clinical trials to include effectiveness in patients, especially for the prevention of diabetes-induced cardiovascular complications and renal diseases. It is also a challenge with potential toxicity, interactions with other medications, and the possibility of adverse effects, like low blood sugar or liver/kidney damage from excessive intake.

References

1. Mooradian AD, Morley JE (1987) Micronutrient status in diabetes mellitus. Am J Clin Nutr 45:877–895. https://doi.org/10.1093/ajcn/45.5.877
2. Failla ML, Kiser RA (1981) Altered tissue content and cytosol distribution of trace metals in experimental diabetes. J Nutr 111:1900–1909. https://doi.org/10.1093/jn/111.11.1900

3. Yale JF, Vigeant C, Nardolillo C et al (1995) In vivo effects of peroxovanadium compounds in BB rats. Mol Cell Biochem 153:181–190. https://doi.org/10.1007/BF01075936
4. Garcia AG, Jurkiewicz A, Jurkiewicz NH (1981) Contractile effect of vanadate and other vanadium compounds on the rat vas deferens. Eur J Pharmacol 70:17–23. https://doi.org/10.1016/0014-2999(81)90427-1
5. Singh S (2023) Chromium picolinate market size, share, growth In: Industry-2034 Market Research Future (Report ID MRFR/F-B & N/24036-HCR). Retrieved October 19, 2025, from https://www.marketresearchfuture.com/reports/chromium-picolinate-market-25675. Reaccessed 19 Oct. 2025
6. Anderson RA (2000) Chromium in the prevention and control of diabetes. Diabetes Metab 26:22–27
7. Achmad R, Budiawan, Auerkari E (2017) Effects of chromium on human body. ARRB 13:1–8. https://doi.org/10.9734/ARRB/2017/33462
8. Kumpulainen JT (1992) Chromium content of foods and diets. Biol Trace Elem Res 32:9–18. https://doi.org/10.1007/BF02784582
9. Wilbur S, Abadin H, Fay M, et al (2012) Toxicological profile for chromium. Agency for Toxic Substances and Disease Registry (US), Atlanta (GA)
10. Chiu A, Shi XL, Lee WKP et al (2010) Review of chromium (VI) apoptosis, cell-cycle-arrest, and carcinogenesis. J Environ Sci Health C 28:188–230. https://doi.org/10.1080/10590501.2010.504980
11. Fenti A, Chianese S, Iovino P et al (2020) Cr(VI) sorption from aqueous solution: a review. Appl Sci 10:6477. https://doi.org/10.3390/app10186477
12. Ghosh D, Bhattacharya B, Mukherjee B et al (2002) Role of chromium supplementation in Indians with type 2 diabetes mellitus. J Nutr Biochem 13:690–697. https://doi.org/10.1016/S0955-2863(02)00220-6
13. Havel PJ (2004) A scientific review: the role of chromium in insulin resistance. Diabetes Educ Suppl:2–14
14. Chen Y, Watson HM, Gao J et al (2011) Characterization of the organic component of low-molecular-weight chromium-binding substance and its binding of chromium. J Nutr 141:1225–1232. https://doi.org/10.3945/jn.111.139147
15. Vincent JB (1999) Mechanisms of chromium action: low-molecular-weight chromium-binding substance. J Am Coll Nutr 18:6–12. https://doi.org/10.1080/07315724.1999.10718821
16. Deng G, Dyroff SL, Lockart M et al (2016) The effects of the glycation of transferrin on chromium binding and the transport and distribution of chromium in vivo. J Inorg Biochem 164:26–33. https://doi.org/10.1016/j.jinorgbio.2016.08.008
17. Wang H, Kruszewski A, Brautigan DL (2005) Cellular chromium enhances activation of insulin receptor kinase. Biochemistry 44:8167–8175. https://doi.org/10.1021/bi0473152
18. Love ST, Di Bona KR, Sinha SH et al (2013) Urinary chromium excretion in response to an insulin challenge is not a biomarker for chromium status. Biol Trace Elem Res 152:57–65. https://doi.org/10.1007/s12011-012-9594-3
19. DiSilvestro RA, Dy E (2007) Comparison of acute absorption of commercially available chromium supplements. J Trace Elem Med Biol 21:120–124. https://doi.org/10.1016/j.jtemb.2007.01.004
20. Laschinsky N, Kottwitz K, Freund B et al (2012) Bioavailability of chromium(III)-supplements in rats and humans. Biometals 25:1051–1060. https://doi.org/10.1007/s10534-012-9571-5
21. Shinde Urmila A, Sharma G, Xu Yan J et al (2004) Anti-diabetic activity and mechanism of action of chromium chloride. Exp Clin Endocrinol Diabetes 112:248–252. https://doi.org/10.1055/s-2004-817971
22. Sahin K, Onderci M, Tuzcu M et al (2007) Effect of chromium on carbohydrate and lipid metabolism in a rat model of type 2 diabetes mellitus: the fat-fed, streptozotocin-treated rat. Metabolism 56:1233–1240. https://doi.org/10.1016/j.metabol.2007.04.021

23. Dogukan A, Tuzcu M, Juturu V et al (2010) Effects of chromium histidinate on renal function, oxidative stress, and heat-shock proteins in fat-fed and Streptozotocin-treated rats. J Ren Nutr 20:112–120. https://doi.org/10.1053/j.jrn.2009.04.009
24. Anderson RA (1995) Chromium and parenteral nutrition. Nutrition 11:83–86
25. Goldberg RB, Jacobson TA (2008) Effects of niacin on glucose control in patients with dyslipidemia. Mayo Clin Proc 83:470–478. https://doi.org/10.4065/83.4.470
26. Kleefstra N, Houweling ST, Jansman FGA et al (2006) Chromium treatment has no effect in patients with poorly controlled, insulin-treated type 2 diabetes in an obese Western population. Diabetes Care 29:521–525. https://doi.org/10.2337/diacare.29.03.06.dc05-1453
27. Fazelian S, Rouhani MH, Bank SS, Amani R (2017) Chromium supplementation and polycystic ovary syndrome: a systematic review and meta-analysis. J Trace Elem Med Biol 42:92–96. https://doi.org/10.1016/j.jtemb.2017.04.008
28. Tang X, Sun Z, Gong L (2018) Chromium supplementation in women with polycystic ovary syndrome: systematic review and meta-analysis. J of Obstet and Gynaecol 44:134–143. https://doi.org/10.1111/jog.13462
29. Saadati N, Haidari F, Barati M et al (2021) The effect of low glycemic index diet on the reproductive and clinical profile in women with polycystic ovarian syndrome: a systematic review and meta-analysis. Heliyon 7:e08338. https://doi.org/10.1016/j.heliyon.2021.e08338
30. Asbaghi O, Naeini F, Ashtary-Larky D et al (2021) Effects of chromium supplementation on lipid profile in patients with type 2 diabetes: a systematic review and dose-response meta-analysis of randomized controlled trials. J Trace Elem Med Biol 66:126741. https://doi.org/10.1016/j.jtemb.2021.126741
31. Jain SK, Rains JL, Croad JL (2007) Effect of chromium niacinate and chromium picolinate supplementation on lipid peroxidation, TNF-alpha, IL-6, CRP, glycated hemoglobin, triglycerides, and cholesterol levels in blood of streptozotocin-treated diabetic rats. Free Radic Biol Med 43:1124–1131. https://doi.org/10.1016/j.freeradbiomed.2007.05.019
32. Institute of Medicine (U.S.) (2010) DRI: dietary reference intakes for vitamin A, vitamin K, arsenic, boron, chromium, copper, iodine, iron, manganese, molybdenum, nickel, silicon, vanadium, and zinc: a report of the Panel on Micronutrients ... and the Standing Committee on the Scientific Evaluation of Dietary Reference Intakes, Food and Nutrition Board, Institute of Medicine. National Academy Press, Washington, D.C
33. Vincent JB (2015) Is the pharmacological mode of action of chromium (III) as a second messenger? Biol Trace Elem Res 166:7–12. https://doi.org/10.1007/s12011-015-0231-9
34. Vincent JB (2017) New evidence against chromium as an essential trace element. J Nutr 147:2212–2219. https://doi.org/10.3945/jn.117.255901
35. Nielsen FH (2019) Summary: The metabolism, nutritional essentiality, and clinical importance of chromium—Clarity emerging after 60 years of research. In: The Nutritional Biochemistry of Chromium (III). Elsevier, pp 361–370
36. Agostoni C, Roberto Berni C, Susan F-T (2014) Scientific opinion on dietary reference values for chromium. EFSA J. https://doi.org/10.2903/j.efsa.2014.3845
37. Costello RB, Dwyer JT, Merkel JM (2019) Chromium supplements in health and disease. In: The Nutritional Biochemistry of Chromium (III). Elsevier, pp 219–249
38. Vincent JB, Lukaski HC (2018) Chromium. Adv Nutrit 9:505–506. https://doi.org/10.1093/advances/nmx021
39. Vincent JB, Brown S (2019) Introduction: A history of chromium studies (1955–2007). In: The Nutritional Biochemistry of Chromium (III). Elsevier, pp 1–58
40. Fessler TA (2013) Trace elements in parenteral nutrition: a practical guide for dosage and monitoring for adult patients. Nut in Clin Prac 28:722–729. https://doi.org/10.1177/0884533613506596
41. Wang ZQ, Cefalu WT (2010) Current concepts about chromium supplementation in type 2 diabetes and insulin resistance. Curr Diab Rep 10:145–151. https://doi.org/10.1007/s11892-010-0097-3

42. Landman GW, Bilo HJ, Houweling ST, Kleefstra N (2014) Chromium does not belong in the diabetes treatment arsenal: current evidence and future perspectives. WJD 5:160. https://doi.org/10.4239/wjd.v5.i2.160
43. Costello RB, Dwyer JT, Bailey RL (2016) Chromium supplements for glycemic control in type 2 diabetes: limited evidence of effectiveness. Nutr Rev 74:455–468. https://doi.org/10.1093/nutrit/nuw011
44. Trumbo PR, Ellwood KC (2006) Chromium picolinate intake and risk of type 2 diabetes: an evidence-based review by the United States Food and Drug Administration. Nutr Rev 64:357–363. https://doi.org/10.1111/j.1753-4887.2006.tb00220.x
45. Gargas ML, Norton RL, Paustenbach DJ, Finley BL (1994) Urinary excretion of chromium by humans following ingestion of chromium picolinate. Implications for biomonitoring. Drug Metab Dispos 22:522–529
46. Andrea L, Germano T, Elisa B (2020) Solid form formulation comprising chromium or copper or a vitamin, compositions and uses thereof
47. LFA Machines (2025) The top 5 problems encountered in developing your formula for tablets and capsules

Chapter 7
The Role of Antioxidant-Associated Trace Elements in the Development Diabetic Cardiomyopathy

Fajar Shodiq Permata, Crystal M. D'Souza, Jaipaul Singh, Frank Christopher Howarth, and Ernest A. Adeghate

Abstract Diabetic cardiomyopathy (DCM) is a distinct myocardial complication of diabetes mellitus, characterized by structural and functional cardiac alterations independent of hypertension and coronary artery disease. Its pathogenesis is complex and multifactorial, with oxidative stress recognized as a central mechanism that interacts with metabolic dysregulation, mitochondrial dysfunction, and chronic inflammation. This chapter synthesizes evidence from experimental and clinical studies to examine the role of antioxidant-associated trace elements in the development and management of DCM. Eight essential trace elements—iron, zinc, copper, selenium, iodine, manganese, molybdenum, and chromium—are highlighted for their involvement as cofactors of key antioxidant enzymes, including superoxide

F. S. Permata
Department of Anatomy, College of Medicine and Health Sciences, United Arab Emirates University, Al Ain, United Arab Emirates

Faculty of Veterinary Medicine, Universitas Brawijaya, Malang, Indonesia

C. M. D'Souza
Department of Anatomy, College of Medicine and Health Sciences, United Arab Emirates University, Al Ain, United Arab Emirates

J. Singh
School of Pharmacy and Biomedical Sciences, University of Central Lancashire, Preston, UK

F. C. Howarth
Department of Physiology, College of Medicine and Health Sciences, United Arab Emirates University, Al Ain, United Arab Emirates

E. A. Adeghate (✉)
Department of Anatomy, College of Medicine and Health Sciences, United Arab Emirates University, Al Ain, United Arab Emirates

Zayed Bin Sultan Foundation for Health Sciences, United Arab Emirates University, Al Ain, United Arab Emirates
e-mail: eadeghate@uaeu.ac.ae

N. S. Dhalla et al. (eds.), *Functional Biochemistry of Metallic Elements*, Advances in Biochemistry in Health and Disease 36,
https://doi.org/10.1007/978-3-032-14445-4_7

dismutase, catalase, and glutathione peroxidase. Disturbances in the homeostasis of these elements weaken antioxidant defenses, increase reactive oxygen and nitrogen species, and accelerate myocardial injury, whereas adequate levels enhance antioxidant activity, preserve mitochondrial function, reduce inflammation, and limit adverse cardiac remodeling. The dual role of trace elements as both inducers and protectors underscores their importance in understanding DCM pathogenesis and highlights their potential as therapeutic modulators. Future strategies should focus on defining optimal micronutrient ranges, developing evidence-based supplementation protocols, and integrating trace element management with conventional treatments to reduce the burden of diabetic cardiomyopathy.

Keywords Antioxidant · Diabetic cardiomyopathy · Development · Management · Micronutrients

Introduction

Diabetes Mellitus

According to the International Diabetes Federation, an estimated 589 million people were living with diabetes mellitus (DM) in 2024, and this figure is projected to rise to over 853 million by 2050 [1]. The growing prevalence of DM inevitably implies a parallel increase in diabetes-related complications in the years ahead.

DM can be broadly classified into four major categories. Type 1 diabetes mellitus (T1DM) is characterized by immune-mediated destruction of pancreatic β-cells, leading to absolute insulin deficiency [2]. It is most frequently diagnosed in children and adolescents and accounts for approximately 5–10% of all DM cases [3]. In contrast, type 2 diabetes mellitus (T2DM), which represents 80–90% of all cases, is primarily associated with insulin resistance. In this condition, skeletal muscle, liver, and adipose tissue become less responsive to insulin, impairing glucose uptake and utilization. Persistent hyperglycemia further aggravates insulin resistance, perpetuating a vicious cycle of metabolic dysfunction [4].

Gestational diabetes mellitus (GDM) constitutes the third main category and occurs during pregnancy. While often transient and resolving after delivery, GDM significantly increases the risk of developing T2DM later in life. The global prevalence of GDM ranges from 6% to 25%, depending on population characteristics and maternal age. For example, in the United States, the prevalence in 2021 was reported at 8.3% overall but rose to 15.6% among women aged 40 years and older [5].

Finally, several less common forms of DM also exist. These include pancreatogenic diabetes, which develops secondary to diseases of the exocrine pancreas such as chronic pancreatitis, and steroid-induced diabetes, which may occur after organ transplantation. In addition, monogenic forms of diabetes, such as *maturity-onset diabetes of the young (MODY)*, result from single-gene defects affecting pancreatic β-cell function [6]. Although these types are less frequent, they contribute to the overall heterogeneity of DM.

How Do Complications of Diabetes Mellitus Occur?

Hyperglycemia, the hallmark feature of diabetes mellitus (DM), induces glucotoxicity and promotes the excessive generation of reactive oxygen species (ROS) and reactive nitrogen species (RNS). Elevated levels of these free radicals disrupt the redox equilibrium, and when endogenous antioxidant defenses are overwhelmed, oxidative stress ensues. This imbalance causes structural and functional damage to DNA, proteins, and lipids, thereby impairing cellular integrity and organ function [7]. Such biochemical insults are most evident in the vasculature supplying vital organs, including the heart, kidney, retina, nerves, and brain. Consequently, patients with poorly controlled DM are predisposed to a range of chronic complications, including diabetic cardiomyopathy, cardiovascular disease, nephropathy, retinopathy, and neuropathy [8].

Under physiological conditions, oxidative stress is counterbalanced by endogenous antioxidant systems. Key antioxidant enzymes, such as superoxide dismutase, catalase, and glutathione peroxidase, require trace elements—including zinc, copper, manganese, selenium, and iron—as essential cofactors [9–11]. Hyperglycemia promotes glucotoxic pathways, including mitochondrial overactivity, NADPH oxidase activation, and AGE–RAGE/PKC signaling, leading to excess production of reactive oxygen and nitrogen species (ROS/RNS). The resulting oxidative stress damages DNA, proteins, and lipids, thereby initiating vascular and organ complications [12]. Endogenous antioxidant defenses, particularly superoxide dismutase, catalase, and glutathione peroxidase, act as checkpoints against radical accumulation. These enzymes depend on essential trace elements, such as zinc, copper, manganese, selenium, and iron [11]. When trace element homeostasis is disrupted, antioxidant capacity declines, amplifying oxidative injury and predisposing the diabetic heart to cardiomyopathy. These trace element-dependent checkpoints play a pivotal role in detoxifying free radicals and maintaining redox homeostasis. When the balance of these micronutrients is disrupted in diabetes, antioxidant capacity declines, amplifying oxidative injury and accelerating the onset of complications [13, 14] (Fig. 7.1).

Diabetic Cardiomyopathy (DCM)

Diabetic cardiomyopathy (DCM) is a specific cardiac complication of diabetes mellitus characterized by structural and functional alterations of the myocardium independent of hypertension, valvular heart disease, or coronary artery disease. It is

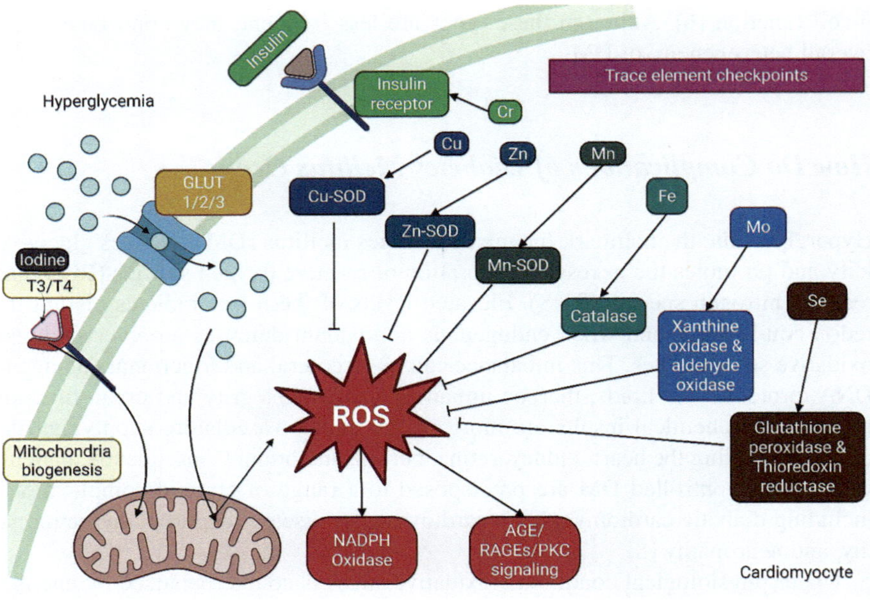

Fig. 7.1 From hyperglycemia to oxidative stress with trace element-dependent antioxidant checkpoints. Hyperglycemia induces glucotoxic pathways that increase ROS production, leading to oxidative stress and biomolecular damage via NADPH oxidase and AGE/RAGEs/PKC signaling. Endogenous antioxidant enzymes, such as Cu/Mn/Zn-SOD, catalase, xanthine oxidase, aldehyde oxidase, glutathione peroxidase, and thioredoxin reductase, dependent on trace elements, act as checkpoints that detoxify radicals and maintain redox balance. Iodine is involved in hormone thyroid (T3/T4) synthesis to support mitochondria biogenesis, and chromium enhances insulin receptor sensitivity

increasingly recognized as a major contributor to heart failure in diabetic patients, with prevalence rates estimated to be significantly higher among individuals with type 2 diabetes [15, 16]. The clinical manifestations of DCM range from early diastolic dysfunction to systolic impairment, ultimately leading to heart failure with preserved or reduced ejection fraction [17, 18].

The pathogenesis of DCM is multifactorial and involves several interrelated mechanisms. Chronic hyperglycemia induces the accumulation of advanced glycation end-products (AGEs), which cross-link with myocardial proteins and impair ventricular compliance [19–21]. Insulin resistance further contributes to altered myocardial metabolism, shifting cardiac energy utilization from glucose to fatty acids, thereby increasing oxygen consumption and mitochondrial workload [22–26]. In addition, low-grade inflammation and activation of pro-fibrotic signaling pathways promote extracellular matrix remodeling and myocardial fibrosis [27, 28].

Oxidative stress has been identified as a central mechanism in the progression of DCM. Excessive production of reactive oxygen species (ROS) such as hydrogen peroxide (OH^-), superoxide (O_2^-), and superoxide dianion (O_2^{2-}), driven by

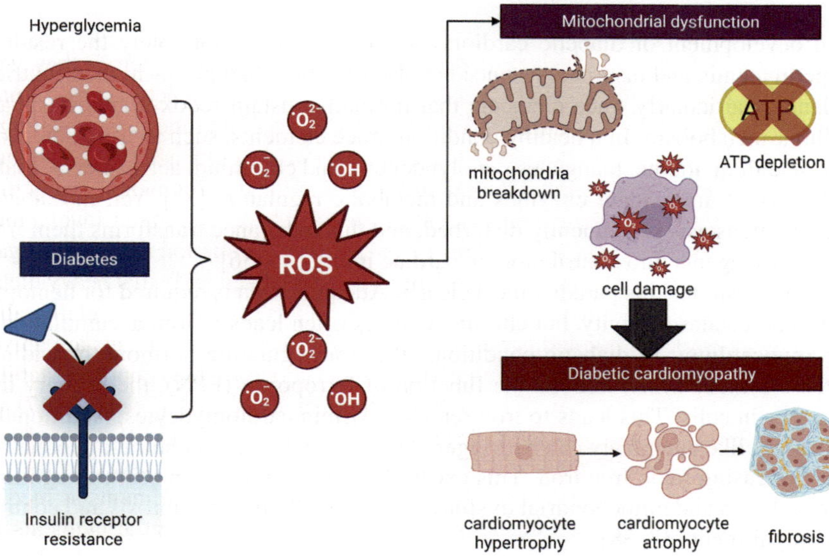

Fig. 7.2 Oxidative stress and mitochondrial dysfunction in the pathogenesis of diabetic cardiomyopathy. Hyperglycemia and insulin resistance promote excessive production of reactive oxygen species such as hydrogen peroxide (OH^-), superoxide (O_2^-), and superoxide dianion (O_2^{2-}), overwhelming antioxidant defenses and causing damage to cell components (DNA, proteins, and lipids). Concurrently, mitochondrial dysfunction impairs electron transport chain efficiency, reduces ATP generation, and disrupts calcium homeostasis, thereby amplifying oxidative injury to the mitochondrial membrane. The synergistic effect of these processes accelerates cardiomyocyte hypertrophy, apoptosis, and fibrosis, ultimately leading to the structural and functional changes characteristic of diabetic cardiomyopathy

mitochondrial dysfunction, NADPH oxidase activation, and impaired antioxidant defenses, leads to damage of cells (their lipids, proteins, and DNA) within cardiomyocytes [29, 30]. This redox imbalance disrupts calcium homeostasis, induces apoptosis, and exacerbates myocardial injury. Alongside oxidative stress, lipotoxicity and impaired autophagy also contribute to structural and functional deterioration of the diabetic heart [31–33]. This autophagy induces fibrotic tissue formation within the heart.

These mechanisms underscore that DCM arises from the interplay of metabolic dysregulation, oxidative stress, and mitochondrial dysfunction (Fig. 7.2). Understanding these causative mechanisms is essential to developing targeted interventions aimed at preventing or slowing the progression of this increasingly prevalent diabetic complication.

Role of Trace Elements in the Development of DCM

The development of diabetic cardiomyopathy (DCM) is not solely the result of hyperglycemia and insulin resistance but also of subtle disruptions in micronutrient balance, particularly trace elements that normally sustain redox homeostasis and cellular metabolism. In a healthy condition, trace elements, such as iron, zinc, copper, selenium, iodine, manganese, molybdenum, and chromium act as indispensable cofactors of antioxidant enzymes and metabolic regulators [34]. Yet, in diabetes, their homeostasis is frequently disturbed, and this imbalance transforms them from protective agents into contributors to cardiac injury [35, 36].

Iron illustrates this paradox most clearly. Adequate iron is required for hemoproteins and catalase activity, but chronic diabetes often leads to iron accumulation in the myocardium. In diabetic conditions, the iron-regulating hormone hepcidin is often elevated, which inhibits the function of ferroportin (FPN), the primary iron exporter in cells. This leads to iron retention within cardiomyocytes. Additionally, processes like autophagy can be triggered that degrade the iron-storage protein ferritin, releasing toxic free iron. This excess iron promotes ferroptosis through lipid peroxidation and mitochondrial dysfunction, gradually fueling fibrosis and contractile impairment [37, 38].

Zinc, by contrast, becomes problematic when deficient. Zinc is a critical cofactor for the antioxidant enzyme Cu/Zn-SOD that safeguards cells against oxidative stress, which converts harmful superoxide radicals (O_2^-) into less damaging hydrogen peroxide (H_2O_2) and oxygen [39]. A zinc deficiency severely impairs SOD activity, allowing reactive oxygen species (ROS) to accumulate and cause widespread oxidative damage to lipids, proteins, and DNA in cardiomyocytes [40]. Moreover, zinc is essential for the proper synthesis, storage, and release of insulin by pancreatic β-cells. Insufficient zinc leads to reduced insulin secretion and impaired glucose tolerance, hallmarks of diabetes [41]. Thus, zinc deficiency amplifies inflammatory responses, further exacerbating myocardial vulnerability [42].

Copper plays a similarly ambivalent role. In chronic diabetes, defective copper metabolism often leads to copper accumulation, particularly of the highly redox-active divalent copper ion in the cardiac extracellular matrix and/or within myocardial cells [43]. Copper overload, which can accompany hyperglycemic states, acts as a catalyst for redox cycling and hydroxyl radical formation. The result is mitochondrial DNA damage, protein oxidation, and eventual cardiomyocyte apoptosis [44, 45].

Selenium is a core component of glutathione peroxidase (GPx) enzymes. Without adequate selenium, the activity of these enzymes is significantly reduced, impairing their ability to neutralize harmful hydrogen peroxides and other reactive oxygen species (ROS). In addition, thioredoxin reductases are also selenoproteins that are crucial for regulating intracellular redox balance. Selenium deficiency compromises the activity of these enzymes, further weakening the cell's ability to cope with oxidative stress. Thus, selenium deficiency compounds by undermining GPx and thioredoxin reductase activity. Without these defenses, peroxides accumulate, mitochondria are damaged, and fibrotic pathways are activated, a cascade strongly associated with adverse cardiac function [46–48]. Even though there is no report about selenium's direct association to diabetic cardiomyopathy, many reports have highlighted that selenium deficiency can induce cardiomyopathy; therefore it can cause more severe DCM.

Other trace elements influence DCM more indirectly, but their impact is no less significant. Iodine, through its role in thyroid hormone synthesis, regulates metabolic rate and mitochondrial activity. Both deficiency and excess iodine can disturb thyroid homeostasis, driving metabolic stress that predisposes the diabetic heart to maladaptive remodeling [49, 50]. Manganese, as a cofactor of mitochondrial Mn-SOD, protects against mitochondrial superoxide accumulation; when manganese availability is insufficient, oxidative stress escalates, ATP production falls, and calcium handling deteriorates, thereby accelerating cardiac dysfunction [51]. People with diabetes showed lower significant manganese blood level associated with ischemic heart failure [52].

The association of molybdenum with diabetes is not well known recently, but some large population studies have found that higher molybdenum levels are associated with a greater prevalence of diabetes and insulin resistance, even at relatively low exposure levels found in the US population [53]. However, molybdenum has a key role in regulating enzymes, such as xanthine oxidase. There is report that disordered glucose levels are associated with xanthine oxidase activity in overweight type 2 diabetic women [54]. Dysregulated molybdenum metabolism may increase oxidase activity, inadvertently heightening ROS production and aggravating oxidative stress in cardiomyocytes [55, 56]. Lastly, chromium, long recognized for its role in potentiating insulin receptor activity, is central to glucose homeostasis. Its deficiency worsens hyperglycemia and lipotoxicity, thereby linking systemic metabolic imbalance directly to myocardial injury in diabetes [57, 58].

Taken together, these eight trace elements shape a complex network of interactions in which imbalance—whether deficiency or overload—contributes to the development of DCM. Through overlapping mechanisms of oxidative stress, mitochondrial dysfunction, inflammatory signaling, and metabolic dysregulation, they transform the diabetic myocardium into a vulnerable substrate for fibrosis, apoptosis, and contractile failure. Recognizing their role not only deepens our understanding of disease pathogenesis but also provides a rationale for considering micronutrient homeostasis in future preventive and therapeutic strategies. For clarity, their physiological roles, mechanisms of disruption, and consequences are summarized in Table 7.1 and Fig. 7.3.

Trace Element in the Management of DCM

Although disturbances in trace element homeostasis can promote the onset and progression of diabetic cardiomyopathy (DCM), adequate levels of these micronutrients exert protective effects that may be harnessed in disease management. Rather

Table 7.1 Summary of trace element imbalances and their consequences in DCM

No	Trace Eeement	Physiological role	Pathogenic disruption	Consequence in DCM
1	Iron (Fe)	Cofactor for hemoproteins and catalase; oxygen transport	Overload leads to ROS generation and ferroptosis	Myocardial iron deposition; fibrosis; apoptosis
2	Zinc (Zn)	Cofactor for Cu/ Zn-SOD; insulin signaling; anti-inflammatory	Deficiency reduces SOD activity; impairs insulin signaling	Oxidative stress; inflammation; cardiac dysfunction
3	Copper (Cu)	Cofactor for SOD, cytochrome c oxidase; redox balance	Excess promotes hydroxyl radical formation and mitochondrial injury	DNA/protein oxidation; cardiomyocyte apoptosis
4	Selenium (Se)	Cofactor for GPx and thioredoxin reductase; peroxide detoxification	Deficiency impairs peroxide detoxification; increases fibrosis	Mitochondrial injury; maladaptive remodeling
5	Iodine (I)	Essential for thyroid hormone synthesis and metabolic regulation	Deficiency/excess disturbs thyroid hormone balance; metabolic stress	Metabolic and mitochondrial stress; cardiac remodeling
6	Manganese (Mn)	Cofactor for Mn-SOD; mitochondrial antioxidant defense	Deficiency reduces Mn-SOD activity; increases mitochondrial ROS	Calcium mishandling; ATP reduction; cell death
7	Molybdenum (Mo)	Cofactor for xanthine oxidase and aldehyde oxidase; purine metabolism	Excess activity of xanthine oxidase enhances ROS production	Enhanced oxidative stress; cardiomyocyte injury
8	Chromium (Cr)	Enhances insulin receptor activity; regulates glucose and lipid metabolism	Deficiency worsens hyperglycemia and lipotoxicity	Metabolic dysregulation; myocardial injury

than acting as harmful agents, balanced concentrations of trace elements contribute to antioxidant defense, stabilization of mitochondrial function, and attenuation of inflammatory and fibrotic processes. This duality highlights their potential as important modulators in the pathogenesis and treatment of DCM.

Zinc is a clear example of this protective role. By serving as a cofactor for Cu/Zn-superoxide dismutase, zinc facilitates the removal of superoxide radicals before they inflict damage on cardiomyocytes. Adequate zinc levels not only restore antioxidant capacity but also improve insulin sensitivity and dampen inflammatory signaling, thereby reducing both metabolic and oxidative stress in the diabetic heart [59–61]. Selenium offers another well-established cardioprotective mechanism. Selenium supplementation can restore glutathione peroxidase activity in animal models of DbCM, strengthening the cellular antioxidant shield and preventing

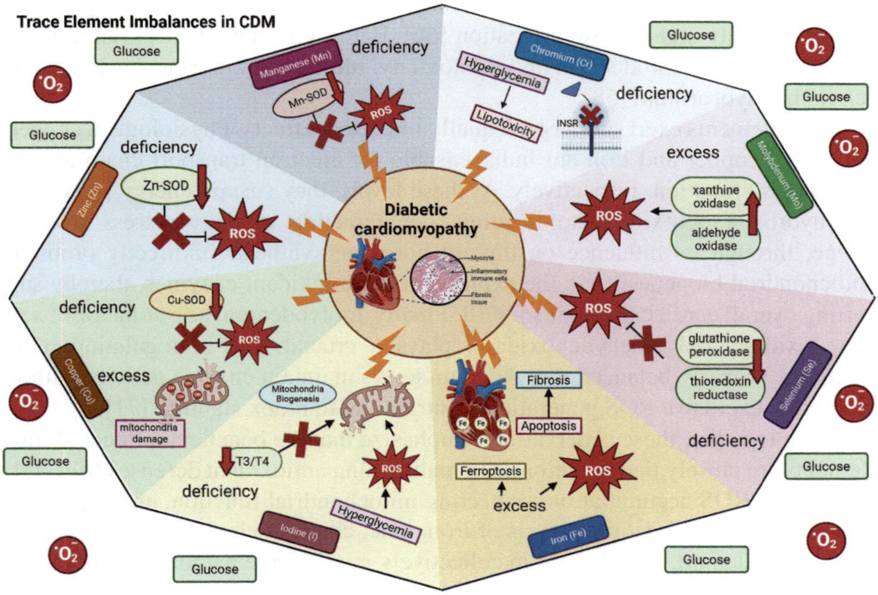

Fig. 7.3 Role of trace elements in the development of DCM. Imbalance of essential trace elements—including iron, zinc, copper, selenium, iodine, manganese, molybdenum, and chromium—contributes to the pathogenesis of DCM in a hyperglycemic environment. Iron overload promotes ferroptosis and lipid peroxidation, while copper excess triggers hydroxyl radical formation. Zinc and selenium deficiency weaken antioxidant defenses by reducing the activity of Cu/Zn-superoxide dismutase and glutathione peroxidase, respectively. Inadequate manganese impairs mitochondrial superoxide dismutase, leading to mitochondrial dysfunction, while iodine imbalance disrupts thyroid hormone-mediated regulation of cardiac metabolism. Chromium deficiency exacerbates insulin resistance and hyperglycemia, and molybdenum dysregulation alters xanthine oxidase activity, further enhancing ROS generation. Collectively, these disruptions converge on oxidative stress, mitochondrial impairment, and inflammation, which drive myocardial fibrosis, apoptosis, hypertrophy, and ultimately diabetic cardiomyopathy

oxidative damage. Incorporated into glutathione peroxidase and thioredoxin reductase, selenium strengthens peroxide detoxification and preserves mitochondrial integrity. Supplementation in experimental models has been associated with reduced myocardial fibrosis and improved ventricular function, underscoring its therapeutic promise [62–64].

Manganese also plays a central role through its action being indispensable cofactor for mitochondrial superoxide dismutase (Mn-SOD), the primary antioxidant enzyme that neutralizes the highly reactive superoxide radical within heart cell mitochondria. Adequate manganese availability ensures the efficient clearance of mitochondrial ROS, thereby sustaining ATP production and preventing calcium overload that could trigger contractile dysfunction. In this way, manganese fortifies mitochondrial resilience under diabetic stress to reduce CDM [65]. Chromium, although not a classical antioxidant, contributes indirectly by improving insulin receptor activity and glucose utilization [66]. Better glycemic control reduces glucose-induced ROS production and lipotoxicity, ultimately lowering oxidative burden in the myocardium [58].

Other elements exert subtler but equally important effects. Physiological concentrations of copper and iron are indispensable for electron transport chain activity and catalase function, respectively. At these levels, they sustain energy production and hydrogen peroxide detoxification, counteracting oxidative stress [67–69]. Iodine, through its influence on thyroid hormone synthesis, indirectly promotes mitochondrial biogenesis and the expression of antioxidant enzymes, thereby supporting overall cardiac metabolism [70]. Finally, molybdenum, as a cofactor of xanthine oxidase and aldehyde oxidase, plays a crucial role in regulating redox reactions. Properly balanced molybdenum levels maintain enzyme activity without promoting excessive ROS generation, contributing to redox stability [71].

Taken together, these observations emphasize that the beneficial effects of trace elements are closely tied to their role in maintaining antioxidant defenses (Fig. 7.4). By limiting ROS accumulation, protecting mitochondrial function, and supporting metabolic balance, adequate levels of iron, zinc, copper, selenium, iodine, manganese, molybdenum, and chromium collectively mitigate the pathological processes underlying DCM. This reinforces the idea that nutritional strategies and targeted supplementation, when carefully monitored, may serve as valuable adjuncts to conventional therapies for diabetic cardiomyopathy. The antioxidant-associated roles of trace elements for managing DCM are summarized in Table 7.2 and Fig. 7.4.

Conclusion

Diabetic cardiomyopathy (DCM) arises from a complex interplay of metabolic dysregulation, mitochondrial dysfunction, inflammation, and, most critically, oxidative stress. Persistent hyperglycemia drives excessive production of reactive oxygen and nitrogen species, overwhelming endogenous defenses and initiating molecular and structural damage within the myocardium.

Antioxidant-associated trace elements—including iron, zinc, copper, selenium, iodine, manganese, molybdenum, and chromium—occupy a pivotal position in this process. As essential cofactors of enzymes, such as superoxide dismutase, catalase, and glutathione peroxidase, these elements safeguard redox equilibrium and limit oxidative injury. When their homeostasis is disrupted, antioxidant defenses weaken, amplifying oxidative stress and accelerating the onset of DCM. Conversely, maintaining optimal levels of these elements strengthens antioxidant capacity, preserves mitochondrial integrity, reduces inflammation, and mitigates adverse cardiac remodeling.

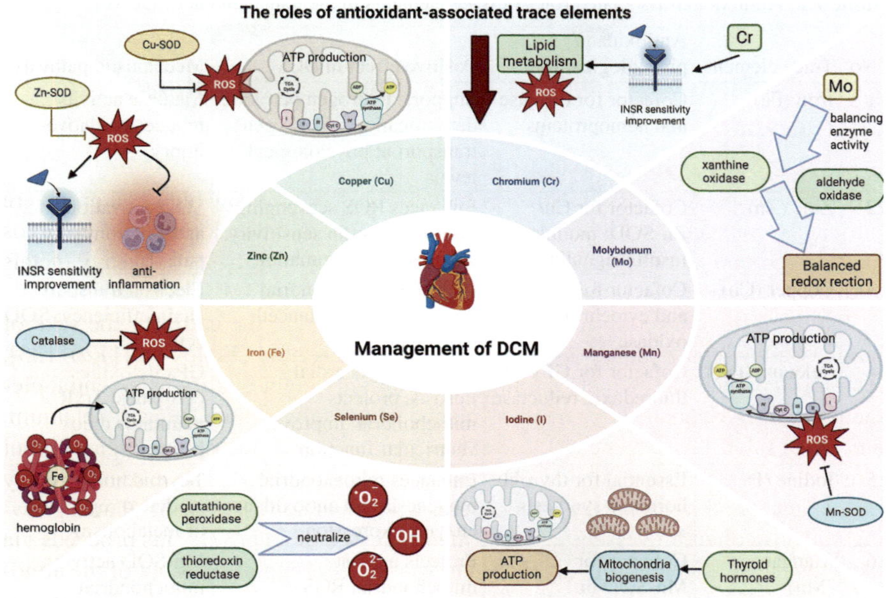

Fig. 7.4 Role of trace elements in management of DCM. Adequate levels of essential trace elements support endogenous antioxidant systems and mitigate diabetic cardiac injury. Zinc enhances Zn-superoxide dismutase activity, improving ROS scavenging, insulin receptor sensitivity and reducing inflammation. Selenium strengthens glutathione peroxidase and thioredoxin reductase to neutralize free radical. Manganese sustains mitochondrial superoxide dismutase activity, maintaining ATP production. Similar to copper, copper enhances Cu-superoxide dismutase for inhibiting ROS and helping ATP production. Chromium improves insulin sensitivity and glycemic control, thereby indirectly reducing lipotoxicity. Physiological concentrations of iron are required for catalase and hemoglobin function, ensuring efficient electron transport for ATP production and ROS detoxification. Iodine contributes through thyroid hormone-mediated regulation of metabolism to induce mitochondria biogenesis, while molybdenum helps maintain redox balance via controlled activity of xanthine oxidase. Collectively, these trace elements strengthen antioxidant defenses, preserve mitochondrial integrity, and attenuate adverse oxidative stress, thereby exerting a cardioprotective influence in diabetic cardiomyopathy

Collectively, the evidence underscores that antioxidant-associated trace elements are not peripheral but central determinants of both the pathogenesis and management of diabetic cardiomyopathy. Future research should aim to define precise micronutrient thresholds, establish evidence-based supplementation strategies, and integrate antioxidant-focused trace element management with conventional therapies. Such approaches hold promise for reducing the burden of oxidative stress, slowing disease progression, and improving outcomes in patients with diabetes.

Table 7.2 Antioxidant-associated roles of trace elements in the management of DCM

No	Trace element	Antioxidant/ physiological role	Positive effect in DCM	Mechanistic pathway
1	Iron (Fe)	Cofactor for catalase and hemoproteins	Supports hydrogen peroxide detoxification and oxygen transport at physiological levels	Catalase activity reduces oxidative injury
2	Zinc (Zn)	Cofactor for Cu/ Zn-SOD; modulates insulin signaling	Enhances ROS scavenging, improves insulin sensitivity, suppresses inflammation	SOD activation; anti-inflammatory pathways
3	Copper (Cu)	Cofactor for SOD and cytochrome c oxidase	Promotes mitochondrial respiration and balanced redox state	Electron transport chain efficiency; SOD activity
4	Selenium (Se)	Cofactor for GPx and thioredoxin reductase	Reduces myocardial fibrosis, protects mitochondria, improves ventricular function	GPx detoxifies peroxides; TrxR maintains redox balance
5	Iodine (I)	Essential for thyroid hormone synthesis	Enhances mitochondrial biogenesis and antioxidant enzyme expression	Thyroid hormone-mediated regulation of metabolism
6	Manganese (Mn)	Cofactor for Mn-SOD in mitochondria	Protects against mitochondrial ROS, maintains ATP production, prevents Ca^{2+} overload	Mn-SOD activity; mitochondrial homeostasis
7	Molybdenum (Mo)	Cofactor for xanthine oxidase and aldehyde oxidase	Maintains controlled enzyme activity without excessive ROS	Balanced redox enzyme regulation
8	Chromium (Cr)	Enhances insulin receptor activity	Improves glycemic control, reduces ROS and lipotoxicity	Insulin signaling; reduced glucose-induced ROS

References

1. International Diabetes Federation. IDF Atlas 11th Edition 2025 [Internet]. 11th ed. International Diabetes Federation; 2025. 130 p. Available from: diabetesatlas.org
2. Zou H, Xie L, Hu J, Zhang R, Wang Y, Xu A et al (2025) Fatty acid binding protein 4 regulates the antigen-presenting function of dendritic cells resulting in T cell priming in streptozotocin-induced type 1 diabetes mice. J Diabetes 17(7):e70123
3. Adeghate E, Schattner P, Dunn E (2006) An update on the etiology and epidemiology of diabetes mellitus. Ann N Y Acad Sci 1084:1–29
4. Mahgoub MO, Ali II, Adeghate JO, Tekes K, Kalász H, Adeghate EA (2023) An update on the molecular and cellular basis of pharmacotherapy in type 2 diabetes mellitus. Int J Mol Sci 24(11):9328
5. QuickStats (2023) Percentage of Mothers with Gestational Diabetes,* by Maternal Age—National Vital Statistics System, United States, 2016 and 2021. MMWR Morb Mortal Wkly Rep 72(1):16
6. Salguero MV, Arosemena M, Pollin T, Greeley SAW, Naylor RN, Letourneau-Freiberg L et al (2023) Monogenic forms of diabetes. In: Lawrence JM, Casagrande SS, Herman WH, Wexler DJ, Cefalu WT (eds) Diabetes in America [Internet]. National Institute of Diabetes and

Digestive and Kidney Diseases (NIDDK), Bethesda (MD). [cited 2025 Sep 2]. Available from: http://www.ncbi.nlm.nih.gov/books/NBK597414/
7. Chandimali N, Bak SG, Park EH, Lim HJ, Won YS, Kim EK et al (2025) Free radicals and their impact on health and antioxidant defenses: a review. Cell Death Discov 11(1):19
8. Adeghate EA, Kalász H, Al Jaberi S, Adeghate J, Tekes K (2021) Tackling type 2 diabetes-associated cardiovascular and renal comorbidities: a key challenge for drug development. Expert Opin Investig Drugs 30(2):85–93
9. Brigelius-Flohé R, Maiorino M (2013) Glutathione peroxidases. Biochim Biophys Acta BBA—Gen Subj 1830(5):3289–3303
10. Valko M, Jomova K, Rhodes CJ, Kuča K, Musílek K (2016) Redox- and non-redox-metal-induced formation of free radicals and their role in human disease. Arch Toxicol 90(1):1–37
11. Wołonciej M, Milewska E, Roszkowska-Jakimiec W (2016) Trace elements as an activator of antioxidant enzymes. Postępy Hig Med Dośw 70:1483–1498
12. Singh A, Kukreti R, Saso L, Kukreti S (2022) Mechanistic insight into oxidative stress-triggered signaling pathways and type 2 diabetes. Molecules 27(3):950
13. Lin CC, Huang HH, Hu CW, Chen BH, Chong IW, Chao YY et al (2014) Trace elements, oxidative stress and glycemic control in young people with type 1 diabetes mellitus. J Trace Elem Med Biol 28(1):18–22
14. Jia MJ, Chen L (2024) Effect of trace elements and nutrients on diabetes and its complications: a Mendelian randomization study. Front Nutr 11:1439217
15. Oktay AA, Paul TK, Koch CA, Lavie CJ (2000) Diabetes, cardiomyopathy, and heart failure. In: Feingold KR, Ahmed SF, Anawalt B, Blackman MR, Boyce A, Chrousos G et al (eds) Endotext [Internet]. MDText.com, Inc., South Dartmouth (MA). [cited 2025 Sep 2]. Available from: http://www.ncbi.nlm.nih.gov/books/NBK560257/
16. Huo JL, Feng Q, Pan S, Fu WJ, Liu Z, Liu Z (2023) Diabetic cardiomyopathy: early diagnostic biomarkers, pathogenetic mechanisms, and therapeutic interventions. Cell Death Discov 9(1):256
17. Jia G, Hill MA, Sowers JR (2018) Diabetic cardiomyopathy: an update of mechanisms contributing to this clinical entity. Circ Res 122(4):624–638
18. Zhao X, Liu S, Wang X, Chen Y, Pang P, Yang Q et al (2022) Diabetic cardiomyopathy: clinical phenotype and practice. Front Endocrinol 13:1032268
19. Hegab Z (2012) Role of advanced glycation end products in cardiovascular disease. World J Cardiol 4(4):90
20. Pappachan JM, Varughese GI, Sriraman R, Arunagirinathan G (2013) Diabetic cardiomyopathy: pathophysiology, diagnostic evaluation and management. World J Diabetes 4(5):177–189
21. Salvatore T, Pafundi PC, Galiero R, Albanese G, Di Martino A, Caturano A et al (2021) The diabetic cardiomyopathy: the contributing pathophysiological mechanisms. Front Med 8:695792
22. Abel ED, O'Shea KM, Ramasamy R (2012) Insulin resistance: metabolic mechanisms and consequences in the heart. Arterioscler Thromb Vasc Biol 32(9):2068–2076
23. Witteles RM, Fowler MB (2008) Insulin-resistant cardiomyopathy. J Am Coll Cardiol 51(2):93–102
24. Kim JA, Wei Y, Sowers JR (2008) Role of mitochondrial dysfunction in insulin resistance. Circ Res 102(4):401–414
25. Karwi QG, Sun Q, Lopaschuk GD (2021) The contribution of cardiac fatty acid oxidation to diabetic cardiomyopathy severity. Cells 10(11):3259
26. De Las HN, Lahera V (2022) Relevance of mitochondrial dysfunction in heart disease associated with insulin resistance conditions. Pflüg Arch—Eur J Physiol 474(1):21–31
27. Russo I, Frangogiannis NG (2016) Diabetes-associated cardiac fibrosis: cellular effectors, molecular mechanisms and therapeutic opportunities. J Mol Cell Cardiol 90:84–93
28. Ramesh P, Yeo JL, Brady EM, McCann GP (2022) Role of inflammation in diabetic cardiomyopathy. Ther Adv Endocrinol Metab 13:20420188221083530

29. Xiong Z, Liao Y, Zhang Z, Wan Z, Liang S, Guo J (2025) Molecular insights into oxidative-stress-mediated cardiomyopathy and potential therapeutic strategies. Biomolecules. 15(5):670
30. Ye Z, Wang S, Wan Z, Huang B, Guo J (2025) Targeting NADPH oxidase-driven oxidative stress in diabetic cardiomyopathy: mechanisms and therapeutic perspectives. Front Pharmacol 16:1610429
31. Van De Weijer T, Schrauwen-Hinderling VB, Schrauwen P (2011) Lipotoxicity in type 2 diabetic cardiomyopathy. Cardiovasc Res 92(1):10–18
32. Moris D, Spartalis M, Spartalis E, Karachaliou GS, Karaolanis GI, Tsourouflis G et al (2017) The role of reactive oxygen species in the pathophysiology of cardiovascular diseases and the clinical significance of myocardial redox. Ann Transl Med 5(16):326
33. Li J, Xie Y, Zheng S, He H, Wang Z, Li X et al (2024 Jun) Targeting autophagy in diabetic cardiomyopathy: from molecular mechanisms to pharmacotherapy. Biomed Pharmacother 175:116790
34. Wechselberger C, Messner B, Bernhard D (2023) The role of trace elements in cardiovascular diseases. Toxics 11(12):956
35. Hasanato RM (2020) Trace elements in type 2 diabetes mellitus and their association with glycemic control. Afr Health Sci 20(1):287–293
36. Tyczyńska M, Hunek G, Kawecka W, Brachet A, Gędek M, Kulczycka K et al (2024) Association between serum concentrations of (certain) metals and type 2 diabetes mellitus. J Clin Med 13(23):7443
37. Tian M, Huang X, Li M, Lou P, Ma H, Jiang X et al (2024) Ferroptosis in diabetic cardiomyopathy: from its mechanisms to therapeutic strategies. Front Endocrinol 15:1421838
38. Wang M, Mo D, Zhang N, Yu H (2024) Ferroptosis in diabetic cardiomyopathy: advances in cardiac fibroblast-cardiomyocyte interactions. Heliyon. 10(15):e35219
39. Marreiro D d N, Cruz KJC, Morais JBS, Beserra JB, Severo JS, de Oliveira ARS (2017) Zinc and oxidative stress: current mechanisms. Antioxidants (Basel) 6(2):24
40. Lee SR (2018) Critical role of zinc as either an antioxidant or a prooxidant in cellular systems. Oxidative Med Cell Longev 2018:9156285
41. Ahmad R, Shaju R, Atfi A, Razzaque MS (2024) Zinc and diabetes: a connection between micronutrient and metabolism. Cells 13(16):1359
42. Rosenblum H, Wessler JD, Gupta A, Maurer MS, Bikdeli B (2020) Zinc deficiency and heart failure: a systematic review of the current literature. J Card Fail 26(2):180–189
43. Cui X, Wang Y, Liu H, Shi M, Wang J, Wang Y (2022) The molecular mechanisms of defective copper metabolism in diabetic cardiomyopathy. Oxidative Med Cell Longev 2022:5418376
44. Pan M, Cheng Z w, Guang HC, Qing YZ, Jun SL, Chen H et al (2022) Long-term exposure to copper induces mitochondria-mediated apoptosis in mouse hearts. Ecotoxicol Environ Saf 234:113329
45. Wang YM, Feng LS, Xu A, Ma XH, Zhang MT, Zhang J (2024) Copper ions: the invisible killer of cardiovascular disease (review). Mol Med Rep 30(5):210
46. Saliba W, El Fakih R, Shaheen W (2010) Heart failure secondary to selenium deficiency, reversible after supplementation. Int J Cardiol 141(2):e26–e27
47. Munguti CM, Al Rifai M, Shaheen W (2017) A rare cause of cardiomyopathy: a case of selenium deficiency causing severe cardiomyopathy that improved on supplementation. Cureus 9(8):e1627
48. Shimada BK, Alfulaij N, Seale LA (2021) The impact of selenium deficiency on cardiovascular function. Int J Mol Sci 22(19):10713
49. Yamakawa H, Kato TS, Noh JY, Yuasa S, Kawamura A, Fukuda K et al (2021) Thyroid hormone plays an important role in cardiac function: from bench to bedside. Front Physiol 12:606931
50. Inoue K, Guo R, Lee ML, Neverova NV, Ebrahimi R, Currier JW et al (2023) Iodine-induced hypothyroidism and long-term risks of incident heart failure. J Am Heart Assoc 12(20):e030511
51. Li L, Yang X (2018) The essential element manganese, oxidative stress, and metabolic diseases: links and interactions. Oxidative Med Cell Longev 2018:7580707

52. Koh ES, Kim SJ, Yoon HE, Chung JH, Chung S, Park CW et al (2014) Association of blood manganese level with diabetes and renal dysfunction: a cross-sectional study of the Korean general population. BMC Endocr Disord 14(1):24
53. Menke A, Guallar E, Cowie CC (2016) Metals in Urine and Diabetes in U.S. Adults. Diabetes 65(1):164–171
54. Hernandez-Hernandez ME, Torres-Rasgado E, Pulido-Perez P, Nicolás-Toledo L, Martínez-Gómez M, Rodríguez-Antolín J et al (2022) Disordered glucose levels are associated with xanthine oxidase activity in overweight type 2 diabetic women. Int J Mol Sci 23(19):11177
55. Berry CE, Hare JM (2004) Xanthine oxidoreductase and cardiovascular disease: molecular mechanisms and pathophysiological implications. J Physiol 555(Pt 3):589–606
56. Kotozaki Y, Satoh M, Nasu T, Tanno K, Tanaka F, Sasaki M (2023) Human plasma xanthine oxidoreductase activity in cardiovascular disease: evidence from a population-based study. Biomedicine 11(3):754
57. Havel PJ (2004) A scientific review: the role of chromium in insulin resistance. Diabetes Educ 2:2–14
58. Gossa Al-Saadde DL, Haider AM, Ali A, Abdu Musad Saleh E, Turki Jalil A, Abdulelah FM et al (2023) The role of chromium supplementation in cardiovascular risk factors: a comprehensive reviews of putative molecular mechanisms. Heliyon 9(9):e19826
59. Li B, Tan Y, Sun W, Fu Y, Miao L, Cai L (2013) The role of zinc in the prevention of diabetic cardiomyopathy and nephropathy. Toxicol Mech Methods 23(1):27–33
60. MacKenzie S, Bergdahl A (2022) Zinc homeostasis in diabetes mellitus and vascular complications. Biomedicine 10(1):139
61. Cai L, Tan Y, Watson S, Wintergerst K (2023) Diabetic cardiomyopathy—zinc preventive and therapeutic potentials by its anti-oxidative stress and sensitizing insulin signaling pathways. Toxicol Appl Pharmacol 477:116694
62. Ayaz M, Can B, Ozdemir S, Turan B (2002) Protective effect of selenium treatment on diabetes-induced myocardial structural alterations. Biol Trace Elem Res 89(3):215–226
63. Okatan EN, Tuncay E, Turan B (2013) Cardioprotective effect of selenium via modulation of cardiac ryanodine receptor calcium release channels in diabetic rat cardiomyocytes through thioredoxin system. J Nutr Biochem 24(12):2110–2118
64. Zhu H, Wang X, Meng X, Kong Y, Li Y, Yang C et al (2022) Selenium supplementation improved cardiac functions by suppressing DNMT2-mediated GPX1 promoter DNA methylation in AGE-induced heart failure. Oxid Med Cell Longev 2022:5402997
65. Shen X, Zheng S, Metreveli NS, Epstein PN (2006) Protection of cardiac mitochondria by overexpression of MnSOD reduces diabetic cardiomyopathy. Diabetes 55(3):798–805
66. Hua Y, Clark S, Ren J, Sreejayan N (2012) Molecular mechanisms of chromium in alleviating insulin resistance. J Nutr Biochem 23(4):313–319
67. Tavsan Z, Ayar KH (2013) The effect of iron and copper as an essential nutrient on mitochondrial electron transport system and lipid peroxidation in trichoderma harzianum. Appl Biochem Biotechnol 170(7):1665–1675
68. Yan F, Li K, Xing W, Dong M, Yi M, Zhang H (2022) Role of iron-related oxidative stress and mitochondrial dysfunction in cardiovascular diseases. Oxidative Med Cell Longev 2022:5124553
69. Fu Y, Hou L, Han K, Zhao C, Hu H, Yin S (2025) The physiological role of copper: dietary sources, metabolic regulation, and safety concerns. Clin Nutr 48:161–179
70. Weitzel JM, Alexander IK (2011) Coordination of mitochondrial biogenesis by thyroid hormone. Mol Cell Endocrinol 342(1–2):1–7
71. Adamus JP, Ruszczyńska A, Wyczałkowska-Tomasik A (2024) Molybdenum's role as an essential element in enzymes catabolizing redox reactions: a review. Biomolecules 14(7):869

Chapter 8
Selenium and the Heart: Molecular Insights into Cardioprotection: A Review

Frank Christopher Howarth, Shamsa M. Baniyas, Mahra H. Alderei, Dana H. Aldhaheri, and Jaipaul Singh

Abstract Selenium is an essential trace element critical for cardiovascular health through its incorporation into antioxidant selenoproteins, such as glutathione peroxidase and thioredoxin reductase. The aim of this review is to explore the multifaceted role of selenium in maintaining cardiac structure and function, emphasizing its influence on redox homeostasis, ion channel regulation, and myocardial integrity. Selenium deficiency is linked to Keshan disease and other cardiomyopathies, with experimental models demonstrating impaired antioxidant defenses, mitochondrial dysfunction, and contractile failure. Supplementation restores antioxidant capacity, improves cardiac energetics, and reduces ischemia-reperfusion injury. Selenium also modulates calcium and potassium currents in cardiomyocytes, influencing action potential duration and arrhythmogenic risk. Electrocardiographic changes, including QT prolongation and low-voltage QRS complexes, are observed in selenium-deficient states and are reversible with selenium repletion. Studies using isolated perfused hearts and cardiac muscle preparations further elucidate selenium's protective effects against oxidative stress and electrophysiological remodeling. Normal selenium concentrations in serum and myocardium are essential for optimal cardiac performance, while both deficiency and excess can be detrimental. Emerging therapeutic strategies, including selenium-based nanotechnology, highlight its potential in managing cardiovascular disease. Overall, selenium is a vital micronutrient in cardiac physiology, with implications for disease prevention and targeted intervention (200 words).

Keywords Selenium · Heart · Electrophysiology

F. C. Howarth (✉) · S. M. Baniyas · M. H. Alderei · D. H. Aldhaheri
Department of Physiology, College of Medicine & Health Sciences, UAE University, Al Ain, UAE
e-mail: chris.howarth@uaeu.ac.ae

J. Singh
School of Pharmacy and Biomedical Sciences, University of Central Lancashire, Preston, UK

N. S. Dhalla et al. (eds.), *Functional Biochemistry of Metallic Elements*, Advances in Biochemistry in Health and Disease 36,
https://doi.org/10.1007/978-3-032-14445-4_8

Introduction

Selenium is an essential trace element with profound biological significance for cardiovascular health. As a critical cofactor for glutathione peroxidase and other selenoproteins, selenium is integral to maintaining cellular redox homeostasis, protecting cardiomyocytes against oxidative damage, and modulating inflammatory pathways [1–3]. Early epidemiological observations linked selenium deficiency to endemic cardiomyopathy, notably Keshan disease in selenium-deficient regions of China, where high rates of heart failure and characteristic electrocardiographic abnormalities were observed [4]. Experimental studies using isolated perfused heart preparations have demonstrated that selenium deficiency severely impairs glutathione peroxidase activity, diminishes the detoxification of peroxides, and heightens susceptibility to oxidative stress-induced contractile dysfunction [1, 5]. Conversely, selenium supplementation restores antioxidant capacity, preserves myocardial energetics, and attenuates ischemia-reperfusion injury, as evidenced by improved contractile recovery and reduced lipid peroxidation in experimental models [6]. Beyond its antioxidant role, selenium influences myocardial ion channel function and calcium handling, thereby affecting action potential propagation and contractility. High-dose selenium administration has been shown to exert dose-dependent negative inotropic effects in isolated perfused heart, underscoring the narrow therapeutic window between deficiency and toxicity [7]. In clinical contexts, selenium status has been implicated in diverse cardiovascular disorders, including dilated cardiomyopathy, heart failure following myocardial infarction, and atrial fibrillation [8, 9]. These findings highlight selenium's multifaceted role in cardiac structure and function, as well as the potential for targeted selenium supplementation to mitigate cardiac injury in select patient populations. This review summarizes the experimental and clinical evidence on selenium's effects on the heart, with emphasis on its impact on oxidative stress, myocardial energetics, electrophysiology, and cardiovascular outcomes.

Roles of Selenium in the Body

Selenium is an essential trace element that plays critical roles in numerous physiological processes, primarily through its incorporation into selenoproteins, which are vital for antioxidant defense, thyroid hormone metabolism, and immune function (Fig. 8.1). Selenoproteins, such as glutathione peroxidases (GPx), thioredoxin reductases, and iodothyronine deiodinases, require selenium for their enzymatic activity, contributing to the protection of cells from oxidative stress and the regulation of redox homeostasis [3]. Selenium is also involved in immune responses, enhancing both innate and adaptive immunity and influencing the proliferation and differentiation of T cells [10]. Furthermore, it plays a role in thyroid health by regulating thyroid hormone synthesis and conversion [11]. Selenium deficiency has

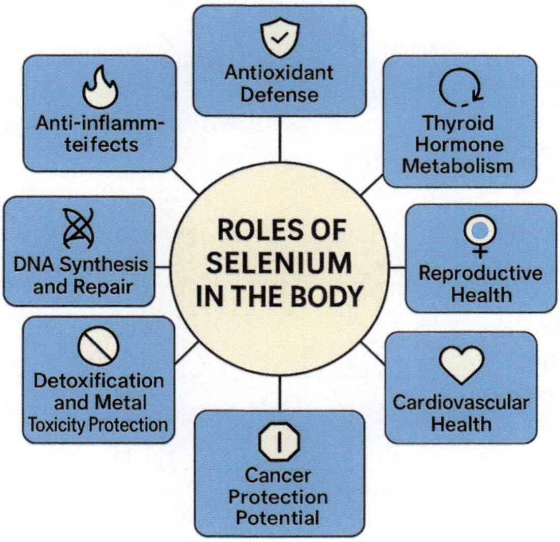

Fig. 8.1 Roles of selenium in the body

been associated with a range of disorders, including Keshan disease, a cardiomyopathy observed in selenium-deficient regions of China, as well as immune dysfunction and cognitive decline [12, 13]. Conversely, excessive selenium intake can lead to selenosis, characterized by gastrointestinal disturbances, hair loss, and neurological abnormalities, underscoring the importance of maintaining optimal selenium levels [14].

Sources of Dietary Selenium

Selenium occurs in a variety of dietary sources, with its concentration largely influenced by soil selenium content, geographic origin, and food processing. Rich natural sources of selenium include Brazil nuts, seafood, organ meats, cereals, and dairy products. Brazil nuts are among the most concentrated sources, containing up to 512 μg/g, while other significant contributors include meats (especially beef kidney), poultry, eggs, fish, and whole grains [15]. Table 8.1 shows the selenium content in different types of foods. In regions with selenium-deficient soils, such as parts of Europe and the Middle East, plant-based foods often contain low levels of selenium, leading to inadequate dietary intake [16]. Animal-derived foods generally offer higher bioavailability due to the predominance of organic selenium forms such as selenocysteine and selenomethionine [13, 15]. Additionally, selenium-fortified foods and supplements, including selenium-enriched yeast and infant formulas, are used to address deficiencies in at-risk populations [16]. The variability in selenium content across food items underscores the importance of considering regional

Table 8.1 Selenium content in selected foods

Source of selenium	Selenium content (μg/g)
Brazil nuts	0.2–512
Yeast	500–4000
Bread	0.09–0.20
Fish	0.06–0.63
Eggs	0.09–0.25
Chicken	0.15
Beef	0.01–0.73
Pork	0.27–0.35
Broccoli	0.012
Milk	0.01–0.06
Chocolate	0.04
Liver	0.3–0.4
Beef kidney	1.45

Adapted from Kieliszek [15]

agricultural and dietary practices when assessing selenium intake. Livestock diets are often supplemented to ensure consistent selenium content in meat and dairy, reducing variability due to environmental factors [15]. Overall, a mixed diet that includes both plant and animal sources can help maintain adequate selenium status, especially in regions with variable soil selenium levels.

Recommended Dietary Allowance (RDA) for Selenium

The Recommended Dietary Allowance (RDA) for selenium varies by age, sex, and geographic region, reflecting differences in environmental selenium availability and dietary patterns. According to the World Health Organization, the general RDA for adults is 55 μg/day, with a tolerable upper intake level of 400 μg/day to prevent toxicity. The Food and Nutrition Board in the United States recommends 40–70 μg/day for men and 45–55 μg/day for women, while the UK suggests slightly higher intakes, 60 μg/day for women and 75 μg/day for men. For children, an intake of 25 μg/day is considered adequate. Selenium intake above 100–200 μg/day has been associated with reduced genetic damage and enhanced immune responses, including increased cytotoxic T cell and natural killer cell activity. Table 8.2 shows the RDAs for selenium in different age groups. The dose that does not appear to show any adverse effect for adults is estimated at 800 μg/day, whereas a dose that can cause the onset of toxicity ranges from 1540 to 1600 μg/day, with symptoms ranging from gastrointestinal disturbances to neurological effects. The selenium content of food varies significantly by geographic region due to differences in soil selenium levels, which underscores the importance of region-specific dietary

Table 8.2 Recommended Daily Allowances (RDAs) for selenium

Age (years)	Males and females selenium (μg/day)
1–3	15–20
4–13	30–40
14–50	55–70
51+	70–100

Adapted from Kieliszek [15]

recommendations [15]. Based on the EFSA 2023 scientific opinion, the tolerable upper intake level (UL) for selenium in adults, including pregnant and lactating women, has been revised to 255 μg/day [17]. This value is derived from a lowest-observed-adverse-effect-level (LOAEL) of 330 μg/day identified in the Selenium and Vitamin E Cancer Prevention Trial (SELECT), where alopecia was observed as a critical adverse effect of excess selenium intake. An uncertainty factor of 1.3 was applied to account for variability and uncertainty in the data. For children, ULs were scaled allometrically based on body weight. Overall, the risk of exceeding the UL through natural dietary sources alone is low in the general European population, except among individuals regularly consuming selenium-rich supplements or large quantities of Brazil nuts. Caution is advised in administering selenium supplements to toddlers and children, and use should be based on individual nutritional needs [17].

Selenium Uptake, Transport, and Homeostasis

Selenium is an essential trace element whose uptake, transport, and homeostasis are still not fully understood. Dietary selenium is ingested primarily in organic forms, such as selenomethionine and selenocysteine, or as inorganic selenite and selenate. Selenium absorption in the gut varies depending on its chemical form. Organic forms like selenomethionine are actively transported, while inorganic forms like selenate are absorbed through passive transport. The duodenum and cecum are major sites of absorption. Absorption occurs mainly in the small intestine via passive diffusion and active transport mechanisms, with selenomethionine showing the highest bioavailability [13]. Once absorbed, selenium is incorporated into selenoproteins through a complex biosynthetic pathway involving selenophosphate synthetase and the unique tRNA synthetase [18]. Selenoprotein P plays a central role in selenium transport, delivering selenium from the liver to peripheral tissues via the bloodstream [19]. The liver is the central organ for selenium regulation, and it produces excretory selenium forms to regulate whole-body selenium. Uptake of selenoprotein P by target cells is mediated by receptors, such as apolipoprotein E receptor 2 and megalin, enabling tissue-specific selenium distribution [20]. Homeostasis is achieved through a balance of dietary intake, hepatic metabolism, tissue redistribution, and renal excretion, with tight regulation ensuring adequate selenium for selenoprotein synthesis while preventing harmful accumulation. Disruption of this balance, as seen in deficiency or excess, can lead to oxidative stress, immune dysfunction, or selenosis, highlighting the importance of understanding the molecular mechanisms underlying selenium homeostasis.

Normal Concentrations of Selenium in Blood and Heart Tissues

Selenium is an essential trace element present in blood and tissues at tightly regulated concentrations necessary for maintaining cellular redox balance, thyroid hormone metabolism, and cardiac function. Selenium concentrations in blood, plasma, and serum serve as key biomarkers for assessing selenium status in human populations. In healthy adults, the reported reference ranges for plasma or serum selenium typically fall between 70 and 150 μg/L, although these values may vary slightly by region, assay method, and population studied. A comprehensive evaluation across several countries identified a global median serum selenium concentration of 78.3 (75.1–81.7) μg/L in men and 80.8 (77.8–83.9) μg in women [21]. In a Chinese population-based study for child-bearing-aged women, the reference range of plasma selenium and whole-blood selenium were 73.81–140.75 μg/L and 81.06–164.72 μg/L, respectively. In clinical contexts, selenium deficiency is often defined as plasma or serum selenium levels below 70 μg/L, which is associated with reduced glutathione peroxidase activity and increased risk of cardiomyopathy, particularly in selenium-deficient endemic regions such as those affected by Keshan disease [3, 22]. Accurate measurement of selenium in plasma or serum is therefore essential for evaluating nutritional adequacy, monitoring supplementation, and identifying populations at risk of deficiency.

Geographic variability in soil selenium content contributes to substantial regional differences in selenium status, as exemplified by endemic Keshan disease regions in China, where selenium levels may fall below 65 μg/L [4]. In animal studies, the protective effect of intracellular GSH against cardiac dysfunction in selenium-deficient neonatal rats and cultured fetal rat myocytes has been demonstrated [23]. Experimental supplementation studies have demonstrated that elevating dietary selenium from marginal levels (0.02 ppm) to higher intake (0.5–1.0 ppm) leads to proportional increases in both serum selenium and cardiac GPx activity, significantly improving resistance to oxidative stress [24]. Similarly, in clinical settings, selenium deficiency in serum has been associated with cardiomyopathy risk, as patients on long-term parenteral nutrition or restrictive diets, such as the ketogenic diet, may develop severe deficiency with serum levels <30 μg/L, predisposing to impaired myocardial function and arrhythmias [25]. Collectively, these data underscore the importance of maintaining adequate selenium concentrations in blood and cardiac tissue to sustain antioxidant defense systems and preserve cardiac health.

Effects of Selenium on the Structure of the Heart

Selenium, an essential trace element incorporated into selenoproteins, exerts significant influences on cardiac structure, particularly through its role in antioxidative defense and maintenance of myocardial cellular integrity. Selenium deficiency has been associated with ultrastructural cardiac alterations, including mitochondrial swelling, myofibrillar degeneration, and interstitial fibrosis, as demonstrated in both animal and human studies of Keshan disease, a cardiomyopathy endemic to selenium-deficient regions [4, 26]. Experimental selenium depletion in rats induces increased lipid peroxidation and reduces glutathione peroxidase activity, ultimately contributing to myocyte injury and impaired structural integrity [23, 27]. Conversely, selenium supplementation has been shown to protect against structural myocardial damage induced by oxidative stressors such as adriamycin, preserving mitochondrial membrane potential, adenosine triphosphate production, and the architecture of sarcomeric proteins [24, 26]. Emerging nanotechnology approaches, including gold-selenium core-shell nanocomposites, further illustrate the promise of selenium in preventing ischemia-reperfusion injury through targeted delivery and protection of mitochondrial structure and function [28]. Collectively, these findings highlight the essential role of selenium in safeguarding cardiac ultrastructure, preventing pathological remodeling, and maintaining functional integrity under both deficiency and oxidative stress conditions.

Effects of Selenium on the Electrocardiogram

Selenium plays a critical role in maintaining cardiac electrophysiological stability, as demonstrated across experimental and clinical studies. Early investigations in selenium-deficient rats revealed characteristic electrocardiographic abnormalities, including progressive T-wave elevation and marked S-T segment displacement, reflecting intrinsic myocardial injury in the absence of major electrolyte disturbances [29] and histopathological evidence of muscle degeneration [30]. Subsequent work has demonstrated that selenium deficiency impairs antioxidant defense by reducing glutathione peroxidase and thioredoxin reductase activity, which in turn promotes lipid peroxidation, mitochondrial dysfunction, and altered repolarization [23, 26]. Selenium deficiency in rats was shown to induce a 62% incidence of electrocardiographic abnormalities, such as sinus arrhythmias or extrasystole [23]. Selenium supplementation in rats resulted in small nonsignificant increases in ST, QT, R wave, R-R, and QRS [26]. Clinically, selenium deficiency has been associated with electrocardiographic changes, including low-amplitude QRS complexes, QT prolongation, and life-threatening arrhythmias, as reported in patients receiving parenteral nutrition or adhering to selenium-poor diets [25, 31]. Selenium supplementation in experimental models of ischemia-reperfusion injury and adriamycin-induced cardiomyopathy significantly improved electrocardiographic parameters and preserved myocardial contractility, supporting the therapeutic potential of selenium as a cardioprotective agent [24, 26, 28]. Overall, these findings underscore selenium's essential role in the preservation of myocardial excitability and conduction, highlighting the electrocardiogram as a sensitive marker of selenium-responsive cardiac dysfunction.

Effects of Selenium in the Isolated Perfused Heart and Cardiac Muscle Preparations

Experimental studies using isolated perfused heart and cardiac muscle preparations have provided detailed mechanistic insights into the cardioprotective effects of selenium and the deleterious consequences of selenium deficiency. In Langendorff-perfused rat hearts, selenium deficiency has been shown to reduce myocardial antioxidant capacity by decreasing glutathione peroxidase and thioredoxin reductase activity, thereby promoting oxidative damage, impairing contractile function, and sensitizing the myocardium to reperfusion injury [23, 26]. Selenium deficiency was shown to cause a significant reduction in left ventricle dP/dt in selenium-deficient rat heart [23]. Selenium supplementation, on the other hand, did not appear to significantly alter left ventricular systolic pressure, left ventricular diastolic pressure, +dP/dt, and -dP/dt [26]. Supplementation with selenium-enriched reperfusion solutions significantly improves postischemic recovery of contractility and preserves mitochondrial membrane potential, adenosine triphosphate levels, and calcium homeostasis [26, 32]. Selenium supplementation in isolated rat hearts exposed to adriamycin reduces mitochondrial dysfunction, attenuates oxidative stress markers, and preserves left ventricular developed pressure and compliance [24, 26]. Novel nanotechnology approaches incorporating selenium into targeted delivery systems have further demonstrated enhanced antioxidant effects and improved recovery of perfused hearts subjected to ischemia-reperfusion protocols [28]. Collectively, findings from isolated perfused heart preparations underscore selenium's critical role in protecting myocardial tissue from oxidative damage, supporting excitation-contraction coupling, and preserving both electrical and mechanical function.

Effects of Selenium on Cardiac Ion Channels

Experimental investigations have shown that selenium exerts significant modulatory effects on ion channels and ionic currents in cardiac myocytes, influencing electrophysiological stability and susceptibility to arrhythmia. In selenium-deficient states, reductions in glutathione peroxidase activity and increased oxidative stress may contribute to alterations in calcium handling and action potential duration [23, 26]. Patch-clamp studies in isolated rat cardiomyocytes have demonstrated that sodium selenite treatment modifies both calcium and potassium currents [33], and related experiments in diabetic rat hearts confirm selenium's role in restoring altered electrical activity and ionic currents [34, 35]. Interestingly, sodium selenite treatment increased rat blood glucose level and lowered plasma insulin level. Selenium treatment resulted in a slight prolongation of the action potential, with no significant effects on spontaneous contraction parameters and intracellular Ca^{2+} transients, and these effects were accompanied by marked alterations in the kinetics of both L-type calcium current (ICa,L) and a significant shift of its half-inactivation potential to more negative values without changes in current density. There was also a significantly faster inactivation of the transient outward potassium current (Ito), no shift in half-inactivation, without any change in its current density. There was also a significant inhibition in the inwardly rectifying potassium current (IK1) density in treated cardiomyocytes [33]. These electrophysiological alterations are functionally relevant, as demonstrated by selenium-induced increases in the total charge carried by calcium currents and decreases in potassium currents, thereby creating a substrate for afterdepolarizations and arrhythmias [34]. In diabetic rat hearts, selenium supplementation has been shown to partially restore impaired Ito and IK currents, shorten action potential duration, and improve calcium homeostasis, supporting its role as an antiarrhythmic micronutrient [34, 36]. Complementary modeling studies further indicate that selenium's effects on Na^+–Ca^{2+} exchanger currents contribute to the normalization of repolarization in diabetic myocardium, highlighting the importance of selenium in maintaining ionic balance and electrical stability in cardiac tissue [36]. Collectively, these findings emphasize that selenium is a critical determinant of cardiomyocyte ion channel function, acting via redox-sensitive mechanisms to regulate both calcium influx and potassium efflux during the cardiac action potential. Interestingly, in streptozotocin-induced diabetic rats, treatment with sodium selenite could reverse the prolongation in both action potential duration and twitch duration of electrically stimulated papillary muscle in diabetic rats and in cardiomyocytes cause significant increases in the diminished amplitudes of Ito and steady state outward K^+ current, suggesting the beneficial effects of sodium selenite treatment on the mechanical and electrical activities of the diabetic rat heart [35].

Selenium modulates cardiac ion channel activity and maintains electrophysiological stability, particularly under pathological conditions such as diabetes. By influencing calcium and potassium currents through redox-sensitive mechanisms, selenium not only preserves action potential characteristics but also mitigates arrhythmogenic risk. These findings underscore its therapeutic potential as an antiarrhythmic micronutrient and highlight the importance of adequate selenium status for optimal cardiac function.

Conclusions

In conclusion, selenium plays a pivotal role in preserving cardiac structure and function through its antioxidant, anti-inflammatory, and electrophysiological effects. Evidence from both experimental and clinical studies highlights the detrimental consequences of selenium deficiency, including oxidative damage, myocardial dysfunction, arrhythmia, and pathological remodeling. Conversely, selenium supplementation demonstrates cardioprotective benefits in models of oxidative stress and ischemia-reperfusion injury. By modulating ion channel activity and maintaining redox homeostasis, selenium supports cardiac excitability, contractility, and structural integrity. These findings underscore the potential of targeted selenium interventions in cardiovascular disease management while also emphasizing the need to avoid both deficiency and excess.

References

1. Xia YM, Hill KE, Burk RF (1985) Effect of selenium deficiency on hydroperoxide-induced glutathione release from the isolated perfused rat heart. J Nutr 115:733–742. https://doi.org/10.1093/jn/115.6.733
2. Hill KE, Taylor MA, Burk RF (1987) Influence of selenium deficiency on glutathione disulfide metabolism in isolated perfused rat heart. Biochim Biophys Acta 923:431–435. https://doi.org/10.1016/0304-4165(87)90051-1
3. Rayman MP (2012) Selenium and human health. Lancet 379:1256–1268. https://doi.org/10.1016/S0140-6736(11)61452-9
4. Lei C, Niu X, Ma X, Wei J (2011) Is selenium deficiency really the cause of Keshan disease? Environ Geochem Health 33:183–188. https://doi.org/10.1007/s10653-010-9331-9
5. Ytrehus K, Ringstad J, Myklebust R, Norheim G, Mjøs OD (1988) The selenium-deficient rat heart with special reference to tolerance against enzymatically generated oxygen radicals. Scand J Clin Lab Invest 48:289–295. https://doi.org/10.3109/00365518809167497
6. Turan B, Saini HK, Zhang M, Prajapati D, Elimban V, Dhalla NS (2005) Selenium improves cardiac function by attenuating the activation of NF-kappaB due to ischemia-reperfusion injury. Antioxid Redox Signal 7:1388–1397. https://doi.org/10.1089/ars.2005.7.1388
7. Franconi F, Manghi N, Giotti A, Martini F, Dini G (1980) Effect of selenium on the contractile force of isolated and perfused Guinea-pig heart. Acta Pharmacol Toxicol (Copenh) 46:98–104. https://doi.org/10.1111/j.1600-0773.1980.tb02427.x
8. Ardahanli I, Ozkan HI (2022) Comparison of serum selenium levels between patients with newly diagnosed atrial fibrillation and normal controls. Biol Trace Elem Res 200:3925–3931. https://doi.org/10.1007/s12011-022-03281-9
9. Thiele R, Wagner D, Gassel M, Winnefeld K, Pleissner J, Pfeifer R (1997) Selenium substitution in acute myocardial infarct. Med Klin (Munich) 92(Suppl 3):26–28. https://doi.org/10.1007/BF03041957
10. Hoffmann PR, Berry MJ (2008) The influence of selenium on immune responses. Mol Nutr Food Res 52:1273–1280. https://doi.org/10.1002/mnfr.200700330
11. Ventura M, Melo M, Carrilho F (2017) Selenium and thyroid disease: from pathophysiology to treatment. Int J Endocrinol 2017:1297658. https://doi.org/10.1155/2017/1297658
12. Hurst R, Hooper L, Norat T, Lau R, Aune D, Greenwood DC, Vieira R, Collings R, Harvey LJ, Sterne JA, Beynon R, Savović J, Fairweather-Tait SJ (2012) Selenium and prostate cancer:

systematic review and meta-analysis. Am J Clin Nutr 96:111–122. https://doi.org/10.3945/ajcn.111.033373
13. Fairweather-Tait SJ, Bao Y, Broadley MR, Collings R, Ford D, Hesketh JE, Hurst R (2011) Selenium in human health and disease. Antioxid Redox Signal 14:1337–1383. https://doi.org/10.1089/ars.2010.3275
14. Kieliszek M, Błażejak S (2013) Selenium: significance, and outlook for supplementation. Nutrition 29:713–718. https://doi.org/10.1016/j.nut.2012.11.012
15. Kieliszek M (2019) Selenium⁻Fascinating microelement, properties and sources in food. Molecules 24. https://doi.org/10.3390/molecules24071298
16. Stoffaneller R, Morse NL (2015) A review of dietary selenium intake and selenium status in Europe and the Middle East. Nutrients 7:1494–1537. https://doi.org/10.3390/nu7031494
17. Turck D, Bohn T, Castenmiller J, de Henauw S, Hirsch-Ernst KI, Knutsen HK, Maciuk A, Mangelsdorf I, McArdle HJ, Peláez C, Pentieva K, Siani A, Thies F, Tsabouri S, Vinceti M, Aggett P, Crous Bou M, Cubadda F, Ciccolallo L, de Sesmaisons Lecarré A, Fabiani L, Titz A, Naska A, EFSA Panel on Nutrition NvFaFAN (2023) Scientific opinion on the tolerable upper intake level for selenium. EFSA J 21:e07704. https://doi.org/10.2903/j.efsa.2023.7704
18. Labunskyy VM, Hatfield DL, Gladyshev VN (2014) Selenoproteins: molecular pathways and physiological roles. Physiol Rev 94:739–777. https://doi.org/10.1152/physrev.00039.2013
19. Schomburg L (2022) Selenoprotein P - selenium transport protein, enzyme and biomarker of selenium status. Free Radic Biol Med 191:150–163. https://doi.org/10.1016/j.freeradbiomed.2022.08.022
20. Burk RF, Hill KE (2015) Regulation of selenium metabolism and transport. Annu Rev Nutr 35:109–134. https://doi.org/10.1146/annurev-nutr-071714-034250
21. Hughes DJ, Fedirko V, Jenab M, Schomburg L, Méplan C, Freisling H, Bueno-de-Mesquita HB, Hybsier S, Becker NP, Czuban M, Tjønneland A, Outzen M, Boutron-Ruault MC, Racine A, Bastide N, Kühn T, Kaaks R, Trichopoulos D, Trichopoulou A, Lagiou P, Panico S, Peeters PH, Weiderpass E, Skeie G, Dagrun E, Chirlaque MD, Sánchez MJ, Ardanaz E, Ljuslinder I, Wennberg M, Bradbury KE, Vineis P, Naccarati A, Palli D, Boeing H, Overvad K, Dorronsoro M, Jakszyn P, Cross AJ, Quirós JR, Stepien M, Kong SY, Duarte-Salles T, Riboli E, Hesketh JE (2015) Selenium status is associated with colorectal cancer risk in the European prospective investigation of cancer and nutrition cohort. Int J Cancer 136:1149–1161. https://doi.org/10.1002/ijc.29071
22. Xia Y, Hill KE, Byrne DW, Xu J, Burk RF (2005) Effectiveness of selenium supplements in a low-selenium area of China. Am J Clin Nutr 81:829–834. https://doi.org/10.1093/ajcn/81.4.829
23. Okamoto T, Mizuta K, Takahashi T, Kishi T, Kitahara S, Komori S, Hashimoto K, Goshima K (1999) Protective effect of gamma-glutamylcysteinylethyl ester on dysfunction of the selenium-deficient rat heart. Biochem Pharmacol 57:955–963. https://doi.org/10.1016/s0006-2952(98)00361-x
24. Jotti A, Maiorino M, Paracchini L, Piccinini F, Ursini F (1994) Protective effect of dietary selenium supplementation on delayed cardiotoxicity of adriamycin in rat: is PHGPX but not GPX involved? Free Radic Biol Med 16:283–288. https://doi.org/10.1016/0891-5849(94)90154-6
25. Bank IM, Shemie SD, Rosenblatt B, Bernard C, Mackie AS (2008) Sudden cardiac death in association with the ketogenic diet. Pediatr Neurol 39:429–431. https://doi.org/10.1016/j.pediatrneurol.2008.08.013
26. Dursun N, Taşkın E, Yerer Aycan MB, Sahin L (2011) Selenium-mediated cardioprotection against adriamycin-induced mitochondrial damage. Drug Chem Toxicol 34:199–207. https://doi.org/10.3109/01480545.2010.538693
27. Matsuda A, Kimura M, Itokawa Y (1998) Influence of selenium deficiency on vital functions in rats. Biol Trace Elem Res 61:287–301. https://doi.org/10.1007/BF02789089
28. Sun Y, Zhang P, Li Y, Hou Y, Yin C, Wang Z, Liao Z, Fu X, Li M, Fan C, Sun D, Cheng L (2022) Light-activated gold-selenium Core-Shell nanocomposites with NIR-II photoacoustic imaging performances for heart-targeted repair. ACS Nano 16:18667–18681. https://doi.org/10.1021/acsnano.2c07311

29. Godwin KO (1965) Abnormal electrocardiograms in rats fed a low selenium diet. Q J Exp Physiol Cogn Med Sci 50:282–288. https://doi.org/10.1113/expphysiol.1965.sp001794
30. Godwin KO, Fraser FJ (1966) Abnormal electrocardiograms, blood pressure changes, and some aspects of the histopathology of selenium deficiency in lambs. Q J Exp Physiol Cogn Med Sci 51:94–102. https://doi.org/10.1113/expphysiol.1966.sp001844
31. Ryuge A, Kim H, Suzuki Y, Okazaki M, Kojima H, Saito S, Kato N, Ishimoto T, Kosugi T, Maruyama S, Mizuno M (2021) Refractory hypotension caused by selenium deficiency in a patient on peritoneal dialysis. Intern Med 60:2461–2464. https://doi.org/10.2169/internalmedicine.6632-20
32. Sinci V, Yildirim S, Turan B (1998) Effects of selenium-enriched reperfusion solutions on isolated Guinea pig hearts. Biol Trace Elem Res 65(3):243–259
33. Ayaz M, Ozdemir S, Yaras N, Vassort G, Turan B (2005) Selenium-induced alterations in ionic currents of rat cardiomyocytes. Biochem Biophys Res Commun 327:163–173. https://doi.org/10.1016/j.bbrc.2004.12.003
34. Ayaz M, Ozdemir S, Yaras N, Vassort G, Turan B (2004) Protective effects of selenium on diabetic heart: mechanical and electrical activity. Biochim Biophys Acta 1675:61–69
35. Ayaz M, Ozdemir S, Ugur M, Vassort G, Turan B (2004) Effects of selenium on altered mechanical and electrical cardiac activities of diabetic rat. Arch Biochem Biophys 426:83–90. https://doi.org/10.1016/j.abb.2004.03.030
36. Yaras N, Ugur M, Ozdemir S, Gurdal H, Purali N, Lacampagne A, Vassort G, Turan B (2005) Effects of diabetes on ryanodine receptor Ca release channel (RyR2) and Ca2+ homeostasis in rat heart. Diabetes 54:3082–3088. https://doi.org/10.2337/diabetes.54.11.3082

Chapter 9
The Role of Cadmium in the Development and Progression of Liver Cancer

Phoka C. Rathebe and Mota X. Kholopo

Abstract Liver cancer, particularly hepatocellular carcinoma (HCC), is a major cause of cancer-related mortality worldwide. Cadmium (Cd), a nonessential toxic metal, has been identified as a significant risk factor due to its widespread presence in industrial emissions, contaminated food, and cigarette smoke. With a long biological half-life, cadmium accumulates in the liver, inducing oxidative stress, DNA damage, and epigenetic modifications, leading to carcinogenesis. The International Agency for Research on Cancer (IARC) classifies cadmium as a Group 1 carcinogen due to its established link with various malignancies, including liver cancer. Cadmium exposure generates reactive oxygen species (ROS), leading to lipid peroxidation, protein oxidation, and DNA damage. It also inhibits DNA repair mechanisms and alters key molecular pathways such as mitogen-activated protein kinase (MAPK) and phosphoinositide 3-kinase (PI3K)/Akt, promoting uncontrolled cell proliferation and inhibiting apoptosis. Hepatocytes absorb cadmium via divalent metal transporter 1 (DMT1) and metallothioneins (MTs), but excessive exposure overwhelms detoxification mechanisms, resulting in hepatocellular injury and fibrosis, which contribute to liver cancer progression. Understanding cadmium's role in liver carcinogenesis is crucial for developing biomarkers, targeted therapies, and regulatory policies to limit exposure. This review explores cadmium exposure sources, toxicokinetics, and molecular mechanisms in liver cancer, emphasizing the need for further research and public health interventions to mitigate its impact.

Keywords Cadmium · Liver cancer · Hepatocellular carcinoma · Oxidative stress · Heavy metal toxicity

P. C. Rathebe (✉) · M. X. Kholopo
Department of Environmental Health, Faculty of Health Sciences, Doornfontein Campus, University of Johannesburg, Johannesburg, South Africa
e-mail: prathebe@uj.ac.za

N. S. Dhalla et al. (eds.), *Functional Biochemistry of Metallic Elements*, Advances in Biochemistry in Health and Disease 36,
https://doi.org/10.1007/978-3-032-14445-4_9

Introduction

Liver cancer is one of the leading causes of cancer-related mortality worldwide, with hepatocellular carcinoma (HCC) being the most prevalent form. Environmental and occupational exposure to heavy metals such as cadmium (Cd) has been increasingly recognized as a significant risk factor in liver carcinogenesis. Cadmium, a non-essential toxic metal, is widely present in industrial emissions, contaminated food, and cigarette smoke. Due to its long biological half-life, cadmium accumulates in vital organs, including the liver, where it exerts its toxic effects over prolonged periods [1]. Studies have shown that chronic exposure to cadmium is associated with oxidative stress, DNA damage, epigenetic alterations, and disruption of key signaling pathways involved in cell proliferation and apoptosis, ultimately contributing to the initiation and progression of liver cancer [2, 3]. The International Agency for Research on Cancer (IARC) has classified cadmium as a Group 1 carcinogen due to its established link with various malignancies, including liver cancer [4]. Epidemiological studies indicate that populations exposed to cadmium, particularly through occupational settings, such as mining, battery production, and smelting, have a higher risk of developing liver-related pathologies, including fibrosis, cirrhosis, and malignancies [5]. Additionally, dietary sources, such as rice, shellfish, and leafy vegetables, grown in contaminated soil contribute significantly to cadmium accumulation in human tissues [6]. Understanding the mechanisms by which cadmium induces liver cancer is crucial for developing targeted preventive and therapeutic strategies.

One of the primary mechanisms through which cadmium exerts its carcinogenic effects is through oxidative stress. Cadmium disrupts the cellular redox balance by depleting glutathione and inhibiting antioxidant enzymes, leading to excessive production of reactive oxygen species (ROS) [7]. This oxidative stress damages cellular macromolecules, including lipids, proteins, and DNA, thereby triggering mutagenesis and genomic instability [8, 9]. Additionally, cadmium-induced ROS production activates pro-inflammatory pathways, such as nuclear factor kappa B (NF-κB), which plays a significant role in promoting chronic inflammation and tumor progression [10, 11]. Furthermore, cadmium has been shown to interfere with DNA repair mechanisms, leading to an accumulation of genetic mutations. It inhibits critical DNA repair enzymes, such as nucleotide excision repair (NER) and base excision repair (BER), thus increasing the likelihood of carcinogenic transformations [12]. Moreover, cadmium-induced epigenetic modifications, including DNA methylation and histone alterations, can silence tumor suppressor genes and promote the activation of oncogenes, further enhancing the risk of liver cancer development [13]. Apart from its direct genotoxic effects, cadmium disrupts key signaling pathways involved in cellular proliferation and apoptosis. The metal has been shown to activate mitogen-activated protein kinase (MAPK) and phosphoinositide 3-kinase (PI3K)/Akt pathways, which contribute to uncontrolled cell division and survival [14]. Concurrently, cadmium inhibits apoptotic pathways by dysregulating the expression of proapoptotic and antiapoptotic proteins, such as Bcl-2 and Bax, thereby enabling the persistence of damaged cells that should otherwise undergo programmed cell death [15].

The liver is particularly vulnerable to cadmium toxicity due to its role in detoxification and metabolism. Hepatocytes actively take up cadmium through divalent metal transporter 1 (DMT1) and metallothioneins (MTs), proteins that bind and sequester heavy metals [16]. However, excessive cadmium exposure overwhelms the detoxification capacity of the liver, leading to hepatocellular injury and fibrosis, which can serve as a precursor to malignant transformation [17]. Additionally, cadmium-induced fibrosis alters the hepatic microenvironment, creating a pro-tumorigenic niche that facilitates cancer progression [18]. Given the global burden of liver cancer and the increasing evidence of cadmium's role in hepatocarcinogenesis, further research is necessary to elucidate the precise molecular mechanisms underlying cadmium-induced liver cancer. Identifying biomarkers for early detection and potential therapeutic targets remains a crucial area of investigation. Moreover, regulatory measures to limit cadmium exposure, particularly in occupational and dietary settings, should be reinforced to mitigate the risk of liver cancer associated with this toxic metal. In this review, we will explore the sources of cadmium exposure, its toxicokinetics, and its molecular mechanisms of action in liver cancer. We will also discuss the latest advancements in cadmium-related cancer research and potential strategies for prevention and intervention. Understanding the interplay between cadmium toxicity and liver carcinogenesis is essential for developing effective public health policies and therapeutic approaches to combat this growing health concern.

Sources of Cadmium Exposure

Cadmium is a toxic heavy metal that poses significant health risks to humans. It is widely distributed in the environment due to both natural and anthropogenic activities. The primary sources of cadmium exposure include occupational exposure, environmental contamination, dietary intake, and lifestyle factors such as smoking. Understanding these sources is crucial in assessing the health risks associated with cadmium and implementing measures to mitigate exposure. Table 9.1 summarizes the major sources, routes of exposure, and associated health risks of cadmium toxicity.

Occupational Exposure

One of the primary sources of cadmium exposure is occupational settings, particularly in industries such as mining, smelting, battery manufacturing, and pigment production. Workers in these industries are at a high risk of cadmium inhalation due to exposure to cadmium dust and fumes. Metal refining, welding, and soldering also

Table 9.1 Sources of cadmium exposure and associated risks

Sources of exposure	Description	Routes of exposure	Health risks	References
Occupational exposure	Common industries such as mining, smelting, battery production, and welding. Workers are exposed to cadmium dust and fumes	Inhalation, dermal contact	Increased risk of lung and liver cancer, kidney damage, and respiratory diseases	[19]
Environmental contamination	Industrial emissions, the combustion of fossil fuels, and the improper disposal of electronic waste contribute to cadmium pollution in soil, air, and water	Inhaling and ingesting contaminated water and food	Kidney dysfunction, cardiovascular disease, liver toxicity	[20]
Dietary intake	Cadmium enters the food chain through crops grown in contaminated soil, seafood (shellfish), and animal organs (liver, kidney)	Ingestion (contaminated food)	Osteoporosis, kidney damage, increased cancer risk	[21]
Tobacco smoking	Cadmium accumulates in tobacco leaves and is inhaled directly into the bloodstream by smokers and secondhand smokers	Inhalation (smoking and passive exposure)	Increased cadmium body burden, higher risks of lung, kidney and liver diseases	[22]
Drinking water	Contaminated groundwater and industrial waste leaching introduce cadmium into water supplies	Ingestion (drinking water)	Chronic toxicity, kidney damage, liver dysfunction	[23]

contribute to increased cadmium exposure, leading to potential health risks such as renal dysfunction and respiratory diseases [18]. Studies have shown that long-term occupational exposure to cadmium increases the risk of developing lung and prostate cancer due to its carcinogenic properties [3]. Inhalation remains the most common route of exposure in industrial settings, as cadmium is efficiently absorbed into the bloodstream via the lungs [24].

Environmental Contamination

Environmental contamination is another major source of cadmium exposure. Cadmium is released into the environment through industrial emissions, fossil fuel combustion, and improper disposal of electronic waste. Soil contamination occurs when cadmium-laden waste and fertilizers containing cadmium are introduced into

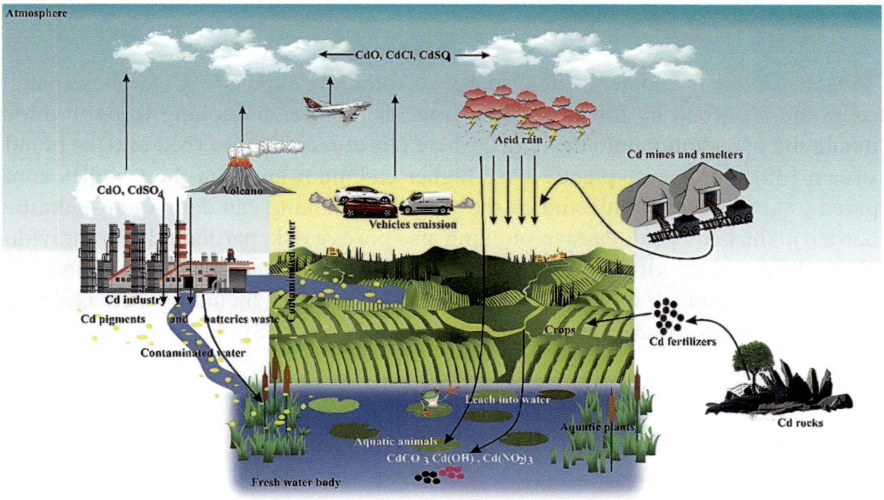

Fig. 9.1 Cadmium sources and bioavailability in the environment

agricultural lands [25]. The use of phosphate fertilizers, which contain trace amounts of cadmium, significantly contributes to soil contamination, leading to the uptake of cadmium by crops [26]. Additionally, cadmium can be present in water sources due to leaching from industrial waste and contaminated soil, posing a risk to communities reliant on such water supplies for drinking and irrigation purposes [27]. Figure 9.1 illustrates the main sources of cadmium contamination, including industrial emissions, agriculture, and waste disposal, along with its pathway into soil, water, and the food chain.

Dietary Intake

Dietary exposure is a significant non-occupational source of cadmium for the general population. Cadmium enters the food chain through contaminated soil and water, affecting staple crops such as rice, wheat, and leafy vegetables [28]. Studies indicate that rice accumulates higher levels of cadmium compared to other crops, particularly in regions where cadmium-contaminated water is used for irrigation [29, 30]. Seafood, particularly shellfish, can also be a source of cadmium due to bioaccumulation in marine organisms [31]. Additionally, animal liver and kidney, being organs that store cadmium, pose a risk when consumed in large quantities [32, 33]. Chronic dietary exposure to cadmium has been linked to kidney damage and an increased risk of osteoporosis [34].

Tobacco Smoking

Tobacco smoke is another critical source of cadmium exposure. Cigarettes contain cadmium absorbed by tobacco plants from the soil, and smoking leads to direct inhalation of cadmium into the lungs, where it is efficiently absorbed into the bloodstream [35]. Smokers typically have higher cadmium levels in their bodies compared to nonsmokers, with studies showing that smoking can double the cadmium burden in the body [36]. Passive smoking also poses a risk, particularly for individuals in close contact with smokers [37]. Given its cumulative nature, long-term exposure through smoking can lead to significant toxic effects, including increased risks of cancer, kidney disease, and cardiovascular disorders [38].

Toxicokinetics of Cadmium and Liver Cancer

Cadmium (Cd) is a toxic heavy metal widely recognized for its adverse health effects, including carcinogenesis. Chronic exposure to cadmium has been implicated in liver cancer due to its ability to accumulate in the liver, disrupt cellular function, and induce oxidative stress [3]. The toxicokinetics of cadmium involve its absorption, distribution, metabolism, and excretion, which collectively influence its persistence and toxicity in the body [39]. Figure 9.2 illustrates the toxicokinetics of cadmium, showing its absorption through the gastrointestinal tract and lungs; distribution and metabolism via blood circulation, liver, kidneys, and tissues; and elimination through urine, bile, and expired air. The liver plays a central role in cadmium storage detoxification, but excess accumulation can lead to toxicity. Understanding the toxicokinetics of cadmium is crucial in elucidating its carcinogenic potential in the liver.

Absorption of Cadmium

Cadmium enters the human body primarily through inhalation and ingestion. Occupational exposure, such as in mining, smelting, and battery manufacturing, contributes significantly to cadmium inhalation, while food and cigarette smoke are major sources of dietary intake [6]. Inhaled cadmium is absorbed through the respiratory tract, with approximately 10–50% retention depending on particle size and solubility [40, 41]. The gastrointestinal absorption of cadmium is influenced by dietary factors, such as iron deficiency, which enhances cadmium uptake due to competition for the divalent metal transporter 1 (DMT1) [12].

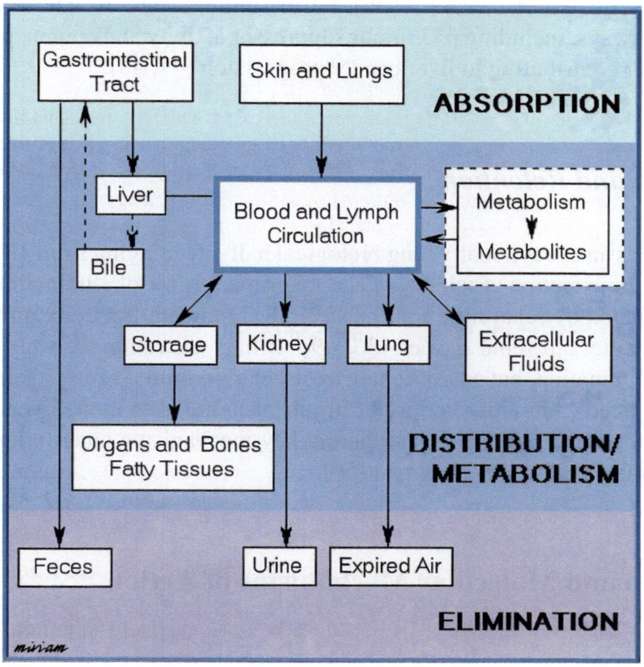

Fig. 9.2 Toxicokinetics of cadmium: absorption, distribution, metabolism, and elimination pathways

Distribution in the Liver

Once absorbed, cadmium is transported via the blood circulation and primarily accumulates in the liver and kidneys. The liver plays a crucial role in cadmium metabolism by binding it to metallothioneins (MTs), which are cysteine-rich proteins that regulate metal homeostasis and detoxification [2]. Cadmium has a high affinity for sulfhydryl groups in proteins, leading to its retention in hepatocytes and subsequent disruption of normal cellular functions [42]. Chronic exposure leads to hepatic accumulation, promoting liver damage and carcinogenesis.

Metabolism and Biotransformation

Unlike organic xenobiotics, cadmium does not undergo classical metabolism but forms stable complexes with metallothioneins. Hepatic cadmium exposure upregulates the expression of MTs, which act as a protective mechanism to sequester cadmium and limit its toxicity [43]. However, excessive cadmium burden leads to mitochondrial dysfunction, endoplasmic reticulum stress, and DNA damage, key events in liver carcinogenesis [44–46]. Furthermore, cadmium interferes with key cellular pathways, including p53 tumor suppressor activity and epigenetic modifications, further contributing to liver cancer progression [47, 48].

Excretion and Retention

Cadmium has an exceptionally long biological half-life, ranging from 10 to 30 years, due to its poor excretion efficiency [17]. The primary route of elimination is through urine, facilitated by renal filtration; however, a significant fraction remains sequestered in hepatic and renal tissues [36, 49]. Biliary excretion also plays a role in cadmium elimination, but enterohepatic recirculation limits its effectiveness, resulting in prolonged cadmium retention. Chronic accumulation in the liver exacerbates oxidative stress and chronic inflammation, key mechanisms underlying hepatocellular carcinoma (HCC) development [50].

Cadmium and Molecular Mechanisms of Action in Liver Cancer

Cadmium (Cd) is a nonessential heavy metal with significant toxicological properties, particularly in relation to carcinogenesis. Chronic cadmium exposure has been strongly associated with liver cancer development, primarily through its ability to induce oxidative stress, disrupt cellular signaling pathways, and interfere with DNA repair mechanisms. Unlike some carcinogens that directly damage DNA, cadmium exerts its effects largely through epigenetic modifications and the dysregulation of cellular processes, promoting tumorigenesis and metastasis in hepatocytes [7, 51, 52].

Oxidative Stress and Liver Cancer

One of the primary molecular mechanisms by which cadmium contributes to liver cancer development is through the induction of oxidative stress. Cadmium exposure leads to excessive reactive oxygen species (ROS) production, which can cause lipid peroxidation, protein oxidation, and DNA damage [2, 53]. The liver, being a major detoxifying organ, accumulates cadmium, leading to mitochondrial dysfunction and increased ROS generation [8]. Oxidative stress not only damages cellular components but also activates oncogenic signaling pathways, such as nuclear factor kappa B (NF-κB) and mitogen-activated protein kinases (MAPKs), which promote inflammation, cell proliferation, and survival of malignant hepatocytes [54].

Epigenetic Modifications

Cadmium-induced liver carcinogenesis is also facilitated by epigenetic modifications, including DNA methylation, histone modification, and microRNA (miRNA) dysregulation. Studies have shown that cadmium exposure leads to global DNA hypomethylation, which can activate oncogenes and promote tumor formation [55]. At the same time, hypermethylation of tumor suppressor genes, such as p16INK4a and RASSF1A, results in the silencing of critical pathways involved in cell cycle control and apoptosis [56]. Additionally, cadmium exposure disrupts miRNA expression patterns, particularly those involved in tumor suppression (e.g., miR-34 and miR-200), further contributing to hepatocarcinogenesis [57].

Inhibition of DNA Repair Mechanisms

Cadmium interferes with essential DNA repair processes, leading to genomic instability and increased mutation rates. The metal has been reported to inhibit nucleotide excision repair (NER) and base excision repair (BER) pathways by displacing essential zinc ions in DNA repair enzymes, such as poly (ADP-ribose) polymerase (PARP) and XPA, a crucial NER factor [58, 59]. This inhibition prevents effective correction of DNA damage, increasing susceptibility to mutations and carcinogenesis.

Dysregulation of Apoptosis and Cell Proliferation

Cadmium has been shown to promote hepatocyte survival by interfering with apoptotic pathways. It inhibits the expression of proapoptotic proteins, such as Bax, and enhances antiapoptotic proteins like Bcl-2, allowing damaged hepatocytes to evade programmed cell death and continue proliferating uncontrollably [60, 61]. Furthermore, cadmium can activate the PI3K/Akt and Wnt/β-catenin pathways, both of which are crucial for cell proliferation and cancer progression [62–64].

Disruption of Metal Homeostasis

Cadmium disrupts normal metal homeostasis, particularly concerning iron, zinc, and calcium, which are essential for cellular function. By replacing zinc in metalloproteins, such as p53, cadmium inhibits the tumor-suppressive activity of p53, leading to uncontrolled cell growth and division [65]. Similarly, cadmium-induced calcium imbalance disrupts normal hepatocyte signaling, further contributing to carcinogenesis [66].

Conclusion

Cadmium exposure represents a significant environmental and occupational hazard that contributes to the rising incidence of liver cancer, particularly hepatocellular carcinoma (HCC). Due to its long biological half-life, cadmium accumulates in the liver, where it induces oxidative stress, DNA damage, epigenetic modifications, and dysregulation of key signaling pathways involved in cell proliferation and apoptosis. These mechanisms collectively contribute to liver carcinogenesis by promoting genetic instability, chronic inflammation, and tumor progression. Given its classification as a Group 1 carcinogen by the International Agency for Research on Cancer (IARC), the need for stringent regulatory policies to minimize cadmium exposure is imperative. Industrial emissions, contaminated food sources, and cigarette smoke remain major routes of cadmium entry into human systems, underscoring the importance of preventive measures, including improved workplace safety, dietary regulations, and public awareness campaigns.

Future research should focus on elucidating the precise molecular pathways by which cadmium induces liver cancer to develop targeted therapies and early detection biomarkers. Additionally, advancements in detoxification strategies and novel therapeutic interventions may help mitigate cadmium-induced hepatotoxicity. Addressing this growing public health concern requires a multidisciplinary approach integrating toxicology, oncology, environmental science, and policymaking to limit cadmium exposure and reduce liver cancer incidence globally.

References

1. IARC Working Group on the Evaluation of Carcinogenic Risks to Humans (1987) Cadmium and cadmium compounds (Group 2A). In: Overall evaluations of carcinogenicity: an updating of IARC Monographs Volumes 1 to 42. International Agency for Research on Cancer
2. Thévenod F (2009) Cadmium and cellular signaling cascades: to be or not to be? Toxicol Appl Pharmacol 238:221–239. https://doi.org/10.1016/j.taap.2009.01.013
3. Waalkes MP (2003) Cadmium Carcinogenesis. Mutat Res 533:107–120. https://doi.org/10.1016/j.mrfmmm.2003.07.011
4. Branca JJV, Fiorillo C, Carrino D, Paternostro F, Taddei N, Gulisano M, Pacini A, Becatti M (2020) Cadmium-induced oxidative stress: focus on the central nervous system. Antioxidants 9:492. https://doi.org/10.3390/antiox9060492
5. Joseph P (2009) Mechanisms of cadmium carcinogenesis. Toxicol Appl Pharmacol 238:272–279. https://doi.org/10.1016/j.taap.2009.01.011
6. Satarug S, Garrett SH, Sens MA, Sens DA (2011) Cadmium, environmental exposure, and health outcomes. Ciencia saude coletiva 16:2587–2602

7. Waisberg M, Joseph P, Hale B, Beyersmann D (2003) Molecular and cellular mechanisms of cadmium carcinogenesis. Toxicology 192:95–117. https://doi.org/10.1016/S0300-483X(03)00305-6
8. Liu J, Qu W, Kadiiska M (2009) Role of oxidative stress in cadmium toxicity and carcinogenesis. Toxicol Appl Pharmacol 238(3):209–214. https://doi.org/10.1016/j.taap.2009.01.029
9. Nzengue Y, Steiman R, Garrel C, Lefèbvre E, Guiraud P (2008) Oxidative stress and DNA damage induced by cadmium in the human keratinocyte HaCaT cell line: role of glutathione in the resistance to cadmium. Toxicology 243(1–2):193–206. https://doi.org/10.1016/J.TOX.2007.10.005
10. Li Z, Chi H, Zhu W, Yang G, Song J, Mo L, Zhang Y, Deng Y, Xu F, Yang J et al (2021) Cadmium induces renal inflammation by activating the NLRP3 inflammasome through ROS/MAPK/NF-κB pathway in vitro and in vivo. Arch Toxicol 95:3497–3513. https://doi.org/10.1007/s00204-021-03157-2
11. Lian S, Xia Y, Khoi PN, Ung TT, Yoon H, Kim N-H, Kim K, Jung Y (2015) Cadmium induces matrix metalloproteinase-9 expression via ROS-dependent EGFR, NF-κB, and AP-1 pathways in human endothelial cells. Toxicology 338:104–116. https://doi.org/10.1016/j.tox.2015.10.008
12. Park JD, Cherrington NJ, Klaassen CD (2002) Intestinal absorption of cadmium is associated with divalent metal transporter 1 in rats. Toxicol Sci 68:288–294
13. Hartwig, A. Cadmium and Cancer. In Cadmium: From toxicity to essentiality; Sigel, A., Sigel, H., Sigel, R.K., Eds.; Springer Netherlands: Dordrecht, 2013; pp. 491–507 ISBN 978–94–007-5179-8
14. Xiao-l W (2013) Cadmium-induced apoptosis on LLC-PK_1 cells by activating PI3K / Akt and ERK1 / 2 pathway. J Trop Med
15. Genchi G, Sinicropi MS, Lauria G, Carocci A, Catalano A (2020) The effects of cadmium toxicity. Int J Environ Res Public Health 17:3782
16. Tellez-Plaza M, Navas-Acien A, Menke A, Crainiceanu CM, Pastor-Barriuso R, Guallar E (2012) Cadmium exposure and all-cause and cardiovascular mortality in the U.S. general population. Environ Health Perspect 120:1017–1022. https://doi.org/10.1289/ehp.1104352
17. Rani A, Kumar A, Lal A, Pant M (2014) Cellular mechanisms of cadmium-induced toxicity: a review. Int J Environ Health Res 24:378–399. https://doi.org/10.1080/09603123.2013.835032
18. Bernard A (2008) Cadmium & its adverse effects on human health. Indian J Med Res 128:557–564
19. Lombaert N, Gilles M, Verougstraete V (2023) Cadmium monitoring at the workplace: effectiveness of a combination of air- and biomonitoring. Toxics 11:354. https://doi.org/10.3390/toxics11040354
20. Peana M, Pelucelli A, Chasapis CT, Perlepes SP, Bekiari V, Medici S, Zoroddu MA (2022) Biological effects of human exposure to environmental cadmium. Biomolecules 13:36. https://doi.org/10.3390/biom13010036
21. Zhao D, Wang P, Zhao F-J (2023) Dietary cadmium exposure, risks to human health and mitigation strategies. Crit Rev Environ Sci Technol 53:939–963. https://doi.org/10.1080/10643389.2022.2099192
22. Matt GE, Quintana PJE, Hoh E, Dodder NG, Mahabee-Gittens EM, Padilla S, Markman L, Watanabe K (2021) Tobacco smoke is a likely source of lead and cadmium in settled house dust. J Trace Elem Med Biol 63:126656. https://doi.org/10.1016/j.jtemb.2020.126656
23. Yang S, Yang D, Taylor D, He M, Liu X, Xu J (2022) Tracking cadmium pollution from source to receptor: a health-risk focused transfer continuum approach. Sci Total Environ 867:161574
24. Satarug S, Moore MR (2004) Adverse health effects of chronic exposure to low-level cadmium in foodstuffs and cigarette smoke. Environ Health Perspect 112:1099–1103. https://doi.org/10.1289/ehp.6751
25. Järup L, Åkesson A (2009) Current status of cadmium as an environmental health problem. Toxicol Appl Pharmacol 238:201–208. https://doi.org/10.1016/j.taap.2009.04.020

26. Loganathan P, Hedley MJ, Grace ND (2008) Pasture soils contaminated with fertilizer-derived cadmium and fluorine: livestock effects. In: Whitacre DM (ed) Reviews of environmental contamination and toxicology, vol 192. Springer New York, New York, NY, pp 29–66. ISBN 978-0-387-71723-4
27. Roberts SM, James RC, Williams PL (2022) Principles of toxicology: environmental and industrial applications. John Wiley & Sons
28. Sharma RK, Agrawal M (2005) Biological effects of heavy metals: an overview. J Environ Biol 26:301–313
29. Gao Y, Duan Z, Zhang L, Sun D, Li X (2022) The status and research Pprogress of cadmium pollution in rice- (Oryza Sativa L.) and wheat- (Triticum Aestivum L.) cropping Ssystems in China: a critical review. Toxics 10. https://doi.org/10.3390/toxics10120794
30. Jahiruddin M, Xie Y, Ozaki A, Islam M, Nguyen T, Kurosawa K (2017) Arsenic, cadmium, Lead and chromium concentrations in irrigated and rain-fed rice and their dietary intake implications. Aust J Crop Sci 11:806–812. https://doi.org/10.21475/AJCS.17.11.07.PNE408
31. Burger J, Gochfeld M (2005) Heavy metals in commercial fish in New Jersey. Environ Res 99:403–412
32. Darwish W, Hussein M, Eldesoky KI, Ikenaka Y, Nakayama S, Mizukawa H, Ishizuka M (2015) Incidence and public health risk assessment of toxic metal residues (cadmium and lead) in Egyptian cattle and sheep meats. Int Food Res J 22:1719–1726
33. Zenad W, Benatallah A, Zaouani M, Boudjellaba S, Ainouz L, Mahdi MHB, Benouadah A (2020) Incidence and public health risk assessment of toxic metal residues (cadmium and lead) in Egyptian cattle and sheep meats. Bull Univ Agric Sci Vet Med Cluj-Napoca Vet Med 77. https://doi.org/10.15835/buasvmcn-vm:2020.0002
34. Nordberg G (1992) Cadmium in the human environment: toxicity and carcinogenicity, vol 118. IARC Scientific Publications, pp 1–470
35. Mathew B, Sharma H, Rawal N (2015) Cadmium: toxicity effect and its mechanism. Res J Pharmacol Toxicol 1:01–05
36. Godt J, Scheidig F, Grosse-Siestrup C, Esche V, Brandenburg P, Reich A, Groneberg DA (2006) The toxicity of cadmium and resulting hazards for human health. J Occup Med Toxicol 1:22. https://doi.org/10.1186/1745-6673-1-22
37. Rahayuningsih CK, Puspitasari D, Puspitasari A (2022) Differences in cadmium (Cd) levels in the blood of active smokers and passive smokers with type 2 diabetic in Puskesmas Kedungdoro Surabaya. j.teknokes 15:69–75. https://doi.org/10.35882/jteknokes.v15i2.198
38. Järup L, Berglund M, Elinder CG, Nordberg G, Vanter M (1998) Health effects of cadmium exposure—a review of the literature and a risk estimate. Scand J Work Environ Health 24:1–51
39. Berlin M, Zalups RK, Fowler BA (2007) Mercury. Handbook on the toxicology of metals, vol 10. Elsevier, B978-012369413
40. Cassee FR, Muijser H, Duistermaat E, Freijer JJ, Geerse KB, Marijnissen JC, Arts JH (2002) Particle size-dependent total mass deposition in lungs determines inhalation toxicity of cadmium chloride aerosols in rats. Application of a multiple path dosimetry model. Arch Toxicol 76:277–286. https://doi.org/10.1007/s00204-002-0344-8
41. Zhou Y, Zeng R, Zhang M, Guo L (2020) Treatment of an acute severe cadmium poisoning patient combined with multiple organ dysfunction syndromes by integrated Chinese and western medicines: a case report. Chin J Integr Med 26:853–856. https://doi.org/10.1007/s11655-020-3089-4
42. Liu Z, Yu X, Shaikh ZA (2008) Rapid activation of ERK1/2 and AKT in human breast cancer cells by cadmium. Toxicol Appl Pharmacol 228:286–294
43. Klaassen CD, Liu J, Choudhuri S (1999) METALLOTHIONEIN: an intracellular protein to protect against cadmium toxicity. Annu Rev Pharmacol Toxicol 39:267–294. https://doi.org/10.1146/annurev.pharmtox.39.1.267
44. Bai H, Yang F, Jiang W, Hu A, Chang H-M, Zhang Y, Jiang L, Lin S, Lu Z, Zhang C et al (2021) Molybdenum and cadmium co-induce mitophagy and mitochondrial dysfunction via

ROS-mediated PINK1/Parkin pathway in hepa1-6 cells. Ecotoxicol Environ Saf 224. https://doi.org/10.1016/j.ecoenv.2021.112618
45. Skipper A, Sims J, Yedjou C, Tchounwou P (2016) Cadmium chloride induces DNA damage and apoptosis of human liver carcinoma cells via oxidative stress. IJERPH 13:88. https://doi.org/10.3390/ijerph13010088
46. Zhu MK, Li HY, Bai LH, Wang LS, Zou XT (2020) Histological changes, lipid metabolism, and oxidative and endoplasmic reticulum stress in the liver of laying hens exposed to cadmium concentrations. Poult Sci 99:3215–3228. https://doi.org/10.1016/j.psj.2019.12.073
47. Mishra A, Brat DJ, Verma M (2015) P53 tumor suppression network in cancer epigenetics. In: Verma M (ed) Cancer epigenetics: risk assessment, diagnosis, treatment, and prognosis. Springer, New York, pp 597–605. ISBN 978-1-4939-1804-1
48. Suzuki M, Takeda S, Teraoka-Nishitani N, Yamagata A, Tanaka T, Sasaki M, Yasuda N, Oda M, Okano T, Yamahira K et al (2017) Cadmium-induced malignant transformation of rat liver cells: potential key role and regulatory mechanism of altered apolipoprotein E expression in enhanced invasiveness. Toxicology 382:16–23. https://doi.org/10.1016/j.tox.2017.03.014
49. Akerstrom M, Barregard L, Lundh T, Sallsten G (2013) The relationship between cadmium in kidney and cadmium in urine and blood in an environmentally exposed population. Toxicol Appl Pharmacol 268:286–293. https://doi.org/10.1016/j.taap.2013.02.009
50. Sabolić I (2006) Common mechanisms in nephropathy induced by toxic metals. Nephron Physiol 104:p107–p114
51. Waalkes MP (2000) Cadmium carcinogenesis in review. J Inorg Biochem 79:241–244. https://doi.org/10.1016/S0162-0134(00)00009-X
52. Zhao L, Islam R, Wang Y, Zhang X, Liu L-Z (2022) Epigenetic regulation in chromium-, nickel- and cadmium-induced carcinogenesis. *Cancers (Basel)* 14:5768. https://doi.org/10.3390/cancers14235768
53. Yan L-J, Allen DC (2021) Cadmium-induced kidney injury: oxidative damage as a unifying mechanism. Biomolecules 11:1575. https://doi.org/10.3390/biom11111575
54. Valko M, Rhodes CJB, Moncol J, Izakovic MM, Mazur M (2006) Free radicals, metals and antioxidants in oxidative stress-induced cancer. Chem Biol Interact 160:1–40
55. Arita A, Costa M (2009) Epigenetics in metal carcinogenesis: nickel, arsenic, chromium and cadmium. Metallomics 1:222–228
56. Atta MM, Aly A, Gad M, El-Hefnawy T, El-Shewi ME, Ahmad EYA (2016) Tumor suppressor genes (P16 and RASSF1A) hypermethylation in hepatocellular carcinoma and chronic hepatitis C patients. Med J Cairo Univ 84:33–62
57. Zhang YM, Wu QM, Chang LY, Liu JC (2020) miR-34a and miR-125a-5p inhibit proliferation and metastasis but induce apoptosis in hepatocellular carcinoma cells via repressing the MACC1-mediated PI3K/AKT/mTOR pathway. neo 67:1042–1053. https://doi.org/10.4149/neo_2020_191019N1062
58. Beyersmann D, Hartwig A (2008) Carcinogenic metal compounds: recent insight into molecular and cellular mechanisms. Arch Toxicol 82:493–512. https://doi.org/10.1007/s00204-008-0313-y
59. Hartwig A (2001) Zinc finger proteins as potential targets for toxic metal ions: differential effects on structure and function. Antioxid Redox Signal 3:625–634. https://doi.org/10.1089/15230860152542970
60. Jin-jun W (2009) Experimental study of the effects of cadmium on the apoptosis and the expression of Bcl-2,Bax in NRK. Cells
61. Ye J-L, Mao W-P, Wu A-L, Zhang N-N, Zhang C, Yu Y-J, Zhou L, Wei C-J (2007) Cadmium-induced apoptosis in human normal liver L-02 cells by acting on mitochondria and regulating Ca2+ signals. Environ Toxicol Pharmacol 24:45–54. https://doi.org/10.1016/j.etap.2007.01.007
62. Kulkarni P, Dasgupta P, Bhat NS, Hashimoto Y, Saini S, Dar AA, Shahryari V, Shiina M, Yamamura S, Tanaka Y et al (2019) Abstract 5040: involvement of PI3K/Akt pathway in cadmium triggered aggressive prostate cancer. Cancer Res 79:5040. https://doi.org/10.1158/1538-7445.AM2019-5040

63. Peng L, Huang Y-T, Zhang F, Chen J-Y, Huo X (2018) Chronic cadmium exposure aggravates malignant phenotypes of nasopharyngeal carcinoma by activating the Wnt/β-catenin signaling pathway via hypermethylation of the casein kinase 1α promoter. CMAR 11:81–93. https://doi.org/10.2147/CMAR.S171200
64. Son Y-O, Wang L, Poyil P, Budhraja A, Hitron JA, Zhang Z, Lee J-C, Shi X (2012) Cadmium induces carcinogenesis in BEAS-2B cells through ROS-dependent activation of PI3K/AKT/GSK-3β/β-catenin signaling. Toxicol Appl Pharmacol 264:153–160. https://doi.org/10.1016/j.taap.2012.07.028
65. Bertin G, Averbeck D (2006) Cadmium: cellular effects, modifications of biomolecules, modulation of DNA repair and genotoxic consequences (a review). Biochimie 88:1549–1559
66. Thévenod F, Lee W-K (2013) Toxicology of cadmium and its damage to mammalian organs. In: Sigel A, Sigel H, Sigel RK (eds) Cadmium: From toxicity to essentiality, Metal ions in life sciences, vol 11. Springer Netherlands, Dordrecht, pp 415–490. ISBN 978-94-007-5178-1

Chapter 10
Functional Biochemistry of Trace Elements in Cancer: Mechanisms, Biomarkers, and Therapeutic Potentials

Nicholas Sagadaya and Ferial Pestano-Gupta

Abstract Trace elements are minerals that the human body needs in very small amounts; however, they are very important for supporting different functions such as immune responses, enzyme activity, and redox balance. These elements, such as zinc, copper, iron, and selenium, can lead to several health problems, including cancer, when they are not properly balanced in the body.

This chapter discusses the roles and functions that zinc, copper, iron, and selenium play in the development, diagnosis, and possible treatment of cancer. Also discussed are how these elements affect signaling pathways in the body, how they may serve as cancer biomarkers, and how they are being looked at in cancer therapy, including using nanoparticles and supplementation. Overall, this paper explains how small elements can play big roles in the body and how they may be used to help manage cancer in the future.

Keywords Trace elements · Cancer · Zinc · Copper · Iron · Selenium · Biomarkers · Biological functions

Introduction

Trace elements (or trace minerals) are defined as micronutrients required in minute quantities but are essential for normal physiological functions and biochemical processes in the human body [1, 2]. Additionally, these elements are required in amounts between 1 and 100 mg/day by adults or make up less than 0.01% of the total body weight [3]. Among them, zinc (Zn), copper (Cu), iron (Fe), and selenium (Se) play critical roles, such as cellular signaling, gene expression, enzymatic catalysis, immune modulation, and redox balance [4, 5].

N. Sagadaya (✉) · F. Pestano-Gupta
Division of Natural Sciences, Berbice Campus, University of Guyana, Tain, Guyana
e-mail: nicholas.sagadaya@uog.edu.gy; ferial.pestano@uog.edu.gy

N. S. Dhalla et al. (eds.), *Functional Biochemistry of Metallic Elements*,
Advances in Biochemistry in Health and Disease 36,
https://doi.org/10.1007/978-3-032-14445-4_10

Advancements in systems biochemistry, metallomics, and redox biology have drawn scientific attention to the complex role of trace elements in cancer [6–8]. For instance, zinc has been linked to genomic instability (such as chromosome breaks and oxidative DNA damage) and an increased risk of cancer, but zinc is also known to stabilize tumor suppressor proteins like p53 (a transcription factor) and modulate apoptosis [9]. Conversely, angiogenesis, the formation of new blood vessels, is facilitated by copper. Copper is also essential for the growth of tumors by influencing various angiogenic factors such as vascular endothelial growth factor (VEGF) and hypoxia-inducible factor 1-alpha (HIF-1α) [10]. Iron is another necessary nutrient that supports fundamental cellular processes like DNA synthesis and mitochondrial respiration. However, when it is present in excess, it causes oxidative DNA damage and carcinogenesis by catalyzing Fenton reactions that form reactive oxygen species (ROS). Additionally, selenium functions as an essential antioxidant and is integrated into selenoproteins. Supraphysiological doses of selenium can also be cytotoxic, and a lack of it may increase vulnerability to prostate and colorectal cancers [11].

Further investigation into trace element-cancer interactions is necessary due to the global cancer burden. As per GLOBOCAN 2018, colorectal, prostate, and breast cancers are among the most common cancers globally, and they are all known to be linked to trace element imbalances [12]. By focusing on the molecular mechanisms, interactions with tumor biology, potential as diagnostic biomarkers, and significance in targeted therapy, this chapter offers a very critical overview of the functional biochemistry of zinc, copper, iron, and selenium in cancer.

Zinc in Cancer Biology

Zinc is a molecular scaffold and functional regulator essential for genomic integrity. The two symbolic zinc-dependent DNA repair proteins are p53 and PARP1, highlighting the metal's role in cancer prevention. The tumor suppressor p53 contains a structural zinc-binding motif that stabilizes the β-sandwich fold necessary for precise DNA recognition [13]. The distortion of this zinc-binding site compromises its transcriptional fidelity. Zinc deficiency causes misfolding of p53 even in the absence of mutation, inactivating its tumor-suppressive functions. The reconstitution of zinc can reactivate conformationally unstable p53 mutants, which make up nearly 30% of all p53 alterations in human cancer [14].

Zn^{2+} does not participate in redox reactions (unlike Fe or Cu), making it uniquely suitable to preserve the redox neutrality required in DNA-protein interfaces without producing ROS. Poly (ADP-ribose) polymerase-1 (PARP1) detects and binds to single-strand DNA breaks via its zinc-finger domains. Zinc loss leads to unfolding of these domains, disabling DNA binding and catalytic activation [15].

Approximately 2–3% of the human genome encodes zinc-finger proteins, many of which belong to the Cys2-His2 class, which coordinate zinc to enable sequence-specific DNA binding. Disruption of zinc homeostasis alters binding kinetics and promoter occupancy of these TFs, perturbing gene regulatory networks central to cancer progression [16]. For instance, SP1 (specificity protein 1) regulates GC-rich promoters, such as VEGF and p21, while EGR1 (early growth response 1) and RREB1 (Ras-responsive element binding protein 1) integrate zinc signals in oncogenic pathways [17].

Zinc cellular homeostasis is tightly controlled by ZIP (Zrt/Irt-like protein) and ZnT (zinc transporter) families [17]. ZIP1 is abundantly expressed in healthy prostate epithelial cells, allowing zinc accumulation (~800–1500 μM) leading to inhibition of mitochondrial aconitase, disrupting the TCA cycle, limiting ATP generation, and apoptosis through Bax activation. In prostate adenocarcinoma, ZIP1 expression is epigenetically repressed, contributing to metabolic reprogramming. Restoring ZIP1 reverses tumor phenotypes, induces apoptosis, and suppresses NF-κB signaling [18].

ZnT1 is the only ZnT located at the plasma membrane, which shows variable expression in prostate cancer. Its dysregulation may modulate cellular zinc buffering, influencing oxidative stress responses and survival. In breast cancer, other ZIPs like ZIP6, ZIP7, and ZIP9 are more dominant, but ZnT1 may still affect systemic homeostasis and metastatic behavior. Ultimately, zinc dysregulation shapes tumor initiation, progression, and therapeutic resistance, particularly in zinc-sensitive cancers such as prostate and breast carcinoma [18].

Copper and Angiogenesis in Tumor Progression

Copper (Cu), a redox-active transition metal, plays a crucial role in tumor angiogenesis by acting as a catalytic and structural cofactor in a network of pro-angiogenic signaling pathways, including VEGF, FGF, and LOX-dependent ECM remodeling [19]. Copper facilitates VEGFR-2 phosphorylation and stabilizes VEGF ligand-receptor interactions, enhancing downstream signaling [20]. It also activates transcription factors like hypoxia-inducible factor-1α (HIF-1α), promoting VEGF gene expression. Copper chelation suppresses HIF-1α expression, thus indirectly silencing VEGF transcription [21].

In redox signaling, copper acts via Fenton-like reactions to generate ROS, which serve as secondary messengers in angiogenic signaling, augmenting VEGF gene expression and increasing endothelial permeability [21]. Copper is also vital for FGF-1 and FGF-2 signaling, stabilizing their interaction with heparan sulfate proteoglycan and enhancing receptor activation. Additionally, angiogenesis induced by FGF-2 is significantly impaired in copper-depleted models [22].

Beyond signaling pathways, copper is an essential cofactor for LOX and LOX-like (LOXL) enzymes, which catalyze the crosslinking of collagen and elastin in the extracellular matrix (ECM). LOX activity is necessary for tumor matrix stiffening and vascular invasion [22]. In breast cancer, LOX promotes hypoxia-mediated metastasis, and its overexpression predicts poor prognosis. Importantly, LOX enzymatic activity is abolished in the absence of copper, making it a direct pharmacological target in antiangiogenic therapy [19].

Ceruloplasmin is the principal copper-binding protein in plasma, is elevated in breast and colorectal cancers, and correlates with increased vascular density (a surrogate for angiogenesis), elevated oxidative stress, cancer stage, and lymphovascular invasion. It also exhibits ferroxidase activity, converting Fe^{2+} to Fe^{3+}, indirectly affecting tumor iron metabolism, another axis of oncogenic growth [23].

ATOX1 is a copper chaperone that shuttles Cu (I) from the cytoplasm to ATP7A/B and cuproenzymes like LOX. It also functions as a transcription factor, upregulating genes like cyclin D1. ATOX1 expression is significantly elevated in colorectal and breast cancers and correlates with metastatic potential, indicating it may act as a copper-dependent oncogene [23].

Elevated tissue and plasma copper levels are consistently documented in cancer patients, particularly those with breast and colorectal malignancies. High Cu levels are positively associated with tumor size, grade, and vascular invasion [24]. Cu/Zn ratio, which is a more sensitive biomarker, reflects trace element imbalance, oxidative stress, and correlates with both cancer aggressiveness and survival outcomes. These findings position copper as a potential diagnostic, prognostic, and predictive biomarker [25].

Iron in Oxidative Stress and Ferroptosis

Iron (Fe) is a transition metal critical to cellular respiration, DNA synthesis, and electron transfer. However, its redox activity renders it a double-edged sword. Ferrous iron (Fe^{2+}) catalyzes the production of reactive oxygen species (ROS) via Fenton and Haber-Weiss chemistry, promoting oxidative stress, DNA damage, lipid peroxidation, and, ultimately, cell death (ferroptosis) [26].

Cancer cells often exhibit dysregulated iron metabolism, marked by increased uptake (via transferrin receptor 1, TfR1) and reduced storage or export. This iron dependency not only supports proliferation but also renders them vulnerable to iron-catalyzed ROS damage [27]. Moderate oxidative stress can activate survival pathways, such as NF-κB and HIF-1α [28], while excessive ROS triggers genomic instability, protein unfolding, lipid peroxidation, and mitochondrial dysfunction hallmarks that can be exploited therapeutically [29].

Hydroxyl radicals generated in the Fenton reaction attack nucleic acids and lipids. They react with deoxyribose sugars, purines, and pyrimidines, forming adducts like 8-oxo-2′-deoxyguanosine (8-oxo-dG), a mutagenic lesion that mispairs with adenine, promoting mutation [30]. Mitochondria are central regulators of iron metabolism via iron-sulfur cluster (Fe-S) biogenesis, heme synthesis, and electron transport. Excess Fe catalyzes superoxide and peroxynitrite generation in the mitochondrial matrix, collapsing membrane potential, releasing cytochrome c, and triggering intrinsic apoptosis [31].

To manage iron, cancer cells modulate regulatory proteins: Transferrin receptor 1 (TfR1) is often overexpressed, while ferritin (FtH/FtL) sequesters excess iron [32]. In tumors, ferritinophagy via NCOA4 is suppressed, promoting intracellular iron retention. In colorectal cancer (CRC), TfR1 is elevated while ferroportin (FPN) is downregulated, leading to iron accumulation and enhanced tumor progression [33].

Ferroptosis has therapeutic relevance in iron-addicted and apoptosis-resistant cancers, such as triple-negative breast cancer (TNBC), hepatocellular carcinoma, and KRAS-mutant CRC. These cancers are often sensitive to glutathione peroxidase 4 (GPX4) inhibitors and other ferroptosis inducers. Furthermore, p53 suppresses or promotes ferroptosis contextually, depending on downstream effectors (e.g., SAT1, SLC7A11) [33].

Selenium in Antioxidant Defense and Cancer Prevention

Selenium (Se) is an essential micronutrient distinguished by its incorporation as the amino acid selenocysteine (Sec) into a family of proteins known as selenoproteins. It is directly encoded in the genetic code via a UGA codon reprogrammed by a specialized tRNA and the selenocysteine insertion sequence (SECIS) element. Selenium's role in cancer is complex, straddling protective, therapeutic, and potentially toxic territories, depending on dosage, chemical form, and baseline selenium status [34, 35].

The most extensively studied selenoproteins in oncology include glutathione peroxidases (GPxs), thioredoxin reductases (TrxR1 and TrxR2), and selenoprotein P (SELENOP). These selenoenzymes form the biochemical backbone of the cellular antioxidant defense system, acting as sentinels against reactive oxygen species (ROS), lipid hydroperoxides, and electrophilic stress. GPx1, a ubiquitously isoform, catalyzes the reduction of H_2O_2 and organic hydroperoxides using glutathione (GSH), while GPx2, expressed in the gastrointestinal epithelium, is critical for maintaining redox balance in the colon crypts, where proliferative stem cells reside. Loss of GPx2 has been associated with increased susceptibility to colorectal carcinogenesis [35].

Thioredoxin reductases (TrxRs) catalyze the NADPH-dependent reduction of thioredoxin (Trx), which in turn influences redox-sensitive processes and DNA synthesis. TrxR1, often overexpressed in many solid tumors, including lungs, liver, and prostate, is paradoxically required for redox stability. Inhibition of TrxR induces oxidative collapse and apoptosis in cancer cells [36].

Selenium deficiency has been associated with increased colorectal cancer (CRC) risk. Low selenium levels (<80 μg/L) have been reported in patients with CRC, particularly in selenium-depleted regions such as parts of Europe and China [37]. Low selenium status reduces expression of selenoproteins, exacerbates oxidative stress, and disrupts p53 signaling and inflammation regulation. GPX1 and SELENOP gene polymorphisms may modulate individual cancer susceptibility [38].

The chemopreventive potential of selenium was tested in clinical trials like Selenium and Vitamin E Cancer Prevention Trial (SELECT), which showed that selenium did not reduce prostate cancer risk and suggested that vitamin E supplementation alone slightly increased prostate cancer incidence. These findings have led to a paradigm shift: selenium may be protective only in deficient populations and could be harmful or inert when provided in excess [39]. Selenium exhibits a narrow therapeutic window, and its effects follow a U-shaped dose-response curve. Deficiency (<60 μg/day) impairs redox defenses, thyroid hormone metabolism, and immune function. Excess (>400 μg/day) may lead to selenosis and paradoxically promote oxidative stress and insulin resistance—factors implicated in tumorigenesis [39].

Interplay of Trace Elements in Cancer Signaling Pathways

These trace metals interact with cellular signaling pathways that regulate critical cell functions like inflammation, angiogenesis, apoptosis, and immune responses. The dysregulation of these micronutrients affects gene expression by affecting molecular regulators, such as kinase cascades, metalloproteins, and transcription factors. Moreover, micronutrient imbalance can also cause changes in the intracellular redox state. Their effects vary depending on the type of cancer, concentration, and cellular compartmentalization.

p53 and Apoptotic Signaling

Zinc is essential in maintaining the structural integrity and DNA-binding ability of the p53 protein. A deficiency in zinc destabilizes p53 conformation, impairing its transcriptional activity and compromising apoptotic responses [40]. In breast and prostate cancer, zinc transporter protein (e.g., ZIP1) downregulation is associated with the inactivation of p53 and increased cell proliferation and decreased apoptosis, characteristics commonly associated with tumor progression [41]. Through the activity of GPx and TrxR, selenium supports p53-mediated apoptosis and contributes to redox homeostasis. These enzymes help to maintain the redox status of cells and to prevent oxidative damage to DNA and proteins, maintaining the integrity of p53 signaling in selenium-replete conditions [11].

NF-κB Pathway and Inflammation

NF-κB is a widely expressed transcription factor that is involved in many processes, including inflammation and cancer. Both iron and copper influence NF-κB activation; iron-induced oxidative stress activates proapoptotic JUN N-terminal kinase

(JNK) signaling [42], while copper-linked proteins such as MEMO1 increase reactive oxygen species production upon nitric oxide stimulation, and COMMD1 modulates NF-κB mediated gene expression [43]. Copper and iron levels have been found to be higher in aggressive breast and colorectal tumors [43], and sustained NF-κB activation encourages angiogenesis, cytokine release (e.g., IL-6, TNF-α), and resistance to apoptosis [44]. In contrast, zinc negatively regulates NF-κB signaling, in part through direct inhibition of IκB kinase (IKK), a key upstream activator of NF-κB [45].

MAPK and PI3K/AKT Pathways

The MAPK and PI3K/AKT pathways are influenced by Fe and Cu, which are central to cancer growth and survival. Fe^{2+} produces ROS, especially hydroxyl radicals. These reactive species strongly activate intracellular signaling pathways, for instance, the phosphorylation of transcription factors downstream of the MAPK cascade. This illustrates how oxidative stress triggers oncogenic transformation through the activation of transcription factors [46]. Copper similarly enhances MAPK signaling and increases the invasiveness of breast cancer cells. This was demonstrated by findings that MCF-7 cells exposed to $CuSO_4$ also showed increased expression of EMT markers vimentin and fibronectin, as well as increased levels of MMP2 and MMP9 [43]. Zinc inhibits the activity of a PI3K-antagonizing lipid phosphatase, PTEN, a phosphatase that negatively controls Akt activity. This occurs presumably through binding of zinc ions to cysteine thiols, which are essential for the catalytic activity of PTEN [47]. Selenium, on the other hand, suppresses tumor-promoting signaling by regulating the MAPK pathway; for example, SeY can protect against Cr (VI)-induced dysfunction and apoptosis by regulating the mitogen-activated protein kinase pathway [48].

Epigenetic Modifications

Numerous zinc and selenium-dependent enzymes play direct roles in epigenetic modifications relevant to cancer. Several key enzymes and zinc finger proteins with zinc atom(s) in the reactive center and binding site play important roles in DNA methylation and histone modifications. As such, zinc deficiency may disrupt the functions of these enzymes and proteins and result in epigenetic dysregulation [49]. Additionally, selenium compounds have been shown to reduce global DNA methylation, enhance acetylation of histones H3 and H4, and reactivate silenced tumor suppressor genes in prostate and breast cancer models by targeting DNA methyltransferases and histone deacetylases [50].

Immune Modulation and Tumor Microenvironment

The trace elements Zn, Cu, Fe, and Se significantly shape the redox environment of the tumor microenvironment. Iron, copper, and selenium are critical for redox metabolism, and their importance is heightened during oncogene-driven perturbations in cancer [51]. To maintain redox homeostasis, it is important to maintain the elements copper and zinc and their ratio (copper/zinc) [52], while selenium prevents DNA oxidative stress by incorporating itself into selenoproteins that mediate redox balance, and zinc acts as an antioxidant [53]. Imbalances of metal ions, such as zinc, iron, selenium, and copper, are often seen in cancer patients, both at intratumoral and systemic level, reflecting their influence on oxidative stress, inflammation, and tumor progression [54].

Diagnostic and Therapeutic Implications

Alterations in transport, concentration, and distribution of trace elements drive malignant transformation and serve as potential biomarkers for early detection and prognosis. Additionally, chelation, supplementation, or delivery by nanoparticles has demonstrated therapeutic potential in a variety of cancer models by manipulating trace element levels. Using trace element profiles in clinical decision-making could enhance individualized treatment plans as precision medicine develops.

Trace Elements as Diagnostic and Prognostic Biomarkers

As discussed in several preceding sections, the abnormal serum or tissue levels of trace elements are increasingly recognized as indicators of cancer risk, progression, and therapeutic response. Findings also support the use of trace elements as noninvasive biomarkers for early detection and monitoring of cancer.

Imaging and Metallomics Applications

The spatial mapping of trace elements within tumors at subcellular resolution can be recorded using metallomics imaging techniques, such as laser ablation-inductively coupled plasma-mass spectrometry (LA-ICP-MS), X-ray fluorescence (XRF) microscopy, and synchrotron radiation. The gradients of trace metals and their association with specific cancer subtypes can be defined because of these techniques [55]. In a clinical setting, PET (positron emission tomography) imaging in tumors with high copper metabolism has been employed using radioactive isotopes, such as ^{64}Cu and ^{68}Ga. This has improved early detection and real-time assessment of therapeutic responses [56].

Therapeutic Strategies: Modulating Trace Element Homeostasis

Utilizing trace elements (specifically Zn, Cu, Fe, and Se) in therapeutic modulation is a rapidly evolving field. Some of the approaches used include elemental supplementation, chelation therapy, and nanoparticle delivery.

A. Supplementation Therapy: In some studies, for patients who lack selenium, taking supplements (seleno-based) like selenomethionine or sodium selenite has been found to help prevent colorectal and prostate cancer, but the results depend on the amount taken and the specific situation. In zinc-deficient patients, zinc supplementation has also been studied for its potential to restore immune surveillance and enhance the efficacy of standard chemotherapy [57].
B. Chelation Therapy: Copper chelators, for example, tetrathiomolybdate (TM), have demonstrated efficacy in reducing systemic copper levels, thereby suppressing angiogenesis, and tumors have been suppressed using [58].
C. Nanoparticle-Based Delivery: Nanotechnology is being utilized to deliver trace elements directly to tumor tissues. For example, zinc oxide (ZnO) nanoparticles, copper sulfide nanoparticles, and selenium quantum dots are under investigation for their cytotoxic effects, ROS generation, and selective uptake by tumor cells [59, 60].

Conclusions and Future Directions

In conclusion, trace elements play very critical roles in cancer development and progression (summarized in Table 10.1). Although we need them in minute amounts, trace elements help support important processes. When they are not properly regulated in the body, they can contribute to cancer by either promoting or failing to prevent abnormal cell growth.

Protective proteins like p53 and PARP1 are supported by zinc, new blood vessel formation in tumors is encouraged by copper, oxidative stress can be produced by iron when in excess, and damage through antioxidant enzymes is reduced by selenium. These trace elements also influence signaling pathways, gene expression, and the tumor microenvironment. Moreover, they are being explored as diagnostic markers and treatment options using chelation, supplementation, and even nanoparticle delivery, because of these roles.

Table 10.1 Functional profiles of essential trace elements in oxidative stress, DNA repair, and cell death

Element	Key functions	Oncologic relevance	Mechanistic targets	Disruption consequences	References
Zinc (Zn)	Structural support for DNA-binding domains (e.g., p53)Cofactor for ~2–3% of genome (zinc-finger TFs) DNA repair via PARP1Maintains redox neutrality in DNA–protein interfacesModulates immune and apoptotic signaling	Prostate & breast cancer: ZIP1 downregulation leads to Zn depletionDisrupts mitochondrial aconitase, impairs TCA, lowers ATPSP1, EGR1, RREB1 TF dysregulation promotes oncogenesisNF-κB signaling suppression via Zn restoration	p53 (Cys176/His179/Cys238/Cys242)PARP1 Zn-finger domainsZIP (e.g., ZIP1/6/7/9), ZnT1 transportersTranscriptional regulators (SP1, EGR1, RREB1)	p53 misfolding/inactivationLoss of DNA repair fidelityChromosomal instabilityDeregulated apoptosis and cell cycle arrest	[13–18]
Copper (Cu)	Catalytic cofactor in VEGF/FGF angiogenesisRedox signaling via Fenton-like ROSStructural cofactor for LOX/LOXL (ECM remodeling)Required for ceruloplasmin & ATOX1 activity	↑ Serum/tumor Cu in CRC and breast cancerDrives HIF-1α/VEGF axis in hypoxiaPromotes LOX-dependent metastasis and stiffnessATOX1: nuclear copper-dependent oncogene (↑ cyclin D1)	VEGFR2 phosphorylationFGF2–FGFR interaction via HSPGLOX/LOXL enzymatic activityCeruloplasmin (CP), ATOX1HIF-1α, MAPK, NF-κB	Blocked angiogenesis with Cu chelatorsLoss of LOX activity impairs ECMATOX1 knockdown reduces tumor proliferationCP elevation correlates with metastasis	[19–24]
Iron (Fe)	Cofactor for ribonucleotide reductase, cytochromesDrives ROS via Fenton reactionEssential for mitochondrial Fe-S cluster biogenesisCentral to ferroptosis	CRC & HCC show ↑ TfR1, ↓ ferroportinFerritinophagy suppression iron overloadFerroptosis sensitivity in TNBC, HCC, KRAS+ CRCMitochondrial damage, DNA adducts (8-oxo-dG)	Transferrin/TfR1-mediated uptakeFerritin (FtH/FtL) storage, NCOA4-mediated releaseFerroptosis mediators: GPX4, SAT1, SLC7A11p53 (regulates SLC7A11)	ROS-induced DNA damageMitochondrial dysfunction & apoptosisGenome instabilityIron addiction exploited via ferroptosis	[26, 27, 29–31, 33]
Selenium (Se)	Incorporated into GPx, TrxR enzymesMaintains cellular redox toneDetoxifies ROS & lipid peroxidesModulates immune and metabolic pathways	Se deficiency linked to CRC (↓ GPx2 in colon crypts)SELECT trial: no benefit in Se-replete populationsSELENOP, GPX1 polymorphisms affect riskSe mimetics induce apoptosis in tumors	GPx1/2/3 reduce H_2O_2 and lipid peroxidesTrxR1/2 regulate thiol–disulfide balanceSELENOP transports Se to tissuesModulates HIF, STAT, redox-responsive genes	Se deficiency → ↑ ROS, DNA damageExcess Se → selenosis, insulin resistanceU-shaped dose-risk curve complicates therapy	[34–39]

However, while the benefits are promising, there are still some important challenges. Trace elements have a narrow therapeutic range, so too much or too little can be harmful. People absorb and respond to minerals differently; even the tumor environment can change how treatment works. To improve outcomes, doctors may need to use personalized tools like genomic and metabolic profiling to guide trace element-based therapies more safely and effectively.

Overall, trace elements hold great promise in helping to understand, detect, and manage cancer, but more research is still needed to guide their safe and effective use in clinical settings.

References

1. Strachan S (2010) Trace elements. Curr Anaesth Crit Care 21(1):44–48
2. Thomas L (2025) Trace elements. In: Thomas L (ed) Clinical laboratory diagnostics: use and assessment of clinical laboratory results, 3rd edn. BSP Publications, Frankfurt, pp 1–42
3. Aliasgharpour M (2020) Trace elements in human nutrition (II)—an update. Int J Prev Med 11:2
4. Prasad AS (2014) Zinc is an antioxidant and anti-inflammatory agent: its role in human health. Front Nutr 1:14
5. Zoroddu MA, Aaseth J, Crisponi G et al (2019) The essential metals for humans: a brief overview. J Inorg Biochem 195:120–129
6. Lossow K, Schwarz M, Kipp AP (2021) Are trace element concentrations suitable biomarkers for the diagnosis of cancer? Redox Biol 46:101900
7. Maret W (2019) The redox biology of redox-inert zinc ions. Free Radic Biol Med 134:311–326
8. Zhang Y, He J, Jin J, Ren C (2022) Recent advances in the application of metallomics in diagnosis and prognosis of human cancer. Metallomics 14:mfac037
9. Ho E (2004) Zinc deficiency, DNA damage and cancer risk. J Nutr Biochem 15(10):572–578
10. Tang X, Yan Z, Miao Y et al (2023) Copper in cancer: from limiting nutrient to therapeutic target. Front Oncol 13:1209156
11. Rayman MP (2012) Selenium and human health. Lancet 379(9822):1256–1268
12. Bray F, Ferlay J, Soerjomataram I et al (2018) Global cancer statistics 2018: GLOBOCAN estimates of incidence and mortality worldwide. CA Cancer J Clin 68(6):394–424
13. Blanden AR, Yu X, Loh SN, Levine AJ, Carpizo DR (2016) Reactivating mutant p53 using small molecules as zinc metallochaperones: awakening a sleeping giant in cancer. Proc Natl Acad Sci USA 113(9):E1182–E1191
14. Joerger AC, Fersht AR (2010) The tumor suppressor p53: from structures to drug discovery. Cold Spring Harb Perspect Biol 2(6):a000919
15. Eustermann S, Videler H, Yang JC, Cole PT, Gruszka D, Veprintsev D, Neuhaus D (2011) The DNA-binding domain of human PARP-1 interacts with DNA single-strand breaks as a monomer through its second zinc finger. J Mol Biol 407(1):149–170
16. Choi S, Bird AJ (2014) Zinc'ing sensibly: controlling zinc homeostasis at the transcriptional level. Metallomics 6(7):1198–1215
17. Bin BH, Seo J, Kim ST (2018) Function, structure, and transport aspects of ZIP and ZnT zinc transporters in immune cells. Mol Asp Med 66:29–44
18. Bafaro E, Liu Y, Xu Y, Dempski RE (2017) The emerging role of zinc transporters in cellular homeostasis and cancer. Signal Transduct Target Ther 2:17029
19. Nasulewicz A, Mazur A, Opolski A (2004) Role of copper in tumour angiogenesis--clinical implications. Curr Med Chem 11:525–530

20. Duffy AM, Bouchier-Hayes DJ, Harmey JH (2013) Vascular Endothelial Growth Factor (VEGF) and its role in non-endothelial cells: autocrine signalling by VEGF. In: Madri JA (ed) Vascular biology of the placenta. NCBI Bookshelf. https://www.ncbi.nlm.nih.gov/books/NBK6482/
21. Feng W, Ye F, Xue W, Zhou Z, Kang YJ (2008) Copper regulation of hypoxia-inducible factor-1 activity. Mol Pharmacol 75:174–182
22. Wang ZJ, Guan QW, Zhou HH, Mao XY, Chen FH (2022) Mechanistic insight into lysyl oxidase in vascular remodeling and angiogenesis. Front Pharmacol 13:10308135
23. Das A, Sudhahar V, Chen GF, Kim HW, Youn SW, Finney L, Yang J, Kweon J, Surenkhuu B, Ushio-Fukai M, Fukai T (2016) Endothelial Antioxidant-1: a key mediator of copper-dependent wound healing in vivo. Arterioscler Thromb Vasc Biol 36:1909–1919
24. Baszuk P, Marciniak W, Derkacz R, Jakubowska A, Cybulski C, Gronwald J, Dębniak T, Huzarski T, Białkowska K, Pietrzak S, Muszyńska M, Kładny J, Narod SA, Lubniski J, Lener MR (2021) Blood copper levels and the occurrence of colorectal cancer in Poland. Biomedicine 9:1628
25. Stepien M, Jenab M, Freisling H, Becker NP (2017) Pre-diagnostic copper and zinc biomarkers and colorectal cancer risk in the European Prospective Investigation into Cancer and Nutrition cohort. Carcinogenesis 38(7):699–707
26. Torti SV, Manz DH, Paul BT, Blanchette-Farra N, Torti FM (2021) Iron and cancer. Nat Rev Cancer 21:466–480
27. Obeagu EI, Tanko MSM (2025) Iron metabolism in breast cancer: mechanisms and therapeutic implications: a narrative review. J Cancer Biol Res [In press]
28. DeNicola GM, Karreth FA, Humpton TJ, Gopinathan A, Wei C, Frese K, Mangal D, Yu KH, Yeo CJ, Calhoun ES, Scrimieri F, Winter JM, Hruban RH, Iacobuzio-Donahue C, Kimmelman AC, Tuveson DA (2011) Oncogene-induced Nrf2 transcription promotes ROS detoxification and tumorigenesis. Nature 475:106–109
29. Toyokuni S (2009) Role of iron in carcinogenesis: cancer as a ferrotoxic disease. Cancer Sci 100:9–16
30. Cadet J, Wagner JR (2013) DNA base damage by reactive oxygen species, oxidizing agents, and UV radiation. Cold Spring Harb Perspect Biol 5:a012559
31. Gao H, Korn JM, Ferretti S, Monahan JE, Wang Y, Singh M, Zhang C, Schnell C, Yang G, Zhang Y, Balbin OA, Barbe S, Cai H, Casey F, Chatterjee S (2015) High-throughput screening using patient-derived tumor xenografts to predict clinical trial drug response. Nat Med 21:1318–1325
32. Abdukarimov N, Kokabi K, Kunz J (2025) Ferroptosis and iron homeostasis: molecular mechanisms and neurodegenerative disease implications. Antioxidants 14:527
33. Huang L, Li W, Lu Y, Ju Q, Ouyang M (2023) Iron metabolism in colorectal cancer. Int J Mol Sci 24:9992732
34. Hariharan S, Dharmaraj S (2020) Selenium and selenoproteins: its role in regulation of inflammation. Free Radic Biol Med 152:45–52
35. Florian S, Krehl S, Loewinger M, Kipp A, Banning A, Esworthy S, Chu FF, Brigelius-Flohé R (2010) Loss of GPx2 increases apoptosis, mitosis, and GPx1 expression in the intestine of mice. Free Radic Biol Med 49:1694–1702
36. Mustacich D, Powis G (2000) Thioredoxin reductase. Biochem J 346:1–8
37. Chalcarz M, Grabarek BO, de Mezer M, Krokowicz P, Patera J, Czupryna K, Żurawski J (2024) Serum selenium concentration as a potential diagnostic marker for early-stage colorectal cancer: a comparative study. Cancer 16:538
38. Trachootham D, Chen G, Zhang W, Lu W, Zhang H, Liu J, Huang P (2014) Loss of p53 in stromal fibroblasts promotes epithelial cell invasion through redox-mediated ICAM1 signal. Cell 149:61–74
39. Lippman SM, Klein EA, Goodman PJ, Lucia MS, Thompson IM, Ford LG, Parnes HL, Minasian LM, Gaziano JM, Hartline JA, Parsons JK, Bearden JD, Crawford ED, Goodman

GE, Claudio J (2009) Effect of selenium and vitamin E on risk of prostate cancer and other cancers: the Selenium and Vitamin E Cancer Prevention Trial (SELECT). JAMA 301:39–51
40. Ho E, Prasad AS (2014) Zinc in cancer prevention. Nutrients 6:2066–2083
41. Chen W, Feng C, Tang W, Zhao Y (2024) Downregulation of zinc transporter ZIP1 promotes breast and prostate cancer progression via inactivation of p53. J Cell Mol Med 28:451–462
42. Torti SV, Torti FM (2013) Iron and cancer: more ore to be mined. Nat Rev Cancer 13:342–355
43. Blockhuys S, Wittung-Stafshede P (2017) Roles of copper-binding proteins in breast cancer. Int J Mol Sci 18:871
44. Bahrami A, Khalaji A, Najafi MB, Sadati S, Raisi A, Abolhassani A, Eshraghi R, Mahabady MK, Rahimian N, Mirzaei H (2024) NF-κB pathway and angiogenesis: novel insights in cancer biology. Pathol Oncol Res 30:51
45. Haase H, Rink L (2013) Zinc signals and immunity. Biofactors 39:186–192
46. McCubrey JA, Steelman LS, Chappell WH, Abrams SL, Wong EW, Chang F, Lehmann B, Terrian DM, Milella M, Tafuri A, Stivala F, Libra M, Basecke J, Evangelisti C, Martelli AM, Franklin RA (2006) Roles of the Raf/MEK/ERK pathway in cell growth, malignant transformation and drug resistance. Biochim Biophys Acta 1773:1263–1284
47. Kim J, Lee Y (2021) Zinc modulation of PTEN activity in PI3K/Akt pathway regulation. Biol Trace Elem Res 199:1125–1133
48. Zhao F, Liu Y, Li Q, Han S, Liu J (2022) Selenomethionine protects against Cr(VI)-induced apoptosis by modulating MAPK signaling in hepatocytes. J Trace Elem Med Biol 69:126898
49. Gu Y, Zang Y (2017) Zinc finger proteins in epigenetic regulation of gene expression. J Genet Genomics 44:231–243
50. Stybalska D, Król T (2022) Epigenetic effects of selenium compounds in cancer: from mechanisms to therapy. Epigenomics 14:1121–1135
51. Terzi EM, Possemato R (2024) Nutrient sensing and redox homeostasis in cancer. Trends Cell Biol 34:150–163
52. Bengtsson J, Johansson A, Olsson B, Björkhem-Bergman L (2023) The copper/zinc ratio as a biomarker in cancer: recent evidence and clinical relevance. Cancer 15:1815
53. Kim J, Lee H, Lee J (2024) Zinc and selenium: dual modulators of antioxidant defense and immune surveillance in cancer. Redox Biol 70:102645
54. Serra D, Rufini A, Comito G, Galli G (2020) Systemic trace element imbalance in cancer: impact on tumor microenvironment and immune response. Cancer 12:2915
55. Qin Z, Caruso JA, Lai B, Matusch A, Becker JS (2011) Trace metal imaging with high spatial resolution: applications in biomedicine. Metallomics 3:28–37
56. Capriotti G, Piccardo A, Giovannelli E, Signore A (2023) Targeting copper in cancer imaging and therapy: a new theragnostic agent. J Clin Med 12:223
57. Peters U, Takata Y (2008) Selenium and the prevention of prostate and colorectal cancer. Mol Nutr Food Res 52(11):1261–1272
58. Kim KK, Abelman S, Yano N, Ribeiro JR, Singh RK, Tipping M, Moore RG (2015) Tetrathiomolybdate inhibits mitochondrial complex IV and mediates degradation of hypoxia-inducible factor-1α in cancer cells. Sci Rep 5:14296
59. Skrajnowska D, Bobrowska-Korczak B (2019) Role of zinc in immune system and anti-cancer defense mechanisms. Nutrients 11(10):2273
60. Rasmussen JW, Martinez E, Louka P, Wingett DG (2010) Zinc oxide nanoparticles for selective destruction of tumor cells and potential for drug delivery applications. Expert Opin Drug Deliv 7(9):1063–1077

Part II
Heavy Metals and Major Minerals

Chapter 11
Heavy Metals for Treating Disease and Maintaining Health

Rolf Teschke, Axel Eickhoff, and Tran Dang Xuan

Abstract The book chapter provides a brief survey of the clinical role of essential and nonessential heavy metals for treating disease and maintaining health. Both forms of heavy metals have a long history as medicinal products to cure human diseases, while indication for the treatment of specific disorders was found for 27/32 heavy metals, ranging from aluminum to zinc. However, their clinical use is nowadays restricted to a few ones due to a lack of evidence based on therapeutic efficacy or because of substitution by other chemical medicines with a positive ratio of benefits versus risks as based on results of randomized controlled trials (RCTs). Promising are heavy metals like gold, silver, platinum, palladium, titanium, copper, iron, and zinc if used as nanoparticles (NPs) ranging from 1 to 100 nm in size. They can become part of personalized therapy approaches for malignancies in the context of cancer drug resistance. Therapeutic cytotoxic effects of metallic NPs are initiated through the generation of reactive intermediates, which lead to oxidative stress within mitochondria or endoplasmic reticulum, representing the microsomes, DNA damage, cell cycle arrest, and finally to apoptotic/necrotic cell death or autophagy. Their therapeutic efficacy can be enhanced if combined with antibodies, DNA, RNA, peptides, and proteins. Issues of biosensing, catalysis, and bioimaging are strongly associated with the properties on NPs. NPs traffic to the spleen and lymph organs with their relevant immune cells, making them good candidates for the delivery of immunotherapeutic agents. For maintaining health, various essential heavy

R. Teschke (✉) · A. Eickhoff
Department of Internal Medicine II, Division of Gastroenterology and Hepatology, Klinikum Hanau, Hanau, Germany

Academic Teaching Hospital of the Medical Faculty, Goethe University Frankfurt/ Main, Frankfurt/Main, Germany

T. D. Xuan
Center for the Planetary Health and Innovation Science, the IDEC Institute, Smart Agriculture Faculty, Laboratory of Plant Physiology and Biochemistry, Graduate School of Advanced Science and Engineering, Hiroshima University, Higashi-Hiroshima City, Hiroshima, Japan
e-mail: tdxuan@hiroshima-u.ac.jp

N. S. Dhalla et al. (eds.), *Functional Biochemistry of Metallic Elements*, Advances in Biochemistry in Health and Disease 36,
https://doi.org/10.1007/978-3-032-14445-4_11

metals in low amounts are responsible. However, heavy metals are potentially toxic in high amounts in a dose-dependent way, while nonessential ones are injurious even in small concentrations.

Keywords Essential heavy metals · Nonessential heavy metals · Metallic nanoparticles · Health · Disease · Environmental pollution, Occupational disease

Introduction

Heavy metals were generated in the universe via atomic fusion before they were donated to our planet [1], where cells of organisms evolved mechanisms to make use of the heavy metals as integral parts of their structure and functions, resulting in human evolution [2]. Without heavy metals, humans would not exist, and both human life and health are intricately connected with the environmental heavy metals; part of these are recognized as essential elements because of their involvement in important metabolic pathways. This helps sustain the function of human organs and organ systems [3]. In a step forward, humans recognized early that heavy metals found below the earth surface can be used for technical purposes as metallic alloy products manufactured in industrial settings [4, 5]. Today, heavy metals are universally used as alloys in innovative industrial products that provide economic stability and welfare [6], but they may be hazardous to human life if used as weapons [7] and if they cause occupational diseases [8] or pollute the environment [9].

The book chapter provides a brief clinical and environmental survey on the role of heavy metals and their chemical derivatives, including nano elements as medicinal products to cure human diseases, their involvement as essential elements for biological functions in humans, and to be differentiated from their role as health risk factors for humans exposed to high amounts.

Heavy Metals in Use as Medicines to Sustain Health and Cure Disorders

Heavy metals and their compounds attracted much interest as potential remedies for a variety of diseases, starting with traditional medicines and followed nowadays by scholar medicine. The analysis showed that most heavy metals have been used for treating specific diseases, but efficacy was rarely established with certainty (Table 11.1) [10–87].

Table 11.1 Listing of heavy metals used as medicines to sustain health and treat diseases

Heavy metal	Used as medicinal product or supplement	References
Aluminum (Al)	Antiperspirant Adjuvant for vaccination Desensitization procedures Antacid agent	[10]
Antimony (Sb)	Acute promyelocytic leukemia Anthroposophical medicine Leishmaniasis Schistosomiasis Psoriasis Tartar Emetic	[11–13]
Arsenic (As)	Acute promyelocytic leukemia Ayurveda medicine Orpiment Syphilis TCM Realgar	[14–17]
Barium (Ba)	Barium isotopes in nuclear medicine	[18]
Beryllium (Be)	NO	[19]
Bismuth (Bi)	Cancer Dyspepsia Gastric ulcers Helicobacter pylori infections Multidrug resistant microbial infections Viral infections	[20]
Cadmium (Cd)	NO	[21]
Chromium (Cr)	Arterial hypertension Cardiovascular diseases Diabetes mellitus type 2 Hyperlipidemia Overweight	[22–25]
Cobalt (Co)	Cancer Drug resistant herpes simplex infection Drug resistant bacterial infections Hyperchromic anemia due to deficiency of vitamin B_{12} (cobalamin) Melanoma Primary biliary cholangitis	[26–30]
Copper (Cu)	Bacterial infections Cancer Chronic ischemic cerebrovascular disease Fungal infections Menkes syndrome due to copper deficiency Viral infections	[31–35]
Gallium (Ga)	Bladder cancer Breast cancer Cancer-related hypercalcemia COVID-19 infections Multi-drug resistant bacterial infections Non-Hodgkin lymphoma	[36–41]

(continued)

Table 11.1 (continued)

Heavy metal	Used as medicinal product or supplement	References
Germanium (Ge)	AIDS Arthritis Breast cancer Melanoma Refractory cancers Senile osteoporosis Spindle cell carcinoma Virus infections	[42–44]
Gold (Au)	Alzheimer disease Cancers Dementia Juvenile idiopathic arthritis Lupus erythematosus Parasitic diseases Psoriatric arthritis Rheumatoid arthritis	[45–52]
Iron (Fe)	Anemia in pregnancy Anemia due to cancer Inflammatory bowel disease Periodic limb movement disorder Restless leg syndrome Microcytic anemia due to iron deficiency	[53–58]
Indium (In)	NO	[59]
Lead (Pb)	Traditional medicines	[60]
Lithium (Li)	Bipolar disorders	[61]
Manganese (Mn)	Cancers	[62]
Mercury (Hg)	Syphilis Traditional Chinese Medicine	[63]
Molybdenum (Mo)	Cancers	[64]
Nickel (Ni)	Antimicrobial agent Cancers	[65]
Platinum (Pt)	Cancers Leishmaniasis Malaria tropica Metastatic urothelial carcinoma	[66–68]
Selenium (Se)	Arterial hypertension Cancers Immunomodulatory agents	[69–71]
Silver (Ag)	Antimicrobial agents Cancers	[72, 73]
Strontium (Sr)	Osteoporosis	[74]
Tellurium (Te)	Acute myeloid leukemia Bacterial infectious Cancers Chemotherapy-induced thrombocytopenia Dermatitis Genital warts HIV Macular degeneration Psoriasis	[75–77]

(continued)

Table 11.1 (continued)

Thallium (Tl)	NO	[78]
Tin (Sn)	Cancer Onchocerciasis Rosacea SARS-CoV-2 infections Scabies Viral infections	[79, 80]
Titanium (Ti)	Cancer	[81]
Uranium (U)	NO	[82]
Vanadium (V)	Cancer Crohn's disease Diabetes mellitus Parasitic diseases	[83]
Zinc (Zn)	Acrodermatitis enteropathica Crohn's disease Cancer Diabetes mellitus Diaper dermatitis Wilson disease Zinc deficiency	[84–87]

Abbreviations: *AIDS* Acquired Immune Deficiency Syndrome*; COVID-19*, Coronavirus Disease; *HIV* Human Immunodeficiency Virus; *SARS-CoV-2* Severe acute respiratory syndrome coronavirus type 2

Overall, 32 heavy metals were analyzed for their potential use as medicines to cure human diseases (Table 11.1). Only a few heavy metals qualify for good therapeutic efficacy; among these are lithium for bipolar disorders [61] and zinc as a soft paste used for prevention and treatment of diaper dermatitis, known to all parents with newborn children [87]. Promising therapeutics are nanomaterials of heavy metals [72]. Literature search revealed medicinal use by 27/32 of heavy metals (84,4%), with lacking information on medicinal application in five heavy metals, beryllium, cadmium, indium, thallium, and uranium (Table 11.1). However, to qualify for an effective medicine in humans, the benchmark is high and must meet the commonly accepted definitions of evidence-based medicine (EBM) [88]. For applying EBM to clinical practice, five steps are required: (1) definition of a clinically relevant question, (2) search for the best evidence, (3) critical appraisal of the evidence, (4) application of the evidence, and (5) evaluation of the EBM performance. Best results are obtained by using a study protocol of randomized controlled trials (RCTs). As a result, since only a few heavy metals were evaluated for EBM qualification, a cautionary note is warranted to use heavy metals that are not qualified as regulatory-approved medicines. Instead, modern medicine provides many drugs that are effective in treating human diseases.

Problematic and clearly outdated are heavy metals as examples in reference to aluminum as an antacid agent, now substituted by proton pump inhibitors [89]; antimony for schistosomiasis, now replaced by praziquantel [90]; arsenic for syphilis, now supplanted by penicillins [91]; bismuth for helicobacter pylori infections, now upgraded by the triple combination therapy [92]; and copper for bacterial infections, now changed to antibiotics [93]. Improvement was also indicated for gallium regarding treatment ofnon-Hodgkin lymphoma, now upgraded by a combination chemotherapy [94]; germanium for melanoma, now substituted by immune-checkpoint inhibitors [95]; platinum for malaria, now replaced by artemether-lumefantrine-amodiaquine [96]; tellurium for acute myeloid leukemia, now replaced by various new chemotherapeutic agents [97]; tin for rosacea, now substituted by antibiotics [98]; and vanadium for Cohn's disease, now replaced by 5-amiosalicyclic acid, steroids, immunomodulators, or biologics [99]. Problematic is often the small therapeutic window for safe use of heavy metals that may easily lead to toxicity. Unquestionable, for many heavy metals used as medicines (Table 11.1) the ratio of benefit as compared to risks is negative.

Other heavy metals, when used as nanoparticles (NPs), will have an excellent perspective as pharmaceuticals in a modern medicine setting, provided data from additional preclinical can be made available. Among these promising new medicines are, for instance, nanoparticles (NPs) of gold, silver, platinum, palladium, titanium, copper, iron, and zinc that can help cure patients with malignancies and are viewed as promotional candidates to enhance the efficacy of current cancer therapies [72, 100–106]. There is also a significant potential of metallic NPs to be used as drugs for bacterial infections [107]. Metallic NPs are variable regarding their size from 1 to 100 nm [100, 106], are multifunctional under therapeutic conditions [72, 100], and have the advantage of being usable as personalized cancer treatment, tailored to the unique molecular features known for individual tumors and among affected patients [72, 101], and targeting even overexpressed receptors commonly found in cancer cells [101]. The highly appreciated tailoring of specific characteristics of metallic NPs is attributed to their modifying and multifunctional features [72, 100]. Therapeutic effects can be ascribed to the cytotoxic properties of metallic NPs that become evident in the course of reactive intermediates generation, which in turn trigger oxidative stress in the subcellular compartments of mitochondria and the endoplasmic reticulum, known also as microsomal fraction among biochemists, and cause DNA damage, cell cycle arrest, conditions that finally result in apoptotic/necrotic cell death or autophagy [102]. Their positive effects can be further improved via their combination with antibodies, DNA, RNA, peptides, and proteins [103–106]. Issues of biosensing, catalysis, and bioimaging are strongly associated with the features of NPs [104]. NPs arrive via the systemic blood circulation at the spleen and lymph organs harboring relevant immune cells, helping them become perfect candidates for release of immunotherapeutic agents [105]. More importantly, metallic NPs are part of cancer immunotherapy destined to recognize and remove cancer cells [72, 106]. There are good perspectives using metallic NPs as adjuvants for vaccines against cancer, providing synergies for their immunomodulatory properties, aiming to allow for antigen presentation and immune cell activation, with the final perspective to facilitate tumor regression [72]. In this context, current immune checkpoint inhibitors can be supplemented by metallic NPs to be prepared to modify immune checkpoint pathways to facilitate antitumor immune responses [72].

More specific information is available for a few special metallic NPs [72, 100–110]: (1) NPs of gold (AuNPs) are highly appreciated for their excellent biocompatibility, modifiable surface chemistry, and special optical properties [72, 103]. These features help generate reactive radicals [106]. They provide drugs directly to tumor cells of specifically target cancers and reduce thereby organ-specific and systemic toxicity [72, 100, 104]. AuNPs accumulate in the tumor and provide the basis of laser hyperthermia therapy [100]. In the context of vaccines, AuNPs can improve their delivery [105]. (2) NPs of silver (AgNPs) effectively contribute to cancer therapy options [104], with a focus on cancer types of resistant to chemotherapy or radiotherapy [72, 103]. AgNPs are highlighted by their antiproliferative and proapoptotic properties in glioma cells after radiation therapy [100] and appreciated for their antimicrobial efficacy [104] and for improving vaccine release [105]. (3) NPs of platinum (PtNPs) are good therapeutic candidates for breast cancer therapy as evidence in several preclinical studies [101–103]. (4) Palladium NPs (PdNPs) with their excellent catalytic characteristics and high surface area are welcomed nano-radiopharmaceuticals and are viewed as promising candidates for the therapy of patients suffering from breast cancer [101–103]. (5) Titanium dioxide NPs (TiNPs) are used because of their focus on increased photocatalytic activity and improved UV absorption efficiency [100] and are appreciated for their properties to improve vaccine delivery [105]. (6) Copper oxide NPs (CuNPs) are characterized specifically by their effective cytotoxic properties against lung cancer cells, while resulting in apoptosis due to increased production of toxic radicals [100, 106], and have the potential of facilitating vaccine delivery [105]. (7) Iron oxide NPs (FeNPs) are appreciated for their magneto-sensitivity, and as such, they are applied in conjunction with doxorubicin resulting in enhanced anticancer effects via the paramagnetic properties of doxorubicin and the potential for electron transitions into nano-complexes with more free radical formation [100]. In addition, FeNPs are privileged to enhance vaccine release [105]. (8) Zinc oxide NPs (ZnNPs) are taken up by the cancer cells and injure DNA in the sense of genotoxic medicines [100], improve vaccine delivery [105], and are capable of toxic radical generation [106]. (9) Further progress is to be expected for NPs of bioactive heavy metals because their compounds can be combined with antibiotic drug molecules; they may function as effective nanocomposites and have the prospective to become carriers in targeted drug-delivery systems of antibacterial drugs. Under special consideration are NPs of gold, silver, titanium, copper, and iron, which exhibit extraordinary antibacterial potential against gram-negative and gram-positive bacterial strains, with a focus on multiresistant bacterial infections [107, 108]. (10) NPs of titanium commonly part in nanostructured ceramics can reduce friction and wear problems known in connection with artificial joint titanium replacements [109, 110] and can also treat bacterial infections often seen in patients with joint replacements [107]. (12) Progress is also underway by using NPs as adjuvants for vaccines against cancer, facilitating synergies for their immunomodulatory properties to improve antigen presentation and immune cell activation, aiming to tumor regression [72, 108]. NPs of heavy metals generally augment vaccine release through improved antigen uptake by dendritic cells and other antigen-presenting cells (APCs), thus ameliorating the respective antitumor cytotoxic T cell response [105].

As opposed to the excellent therapeutic perspectives of heavy metals as NPs [100–110], heavy metals, such antimony, arsenic, cadmium, chromium, copper, iron, lead, manganese, mercury, nickel, selenium, tin, thallium, and zinc, as detected in herbal medicines, have no future as they are potentially hepatotoxic [111]. As an example, problematic is the presence of arsenic in Ayurvedic medicine [112–117] or in herbal traditional Chinese medicine [118]. The inclusion of arsenic in herbal medicines was promoted under the expectation to increase the efficacy of the herbal products, but this assumption was never verified. Instead, the high risks of liver injury in cases [112–118] were evaluated by the Roussel Uclaf Causality Assessment Method (RUCAM) [119].

Essential Heavy Metals and Their Biological Functions in Humans

Of clinical interest are essential heavy metals at low amounts, which as a group are intricately connected with human evolution, intrauterine development of embryos and fetuses, and human health from birth until adults [1–3]. Essential heavy metals with the biological functions are listed (Table 11.2) [120–132].

Essential heavy metals are a perfect godsend given to the earth for the sake of humans, allowing for their evolution and health preservation (Table 11.2) [120–133]. Among the many heavy metals found on earth, only eight qualify as essential elements, namely, cobalt, copper, iron, manganese, molybdenum, nickel, selenium, and zinc (Table 11.2). None of these can be synthesized in the human organism but must be provided from outside sources of the environment. The essential heavy metals are found in rocks, the atmosphere, water, and soil [134–136]. Plants growing on soils containing essential heavy metals absorb them [137]. This proceeds via their roots and rhizomes, but uptake is also possible through leaves out of the atmosphere or through human activities like watering. As a result, essential heavy metals contained in the plants serve as feed for grazing cattle, providing these to humans via their meat. Such plants are, as vegetables or fruits, also part of the food chain of humans, who eat them. Essential heavy metals also arrive via the drinking water.

Heavy Metal Toxicities

Of note, many nonessential heavy metals are toxic in small and high amounts [133]. As expected, high amounts of essential heavy metals are hazards for human health in the context of genetic predisposition with excess deposition of copper in patients with Wilson disease [86] and of iron accumulation in patients with hemochromatosis [138], or via environmental and occupational exposure [133].

Table 11.2 Listing of essential heavy metals with their biological properties in human

Essential heavy metal	Benefits of environmental essential heavy metals for human evolution and sustained healthy life	References
Cobalt (Co)	Cobalt contributes to angiogenesis and is involved in bone tissue regeneration. Cobalt consumption from environmental sources impedes deficiency causing megaloblastic pernicious anemia because it is a pivotal part of vitamin B_{12} (cobalamin), which is a cofactor of metabolic pathways involved in hematologic or neurologic functions and bone tissue regeneration.	[120–123]
Copper (Cu)	Copper ingested through environmental products contaminated by copper including drinking water helps provide cellular copper homeostasis. This HM plays a crucial role in many enzymes such as cytochrome *c* oxidase, superoxide dismutase 3, Cu-dependent amino-oxidase 1-3, and dopamine-β- hydroxylase, and is a constituent of Cu/Zn-dependent superoxide 1. This allows copper to influence many metabolic pathways required to sustain human health.	[124]
Iron (Fe)	Iron participates in physiological metabolic pathways, including oxygen transport, cellular energy generation, and DNA metabolism and is necessary to prevent severe clinical iron deficiency as evidenced by microcytic anemia, problems during pregnancy, and delays in childhood cognitive development	[125
Manganese (Mn)	Manganese is involved in the normal human development and many energy and related metabolic pathways including those involved in neurotransmitter production and metabolism in the brain. Manganese acts as an cofactor for metalloenzymes, transferases, superoxide dismutase, gluconeogenic enzymes, isocitrate dehydrogenase, pyruvate carboxylase, and various antioxidant enzymes that protect cells from free toxic radical damage.	[126, 127]
Molybdenum (Mo)	Molybdenum is an essential element for human life, acting as a cofactor in enzymes such as the molybdenum-containing enzymes xanthine oxidase (XO), which plays a pivotal role in purine catabolism, generating reactive oxygen species (ROS) capable of inducing oxidative stress and subsequent organ dysfunction. Elevated XO activity is associated with liver pathologies such as metabolic dysfunction-associated steatotic liver disease (MASLD) and hepatocellular carcinoma (HCC). Aldehyde oxidases (AOs) are also molybdenum-containing enzymes that, similar to XO, participate in drug metabolism, with notable roles in the oxidation of various substrates.	[128, 129]
Nickel (Ni)	Nickel is an essential element for humans, as evidenced by its deficiency, which causes reduced growth in intra-uterine development iron adsorption leading to anemia. Urease, hydrogenase, *s*-methyl coenzyme-M reductase, acetyl CoA synthase, nickel- superoxide dismutase, glyoxalase I, and *cis–trans* isomerase are all essential nickel-containing enzymes	[130]

(continued)

Table 11.2 (continued)

Essential heavy metal	Benefits of environmental essential heavy metals for human evolution and sustained healthy life	References
Selenium (Se)	Selenium is incorporated into selenoproteins that have a wide range of pleiotropic effects, ranging from antioxidant and anti-inflammatory effects to the production of active thyroid hormone. In the past 10 years, the discovery of disease-associated polymorphisms in selenoprotein genes has drawn attention to the relevance of selenoproteins to health. Low selenium status has been associated with increased risk of mortality, poor immune function, and cognitive decline. significant positive effect on the prevention and treatment of cardiovascular disease, inflammatory diseases, and cancer	[131, 132]
Zinc (Zn)	Zinc plays a crucial part in cell growth and development, metabolism, cognitive, reproductive, and immune system function. Zn^{2+} has regulatory and catalytic roles as a component in more than 2500 proteins, including enzymes and transcription factors.It plays a regulative role for gene expression, DNA metabolism, chromatin structure, cell proliferation, apoptosis, immunity, ad defenses towards oxidants. In addition, it regulates intracellular signal transduction and is critical for efficient synaptic transmission in the central nervous system. Low nutritional zinc intake and absorption are linked with cancer manifestation, progression and in some cases metastasis.	[84]

This table was modified from a previous open-access report [133]

Conclusion

Selected essential and nonessential heavy metals and their compounds, including NPs, are a promising medicine for disease treatment like malignancies in case of cancer drug resistance, while essential heavy metals in low amounts are beneficial for sustaining health through providing cellular homeostasis. As an example, (1) cobalt is a pivotal part of vitamin B_{12} contributing to erythrocyte generation in the bone marrow, (2) copper is a constituent of the Cu/Zn-dependent superoxide 1, (3) iron is included in the hemoglobin of erythrocytes, (4) manganese is a cofactor for metalloenzymes, (5) molybdenum is a constituent of molybdenum-containing enzymes like xanthine oxidase, (6) nickel is included in the nickel-superoxide dismutase, (7) selenium is incorporated in selenoproteins that have a wide range of pleiotropic effects, and (8) zinc plays a crucial part in cell growth and development, metabolism, cognitive, reproductive, and immune systems, such as cancer drug resistance. Despite some appreciation, nonessential heavy metals may be toxic even in small concentrations, while essential heavy metals are toxic in high amounts.

Conflict of Interest The authors declare that they have no conflict of interest.

Author Contribution RT and AE provided the outline of the article and collected the relevant literature; TDX conceptualized the tables. RT wrote the first draft, which was edited by AE and TDX. All authors agreed on the final version to be submitted for publication.

Funding There was no funding for this invited article.

References

1. Belford R. The origin of the elements. Available online: https://chem.libretexts.org/Courses/University_of_Arkansas_Little_Rock/Chem_1403%3A_General_Chemistry_2/Text/21%3A_Nuclear_Chemistry/21.06%3A_The_Origin_of_the_Elements. Accessed 19 Mar 2025
2. Clarkson T (1995) Health effects of metals: a role for evolution? Environ Health Perspect 103 Suppl 1(Suppl 1):9–12. https://doi.org/10.1289/ehp.95103s19
3. Jomova K, Alomar SY, Nepovimova E et al (2025) Heavy metals: toxicity and human health effects. Arch Toxicol 99:153–209. https://doi.org/10.1007/s00204-024-03903-2
4. Zhang X, Yang H, Sun R, Cui M et al (2022) Evaluation and analysis of heavy metals in iron and steel industrial area. Environ Dev Sustain 24:10997–11010. https://doi.org/10.1007/s10668-021-01893-0
5. Han FX, Banin A, Su Y, Monts DL et al (2002) Industrial age anthropogenic inputs of heavy metals into the pedosphere. Naturwissenschaften 89:497–504. https://doi.org/10.1007/s00114-002-0373-4
6. Raabe D (2023) The materials science behind sustainable metals and alloys. Chem Rev 123:2436–2608. https://doi.org/10.1021/acs.chemrev.2c00799
7. Tin D, Margus C, Ciottone GR (2021) Half-a-century of terrorist attacks: weapons selection, casualty outcomes, and implications for counter-terrorism medicine. Prehosp Disaster Med 36:526–530. https://doi.org/10.1017/S1049023X21000868
8. Zhang L, Yu JM, Shan XY, Shao J et al (2023) Characterization of welding fume and airborne heavy metals in electronic manufacturing workshops in Hangzhou, China: implication for occupational population exposure. Environ Sci Pollut Res Int 30:57398–57409. https://doi.org/10.1007/s11356-023-26569-2
9. Meng F, Liu D, Bu T, Zhang M et al (2025) Assessment of pollution and health risks from exposure to heavy metals in soil, wheat grains, drinking water, and atmospheric particulate matter. J Environ Manag 376:124448. https://doi.org/10.1016/j.jenvman.2025.124448
10. Klotz K, Weistenhöfer W, Neff F, Hartwig A et al (2017) The health effects of aluminum exposure. Dtsch Arztebl Int 114:653–659. https://doi.org/10.3238/arztebl.2017.0653
11. Periferakis A, Caruntu A, Periferakis AT et al (2022) Availability, toxicology and medical significance of antimony. Int J Environ Res Public Health 19:4669. https://doi.org/10.3390/ijerph19084669
12. Sharma P, Perez D, Cabrera A, Rosas N et al (2008) Perspectives of antimony compounds in oncology. Acta Pharmacol Sin 29:881–890. https://doi.org/10.1111/j.1745-7254.2008.00818.x
13. Gendrisch F, Haarhaus B, Schempp CM, Wölfle U (2021) Anti-psoriatic effects of antimony compounds in vitro. Molecules 26:5814. https://doi.org/10.3390/molecules26195814
14. Liu F, Deng Y, Wang A, Yang T et al (2024) Harness arsenic in medicine: current status of arsenicals and recent advances in drug delivery. Expert Opin Drug Deliv 21:867–880. https://doi.org/10.1080/17425247.2024.2372363
15. Ernst E (2002) Toxic heavy metals and undeclared drugs in Asian herbal medicines. Trends Pharmacol Sci 23:136–139
16. Paul NP, Galván AE, Yoshinaga-Sakurai K et al (2023) Arsenic in medicine: past, present and future. Biometals 36:283–301. https://doi.org/10.1007/s10534-022-00371-y
17. Yang Y, Li Y, Li R, Wang Z (2024) Research progress on arsenic, arsenic-containing medicinal materials, and arsenic-containing preparations: clinical application, pharmacological effects, and toxicity. Front Pharmacol 15:1338725. https://doi.org/10.3389/fphar.2024.1338725
18. Reissig F, Kopka K, Mamat C (2021) The impact of barium isotopes in radiopharmacy and nuclear medicine - from past to presence. Nucl Med Biol 98–99:59–68. https://doi.org/10.1016/j.nucmedbio.2021.05.003
19. Cooper RG, Harrison AP (2009) The uses and adverse effects of beryllium on health. Indian J Occup Environ Med 13:65–76. https://doi.org/10.4103/0019-5278.55122

20. Griffith DM, Li H, Werrett MV et al (2021) Medicinal chemistry and biomedical applications of bismuth-based compounds and nanoparticles. Chem Soc Rev 50:12037–12069. https://doi.org/10.1039/d0cs00031k
21. Rafati Rahimzadeh M, Rafati Rahimzadeh M, Kazemi S, Moghadamnia AA (2017) Cadmium toxicity and treatment: an update. Caspian J Intern Med 8:135–145. https://doi.org/10.22088/cjim.8.3.135
22. Gossa Al-Saadde DL, Haider AM, Ali A et al (2023) The role of chromium supplementation in cardiovascular risk factors: a comprehensive review of putative molecular mechanisms. Heliyon 9:e19826. https://doi.org/10.1016/j.heliyon.2023.e19826
23. Asbaghi O, Naeini F, Ashtary-Larky D et al (2021) Effects of chromium supplementation on blood pressure, body mass index, liver function enzymes and malondialdehyde in patients with type 2 diabetes: a systematic review and dose-response meta-analysis of randomized controlled trials. Complement Ther Med 60:102755. https://doi.org/10.1016/j.ctim.2021.102755
24. Wan S, He J, Simoes EJ et al (2023) Chromium supplementation to reduce cardiometabolic risk factors: a novel dose-response meta-analysis of randomized clinical trials. JACC Adv 2:100729. https://doi.org/10.1016/j.jacadv.2023.100729
25. Vajdi M, Musazadeh V, Karimi A et al (2023) Effects of chromium supplementation on lipid profile: an umbrella of systematic review and meta-analysis. Biol Trace Elem Res 201:3658–3669. https://doi.org/10.1007/s12011-022-03474-2
26. Kowdley KV, Hirschfield GM, Coombs C et al (2025) COBALT: a confirmatory trial of Obeticholic acid in primary biliary cholangitis with placebo and external controls. Am J Gastroenterol 120:390–400. https://doi.org/10.14309/ajg.0000000000003029
27. Heffern MC, Yamamoto N, Holbrook RJ et al (2013) Cobalt derivatives as promising therapeutic agents. Curr Opin Chem Biol 17:189–196. https://doi.org/10.1016/j.cbpa.2012.11.019
28. Ma Y, Lin W, Ruan Y et al (2022) Advances of cobalt nanomaterials as anti-infection agents, drug carriers, and immunomodulators for potential infectious disease treatment. Pharmaceutics 14:2351. https://doi.org/10.3390/pharmaceutics14112351
29. Yamada K (2013) Cobalt: its role in health and disease. Met Ions Life Sci 13:295–320. https://doi.org/10.1007/978-94-007-7500-8_9
30. Mohamed HRH, Mohamed BA, Hakeem GM et al (2025) Cobalt oxide nanoparticles induce cytotoxicity and excessive ROS mediated mitochondrial dysfunction and p53-independent apoptosis in melanoma cells. Sci Rep 15:2220. https://doi.org/10.1038/s41598-025-85691-y
31. Woźniak-Budych MJ, Staszak K, Staszak M (2023) Copper and copper-based nanoparticles in medicine-perspectives and challenges. Molecules 28:6687. https://doi.org/10.3390/molecules28186687
32. Mo N, Tai C, Yang Y et al (2025) MT2A promotes angiogenesis in chronically ischemic brains through a copper-mitochondria regulatory mechanism. J Transl Med 23:162. https://doi.org/10.1186/s12967-025-06163-5
33. Xie W, Zhenhu G, Zhao L, Wei Y (2023) The copper age in cancer treatment: from copper metabolism to cuproptosis. Progr Mater Sci 138:101145. https://doi.org/10.1016/j.pmatsci.2023.101145
34. Peng G, Huang Y, Xie G, Tang J (2024) Exploring copper's role in stroke: progress and treatment approaches. Front Pharmacol 15:1409317. https://doi.org/10.3389/fphar.2024.1409317
35. Kręcisz P, Stefańska K, Studziński J et al (2025) Radiocopper in radiopharmacy and medical use: current status and perspective. J Med Chem. https://doi.org/10.1021/acs.jmedchem.4c02885. Epub ahead of print
36. de Assis ASJ, Pegoraro GM, Duarte ICS (2022) Evolution of gallium applications in medicine and microbiology: a timeline. Biometals 35:675–688. https://doi.org/10.1007/s10534-022-00406-4
37. Torabi S, Bahreini F, Rezaei N (2022) Gallium components-based drug delivery: a potential treatment for COVID-19. Infect Disord Drug Targets 22:e270122200590. https://doi.org/10.2174/1871526522666220127120617

38. Kurtuldu F, Mutlu N, Boccaccini AR, Galusek D (2022) Gallium containing bioactive materials: a review of anticancer, antibacterial, and osteogenic properties. Bioact Mater 17:125–146. https://doi.org/10.1016/j.bioactmat.2021.12.034
39. Goss CH, Kaneko Y, Khuu L et al (2018) Gallium disrupts bacterial iron metabolism and has therapeutic effects in mice and humans with lung infections. Sci Transl Med 10(460):eaat7520. https://doi.org/10.1126/scitranslmed.aat7520
40. Liu Y, Yang F, Pan Z et al (2023) Gallium-enabled bactericidal medicine. Mater Today 67:548–5565. https://doi.org/10.1016/j.mattod.2023.06.014
41. Collery P, Keppler B, Madoulet C, Desoize B (2002) Gallium in cancer treatment. Crit Rev Oncol Hematol 42:283–296. https://doi.org/10.1016/s1040-8428(01)00225-6
42. Luo X, Sun J, Kong D et al (2023) The role of germanium in diseases: exploring its important biological effects. J Transl Med 21:795. https://doi.org/10.1186/s12967-023-04643-0
43. Goodman S (1988) Therapeutic effects of organic germanium. Med Hypotheses 26:207–215. https://doi.org/10.1016/0306-9877(88)90101-6
44. Ohri LK, Vicari SM, Malone PM (1993) Germanium use and associated adverse effects: a review. J Pharm Technol 9:237–241. https://doi.org/10.1177/875512259300900602
45. Berners-Price SJ, Filipovska A (2011) Gold compounds as therapeutic agents for human diseases. Metallomics 3:863–873. https://doi.org/10.1039/c1mt00062d
46. Silveira GB, Muller AP, Machado-de-Ávila RA et al (2021) Advance in the use of gold nanoparticles in the treatment of neurodegenerative diseases: new perspectives. Neural Regen Res 16:2425–2426. https://doi.org/10.4103/1673-5374.313040
47. Balfourier A, Kolosnjaj-Tabi J, Luciani N et al (2020) Gold-based therapy: from past to present. Proc Natl Acad Sci USA 117:22639–22648. https://doi.org/10.1073/pnas.2007285117
48. Yeo CI, Ooi KK, Tiekink ERT (2018) Gold-based medicine: a paradigm shift in anti-cancer therapy? Molecules 23:1410. https://doi.org/10.3390/molecules23061410
49. Taghizadeh S, Alimardani V, Roudbali PL et al (2019) Gold nanoparticles application in liver cancer. Photodiagn Photodyn Ther 25:389–400. https://doi.org/10.1016/j.pdpdt.2019.01.027
50. Casini A, Sun RW, Ott I (2018) Medicinal chemistry of gold anticancer metallodrugs. Met Ions Life Sci 18:books/9783110470734/9783110470734-007/9783110470734-013.xml. https://doi.org/10.1515/9783110470734-013
51. Faa G, Gerosa C, Fanni D et al (2018) Gold - old drug with new potentials. Curr Med Chem 25:75–84. https://doi.org/10.2174/0929867324666170330091438
52. Kaye AD, Sala KR, Dethloff D et al (2024) The evolving use of gold nanoparticles as a possible reversal agent for the symptoms of neurodegenerative diseases: a narrative review. Cureus 16:e64846. https://doi.org/10.7759/cureus.64846
53. Dickson EA, Ng O, Keeler BD et al (2023) The ICaRAS randomised controlled trial: intravenous iron to treat anaemia in people with advanced cancer - feasibility of recruitment, intervention and delivery. Palliat Med 37:372–383. https://doi.org/10.1177/02692163221145604
54. Kaur T, Upadhyay J, Nandave M et al (2024) Exploring progress in iron supplement formulation approaches for treating iron deficiency anemia through bibliometric and thematic analysis. Heliyon 10:e29058. https://doi.org/10.1016/j.heliyon.2024.e29058
55. Winkelman JW, Berkowski JA, DelRosso LM et al (2025) Treatment of restless legs syndrome and periodic limb movement disorder: an American Academy of Sleep Medicine clinical practice guideline. J Clin Sleep Med 21:137–152. https://doi.org/10.5664/jcsm.11390
56. Gordon M, Sinopoulou V, Iheozor-Ejiofor Z et al (2021) Interventions for treating iron deficiency anaemia in inflammatory bowel disease. Cochrane Database Syst Rev 1:CD013529. https://doi.org/10.1002/14651858.CD013529.pub2
57. Richards T, Breymann C, Brookes MJ et al (2021) Questions and answers on iron deficiency treatment selection and the use of intravenous iron in routine clinical practice. Ann Med 53:274–285. https://doi.org/10.1080/07853890.2020.1867323
58. Skolmowska D, Głąbska D, Kołota A, Guzek D (2022) Effectiveness of dietary interventions to treat iron-deficiency anemia in women: a systematic review of randomized controlled trials. Nutrients 14:2724. https://doi.org/10.3390/nu14132724

59. Ajiboye TO, Amao IO, Adeyemi WJ et al (2024) Overview of medical and biological applications of Indium(III) complexes. Chem Afr 7:1729–1748. https://doi.org/10.1007/s42250-023-00871-5
60. McRae A, Vilcins D, Le HHTC et al (2022) Lead in traditional and complementary medicine: a systematic review. Rev Environ Health 39:111–120. https://doi.org/10.1515/reveh-2022-0079
61. Gitlin M, Bauer M (2024) Lithium: current state of the art and future directions. Int J Bipolar Disord 12:40. https://doi.org/10.1186/s40345-024-00362-7
62. Huang P, Tang Q, Li M et al (2024) Manganese-derived biomaterials for tumor diagnosis and therapy. J Nanobiotechnology 22:335. https://doi.org/10.1186/s12951-024-02629-8
63. Zhao M, Li Y, Wang Z (2022) Mercury and mercury-containing preparations: history of use, clinical applications, pharmacology, toxicology, and pharmacokinetics in traditional Chinese medicine. Front Pharmacol 13:807807. https://doi.org/10.3389/fphar.2022.807807
64. Odularu AT, Ajibade PA, Mbese JZ (2019) Impact of molybdenum compounds as anticancer agents. Bioinorg Chem Appl 2019:6416198. https://doi.org/10.1155/2019/6416198
65. Berhe MG, Gebreslassie YT (2023) Biomedical applications of biosynthesized nickel oxide nanoparticles. Int J Nanomedicine 18:4229–4251. https://doi.org/10.2147/IJN.S410668
66. Mendez-Arriaga JM (2025) Platinum group metals against parasites: state of the art and future perspectives. Med Chem 21:2–10. https://doi.org/10.2174/0115734064324855240806052735
67. Ishizuya Y, Kawashima A, Horibe Y et al (2025) FAN score predicts prognosis after platinum-based first-line chemotherapy in patients with metastatic urothelial carcinoma. Sci Rep 15:4640. https://doi.org/10.1038/s41598-025-86212-7
68. Zhang C, Xu C, Gao X, Yao Q (2022) Platinum-based drugs for cancer therapy and anti-tumor strategies. Theranostics 12:2115–2132. https://doi.org/10.7150/thno.69424
69. Tang P, Huang R, Zhong X et al (2025) A comprehensive review on selenium and blood pressure: recent advances and research perspectives. J Trace Elem Med Biol 88:127607. https://doi.org/10.1016/j.jtemb.2025.127607
70. Sadler RA, Mallard BA, Shandilya UK et al (2024) The immunomodulatory effects of selenium: a journey from the environment to the human immune system. Nutrients 16:3324. https://doi.org/10.3390/nu16193324
71. Guan B, Yan R, Li R, Zhang X (2018) Selenium as a pleiotropic agent for medical discovery and drug delivery. Int J Nanomedicine 13:7473–7490. https://doi.org/10.2147/IJN.S181343
72. Singh P, Pandit S, Balusamy SR et al (2025) Advanced nanomaterials for cancer therapy: gold, silver, and iron oxide nanoparticles in oncological applications. Adv Healthc Mater 14:e2403059. https://doi.org/10.1002/adhm.202403059
73. Żyro D, Sikora J, Szynkowska-Jóźwik MI, Ochocki J (2023) Silver, its salts and application in medicine and pharmacy. Int J Mol Sci 24(21):15723. https://doi.org/10.3390/ijms242115723
74. Kołodziejska B, Stępień N, Kolmas J (2021) The influence of strontium on bone tissue metabolism and its application in osteoporosis treatment. Int J Mol Sci 22:6564. https://doi.org/10.3390/ijms22126564
75. Chiaverini L, Cirri D, Tolbatov I et al (2022) Medicinal hypervalent tellurium prodrugs bearing different ligands: a comparative study of the chemical profiles of AS101 and its Halido replaced analogues. Int J Mol Sci 23:7505. https://doi.org/10.3390/ijms23147505
76. Sári D, Ferroudj A, Semsey D et al (2024) Tellurium and nano-tellurium: medicine or poison? Nanomaterials (Basel) 14:670. https://doi.org/10.3390/nano14080670
77. Ba LA, Döring M, Jamier V, Jacob C (2010) Tellurium: an element with great biological potency and potential. Org Biomol Chem 8:4203–4216. https://doi.org/10.1039/c0Ob00086h
78. Peter AL, Viraraghavan T (2005) Thallium: a review of public health and environmental concerns. Environ Int 31:493–501. https://doi.org/10.1016/j.envint.2004.09.003
79. Tang M, Hu X, Wang Y et al (2021) Ivermectin, a potential anticancer drug derived from an antiparasitic drug. Pharmacol Res 163:105207. https://doi.org/10.1016/j.phrs.2020.105207

80. Kaur B, Blavo C, Parmar MS (2024) Ivermectin: a multifaceted drug with a potential beyond anti-parasitic therapy. Cureus 16:e56025. https://doi.org/10.7759/cureus.56025
81. Rodríguez-Barajas N, Anaya-Esparza LM, Villagrán-de la Mora Z et al (2022) A. Review of therapies using TiO_2 nanomaterials for increased anticancer capability. Anti Cancer Agents Med Chem 22:2241–2254. https://doi.org/10.2174/1871520622666211228112631
82. Zhang S, Wang X, Gao X et al (2025) Radiopharmaceuticals and their applications in medicine. Signal Transduct Target Ther 10(1). https://doi.org/10.1038/s41392-024-02041-6
83. Pessoa JC, Etcheverry S, Gambino D (2015) Vanadium compounds in medicine. Coord Chem Rev 301:24–48. https://doi.org/10.1016/j.ccr.2014.12.002
84. Kiouri DP, Chasapis CT, Mavromoustakos T et al (2025) Zinc and its binding proteins: essential roles and therapeutic potential. Arch Toxicol 99:23–41. https://doi.org/10.1007/s00204-024-03891-3
85. Sturniolo GC, Di Leo V, Ferronato A et al (2001) Zinc supplementation tightens "leaky gut" in Crohn's disease. Inflamm Bowel Dis 7:94–98. https://doi.org/10.1097/00054725-200105000-00003
86. Teschke R, Eickhoff A (2024) Wilson disease: copper-mediated cuproptosis, iron-related ferroptosis, and clinical highlights, with comprehensive and critical analysis update. Int J Mol Sci 25:4753. https://doi.org/10.3390/ijms25094753
87. Gupta M, Mahajan VK, Mehta KS, Chauhan PS (2014) Zinc therapy in dermatology: a review. Dermatol Res Pract 2014:709152. https://doi.org/10.1155/2014/709152
88. Tenny S, Varacallo MA (2025) Evidence-based medicine. In: StatPearls [Internet]. StatPearls Publishing, Treasure Island
89. Shah N, Gossman W (2025) Omeprazole. In: StatPearls [Internet]. StatPearls Publishing, Treasure Island. Available from: https://www.ncbi.nlm.nih.gov/sites/books/NBK539786/
90. de Castro Carvalho Silva L, Cunha Dos Reis LF, Malaquias LCC et al (2025) Impact of nanostructured formulations for schistosomiasis treatment: a systematic review of in vivo preclinical evidence. J Pharm Pharmacol:rgae155. https://doi.org/10.1093/jpp/rgae155. Epub ahead of print
91. Callado GY, Gutfreund MC, Pardo I et al (2024) Syphilis treatment: systematic review and meta-analysis investigating nonpenicillin therapeutic strategies. Open Forum Infect Dis 11:ofae142. https://doi.org/10.1093/ofid/ofae142
92. Chey WD, Howden CW, Moss SF et al (2024) ACG Clinical guideline: treatment of helicobacter pylori infection. Am J Gastroenterol 119:1730–1753. https://doi.org/10.14309/ajg.0000000000002968
93. Li G, Deng H, Xu W et al (2025) Combating antibiotic-resistant bacterial infection using coassembled dimeric antimicrobial peptide-based nanofibers. ACS Nano 19(3):3155–3171. https://doi.org/10.1021/acsnano.4c09347. Epub 2025 Jan 13
94. Ansell SM (2024) Hodgkin lymphoma: 2025 update on diagnosis, risk-stratification, and management. Am J Hematol 99:2367–2378. https://doi.org/10.1002/ajh.27470
95. Boutros A, Croce E, Ferrari M et al (2024) The treatment of advanced melanoma: current approaches and new challenges. Crit Rev Oncol Hematol 196:104276. https://doi.org/10.1016/j.critrevonc.2024.104276. Epub 2024 Jan 29
96. Tarning J, White NJ, Dondorp AM (2025) Dose-optimization of a novel co-formulated triple combination antimalarial therapy: artemether-lumefantrine-amodiaquine. Clin Pharmacol Ther. https://doi.org/10.1002/cpt.3582. Epub ahead of print
97. Shimony S, Stahl M, Stone RM (2025) Acute myeloid leukemia: 2025 update on diagnosis, risk-stratification, and management. Am J Hematol. https://doi.org/10.1002/ajh.27625. Epub ahead of print
98. Moran SK, Wong SC, Taylor SL, Feldman SR (2025) Early investigational agents for the treatment of rosacea: drugs in phase I and II clinical development. Expert Opin Investig Drugs:1–10. https://doi.org/10.1080/13543784.2025.2463093. Epub ahead of print

99. Yan L, Xu Y, Gao J et al (2025) Treatment patterns for Crohn's disease: a multinational analysis from 2011 to 2022 in the United States, Japan, and China. J Crohns Colitis 19 (Supplement 1):1891–1893. https://doi.org/10.1093/ecco-jcc/jjae190.1195
100. Al-Samydai A, Abu Hajleh MN, Al-Sahlawi F et al (2024) Advancements of metallic nanoparticles: a promising frontier in cancer treatment. Sci Prog 107:368504241274967. https://doi.org/10.1177/00368504241274967
101. Alven S, Gandidzanwa S, Ngalo B et al (2024) Platinum group metals nanoparticles in breast cancer therapy. Pharmaceutics 16:1162. https://doi.org/10.3390/pharmaceutics16091162
102. Villalobos Gutierrez PT, Patakfalvi R, Soriano Hernández AD et al (2024) The role of noble metal nanomaterials in cancer therapy. In: Interdisciplinary cancer research. Springer, Cham, pp 1–20. https://doi.org/10.1007/16833_2024_442
103. Villalobos Gutiérrez PT, Muñoz Carrillo JL et al (2023) Functionalized metal nanoparticles in cancer therapy. Pharmaceutics 15:1932. https://doi.org/10.3390/pharmaceutics15071932
104. Georgeous J, AlSawaftah N, Abuwatfa WH, Husseini GA (2024) Review of gold nanoparticles: synthesis, properties, shapes, cellular uptake, targeting, release mechanisms and applications in drug delivery and therapy. Pharmaceutics 16:1332. https://doi.org/10.3390/pharmaceutics16101332
105. Evans ER, Bugga P, Asthana V, Drezek R (2018) Metallic nanoparticles for cancer immunotherapy. Mater Today (Kidlington) 21:673–685. https://doi.org/10.1016/j.mattod.2017.11.022
106. Chen Q, Fang C, Xia F et al (2024) Metal nanoparticles for cancer therapy: precision targeting of DNA damage. Acta Pharm Sin B 14:1132–1149. https://doi.org/10.1016/j.apsb.2023.08.031
107. Shabatina T, Vernaya O, Shumilkin A et al (2022) Nanoparticles of bioactive metals/metal oxides and their nanocomposites with antibacterial drugs for biomedical applications. Materials (Basel) 15:3602. https://doi.org/10.3390/ma15103602
108. Marques Neto LM, Kipnis A, Junqueira-Kipnis AP (2017) Role of metallic nanoparticles in vaccinology: implications for infectious disease vaccine development. Front Immunol 8:239. https://doi.org/10.3389/fimmu.2017.00239
109. Souza JCM, Sordi MB, Kanazawa M et al (2019) Nano-scale modification of titanium implant surfaces to enhance osseointegration. Acta Biomater 94:112–131. https://doi.org/10.1016/j.actbio.2019.05.045
110. Catledge SA, Fries MD, Vohra YK et al (2002) Nanostructured ceramics for biomedical implants. J Nanosci Nanotechnol 2:293–312. https://doi.org/10.1166/jnn.2002.116
111. Quan NV, Xuan TD, Teschke R (2020) Potential hepatotoxins found in herbal medicinal products: a systematic review. Int J Mol Sci 21:5011. https://doi.org/10.3390/ijms21145011
112. Gerke L, Seifert R (2025) Lead and arsenic intoxications by traditional and alternative medicine: men are more sensitive than women. Naunyn Schmiedeberg's Arch Pharmacol 398:799–818. https://doi.org/10.1007/s00210-024-03317-y
113. Karousatos CM, Lee JK, Braxton DR, Fong TL (2021) Case series and review of Ayurvedic medication induced liver injury. BMC Complement Med Ther 21:91. https://doi.org/10.1186/s12906-021-03251-z
114. Pinto B, Goyal P, Flora SJ et al (2014) Chronic arsenic poisoning following ayurvedic medication. J Med Toxicol 10:395–398. https://doi.org/10.1007/s13181-014-0389-0
115. Devarbhavi H (2018) Ayurvedic and herbal medicine-induced liver injury: it is time to wake up and take notice. Indian J Gastroenterol 37:5–7. https://doi.org/10.1007/s12664-018-0820-624696169
116. Philips CA, Paramaguru R, Augustine P (2018) Ayurveda metallic-mineral 'Bhasma'-associated severe liver injury. BMJ Case Rep 2018:bcr2018225590. https://doi.org/10.1136/bcr-2018-225590
117. Philips CA, Augustine P, Paramaguru R, Padsalgi G (2019) Ayurveda herbal medicine-induced liver cirrhosis. Cureus 11:e4122. https://doi.org/10.7759/cureus.4122
118. Teschke R, Zhang L, Long H et al (2015) Traditional Chinese Medicine and herbal hepatotoxicity: a tabular compilation of reported cases. Ann Hepatol 14:7–19

119. Danan G, Teschke R (2016) RUCAM in drug and herb induced liver injury: the update. In: Special Issue: Drug, Herb, and Dietary Supplement Hepatotoxicity, guest editors Rolf Teschke and Raúl J. Andrade. Int J Mol Sci 17:14. https://doi.org/10.3390/ijms17010014
120. Ankar A, Kumar A (2025) Vitamin B12 deficiency. In: StatPearls [Internet]. StatPearls Publishing, Treasure Island
121. Russell MJ (2022) Cobalt: a must-have element for life and livelihood. Proc Natl Acad Sci USA 119:e2121307119. https://doi.org/10.1073/pnas.2121307119
122. Safari-Gezaz M, Parhizkar M, Asghari E (2025) Effect of cobalt ions doping on morphology and electrochemical properties of hydroxyapatite coatings for biomedical applications. Sci Rep 15:149. https://doi.org/10.1038/s41598-024-84055-2
123. Zheng Y, Yag Y, Deng Y (2019) Dual therapeutic cobalt-incorporated bioceramics accelerate bone tissue regeneration. Mater Sci Eng C Mater Biol Appl 99:770–782. https://doi.org/10.1016/j.msec.2019.02.020
124. Lutsenko S, Roy S, Tsvetkov P (2025) Mammalian copper homeostasis: physiological roles and molecular mechanisms. Physiol Rev 105:441–491. https://doi.org/10.1152/physrev.00011.2024
125. Roemhild K, von Maltzahn F, Weiskirchen R et al (2021) Iron metabolism: pathophysiology and pharmacology. Trends Pharmacol Sci 42:640–656. https://doi.org/10.1016/j.tips.2021.05.001
126. Obeng SK, Kulhánek M, Balík J et al (2024) Manganese: from soil to human health-a comprehensive overview of its biological and environmental significance. Nutrients 16:3455. https://doi.org/10.3390/nu16203455
127. Avila DS, Puntel RL, Aschner M (2013) Manganese in health and disease. Met Ions Life Sci 13:199–227. https://doi.org/10.1007/978-94-007-7500-8_7
128. Schwarz G, Belaidi AA (2013) Molybdenum in human health and disease. Met Ions Life Sci 13:415–450. https://doi.org/10.1007/978-94-007-7500-8_13
129. Adamus JP, Ruszczyńska A, Wyczałkowska-Tomasik A (2024) Molybdenum's role as an essential element in enzymes catabolizing redox reactions: a review. Biomolecules 14:869. https://doi.org/10.3390/biom14070869
130. Begum W, Rai S, Banerjee S et al (2022) A comprehensive review on the sources, essentiality and toxicological profile of nickel. RSC Adv 12:9139–9153. https://doi.org/10.1039/d2ra00378c
131. Rayman MP (2012) Selenium and human health. Lancet 379(9822):1256–1268. https://doi.org/10.1016/S0140-6736(11)61452-9
132. Chen C, Yang Z, Ma J, Xie W, Wang Z (2025) Recent research progress on the biological functions, synthesis and applications of selenium nanoparticles. J Chromatogr B Analyt Technol Biomed Life Sci 1252:124448. https://doi.org/10.1016/j.jchromb.2024.124448
133. TeschkeR, Chien NX, Xuan TD. Impact of heavymetals as trace elements on the ecosystem and health. Recent Progress in Nutrition. 2025; 5: 023. https://doi.org/10.21926/rpn.2504023
134. Fatullayeva S, Tagiyev D, Zeynalov N, Tagiyev S (2024) Perspective chapter: heavy metals—sources of releasing into ecosystems, biological importance, toxicity, and sorption methods. IntechOpen. https://doi.org/10.5772/intechopen.100672617
135. Vongdala N, Tran HD, Xuan TD et al (2019) Heavy metal accumulation in water, soil, and plants of municipal solid waste landfill in Vientiane, Laos. Int J Environ Res Public Health 16:22. https://doi.org/10.3390/ijerph16010022
136. Ibrahimov R (2020) The effects of heavy metals on biochemical processes in the human body (review). J Life Sci Biomed 2:38–46. https://doi.org/10.29228/jlsb.40
137. Panda A, Fatnani D, Parida AK (2025) Uptake, impact, adaptive mechanisms, and phytoremediation of heavy metals by plants: role of transporters in heavy metal sequestration. Plant Physiol Biochem 221:109578. https://doi.org/10.1016/j.plaphy.2025.109578
138. Teschke R (2024) Hemochromatosis: Ferroptosis, ROS, gut microbiome, and clinical challenges with alcohol as confounding variable. Int J Mol Sci 25:2668. https://doi.org/10.3390/ijms25052668

Chapter 12
Role of Micronutrients in the Treatment of Metal/Metalloid Toxicity: Mechanisms and Therapeutic Potential

Romi Singh, Jayant Kumar, Mohini Chaurasia, Abhiram Kumar, and S. J. S. Flora

Abstract Heavy metals often mimic or compete with essential micronutrients for absorption and binding sites, resulting in deficiencies and toxic manifestations. Micronutrient interactions are critical to exploring strategies to overcome heavy metal toxicity due to their significant influence on metal absorption, transportation, metabolism, and detoxification. Essential micronutrients such as zinc, iron, calcium, selenium, and vitamins C and E can modulate the toxicity of heavy metals like lead, cadmium, mercury, and arsenic by competing for absorption sites, enhancing metal excretion, and supporting antioxidant defenses. Deficiency in these micronutrients predisposes individuals to increased susceptibility to metal toxicity, while adequate intake can mitigate harmful effects by reducing metal uptake and alleviating oxidative stress. Additionally, micronutrients facilitate binding to target proteins and metabolic sequestration of toxic metals, thus decreasing cellular damage. The complex interplay between micronutrients and heavy metals suggests that dietary supplementation and nutritional interventions can serve as practical, low-cost approaches for preventing and treating heavy metal toxicity in exposed populations. This rationale underscores the importance of understanding micronutrient-heavy metal interactions to develop effective public health strategies.

Keywords Metalloid toxicity · Heavy metal toxicity · Micronutrients

---

R. Singh · M. Chaurasia · S. J. S. Flora (✉)
Era College of Pharmacy, Era University, Lucknow, India

J. Kumar
Department of Pharmaceutical Sciences, HNB Garhwal University (A Central University), Chauras Campus, Tehri Garhwal, Uttarakhand, India

A. Kumar
Laboratoy of Molecular Medicine, Birla Institute of Technology and Sciences, Pilani, Hyderabad Campus, Hyderabad, India

N. S. Dhalla et al. (eds.), *Functional Biochemistry of Metallic Elements*, Advances in Biochemistry in Health and Disease 36,
https://doi.org/10.1007/978-3-032-14445-4_12

Introduction

An Overview of Heavy Metal Toxicity

Heavy metal toxicity is a critical environmental and public health concern due to the widespread presence and persistence of metals like lead, cadmium, arsenic, and mercury in various ecosystems [1]. These metals are naturally occurring elements with high atomic weights and densities; they become toxic when accumulated in biological systems beyond certain limits. Unlike organic toxins, heavy metals do not degrade easily, leading to long-term soil, water, and air contamination, which poses chronic exposure risks to humans and wildlife [2]. Exposure to these metals occurs through inhalation, ingestion of contaminated food and water, and dermal contact. Occupational exposure is prevalent in metal extraction, processing, and recycling industries. Environmental contamination from these metals is also widespread, affecting global air quality, soil, and water bodies [3]. The toxicity of heavy metals manifests through the disruption of vital cellular functions, significantly impacts human health, and damages various organs as shown in Fig. 12.1.

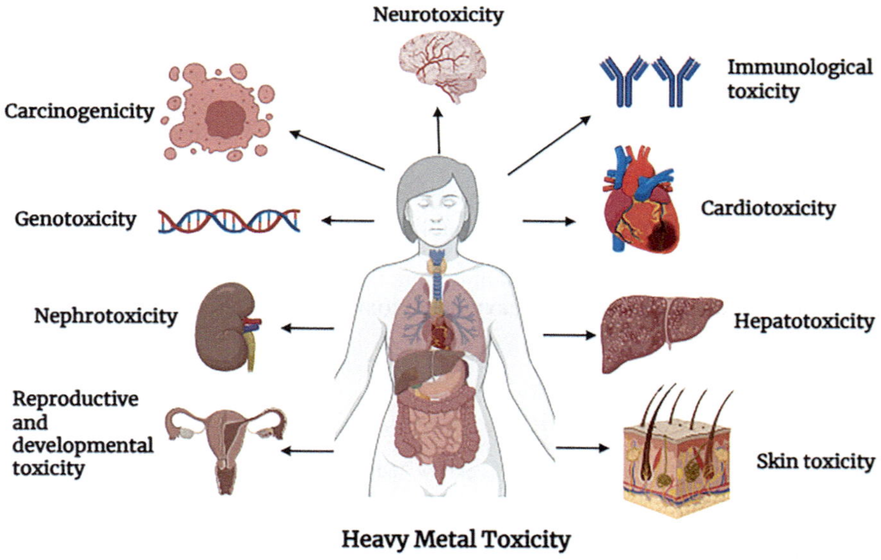

Fig. 12.1 Exposure to heavy metals poses a significant threat to human health, leading to a range of severe toxic effects, which are illustrated in this pictorial view. Heavy metal exposure is associated with serious conditions such as carcinogenicity, increasing cancer risk; nephrotoxicity, harming the kidneys; genotoxicity, causing genetic damage; and neurotoxicity, impairing neurological function [4]

Objective of the Chapter

This chapter comprehensively addresses the complex relationship between heavy metal toxicity and micronutrient interactions, aiming to provide a thorough understanding of their mechanisms, health implications, and potential therapeutic approaches. Beginning with an introduction to key heavy metals such as lead, cadmium, arsenic, and mercury, their sources, exposure routes, and the significant public health burden they impose, the sections provide an in-depth review of heavy metal toxicities and their cellular mechanisms, including oxidative stress, mitochondrial dysfunction, and enzyme inhibition. Further, the chapter delves into an overview of micronutrients, defining and classifying essential trace elements and vitamins, and detailing their dietary sources and physiological roles. A key focus is on mechanistic insights into how micronutrients mitigate heavy metal toxicity through antioxidant defense, chelation, repair of oxidative damage, and modulation of gene expression related to metal metabolism. The chapter further explores the heightened susceptibility to toxicity under micronutrient deficiencies, supported by epidemiological evidence. Detailed interactions between specific micronutrients such as zinc, selenium, iron, calcium, magnesium, copper, and vitamins, and various heavy metals, are examined individually, supported by in vivo and clinical studies. The chapter concludes with discussions on therapeutic and preventive strategies, recent advances, and future directions to enhance protection against heavy metal toxicity, thereby offering a structured and integrative resource for researchers, clinicians, and public health professionals.

Sources and Exposure to Heavy Metals/Metalloids

Heavy metals originate from two primary categories: natural and anthropogenic sources. Natural sources include weathering metal-bearing rocks, volcanic eruptions, and soil-forming processes. Forest fires and the evaporation of oceans also contribute to naturally occurring heavy metals in the environment [5]. In addition, anthropogenic or human activities are a significant contributor to heavy metal pollution. That encompass industrial activities, agriculture, mining, and domestic effluents [6]. Key anthropogenic sources include agricultural practices, fossil fuel combustion, vehicle traffic, and waste disposal [7]. Source of origin of various heavy metals has been discussed in later section where each heavy metal has been discussed individually.

The United States Environmental Protection Agency (USEPA) recommended the model for human exposure to heavy metals. Pollution levels of selected heavy metals could be considered in terms of their ecological effects during dry and wet seasons. Moreover, these contaminations are subcategorized in five different levels based on the lithofacies of the subsurface considered: Layer 5 > Layer 4 > Layer 3 > Layer 2 > Layer 1. According to this report, heavy metals present in decreasing

Table 12.1 Summary of the presence of heavy metals over the lithofacies of the subsurface. The data has been taken from a nature scientific report under license CC 4.0

	Subsurface lithofacies				
Metals	L1	L2	L3	L4	L5
Pb, mg kg^{-1}	163 ± 0.10	174 ± 0.10	174 ± 0.10	192 ± 0.10	202 ± 0.15
Cd, mg kg^{-1}	16 ± 0.10	20 ± 0.10	24 ± 0.15	28 ± 0.20	31 ± 0.10
Hg, mg kg^{-1}	12 ± 0.15	14 ± 0.10	15 ± 0.15	18 ± 0.10	19 ± 0.20
Cu, mg kg^{-1}	26 ± 0.10	29 ± 0.15	32 ± 0.10	34 ± 0.05	35 ± 0.10
As, mg kg^{-1}	14 ± 0.10	16 ± 0.15	19 ± 0.15	22 ± 0.15	25 ± 0.10
Cr, mg kg^{-1}	74 ± 0.15	77 ± 0.10	82 ± 0.15	90 ± 0.15	95 ± 0.05
Zn, mg kg^{-1}	227 ± 0.15	238 ± 0.10	252 ± 0.05	264 ± 0.10	272 ± 0.10
Ni, mg kg^{-1}	187 ± 0.10	198 ± 0.15	212 ± 0.10	226 ± 0.15	232 ± 0.10
Al, mg kg^{-1}	457 ± 0.15	490 ± 0.10	534 ± 0.20	587 ± 0.10	606 ± 0.15

order: Al > Zn > Ni > Pb > Cr > Cu > Cd > As > Hg. Al, Cu, and Zn present maximum allowable limits, and As, Cd, Hg, and Ni are observed above the maximum allowable limits [8]. Table 12.1 below summarizes heavy metals over the lithofacies of the subsurface.

Humans can absorb heavy metals through three main routes: ingestion, inhalation, and dermal contact. Ingestion typically happens through consuming contaminated food or water, such as fish containing methylmercury or vegetables grown in polluted soil. Inhalation occurs mainly via breathing air contaminated with metal particles or dust, which is common in occupational settings like factories or areas near mining and smelting activities. Dermal absorption, though less significant, involves metals entering through skin contact with contaminated soil, water, or products. For instance, lead dust can be absorbed by workers handling old paint. These exposure pathways pose significant health risks, emphasizing the need to control environmental contamination and monitor population exposure to protect public health. However, the above contamination, such as the hazard quotient (HQ) and hazard index (HI), can be identified by following Eq. 12.1 [9].

The hazard quotient (HQ) and hazard index (HI): It calculates the noncarcinogenic health risk and probability of drilling crew members or people in the environment suffering an adverse effect from heavy metals exposure:

$$HQ = \frac{ADD}{R_f D} \tag{12.1}$$

whereas reference dose (R_fD) values were adopted from USEPA followed by the risk-based concentration in the above table, and the ADD is the acceptable oral dose for a heavy metal.

Common Heavy Metals and Mechanisms Responsible for Their Toxic Effects

Mechanisms of Heavy Metal Toxicity

Heavy metal toxicity occurs primarily through their ability to disrupt cellular and molecular functions. These metals can enter cells and bind to proteins, enzymes, and DNA, causing structural and functional damage 10–13]. They induce oxidative stress by generating reactive oxygen species (ROS), damaging lipids, proteins, and nucleic acids. Additionally, heavy metals interfere with essential metal ion homeostasis by displacing them in metalloproteins, leading to enzyme inhibition and disrupted cellular processes [13]. They also affect the cytoskeleton and vesicle trafficking, impairing cellular transport and membrane function. Chronic exposure can cause DNA damage and carcinogenesis via interactions with transcription factors and signaling pathways [14].

Disruption of Cellular Homeostasis

Disruption of cellular homeostasis occurs when the delicate balance of cellular processes is disturbed by internal or external stressors, leading to impaired function and disease. Factors such as oxidative stress, metal ion imbalance, mitochondrial dysfunction, and protein misfolding contribute to this disruption. It triggers stress responses like apoptosis, autophagy, or senescence to restore balance or eliminate damaged cells. Chronic disruption causes metabolic disorders, neurodegeneration, cancer, and inflammation. Mechanisms include altered calcium signaling, redox imbalance, and impaired protein quality control, affecting cellular viability and organ function. Understanding these mechanisms aids in developing therapies targeting homeostatic [11, 15].

Generation of Reactive Oxygen Species (ROS)

Reactive oxygen species (ROS) are chemically reactive molecules containing oxygen that are generated mainly as by-products of cellular metabolism. The primary source of ROS in cells is the mitochondrial electron transport chain, particularly at Complexes I and III, where electrons can prematurely reduce oxygen to form superoxide radicals. Additional sources include NADPH oxidase enzymes, peroxisomes, endoplasmic reticulum, and enzymatic reactions involving xanthine oxidase and cytochrome P450. ROS generation can be triggered by environmental stimuli, cellular stress, or ligand-receptor interactions, serving roles in cell signaling, defense, and homeostasis. However, excessive ROS production leads to oxidative stress, damaging lipids, proteins, and DNA. For instance, microbacterium liquefaciens can remove up to 81% of Cr from wastewater when immobilized on a solid surface. This

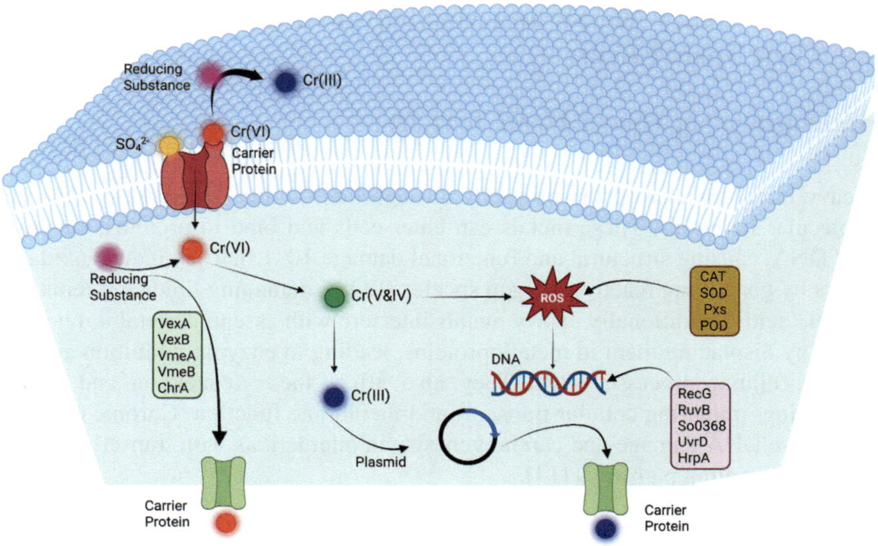

Fig. 12.2 Bacteria play a crucial role in the remediation of hexavalent chromium, a highly toxic heavy metal, through various mechanisms that reduce it to the less harmful trivalent form. This process is considered a favorable technique for environmental cleanup due to its cost-effectiveness compared to physicochemical methods, which can be expensive and inappropriate for low concentrations of Cr [4]

bacterium employs several mechanisms for heavy metal removal. It may utilize heavy metal ions during metabolic processes or detoxify them through soluble enzymes secreted by the cells. The overall bioremediation process of Cr by bacteria is illustrated in Fig. 12.2.

Mitochondrial Dysfunction and Apoptosis

Mitochondrial dysfunction plays a central role in apoptosis by disrupting energy metabolism and triggering the release of pro-apoptotic factors like cytochrome c from the mitochondrial intermembrane space into the cytoplasm. This release initiates caspase activation, leading to programmed cell death. Changes such as loss of mitochondrial membrane potential, increased reactive oxygen species (ROS) production, and outer mitochondrial membrane permeabilization contribute to apoptosis [16]. The Bcl-2 family of proteins tightly regulates this process, balancing pro- and anti-apoptotic signals [17]. Mitochondrial dysfunction and apoptosis are implicated in various diseases, including neurodegeneration, cancer, and cardiovascular [18].

Metal-Induced Enzyme Inhibition

Metal-induced enzyme inhibition is a critical biological phenomenon where metal ions interact with enzymes, altering their activity either by activating or inhibiting them. Metalloenzymes, essential in numerous biological processes, can be inhibited by excess or non-native metal ions that bind to active or allosteric sites, causing conformational changes or competing with natural metal cofactors [12]. For instance, zinc ions competitively inhibit phosphomannose isomerase by binding to the substrate site, causing significant conformational alterations. Metals such as Hg^{2+}, Cd^{2+}, and Zn^{2+} are potent inhibitors of microsomal and soluble epoxide hydrolases through competitive or noncompetitive mechanisms. Furthermore, substrate inhibition by metal ions is observed in enzymes like ferrochelatase, where metals such as Mn^{2+} and Cd^{2+} inhibit activity at higher concentrations, affecting heme biosynthesis. Metal-induced inhibition can be reversible, as chelating agents like EDTA can restore enzyme activity by removing inhibitory metals. Understanding these mechanisms is vital for therapeutic targeting of metalloenzymes and managing metal toxicity (Table 12.2) [33].

Common Approaches for Treatment of Heavy Metal Toxicity

Removing the source of exposure and employing chelation therapy in which substances like EDTA, dimercaprol (BAL), or DMSA bind to metals like lead, mercury, or arsenic to aid in their excretion are the usual treatments for heavy metal poisoning. Additionally, crucial supportive care is rendered, which includes symptomatic management, dietary assistance (such as antioxidants, zinc, and selenium), and hydration. Hospitalization may be necessary in extreme circumstances. To avoid long-term harm to organs like the kidneys, liver, and nervous system, early diagnosis and timely treatments are essential. Details on other approaches are rigorously studied and are available in literature. We will be discussing roles of micronutrients, i.e., Zn, Se, vitamins, etc., in the section ahead (Table 12.3).

Rationale for Exploring Micronutrient Interaction

Micronutrient interactions are critical to exploring strategies to overcome heavy metal toxicity due to their significant influence on metal absorption, metabolism, and detoxification. Essential micronutrients such as zinc, iron, calcium, selenium, and vitamins C and E can modulate the toxicity of heavy metals like lead, cadmium, mercury, and arsenic by competing for absorption sites, enhancing metal excretion, and supporting antioxidant defenses [34]. Deficiency in these micronutrients predisposes individuals to increased susceptibility to metal toxicity, while adequate intake can mitigate harmful effects by reducing metal uptake and alleviating oxidative

Table 12.2 Toxicological profile of common heavy metals: their health impacts and interventions

Metals/ metalloid	Organ affected	Symptoms	Treatments/ therapy	Supportive treatments/micronutrients	Ref.
Pb	Brain, kidneys, liver	Cognitive deficits, abdominal pain, anemia, fatigue	Chelation with $CaNa_2EDTA$, DMSA, penicillamine, DMPS, removal from exposure	Calcium, iron, zinc, magnesium, vitamin C, vitamin E, thiamine (vitamin B1), N-acetylcysteine (NAC)	[19, 20]
Hg	Brain, kidneys, lungs, fetus	Tremors, memory loss, kidney dysfunction, neurotoxicity	Chelation with DMSA, DMPS, D-penicillamine, supportive therapy	Selenium, zinc, vitamin C (ascorbic acid), alpha-lipoic acid (ALA), glutathione, chlorella, garlic (allicin)	[21, 22]
Cd	Kidneys, liver, bones, lungs	Bone demineralization, kidney failure, pulmonary edema	Chelation with $CaNa_2EDTA$, DMSA, D-penicillamine, and antioxidant therapy	Zinc, selenium, curcumin (from turmeric), N-acetylcysteine (NAC), alpha-lipoic acid (ALA), vitamin C, vitamin E	[23, 24]
As	Skin, lungs, liver, bladder	Skin lesions, cardiovascular disease, and cancer	Chelation with $CaNa_2EDTA$, DMSA, penicillamine, DMPS	Selenium, zinc, glutathione (GSH), folic acid (vitamin B9), vitamin C, vitamin E, alpha-lipoic acid (ALA), curcumin	[16, 25]
Tl	Nervous system, GI tract, liver	Hair loss, nerve damage, diarrhea, convulsions	Prussian blue, forced diuresis, activated charcoal, DMSA, DMPS, symptomatic care	Potassium supplementation, selenium, vitamins C and E, N-acetylcysteine (NAC), zinc	[12, 26]
Cr	Lungs, kidneys, liver	Respiratory irritation, lung cancer (Cr VI), skin ulcers, nephrotoxicity	Chelation with $CaNa_2EDTA$ or DMPS, DMSA, N-acetylcysteine, ascorbic acid (vitamin C), antioxidant therapy	Alpha-lipoic acid (ALA), glutathione (GSH), vitamin C, vitamin E, selenium, zinc	[27]
Ni	Skin, lungs, kidneys	Allergic dermatitis, respiratory issues, lung fibrosis	Chelation with $CaNa_2EDTA$ or DMSA, penicillamine, DMPS, symptomatic treatment, corticosteroids	Zinc, selenium, N-acetylcysteine (NAC), alpha-lipoic acid (ALA), vitamin C, vitamin E	[28]

(continued)

Table 12.2 (continued)

Metals/ metalloid	Organ affected	Symptoms	Treatments/ therapy	Supportive treatments/micronutrients	Ref.
Co	Heart, thyroid, lungs	Cardiomyopathy, asthma, hypothyroidism, dermatitis	Chelation with EDTA or DMSA, penicillamine, DMPS, removal from exposure, corticosteroids for inflammation	Curcumin, glutathione (GSH), alpha-lipoic acid (ALA), zinc, selenium, vitamin C, vitamin E	[25, 29]
Ag	Skin, eyes, liver	Argyria (blue-gray skin pigmentation), liver dysfunction	Sodium thiosulfate, DMSA, penicillamine, DMPS, chelation therapy	Zinc, selenium, N-acetylcysteine (NAC), alpha-lipoic acid (ALA), vitamin C, vitamin E	[12, 30]
Mn	Brain, liver, lungs	Manganese (Parkinson-like symptoms), cognitive and motor dysfunction	$CaNa_2EDTA$, Para-aminosalicylic acid, DMPS, DMSA, supportive neurological care	Alpha-lipoic acid (ALA), glutathione (GSH), vitamin C, vitamin E, selenium, zinc	[31, 32]

Abbreviation: *DMSA* Dimercaptosuccinic acid, *EDTA* ethylenediaminetetraacetic acid, *DMPS* 2,3-dimercapto-1-propanesulfonic acid, *IV* intravenous, *IM* intramuscular, *ALA* alpha-lipoic acid, *GSH* glutathione, *NAC* N-acetylcysteine

Table 12.3 Common essential heavy metals and their physiological importance and permissible levels

Metals	Physiological role	Recommended daily intake (adults)	Toxic level/UL (upper limit)
Fe	Hemoglobin synthesis and oxygen transport	8–18 mg/day	>45 mg/day
Zn	Immune function, wound healing, and enzyme activity	8–11 mg/day	>40 mg/day
Cu	Enzyme cofactor and iron metabolism	0.9 mg/day	>10 mg/day
Mn	Antioxidant enzymes and bone formation	1.8–2.3 mg/day	>11 mg/day
Cr	Carbohydrate and lipid metabolism	25–35 μg/day	>1000 μg/day
Mo	Enzyme function (xanthine oxidase, sulfite oxidase)	45 μg/day	>2 mg/day
Co	Component of vitamin B12	0.9 μg/day	>2.4 μg/day
Se	Antioxidant (glutathione peroxidase)	55 μg/day	>400 μg/day

stress. Additionally, micronutrients facilitate binding to target proteins and metabolic sequestration of toxic metals, thus decreasing cellular damage. The complex interplay between micronutrients and heavy metals suggests that dietary supplementation and nutritional interventions can serve as practical, low-cost approaches for preventing and treating heavy metal toxicity in exposed populations. This rationale underscores the importance of understanding micronutrient-heavy metal interactions to develop effective public health strategies [35].

Overview of Micronutrients

Micronutrients are essential chemicals required by organisms in small quantities to perform various biochemical processes and regulate the physiological functions of cells and organs. For humans, micronutrients typically take one of three forms: vitamins, trace elements, and dietary minerals (Tables 12.4 and 12.5) [36].

Micronutrients collectively encompass vitamins and trace elements, substances that the body requires in trace amounts but cannot synthesize sufficiently or at all, necessitating their provision through diet or supplementation. Unlike macronutrients such as carbohydrates, fats, and proteins that provide energy, micronutrients primarily serve as cofactors, coenzymes, structural components, or regulators of physiological and biochemical processes [37]. They are indispensable for enzymatic reactions, antioxidant defense, immune system function, gene expression, and the prevention of metabolic disorders and deficiency. Micronutrient requirements are defined as dietary intake levels sufficient to meet physiological needs, prevent deficiency symptoms, and avoid toxicity from excess intake. Maintaining micronutrient homeostasis is vital as both deficiency and overload can produce pathological manifestations [38].

Micronutrients are primarily subdivided into two categories: trace elements (also called microminerals) and vitamins, each with distinct chemical characteristics, biological roles, and nutritional requirements.

Trace Elements

Trace elements are inorganic chemical elements required in minuscule amounts (usually in the mg or μg range daily) for normal cellular, molecular, and systemic functions. They mostly occur as metals or metalloids and are incorporated as structural or catalytic components of enzymes, cofactors in metabolic pathways, or regulators at the genetic level [39].

Iron acts as cofactor in cytochromes and copper in oxidases, facilitating redox reactions and electron transfer. Zinc regulates gene expression through zinc fingers and metal-responsive transcription factors, while selenium supports antioxidant defense as a key component of glutathione peroxidase. Iodine is vital for thyroid

Table 12.4 Micronutrients in heavy metal detoxification: mechanisms, sources, and marketed products

Micronutrient	Mechanism/role in heavy metal detoxification	Examples of natural sources	Metal-nutrient: interaction and mechanism	Common dosage (supplements of micronutrients)
Zn	Competes with heavy metals (like Cd, Pb) for binding sites; induces metallothionein synthesis, which binds toxic metals	Nuts, seeds, whole grains, legumes, and meat	**Zinc-Pb:** Competes for GI uptake, perhaps on the same MT-like transport system **Zinc-Cd:** Competes for GI absorption; Cd interferes with zinc metabolism	**Tablet:** *Zincovit, Zinconia, Becozinc, zinc lozenges, Zincovit-chewy, Limcee zinc* **Capsule:** *Zincolife, Zincofer* **Syrups/suspensions:** *Zinconia syrup, Z&D Syrup*
Se	Forms selenoenzymes (e.g., glutathione peroxidase) that reduce oxidative stress and bind heavy metals like mercury	Brazil nuts, fish, eggs, and sunflower seeds	**Selenium-Cd:** Selenium shifts Cd binding to higher molecular weight proteins **Selenium-Pb:** Selenium activates Pb-inhibited δ-ALAD **Selenium-Hg:** Complexing of Hg by selenium **Selenium-organic Hg:** Mechanism not clear; selenium may catalyze the reduction of Hg and organic hydroperoxides to products of greater stability **Selenium-Cd:** Selenium shifts Cd binding to higher molecular weight proteins **Selenium-As:** Antagonistic effects	**Tablets:** *Selenium ACE, Selace* **Capsules:** *Selace forte, Healthvit selenium capsules, puritan's pride selenium* **Injection:** *Selenium injection BP*
Fe	Reduces lead absorption by competing for uptake pathways; helps maintain hemoglobin and enzyme functions	Red meat, spinach, lentils, fortified cereals	**Iron-Pb:** Competes for iron transport systems (i.e., ferritin) of the intestine **Iron-Cd:** Cd decreases iron absorption and metabolism (Cd possibly binds with ferritin and transferrin)	**Tablets:** Fefol, Feronia, Orofer, Irovit-C, Autrin-C, Fefol Spansules **Capsules:** *Autrin, Dexorange, Fefol-Z, Livogen, Autrin Softgels* **Syrups:** *Dexorange syrup, Orofer syrup, Haem up syrup* **Drops:** *Orofer drops, Feronia drops* **Suspensions:** Ferplex suspension **Injection:** *Venofer, Orofer-S Ferinject, Encicarb, Imferon, Cosmofer*

(continued)

Table 12.4 (continued)

Micronutrient	Mechanism/role in heavy metal detoxification	Examples of natural sources	Metal-nutrient: interaction and mechanism	Common dosage (supplements of micronutrients)
Ca	Competes with lead and cadmium for absorption; stabilizes cell membranes and reduces oxidative damage	Dairy, leafy greens, almonds	**Calcium-Pb:** Competes for binding sites on intestinal mucosal proteins **Calcium-Cd:** Cd decreases intestinal calcium transport; increases Cd deposits in bone tissue in a calcium-deficient state	**Tablets:** *Shelcal, Calcirol, coral calcium, tums, Riconia, Shelcal-chewy, Limcee calcium, CCM effervescent* **Capsules:** *Coral calcium capsules, Calcimax forte, Shelcal-OS, Coraline-D* **Syrups/suspensions:** *Ostocalcium syrup, Calcimax syrup* **Injection:** *Calcium-Sandoz injection*
Mg	Competes with heavy metals for binding sites and helps reduce oxidative stress	Nuts, seeds, whole grains, green vegetables	**Magnesium-Pb:** Competes with lead for calcium channels, reducing lead uptake into cells**magnesium–Cd:** Mg reduces cd-induced oxidative damage and mitochondrial dysfunction **Magnesium-Hg:** May protect against Hg-induced oxidative stress by enhancing antioxidant defense systems	**Tablets:** *Magne-B6, Magnesol, slow-mag, magnesium citrate chewable tablets, magnesium effervescent tablets (mag-tab)* **Capsules:** *Healthvit magnesium, Magmin capsules magnesium Softgels* **Magnesium sulfate injection:** *Eclampsia*
Cu	A cofactor for antioxidant enzymes like superoxide dismutase (SOD), helping counter oxidative stress	Shellfish, nuts, seeds, whole grains	**Copper-Zn:** Antagonistic at high doses; balance is important for metallothionein function **Copper-Pb:** Adequate copper may help protect against Pb-induced anemia via improved iron mobilization **Copper-Cd:** May counteract Cd effects on oxidative stress enzymes	**Tablets:** *Supradyn, Zincovit* **Capsules:** *Autrin, Becozinc, coral calcium* **Syrups:** *Riconia, Zincofer*

Mn	Acts as an antioxidant enzyme cofactor (Mn-SOD) to protect against metal-induced oxidative damage	Whole grains, nuts, leafy vegetables	**Manganese-Pb:** Mn-SOD activity may protect cells from Pb-induced oxidative damage **Manganese-Cd:** Supports antioxidant defense enzymes impaired by Cd **Manganese-Hg:** May modulate oxidative stress responses in hg exposure	**Tablet:** *Zincovit, Supradyn, Revital* **Capsule:** *Becozinc, Riconia* **Syrups:** *Zincofer syrup, Riconia syrup*
Vitamin C	Chelates metals like lead and enhances excretion; strong antioxidant that reduces free radical damage	Citrus fruits, bell peppers, kiwi	**Vitamin C-Pb:** Enhances urinary excretion of lead by forming soluble complexes **Vitamin C-Ni:** Reduces oxidative damage caused by nickel exposure **Vitamin C-Cr:** Acts as a reducing agent converting toxic Cr (VI) to less harmful Cr (III)	**Tablets:** *Limcee, Celin, Redoxon, Limcee chewable, Celin 500, Limcee plus, C-sip effervescent, redox-C* **Capsules:** *Celin capsules, Healthvit vitamin C, vitamin C Softgels* **Syrups:** *Celin syrup, Limcee syrup* **Injection**: *Celin injection, Ascoril-C injection*
Vitamin E	Lipid-soluble antioxidant that protects cell membranes from oxidative damage caused by metals	Almonds, sunflower seeds, vegetable oils	**Vitamin E-Hg:** Protects cell membranes from mercury-induced lipid peroxidation **Vitamin E-Pb:** Reduces lead-induced oxidative stress in neural and reproductive tissues **Vitamin E-Cd:** Prevents cadmium-induced peroxidative damage in membranes	**Tablets:** *Evion, E cap, Evit, vitamin E chewable* **Capsules:** *Evion 400, Evion LC, E-gen, Evion 600, Healthvit E-Vit* **Injection**: *Alpha-tocopherol Evion injection*
Vitamin B-complex	Supports detoxification enzymes and reduces oxidative stress; B6 and B12 aid in methylation of toxins	Whole grains, eggs, fish, legumes	**B6-Pb:** Reduces lead-induced neurotoxicity **B12-Hg:** Supports methylation to facilitate mercury detox **B1-As:** Helps reduce arsenic-induced oxidative damage	**Tablets:** *Neurobion, Becosules, Riconia, Supradyn, Becosules chewable, Neurobion chewable, Supradyn effervescent, Beplex forte effervescent* **Capsules:** *Becosules Z, Becosules performance, Autrin B-complex, Neurobion forte Softgel, Becozym Softgel* **Syrups:** *Becosules syrup, Riconia syrup, Beplex forte syrup* **Injections:** *Neurobion injection, Becosules injection, Polybion injection (contains B1, B2, B3, B6, and B12)*

Table 12.5 Micronutrients, deficiency, and their role in metal toxicity

Micronutrient	Heavy metal toxicity risk (deficiency state)	Mechanism of protection
Se	Cadmium, mercury, arsenic, lead	Formation of reversible Se-metal complexes (Hg-Se, Cd-Se); cofactor of glutathione peroxidase (GPx) and thioredoxin reductase; reduces oxidative stress and lipid peroxidation
Fe	Lead, cadmium	Iron deficiency upregulates DMT1, enhancing Pb^{2+}/Cd^{2+} absorption; adequate Fe reduces uptake
Ca	Lead, cadmium	Ca^{2+} deficiency enhances Pb^{2+} uptake via Ca channels; adequate Ca competes with Pb, reducing deposition in bones and nerves
Mg	Lead, aluminum	Competes with Pb^{2+}/Al^{3+} at binding sites; stabilizes cell membranes; reduces ROS generation
Vitamin A (retinol)	Cadmium, lead	Deficiency increases oxidative stress and impairs metallothionein regulation; adequate vitamin A maintains antioxidant gene expression and membrane stability
Vitamin E (tocopherol)	Mercury, lead, cadmium	Potent lipid-soluble antioxidant; scavenges free radicals induced by metals; protects membranes from lipid peroxidation
Vitamin K	Lead, cadmium	Involved in γ-carboxylation of glutamate residues, forming Gla proteins that chelate Ca^{2+}; helps maintain calcium homeostasis disrupted by Pb^{2+}/Cd^{2+}; indirectly reduces toxicity
Phosphates (PO_4^{3-})	Lead	Adequate phosphate reduces Pb absorption by forming insoluble Pb-phosphate complexes; it promotes Pb sequestration in bone, reducing free Pb^{2+} circulation
Fe	Lead, cadmium	Iron deficiency upregulates DMT1, enhancing Pb^{2+}/Cd^{2+} absorption; adequate Fe reduces uptake

hormone synthesis, influencing metabolism and development, and iron is crucial for oxygen transport via hemoglobin and myoglobin. Both deficiencies (e.g., anemia from iron, immune impairment from zinc or selenium, developmental issues from iodine) and excesses (e.g., toxicity from lead or cadmium) can cause significant health disorders.

Vitamins

Vitamins are a heterogeneous group of organic compounds required in small amounts as humans cannot synthesize them adequately; they mainly function as enzyme cofactors or precursors vital for biochemical reactions involved in energy metabolism, cellular maintenance, and physiological regulation [40, 41]. Based on solubility, they are classified as water-soluble vitamins, including all B-complex vitamins (B1, B2, B6, B12, folate) and vitamin C, which are hydrophilic, not extensively stored in the body, and need regular replenishment [42], and fat-soluble vitamins (A, D, E, K), which dissolve in lipids, are stored in body fat and the liver, and support functions like vision, bone health, antioxidant defense, and blood clotting.

Dietary Sources and Recommended Daily Allowances

Micronutrients, including vitamins and trace elements, must be obtained primarily through dietary sources because the body cannot produce adequate amounts. Common dietary sources of essential micronutrients include fruits, vegetables, whole grains, dairy products, meat, fish, eggs, and legumes. For example, iron is abundant in red meat and leafy green vegetables, calcium in dairy products, vitamin C in citrus fruits, and vitamin A in dark green leafy vegetables and liver. Animal-source foods generally provide more bioavailable forms of minerals such as iron, zinc, and vitamin B12 than plant-based sources, which may contain absorption inhibitors. Recommended daily allowances (RDAs) are dietary intake levels set to meet the nutritional needs of most healthy individuals and prevent deficiency diseases. RDAs vary by age, sex, and physiological status. For instance, the RDA for vitamin C for adults is approximately 75–90 mg/day, while the RDA for iron ranges from 8 mg/day for adult men to 18 mg/day for menstruating. RDAs serve as a guideline for health professionals to assess and plan diets for populations [38].

General Roles in Physiological and Cellular Processes

Micronutrients play crucial roles in numerous physiological and cellular processes essential for maintaining health and homeostasis. They act primarily as cofactors or components of enzymes, facilitating biochemical reactions such as energy production, DNA synthesis, and repair [43]. Trace elements like zinc and selenium contribute to antioxidant defense by neutralizing reactive oxygen species, protecting cells from oxidative damage. Vitamins, such as the B-complex group, serve as coenzymes in the metabolism of carbohydrates, fats, and proteins, supporting cellular energy and growth. Micronutrients also regulate gene expression and immune responses, enhancing the body's ability to fight infections and maintain tissue integrity [44]. For example, vitamin A influences cell differentiation, while iron is vital for oxygen transport via hemoglobin. They are involved in signaling pathways, cell membrane stabilization, and neurotransmission, affecting neurological and developmental functions. Deficiencies can impair these processes, leading to metabolic disturbances, weakened immunity, and increased disease risk. Thus, adequate micronutrient intake is fundamental for optimal cellular function and overall physiological health [15].

Micronutrients in Protection of Metal Toxicity

Micronutrients and heavy metals share similar mechanisms for absorption, sites at transporter proteins for their distribution, and other physiological actions. Hence, they may use the same mechanism for their entry in the body and to interfere with vital mechanisms, ultimately leading to toxic manifestations [45].

Vitamin E

Vitamin E is a fat-soluble organic micronutrient that helps to preserve human health against heavy metal toxicity. Vegetable oils, such as wheat germ, sunflower, corn germ, soybean, and rapeseed, are the primary dietary source of vitamin E for humans. It is also found in some nuts, fruits, and vegetables, such as almonds, avocados, spinach, and kale. In 1922, vitamin E was first described as an essential dietary factor maintaining fertility in rats. After that, numerous physiological activities of vitamin E have been discovered. Particularly, α-tocopherol neutralizes highly reactive free radicals generated by heavy metals like mercury, cadmium, and lead, which protects biological membranes from lipid peroxidation [46]. Beyond its classical function as an antioxidant, vitamin E is also involved in the modulation of enzyme activities and signaling cascades, as well as gene and protein expression. Many cell responses including inflammatory responses, cell proliferation, programmed cell death, and lipid homeostasis have been shown to be affected by vitamin E (Fig. 12.3) [48, 49]. Despite the convincing findings from in vitro, as well as animal, studies, the conflicting results obtained from studies in humans failed to provide clear evidence for the benefits of vitamin E supplementation for disease prevention and therapy in patients with adequate levels of circulating vitamin E

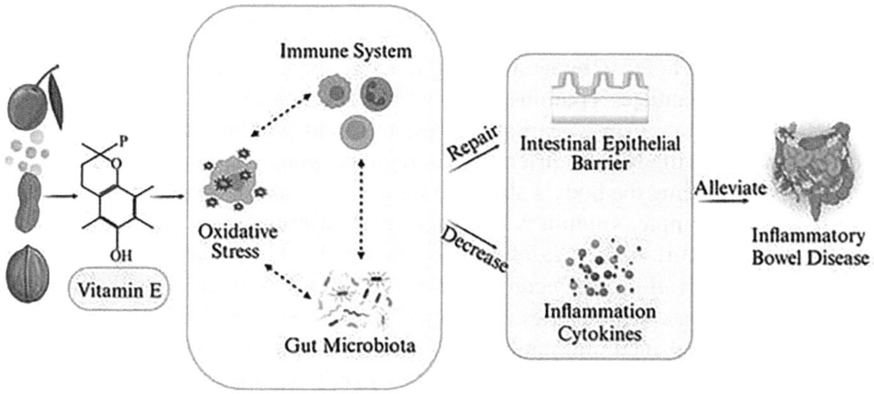

Fig. 12.3 Possible mechanisms underlying the interaction between vitamin E and other micronutrients [47]

[50]. In summary, the cellular and molecular mechanisms by which vitamin E acts as well as its physiological relevance remains to be fully understood. Free radicals such as O_2 (hyperoxide anion) attack the double bonds of the unsaturated fatty acids of membrane phospholipids resulting in the formation of a lipid peroxyl radical ($RO_2^{\bullet}$). Vitamin E (TOHs and T3s) provides its phenolic hydrogen atom to the peroxyl radical and converts it to a hydroperoxide. The resulting vitamin E radical can be reduced by vitamin C that forms then an ascorbyl radical that is regenerated with the help of glutathione (GSH). In this process, two monomers of GSH are oxidized to form a dimer (GSSG). Selenoproteins and the metal ion transporter metallothionein are involved in reducing GSSG to GSH. Pro-vitamin A (β-carotene) regenerates TOH from tocopheryl radicals. The resulting carotenoid radical cation is further regenerated by the vitamin C cycle. Besides that, magnesium ions contribute to the stabilization of the vitamin E radical and vitamin K competes hypothetically with vitamin E for the enzyme CYP4F2 which ω-hydroxylates the side chains of both vitamin K_1 (phylloquinone) and vitamin E. Abbreviations: CYP, cytochrome P450; SOD, superoxide dismutase. Figure 12.3 illustrates possible mechanisms underlying the interaction between vitamin E and other micronutrients.

Vitamin C

Vitamin C is classified as a water-soluble antioxidant and serves as an indispensable cofactor for enzymes involved in numerous biosynthetic pathways. It exerts significant influence on various facets of the immune response. Vitamin C effectively scavenges reactive oxygen species (ROS) produced as a consequence of the actions of heavy metals such as lead, cadmium, mercury, and arsenic, thereby preventing lipid peroxidation and ultimately mitigating cellular damage. It possesses the ability to chelate metal ions including Fe^{3+}, Pb^{2+}, and Cu^{2+}, facilitating their excretion via renal mechanisms and consequently diminishing their toxicological effects. Furthermore, vitamin C is crucial for maintaining proper barrier integrity and facilitating wound healing through the regulation of collagen synthesis. Additionally, vitamin C plays a vital role in the regeneration of antioxidants and modulates redox-sensitive signaling pathways, thereby alleviating oxidative stress during the inflammatory response. Furthermore, vitamin C influences the differentiation and proliferation of immune cells and is actively involved in the modulation of a myriad of immune activities (Fig. 12.4) [51].

Vitamin A

The term vitamin A comprises a group of lipophilic molecules including retinol, retinal, retinoic acid (RA), as well as several carotenoids as the so-called pro-vitamin A. Vitamin A is involved in various processes in our body: It is essential for

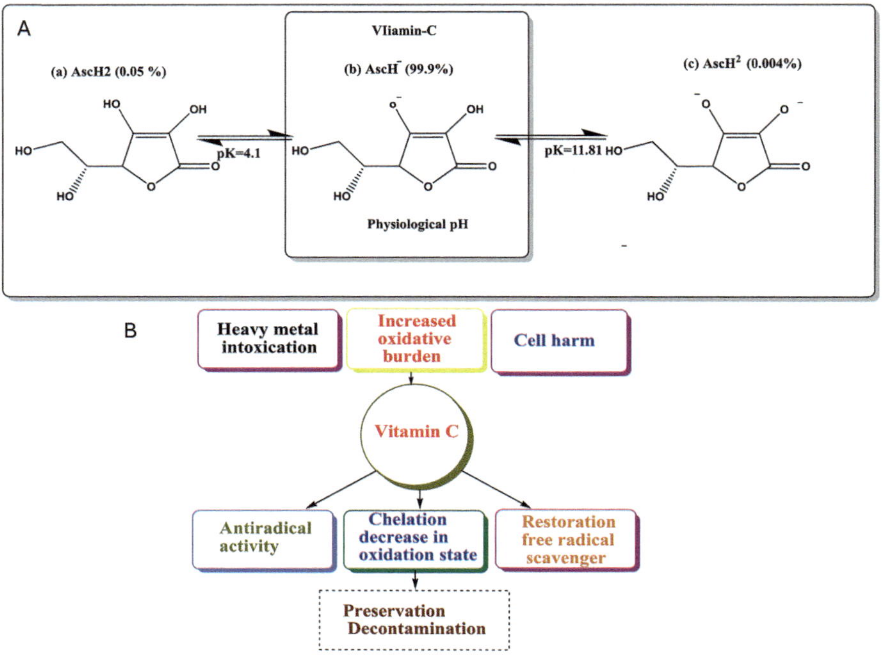

Fig. 12.4 Physicochemical properties and biological antioxidant mechanisms of vitamin C (ascorbic acid). (A) The acid-base equilibrium of vitamin C at physiological pH. The pictorial view shows the speciation between the fully protonated form, ascorbic acid ($AscH_2$), the monoanionic ascorbate ($AscH^-$), and the dianionic form (Asc^{2-}). With acid dissociation constants (pKa) of 4.1 and 11.81, the monoanionic ascorbate is the overwhelmingly predominant species (99.9%) at physiological pH. (B) A schematic diagram illustrating the protective roles of vitamin C against increased oxidative burden. Stressors like heavy metal intoxication and cell harm lead to oxidative stress. Vitamin C mitigates this damage through multiple pathways, including direct antiradical activity, chelation and reduction of metal ions to a less toxic oxidation state, and acting as a potent free radical scavenger. These collective actions result in cellular preservation and decontamination [25]

(i) vision (vitamin A deficiency can lead to night blindness), (ii) growth and development (vitamin A is involved in the genetic regulation of cell and tissue formation), and (iii) immune function. Furthermore, as described by Huang et al., (iv) vitamin A regulates the formation of epithelial tissues, which act as the first barrier against pathogen invasion [52]. Vitamin A also plays a crucial role in the modulation of cell proliferation, differentiation, and programmed cell death of innate immune cells.

Vitamin K

The term vitamin K comprises the molecules phylloquinone (vitamin K1), as well as the menaquinones (vitamin K2 or MKs). Vitamin K functions as a cofactor for the enzyme glutamyl carboxylase that is involved in blood coagulation, bone metabolism, and prevention of vessel mineralization, as well as many other essential cellular functions. Disordered mineralization in blood vessels leads to an increased risk of CVD. Vitamin K deficiency is associated with accelerated coronary artery calcification. Vitamin K is crucial for the carboxylation of glutamate residues in certain proteins, forming gamma-carboxyglutamate (Gla) residues [53]. These Gla residues can chelate calcium ions [54]. While vitamin K doesn't directly chelate toxic heavy metals like lead or cadmium, it influences calcium homeostasis, which can be disrupted by these metals [53]. By promoting proper calcium regulation, vitamin K can help mitigate some of the adverse effects of heavy metal toxicity that involve calcium imbalances. Vitamin K can also influence redox homeostasis by influencing the activity of enzymes involved in antioxidant defense. Additionally, vitamin K can affect the levels of certain antioxidants like glutathione, which plays a crucial role in detoxifying heavy metals. Studies have revealed that vitamin K can protect against methylmercury-induced neuronal death, potentially by influencing glutathione levels or other antioxidant mechanisms. Vitamin K has also been shown to have protective effects against cell death induced by agents that deplete glutathione. Furthermore, vitamin K's role in bone health, through its influence on osteocalcin (a calcium-binding protein), may indirectly contribute to overall health and resilience against heavy metal toxicity according to the National Institutes of Health (NIH) [55].

Zinc

Zinc is most abundant essential micronutrient after iron. It plays catalytical/structural role or binds with more than 3000 proteins of human body, which catalyzes various crucial processes and is responsible for vital physiological manifestations, i.e., proper immune function, suppression of inflammation, growth and development, wound repair, sensory and neurological functions, and general antioxidant defense among other physiological roles that support human health [56–59]. Heavy metals cause imbalance. Zinc, though indirectly, plays a crucial role in maintaining redox balance [60]. Multiple mechanisms such as protection of protein sulfhydryl groups, competition with redox-active metals, actions as a cofactor for antioxidant enzymes such as Cu/Zn superoxide dismutase, and induction of the antioxidant system response underlie zinc's role in antioxidant defense [61].

Zinc finger proteins, a zinc-binding motif, are reflected in numerous essential cellular functions and are characterized by coordination of zinc by four histidine and/or cysteine residues. Toxic heavy metals, like cadmium, cobalt, lead, arsenic, and uranium, interfere with zinc finger proteins either by displacing zinc or causing oxidation of zinc coordinating cysteine residues in this key structure [61–63]. Consequently, affecting the concerned protein function, zinc replenishment prevents/mitigates such toxic effects of heavy metals. Metallothionein is a family of protein that can bind with essential as well as other heavy metals. In the scenario where exposure to toxic heavy metals results in their toxicity, metallothionein binds with these metals and reduces the resulting toxicity [64]. Metallothionein also scavenges reactive oxygen species and plays role in maintaining redox balance caused due to exposure to heavy metals [16]. Zinc helps in mitigation of toxicity caused by heavy metals by playing role in induction of expression of metallothionein [65]. Metal-induced organ-specific toxicity or carcinogenesis is often associated with greater accumulation of metal in a particular target tissue, and transporters play an important role in reducing metal burden in the body along excretory routes [66]. The evidence suggests that for at least some metals, i.e., As and Cd, zinc may decrease metal toxicity by reducing tissue burden of toxic metals based on transport mechanisms [67].

Selenium

Selenium is a potent immune modulator and antioxidant micronutrient. It also boosts the immune system, regulates thyroid function, prevents cancer, protects the heart, and enhances reproductive potential. It is present most abundantly in some nuts (Brazil nut), whole grains, dairy products, animal viscera, and sea animals. Being an essential part of many proteins and enzymes, selenium mainly performs its biological roles in the body as melanoproteins. More than 20 of the more than 30 different kinds of melanoproteins that have been discovered so far contain selenocysteine residues. Among these, iodothyronine deiodinases (DIOs), glutathione peroxidases (GPXs), and thioredoxin reductases (TrxRs) have all been extensively researched. Selenium has many biological benefits, such as boosting the immune system, regulating thyroid function, preventing cancer, protecting the heart, enhancing reproductive potential, and reducing inflammation. The chemical structures of compound are depicted in Fig. 12.5 [69].

This makes them less available to living things and less dangerous. Selenium is part of antioxidant enzymes, such as glutathione peroxidase (GPx). So it boosts the body's natural defenses to fight off the dangerous free radicals that heavy metals make and lower oxidative stress. Selenium supplementation has been demonstrated to diminish cadmium accumulation in many organisms, such as fish and plants, hence alleviating liver damage, enhancing hematological parameters, and counteracting oxidative stress generated by cadmium [70–73]. Selenium can bind to mercury to make stable complexes and help the body get rid of it, which could reduce neurotoxicity and other harmful effects [74]. Selenium can counteract lead-induced

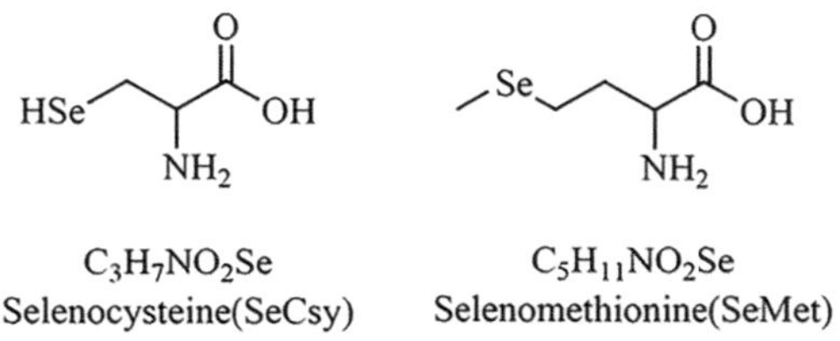

Fig. 12.5 Structural formulas of selenocysteine (SeCys) and selenomethionine Se can interact directly with some heavy metals, such cadmium (Cd) and mercury (Hg), to make complexes, i.e., Cd-Se and Hg-Se/MeHg-Se, that are less poisonous and physiologically inactive (like cadmium selenide) [68]

oxidative stress (NRF2 activation → increased GSH, SOD, CAT; reduced MDA), apoptosis (modulates MAPK (P38/JNK/ERK) signaling to suppress apoptotic/necroptotic gene expression), and necrosis (regulates Bcl-2, Bax, cytochrome c, caspase-3; lowers HSP70/90) in renal cells via altering signaling pathways and enhancing antioxidant defenses [75]. Selenium can help guard against heavy metal poisoning, but it's important to remember that both not getting enough selenium and getting too much can be harmful. The body's antioxidant defense and other important functions depend on having the right amount of selenium (Table 12.5).

Iron

The trace element iron is essential for optimal physiological functioning and overall health and must be derived from dietary food sources and supplements. It is found in hemoglobin in erythrocytes and is also a component of myoglobin, aiding in storing and releasing oxygen within muscle cells. It also contributes to enzymatic reactions in energy production, DNA and amino acid synthesis, and immune function [76]. Iron helps in prevention and mitigation of toxicity caused by heavy metals. It prevents the entry of heavy metal in our body by environmental remediation. Iron-based materials (like iron oxides/hydroxides and zero-valent iron nanoparticles) induces adsorption, reductive precipitation, coprecipitation, and lattice incorporation (Fig. 12.6), thus immobilizing them on surfaces or reducing their solubility through precipitation and entrapment within the iron material's structure, ultimately transforming toxic heavy metal ions into less harmful forms, and preventing them from entering the environment or the body and causing harm. Additionally, cadmium and lead compete with iron in intestine via divalent metal transporter-1. Adequate iron in diet prevents toxicity by these two heavy metals. Conversely, iron deficiency worsens metal's toxic effects [78]. An optimum level of iron is essential for normal health.

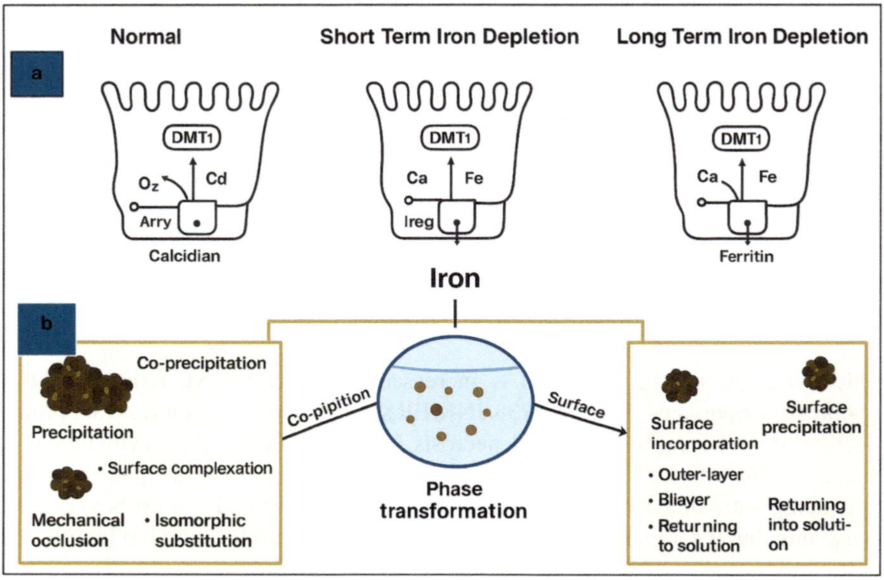

Fig. 12.6 (**a**) Mechanisms involved in heavy metals retention by iron oxyhydroxide. (**b**) Transportation of lead (Pb) and cadmium (Cd) in short-term and long-term iron depletion [77]

Calcium

Calcium is mineral, which is found in abundance in our body and is often thought of as macronutrient. However, it is classified as micronutrient due to its role in physiological functions at cellular levels. Its measure functional routes include formation of bones and tooth, nerve impulse transmission, muscle contraction, and as clotting factor. Few other functions include prevention of colon cancer, formation and progression of kidney stones, and reduction of blood pressure. The daily dietary intake of calcium for both man and woman is 1200–1500 mg. Milk and milk products; green leafy vegetables that includes kale, spinach, etc.; and fortified flour are the major source of calcium [79].

Many toxic heavy metals have a similar charge and ionic radius to calcium, allowing them to interfere with and mimic calcium's metabolic pathways. This can disrupt neural signaling by impostering as calcium for binding with calmodulin. Deposition of lead in bones in place of calcium leads to an overload of calcium ion, which can trigger excitotoxity (overstimulation of nerve cells to the point of cell death) which causes neurological damage. These toxic manifestations worsen when an individual is presented with calcium deficiency. A calcium-rich diet prevents absorption of these heavy metals (Pb and Cd) by preferentially binding with DMT 1 and prevents their entry in blood circulation, ultimately preventing their distribution and toxic effects upon vital organs, i.e., heart, kidney, liver, skeletal system, and nervous system [80].

Magnesium

Magnesium is the fourth most abundant mineral in the body, with the highest content in bones, followed by muscles, soft tissues, and blood. Seeds, nuts, and green leafy vegetables, such as spinach, are rich dietary sources of magnesium. Over 75 years ago, it was first reported that a deficiency of magnesium may be associated with a pro-. After that, more evidence has been provided on the connection between magnesium deficiency and numerous metabolic diseases. Magnesium deficiency is a risk factor for diseases connected to inflammation and oxidative stress, including obesity, atherosclerosis, and T2DM. The plasma concentration of magnesium negatively correlates with the risk of many metabolic diseases which are accompanied by low-grade inflammation [81]. Animal and human studies also support that CVD patients have a significantly higher requirement of both magnesium and vitamin E. Like vitamin E, supplementation of magnesium shows different but beneficial effects in regulating glucose, as well as lipid. For this reason, it has been hypothesized that the combined supplementation could probably provide more benefits than the supplementation of magnesium or vitamin E alone. Four interventional studies were performed in Iran to determine the effects of a supplementation of magnesium and vitamin E on oxidative stress, cardiovascular risk, T2DM, as well as blood lipid profile [82]. All of them used the same study design: The participants received 250 mg/day of magnesium and 268 mg/day of -TOH. Afzali and co-workers reported that this co-supplementation of magnesium and -TOH for 12 weeks reduced the wound size in patients suffering from diabetic ulcer. In addition, taking magnesium and vitamin E led to improved glycemic control via a reduction of fasting plasma glucose and alleviative insulin resistance. The supplementation also slightly decreased triglycerides (TG) and LDL cholesterol as well as reducing high-sensitivity C-reactive protein (hs-CRP) by more than 30%. Maktabi et al. investigated the effects in patients with gestational diabetes and found that the co-supplementation ameliorated the adverse plasma lipid parameters: Plasma TG and the ratio of total cholesterol to high-density lipoprotein (HDL) cholesterol were reduced after 6 weeks [83].

Micronutrients During Chelation of Metal

Coadministration of micronutrients like essential metals, vitamins, amino acids, and phytochemicals during chelation therapies is currently considered the most viable and recommended therapy for toxic metal-exposed animals/patients for whom monotherapy regimens have failed to achieve required level of benefits for the rapid reduction of excess metals and associated clinical disorders including oxidative stress and to avoid permanent organ damage. Combination chelation may provide a more flexible approach for individualizing chelation therapy, thereby improving tolerability, adherence, and quality of life. In principle, chelators can be combined with an infinite number of dosing regimens; these involve simultaneous or sequential exposure to the chelators on the same day or alternating the drugs on different days. Clinical studies have established the safety and efficacy of chelation combinations. However, real-life data with combination therapies still lacks, but it definitely underlines the significance of compliance for a meaningful reduction in metal overload compared to monotherapies [84].

Conclusion and Future Direction

There is lack of studies designed specifically to address the effect of micronutrient status on toxicity from exposure to toxic metals. The ability of micronutrients to modulate the toxic effects of metals is indisputable. Toxic metals and micronutrients interact at several points in the body: absorption and excretion of toxic metals, transport of metals in the body, binding to target proteins, metabolism and sequestration of toxic metals, and finally, in secondary mechanisms of toxicity like oxidative stress. Population consuming a diet deficient in micronutrients will be predisposed to metal toxicity. Numerous finding suggest the fact that supplementation of micronutrients can have a significant impact on metal toxicity. Studies that determine the impact of diet on toxicity will make risk assessment more accurate. These possibilities provided by the interaction of diet with toxic metals are clear. We now need to seriously consider the effect of micronutrients on the health of humans and animals exposed to toxic metals. The typical remedy for toxic metal exposure has been to remove the exposure source or use chelation in depleting the body burden of toxic metals. However, situations such as the arsenicism in India today will arise situations in which millions of people are affected and it is impossible to remove everyone from exposure. In these cases, dietary manipulation may provide the best defense and must be understood to be of use.

References

1. Jayamurali D, Varier KM, Liu W et al (2021) An overview of heavy metal toxicity. In: Metal, metal oxides and metal sulphides for biomedical applications. Springer International Publishing, Cham, pp 323–342
2. Flora SJ (2000) Possible health hazards associated with the use of toxic metals in semiconductor industries. J Occup Health 42(3):105–110
3. Mishra S, Bharagava RN, More N et al (2018) Heavy metal contamination: an alarming threat to environment and human health. In: Environmental biotechnology: for sustainable future. Springer, pp 103–125
4. Singh V, Singh N, Rai SN et al (2023) Heavy metal contamination in the aquatic ecosystem: toxicity and its remediation using eco-friendly approaches. Toxics 11(2):147
5. Kaur R, Sharma S, Kaur H (2019) Heavy metals toxicity and the environment. J Pharmacog Phytochem 1:247–249

6. Ali H, Khan E, Ilahi I (2019) Environmental chemistry and ecotoxicology of hazardous heavy metals: environmental persistence, toxicity, and bioaccumulation. J Chem 1:6730305
7. Das S, Sultana KW, Ndhlala AR et al (2023) Heavy metal pollution in the environment and its impact on health: exploring green technology for remediation. Environ Health Insights 17:11786302231201259
8. Patwa J, Sharma A, Flora S (2022) Arsenic, cadmium, and lead. In: Reproductive and developmental toxicology. Elsevier, pp 547–571
9. Okoro EE, Okolie AG, Sanni SE, Omeje M (2020) Toxicology of heavy metals to subsurface lithofacies and drillers during drilling of hydrocarbon wells. Sci Rep 10(1):6152
10. Limcharoensuk T, Chumsawat W, Siraj U et al (2025) Aqueous extract of Cissus quadrangularis L. alleviates heavy metal toxicity in Saccharomyces cerevisiae by limiting metal uptake and enhancing detoxification mechanisms. Ecotoxicol Environ Saf 299:118408
11. Silva SB, Ruiz TFR, Dos Santos FCA et al (2023) Impacts of heavy metal exposure on the prostate of murine models: mechanisms of toxicity. Reprod Toxicol 120:108448
12. Flora SJ, Pachauri V (2010) Chelation in metal intoxication. Int J Environ Res Public Health 7(7):2745–2788
13. Patwa J, Flora SJS (2020) Heavy metal-induced cerebral small vessel disease: insights into molecular mechanisms and possible reversal strategies. Int J Mol Sci 21(11):3862
14. Kim HS, Kim YJ, Seo YR (2015) An overview of carcinogenic heavy metal: molecular toxicity mechanism and prevention. J Cancer Prev 20(4):232–240
15. Morris AL, Mohiuddin SS (2025) Biochemistry, nutrients. StatPearls Publishing, Treasure Island
16. Flora S, Mittal M, Mehta A (2008) Heavy metal induced oxidative stress & its possible reversal by chelation therapy. Indian J Med Res 128(4):501–523
17. Qian S, Wei Z, Yang W et al (2022) The role of BCL-2 family proteins in regulating apoptosis and cancer therapy. Front Oncol 12:985363
18. Soni JP, Sood R, Kumar J (2024) Reactive oxygen species (ROS) and cancer. In: Mishra N, Kaundal RK (eds) Role of autophagy and reactive oxygen species in cancer treatment: principles and current strategies. Springer Nature Switzerland, pp 27–45
19. Flora SJ, Flora G, Saxena G (2006) Environmental occurrence, health effects and management of lead poisoning. In: Lead. Elsevier, pp 158–228
20. Hsu P-C, Guo YL (2002) Antioxidant nutrients and lead toxicity. Toxicology 180(1):33–44
21. Kosnett MJ (2013) The role of chelation in the treatment of arsenic and mercury poisoning. J Med Toxicol 9(4):347–354
22. Unsal V (2018) Natural phytotherapeutic antioxidants in the treatment of mercury intoxication-a review. Adv Pharm Bull 8(3):365–376
23. Rafati Rahimzadeh M, Rafati Rahimzadeh M, Kazemi S et al (2017) Cadmium toxicity and treatment: an update. Caspian J Intern Med 8(3):135–145
24. Zhai Q, Narbad A, Chen W (2015) Dietary strategies for the treatment of cadmium and lead toxicity. Nutrients 7:552–571
25. Flora SJ, Shrivastava R, Mittal M (2013) Chemistry and pharmacological properties of some natural and synthetic antioxidants for heavy metal toxicity. Curr Med Chem 20(36):4540–4574
26. Kim HS, Kim YJ, Seo YR (2019) General review of titanium toxicity. Int J Implant Dent 5(1):10
27. alali-Mood M, Eizadi-Mood N, Hassanian-Moghaddam et al (2005) Recent advances in the clinical management of intoxication by five heavy metals: mercury, lead, chromium, cadmium and arsenic. Heliyon 11(4)
28. Genchi G, Carocci A, Lauria G et al (2020) Nickel: human health and environmental toxicology. Int J Environ Res Public Health 17(3)
29. Leyssens L, Vinck B, Van Der Straeten C et al (2017) Cobalt toxicity in humans—a review of the potential sources and systemic health effects. Toxicology 387:43–56
30. Takamiya AS, Monteiro DR, Bernabé DG et al (2016) In vitro and in vivo toxicity evaluation of colloidal silver nanoparticles used in endodontic treatments. J Endodont 42(6):953–960

31. Jiang YM, Mo XA, Du FQ et al (2006) Effective treatment of manganese-induced occupational parkinsonism with p-aminosalicylic acid: a case of 17-year follow-up study. J Occup Environ Med 48(6):644–649
32. Kwakye GF, Paoliello MM, Mukhopadhyay S et al (2015) Manganese-induced parkinsonism and Parkinson's disease: shared and distinguishable features. Int J Environ Res Public Health 12(7):7519–7540
33. Sharma SS, Schat H, Vooijs R (1998) In vitro alleviation of heavy metal-induced enzyme inhibition by proline. Phytochem 49(6):1531–1535
34. Kumar A, Sakhare K, Bhattacharya D et al (2022) Communication in non-communicable diseases (NCDs) and role of immunomodulatory nutraceuticals in their management. Front Nutr 9
35. Peraza MA, Ayala-Fierro F, Barber DS et al (1998) Effects of micronutrients on metal toxicity. Environ Health Perspect 106(suppl 1):203–216
36. Alsafran M, Saleem MH, Rizwan M et al (2023) An overview of heavy metals toxicity in plants, tolerance mechanism, and alleviation through lysine-chelation with micro-nutrients—a novel approach. Plant Growth Regul 100(2):337–354
37. Roth B, Al-Shareef D, Ohlsson B (2025) Intake of fiber and micronutrients in patients with IBS. Scand J Gastroenterol 60:1–12
38. Kumar A, Sharma J, Chattopadhyay R et al (2022) EON therapy ameliorates cachexia and quality of life in cancer patients. J Cancer Ther 13(07):379–396
39. Shenkin A, Talwar D, Ben-Hamouda N et al (2024) LLL 44-1 micronutrients in clinical nutrition: trace elements. Clin Nutr ESPEN 61:369–376
40. Karalliedde L, Brooke N (2012) Toxicity of heavy metals and trace elements. In: Essentials of toxicology for health protection: a handbook for field professionals. Oxford University Press, Oxford
41. Patra R, Rautray AK, Swarup D (2011) Oxidative stress in lead and cadmium toxicity and its amelioration. Vet Med Int 2011(1):457327
42. Ofoedu CE, Iwouno JO, Ofoedu EO et al (2021) Revisiting food-sourced vitamins for consumer diet and health needs: a perspective review, from vitamin classification, metabolic functions, absorption, utilization, to balancing nutritional requirements. Peer J 9:e11940
43. Hake G, Mhaske A, Shukla R, Flora SJS (2025) Copper-induced neurodegenerative disorders and therapeutic potential of curcumin-loaded nanoemulsion. Toxics 13(2):108
44. Arigony AL, de Oliveira IM, Machado M et al (2013) The influence of micronutrients in cell culture: a reflection on viability and genomic stability. Biomed Res Int 2013:597282
45. Flora S, Jain V, Behari JR, Tandon S (1982) Protective role of trace metals in lead intoxication. Toxicol Lett 13(1–2):51–56
46. Azzi A (2007) Molecular mechanism of α-tocopherol action. Free Radic Biol Med 43(1):16–21
47. Liao S, Omage SO, Börmel L et al (2022) Vitamin E and metabolic health: relevance of interactions with other micronutrients. Antioxidants 11(9):1785
48. Sharma A, Flora SJS (2021) Positive and negative regulation of ferroptosis and its role in maintaining metabolic and redox homeostasis. Oxid Med Cell Longev 2021(1):9074206
49. Srivastava S, Kumar A, Yadav PK et al (2021) Formulation and performance evaluation of polymeric mixed micelles encapsulated with baicalein for breast cancer treatment. Drug Dev Ind Pharm 47(9):1512–1522
50. Sharma N, Kumar A, Sambhakar S et al (2025) Recent advances in Nanotherapeutics and Theranostics for squamous cell carcinoma: a comprehensive review. Curr Drug Deliv
51. Gonsalves H, Santos AFd, Azevedo LH, Corrêa L (2025) Oral surgery, oral medicine, oral pathology and oral radiology. 139(5):e131
52. Huang Z, Liu Y, Qi G et al (2018) Role of vitamin A in the immune system. J Clin Med 7(9):258
53. Aaseth JO, Alehagen U, Opstad TB, Alexander J (2023) Vitamin K and calcium chelation in vascular health. Biomedicines 11(12):3154
54. Oldenburg J, Bevans CG, Müller CR, Watzka M (2006) Vitamin K epoxide reductase complex subunit 1 (VKORC1): the key protein of the vitamin K cycle. Antioxid Redox Signal 8(3–4):347–353

55. Ivanova D, Zhelev Z, Getsov P et al (2018) Vitamin K: redox-modulation, prevention of mitochondrial dysfunction and anticancer effect. Redox Biol 16:352–358
56. Maywald M, Rink L (2022) Zinc in human health and infectious diseases. Biomolecules 12(12)
57. Prasad AS, Bao B (2019) Molecular mechanisms of zinc as a pro-antioxidant mediator: clinical therapeutic implications. Antioxidants 8(6):164
58. Stiles LI, Ferrao K, Mehta KJ (2024) Role of zinc in health and disease. Clin Exp Med 24(1):38
59. Thompson MW (2022) Regulation of zinc-dependent enzymes by metal carrier proteins. Biometals 35(2):187–213
60. Flora SJ (1991) Influence of simultaneous supplementation of zinc and copper during chelation of lead in rats. Hum Exp Toxicol 10(5):331–336
61. Speer RM, Zhou X, Volk LB et al (2023) Arsenic and cancer: evidence and mechanisms. In: Advances in pharmacology. Elsevier, pp 151–202
62. Abbehausen C (2019) Zinc finger domains as therapeutic targets for metal-based compounds–an update. Metallomics 11(1):15–28
63. Yu H-t, Zhen J, Leng J-y et al (2021) Zinc as a countermeasure for cadmium toxicity. Acta Pharmacol Sin 42(3):340–346
64. Qi Z, Wang Q, Wang H, Tan M (2021) Metallothionein attenuated arsenic-induced cytotoxicity: the underlying mechanism reflected by metabolomics and lipidomics. J Agric Food Chem 69(18):5372–5380
65. Hudson LG, Dashner-Titus EJ, MacKenzie D (2025) Zinc as a mechanism-based strategy for mitigation of metals toxicity. Curr Environ Health Rep 12(1):5
66. Fujishiro H, Himeno S (2019) New insights into the roles of ZIP8, a cadmium and manganese transporter, and its relation to human diseases. Biol Pharm Bull 42(7):1076–1082
67. Banerjee M, Yaddanapudi K, States JC (2022) Zinc supplementation prevents mitotic accumulation in human keratinocyte cell lines upon environmentally relevant arsenic exposure. Toxicol Appl Pharmacol 454:116255
68. Khandelwal S, Flora S, Tandon S (1985) Possible role of nickel induced metallothionein (MT) in protection against cadmium nephrotoxicity. Zeitschrift fur die gesamte Hygiene und ihre Grenzgebiete 31(7):400–403
69. Bai S, Zhang M, Tang S et al (2024) Effects and impact of selenium on human health, a review. Molecules 30(1):50
70. Hashtjin YA, Raeeszadeh M, Khanghah AP (2025) Interaction of heavy metals (cadmium and selenium) in an experimental study on goldfish: Hematobiochemical changes and oxidative stress. J Xenobio 15(2):57
71. Sahito JH, Ma C, Zhang J et al (2025) Selenium and its mechanisms mitigate cadmium toxicity in plants: promising role and future potentials. Ecotoxico Environ Saf 300:118422
72. Spiller HA (2018) Rethinking mercury: the role of selenium in the pathophysiology of mercury toxicity. Clin Toxicol 56(5):313–326
73. Zwolak I (2020) The role of selenium in arsenic and cadmium toxicity: an updated review of scientific literature. Biol Trace Elem Res 193(1):44–63
74. Hong W, Liu Y, Liang J et al (2023) Molecular mechanisms of selenium mitigating lead toxicity in chickens via mitochondrial pathway: selenoproteins, oxidative stress, HSPs, and apoptosis. Toxics. 11(9):734
75. Miao Z, Z. Miao, Shi X et al (2022) The antagonistic effect of selenium on lead-induced apoptosis and necroptosis via P38/JNK/ERK pathway in chicken kidney. Ecotoxicol Environ Saf 231:113176
76. Moustarah F, Daley SF (2025) Dietary iron. StatPearls Publishing, Treasure Island
77. Bressler JP, Olivi L, Cheong JH et al (2004) Divalent metal transporter 1 in lead and cadmium transport. Ann N Y Acad Sci 1012(1):142–152
78. Li H-B, Xue R-Y, Chen X-Q et al (2022) Ca minerals and oral bioavailability of Pb, Cd, and As from indoor dust in mice: mechanisms and health implications. Environ Health Perspect 130(12):127004
79. Cormick G, Belizán JM (2019) Calcium intake and health. Nutrients 11(7):1606

80. Carmona FJ, Guagliardi A, Masciocchi N (2022) Nanosized calcium phosphates as novel macronutrient nano-fertilizers. Nanomaterials (Basel) 12(15)
81. Jiao J, Wei L, Shi S et al (2025) Response of calcium-dependent protein kinase genes' expression in 'Feizixiao' litchi pulp to foliar nutrient treatment of calcium-magnesium mixed solution and their regulation of sugar transformation. Plants (Basel) 14(11)
82. Kumar A, Maurya P, Narayan KP (2025) Fabrication of polymeric matrix type transdermal patch containing glimepiride for therapeutic management of type-2 diabetes. Ind J Pharm Educ Res 59(3):936–948
83. Maktabi M, Jamilian M, Amirani E et al (2018) The effects of magnesium and vitamin E co-supplementation on parameters of glucose homeostasis and lipid profiles in patients with gestational diabetes. Lipids Health Dis 17(1):163
84. Zhai Q, Narbad A, Chen W (2015) Dietary strategies for the treatment of cadmium and lead toxicity. Nutrients 7(1):552–571

Chapter 13
Development of Hypertension in Response to Chronic Exposure to Mercury

Elle Garriock, Shawn D. George, Paramjit S. Tappia, and Bram Ramjiawan

Abstract Hypertension is believed to be the leading modifiable risk factor for cardiovascular morbidity and mortality worldwide. While obesity, high sodium intake, and sedentary lifestyle are well-established pathological factors for hypertension, environmental toxicants including mercury (Hg) are increasingly being recognized as contributors to elevated blood pressure. Exposure to Hg occurs through occupational sources, pollution, and dietary intake, particularly from methylmercury-contaminated fish. Several lines of accumulating evidence suggest Hg induces oxidative stress, endothelial dysfunction, and renal injury, as well as causing a disturbance of the renin–angiotensin system and the sympathetic nervous system. Accordingly, this review describes some of the experimental and clinical lines of evidence for Hg exposure and the induction of hypertension as well as the underlying mechanisms of action of Hg in the pathogenesis of hypertension in susceptible populations as well as the implications for public health.

Keywords Mercury · Hypertension · Oxidative stress · Neurohormonal systems · Endothelial dysfunction, Kidney injury

E. Garriock · S. D. George · P. S. Tappia (✉)
Asper Clinical Research Institute and St. Boniface Hospital Albrechtsen Research Centre, Winnipeg, MB, Canada
e-mail: ptappia@sbrc.ca

B. Ramjiawan
Department of Pharmacology and Therapeutics, Max Rady College of Medicine, University of Manitoba, Winnipeg, MB, Canada

N. S. Dhalla et al. (eds.), *Functional Biochemistry of Metallic Elements*, Advances in Biochemistry in Health and Disease 36,
https://doi.org/10.1007/978-3-032-14445-4_13

Introduction

Hypertension affects over 1 billion people globally and accounts for more deaths than any other modifiable risk factor [1]. Despite advances in pharmacological therapy targeting the sympathetic nervous system and RAS, treatment resistance persists in approximately 40% of cases [2]. Emerging evidence suggests that environmental contaminants, including heavy metals such as mercury (Hg), play a role in the etiology of hypertension [3]. Indeed, Hg, in particular, remains a global concern due to its persistence in the environment and ability to bioaccumulate in the food chain.

There are three main forms of Hg: elemental, inorganic, and organic [4]. Moreover, there are two principal forms of mercury that humans are exposed to, namely, Hg vapor and methylmercury compounds [5]. Human exposure arises from industrial emissions, coal combustion, artisanal gold mining, and waste incineration, accounting for up to 90% of environmental Hg release [6]. In aquatic systems, inorganic Hg is converted by microorganisms into methylmercury (MeHg), a form that biomagnifies through the food chain, reaching highest levels in predatory fish such as shark, swordfish, and king mackerel [7].

The Minamata disaster in the 1950s in Japan, where MeHg-contaminated seafood caused severe neurotoxicity, highlighted mercury's long-term health risks [8]. However, several human studies have demonstrated associations between Hg exposure and elevated blood pressure. Epidemiological syntheses and recent systematic appraisals report a consistent association between higher Hg burden and increased blood pressure indices and hypertension risk, particularly when considered alongside co-exposures to other metals and cardiometabolic risk factors. These population data support a biologically plausible link between Hg exposure and impaired vascular homeostasis [9, 10].

Occupational studies examining Hg miners showed higher prevalence of hypertension and increased cardiovascular and renal disease compared with controls [11]. The Finnish Kuopio Ischemic Heart Disease Risk Factor Study reported a twofold increased risk of hypertension in men with high hair Hg levels [12]. The US NHANES data has revealed that blood Hg concentrations were positively associated with systolic blood pressure in women, particularly in non-fish consumers [13]. Notably, in the Faroese and Inuit Indigenous populations consuming Hg-contaminated fish and marine mammals, elevated carotid intima-media thickness and blood pressure have been reported [14]. It should be mentioned that while fish consumption provides cardioprotective omega-3 fatty acids, Hg contamination counteracts these benefits through oxidative stress and endothelial dysfunction [15].

Populations consuming low-Hg fish still experience net cardiovascular benefits, underscoring the importance of targeted dietary guidance [16]. Several mechanisms underlying hypertension associated with Hg exposure have been proposed including oxidative stress, impaired nitric oxide (NO) signaling, modified vascular response to neurotransmitters and disturbed vascular muscle Ca^{2+} signaling, renal damage, and interference with the RAS [17]. Accordingly, this review describes some of the experimental and clinical findings on Hg exposure and hypertension

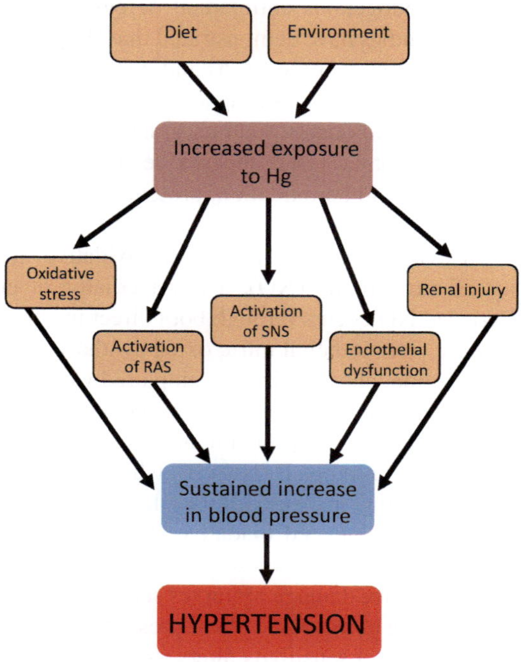

Fig. 13.1 Primary mechanisms associated with Hg-induced hypertension. Hg mercury; RAS renin–angiotensin system; SNS sympathetic nervous system

and highlights the primary mechanisms (summarized in Fig. 13.1) associated with Hg-induced hypertension in susceptible populations and its implications for public health awareness.

Pathophysiological Mechanisms Linking Mercury and Hypertension

It is now known that multiple interrelated mechanisms involving oxidative stress, disruption of neurohormonal systems, endothelial dysfunction, and renal injury contribute to the development of hypertension in response to chronic Hg exposure and are described as follows.

Role of Mercury-Induced Oxidative Stress in the Development of Hypertension

Chronic exposure to Hg has been identified as a contributing factor to the development of hypertension, primarily through mechanisms involving oxidative stress. In this regard, Hg has been reported to promote the formation of reactive oxygen species (ROS), increase lipid peroxidation, induce glutathione depletion, as well as deplete endogenous antioxidant defense systems [18]. Indeed, experimental and clinical studies have demonstrated that Hg promotes the generation of ROS, thereby disrupting redox homeostasis and impairing vascular function [19, 20]. It is pointed out that Hg exhibits a strong affinity for sulfhydryl groups, resulting in depletion of essential antioxidants such as glutathione and inhibition of antioxidant enzymes including catalase and superoxide dismutase [21–23]. This oxidative imbalance facilitates lipid peroxidation, protein oxidation, and DNA damage, leading to compromised vascular integrity and cellular homeostasis.

In vascular tissues, Hg-induced oxidative stress leads to decreased bioavailability of nitric oxide (NO), a key mediator of endothelial-dependent vasodilation. Reduced NO occurs through both direct inactivation by superoxide and diminished activity of endothelial nitric oxide synthase (eNOS) [24]. The resultant endothelial dysfunction, characterized by impaired vasorelaxation and increased vascular tone, is a central mechanism in hypertension development. Additionally, Hg-evoked production of ROS is reported to activate redox-sensitive transcription factors, such as nuclear factor-kappa B (NF-κB), which amplify pro-inflammatory signaling, vascular remodeling, and smooth muscle cell proliferation [25, 26].

While Hg-induced oxidative stress exerts vascular effects, the oxidative stress also disrupts renal function and impairs autonomic regulatory systems. For example, renal oxidative damage promotes sodium retention and extracellular fluid expansion, contributing to increased blood pressure [27]. Moreover, ROS-mediated impairment of baroreceptor sensitivity alters autonomic regulation of vascular resistance, further perpetuating hypertension [24]. It should be mentioned that the Hg-induced oxidative stress increases the expression of angiotensin-converting enzyme (ACE) and angiotensin II type 1 receptors (AT1R) in vascular tissues [26, 28]. The consequence of elevated angiotensin II (Ang II) levels, acting via AT1R, is the promotion of vasoconstriction, vascular smooth muscle hypertrophy, and pro-inflammatory gene expression while simultaneously suppressing NO bioavailability. This dual effect exacerbates vascular stiffness and increases peripheral resistance. Taken together, these findings highlight oxidative stress as a critical mechanism associated with Hg exposure in the pathogenesis of hypertension. Furthermore, the interplay of reduced antioxidant defense, diminished NO bioavailability, endothelial dysfunction, and pro-inflammatory activation provides a foundation for understanding Hg-induced cardiovascular risk as well as the potential for therapeutic strategies that target oxidative stress in mitigating the hypertensive effects of chronic Hg exposure.

Dysregulation of the Renin–Angiotensin System due to Mercury Exposure and Hypertension

The RAS is a critical regulator of vascular tone, sodium balance, and fluid homeostasis. Mercury exposure not only provokes oxidative stress but also disrupts neurohumoral pathways central to blood pressure regulation, particularly the renin–angiotensin system (RAS), and its dysregulation is strongly implicated in the pathogenesis of hypertension [29]. A dysregulation of the RAS has been reported to occur in response to Hg exposure resulting in an increase in vascular tone [30]. Exposure to Hg dysregulates systemic and tissue-specific components of the RAS and thus favors vasoconstriction, sodium retention, vascular remodeling, and subsequent development of hypertension. There is now accumulating preclinical and translational data indicating that Hg also modulates key RAS elements (ACE, Ang II, AT1R) and may disrupt compensatory ACE2/Ang-(1–7)/Mas axis, shifting the homeostatic balance toward the classical, pressor arm of the RAS [9, 31]. Mechanistic evidence from animal and ex vivo studies have demonstrated that Hg increases ACE activity and augments Ang II signaling in vascular and renal tissues. The upregulation of ACE and AT1R expression with ROS-driven NADPH oxidase activation amplifies Ang II-mediated vasoconstriction, smooth muscle hypertrophy, and inflammatory gene expression while simultaneously diminishing NO bioavailability. These convergent actions increase systemic vascular resistance and facilitate structural arterial changes that sustain elevated blood pressure. Such actions of Hg on the RAS produce endothelial dysfunction that precedes hypertension in experimental models [32].

It has been demonstrated that Hg exposure promotes intrarenal RAS activation, tubular injury, and sodium retention resulting in an expansion of intravascular volume and an increase in arterial pressure [33]. Renal oxidative injury also impairs autoregulatory responses and sensitizes the kidney to profibrotic and pro-inflammatory effects of Ang II, thereby linking Hg-induced nephrotoxicity to chronic blood pressure elevation [32]. It should be mentioned that ACE2, which degrades Ang II to Ang-(1–7) counterbalances classical RAS signaling [34]. Although direct experimental studies specifically demonstrating Hg-induced down-regulation of ACE2 are very limited, based on Hg-induced oxidative stress, NF-κB-driven inflammation, and protease activation, which are known to reduce membrane ACE2 and/or increase ACE2 shedding, it is possible that exposure to Hg could blunt the protective ACE2/Ang-(1–7) axis and thereby exaggerate Ang II-dependent hypertensive signaling; however, this remains to be fully established [35].

In view of the Hg-driven RAS activation, endothelial dysfunction, and renal impairment, it is thus suggested that interventions targeting the RAS may be particularly effective in populations with significant and detrimental exposure to Hg. While pharmacological inhibitors of the RAS remain an appropriate therapeutic approach, experimental strategies that preserve or restore ACE2/Ang-(1–7) signaling are potentially of benefit. At the population level, reducing environmental and dietary exposure to Hg would be seen as a primary preventive approach. Importantly, given the recent interest in ACE2 and cardiometabolic disease, exploration into the relationship between Hg exposure, ACE/ACE2 balance, and clinical cardiovascular endpoints is warranted [35, 36]. Together, these findings highlight the RAS as a pivotal mediator of mercury-induced hypertension. Mercury amplifies classical Ang II/AT1R signaling while suppressing counter-regulatory pathways, contributing to sustained vasoconstriction, vascular remodeling, renal sodium retention, and systemic hypertension. Interventions targeting RAS activation—such as ACE inhibitors, AT1R blockers, or therapies enhancing ACE2/Ang-(1–7) pathways—may hold therapeutic promise in mitigating the cardiovascular burden of mercury exposure.

Mercury-Induced Hypertension due to Activation of the Sympathetic Nervous System

Mercury exposure contributes to the development and maintenance of hypertension not only through oxidative stress, endothelial dysfunction, and activation of the RAS but also by disrupting autonomic regulation, particularly the sympathetic nervous system (SNS). Sympathetic overactivity is a well-established driver of elevated blood pressure, increasing heart rate, cardiac output, peripheral resistance, and vascular remodeling [37]. Mercury exposure activates the sympathetic nervous system (SNS), compounding its hypertensive impact by increasing sympathetic outflow, impairing autonomic reflexes, and disrupting catecholamine metabolism, creating a sustained pressor environment. Recent epidemiological, experimental, and autopsy-based evidence underscores the relevance of these pathways in human and animal models. In a controlled animal study, chronic low-dose exposure to inorganic mercury ($HgCl_2$) induced marked alterations in autonomic control of cardiovascular function. Rats exhibited elevated mean arterial pressure and heart rate, diminished baroreflex sensitivity, blunted reflexive hemodynamic responses (e.g., to the Von Bezold–Jarisch reflex), and a shift in heart rate variability (HRV) consistent with sympathetic predominance. Pharmacological blockade confirmed heightened sympathetic tone and impaired vagal (parasympathetic) regulation in mercury-exposed animals [38].

In adults environmentally exposed to methylmercury—such as Inuit populations—higher blood mercury levels correlated with reduced HRV metrics, particularly parameters representing parasympathetic activity. This autonomic dysfunction was accompanied by elevations in systolic blood pressure and pulse pressure, aligning with mercury's association with hypertension in human cohorts [39]. Another human study assessing heart rate recovery (HRR) after maximal exercise observed blunted recovery in mercury-exposed individuals. Given that HRR dynamics reflect parasympathetic reactivation and sympathetic withdrawal, these findings further point to mercury-mediated autonomic dysregulation [40]. Autopsy studies reveal that mercury accumulates with age in the human adrenal medulla. Mercury interferes with S-adenosyl-l-methionine (SAM), a crucial cofactor for catecholamine-degrading enzymes, and impairs both catechol-O-methyltransferase (COMT) and phenylethanolamine N-methyltransferase (PNMT) function. These disruptions elevate noradrenaline levels and reduce adrenaline synthesis, potentially contributing to sympathetic overdrive in vivo [41]. Pediatric cases of elemental mercury intoxication have been documented in which children presented with severe hypertension and elevated catecholamine levels, initially suggestive of pheochromocytoma.

Clinical evaluation revealed no adrenal tumors; rather, mercury exposure explained the hypertensive and adrenergic presentation [42].

A major mechanism involves Hg-induced oxidative stress and neuroinflammation within central autonomic nuclei; Hg accumulates in the central nervous system, where it enhances ROS production and activates microglia, leading to neuronal dysfunction in areas that regulate sympathetic outflow, such as the hypothalamic paraventricular nucleus and brainstem centers [28]. The resulting activation of sympathetic premotor neurons promotes heightened efferent sympathetic activity, thereby increasing vasoconstriction and arterial pressure. In addition, Hg also indirectly influences SNS activity through interaction with the RAS. In this regard, angiotensin II, which is upregulated by Hg exposure, acts centrally to stimulate sympathetic nerve firing and peripherally to sensitize vascular smooth muscle to sympathetic neurotransmitters [29]. This crosstalk between mercury, RAS, and SNS establishes a feed-forward loop that sustains hypertension. Experimental studies have demonstrated increased plasma catecholamine levels and enhanced vascular responsiveness to norepinephrine following Hg exposure [26, 43]. Furthermore, baroreflex dysfunction, partly mediated by oxidative damage to arterial baroreceptor afferents and central integration pathways, reduces inhibitory feedback on sympathetic drive, further augmenting pressor responses [24].

Taken together, these findings suggest that Hg contributes to hypertension through SNS overactivation at both central and peripheral levels. Central neurotoxicity enhances sympathetic outflow, while peripheral sensitization increases vascular responsiveness to adrenergic stimuli. These mechanisms, along with oxidative stress and RAS activation, highlight the role of the SNS in Hg-induced hypertension. Thus, targeting sympathetic overactivity either through lifestyle modifications, sympatholytic drugs, or neuromodulatory approaches may represent an adjunctive strategy in reducing Hg-related cardiovascular risk.

Role of Endothelial Dysfunction in Mercury-Induced Hypertension

Mercury exposure has been implicated in the pathogenesis of hypertension through its detrimental effects on vascular endothelium. The vascular endothelium plays a critical role in the regulation of vascular tone, hemostasis, and inflammation, with endothelial dysfunction representing an early and central mechanism in the development of hypertension and cardiovascular disease [44]. Mercury exerts direct toxic effects on endothelial cells by increasing oxidative stress, reducing nitric oxide (NO) bioavailability, and promoting a pro-inflammatory and pro-thrombotic endothelial phenotype [19, 26]. One of the primary mechanisms by which mercury induces endothelial dysfunction is through interference with NO signaling.

Mercury exposure reduces the activity and expression of endothelial nitric oxide synthase (eNOS) and promotes the interaction of superoxide anion with NO to form peroxynitrite, thereby decreasing the bioavailability of NO, a potent vasodilator [24, 26]. This impairment of NO-dependent vasodilation leads to increased vascular tone and peripheral resistance, key determinants of elevated blood pressure. In addition, mercury enhances the production of endothelin-1 (ET-1), a potent vasoconstrictor peptide secreted by endothelial cells, thereby contributing further to vascular dysfunction [25]. At the cellular level, mercury targets the vascular endothelium through multiple, convergent pathways. Mercury species have a high affinity for thiol groups on proteins and low-molecular-weight thiols, leading to depletion of glutathione and dysregulation of redox buffering systems; these drive sustained production of reactive oxygen species (ROS) and promote eNOS uncoupling. The consequence is a marked reduction in nitric oxide (NO) bioavailability, through both decreased eNOS activity/expression and chemical quenching of NO by superoxide, which directly impairs endothelium-dependent vasorelaxation and predisposes to increased vascular tone [45].

Mercury-induced activation of redox-sensitive signaling pathways, such as nuclear factor-kappa B (NF-κB), also results in upregulation of adhesion molecules and pro-inflammatory cytokines, which not only exacerbate endothelial dysfunction but also foster vascular remodeling and stiffness [20, 26]. Furthermore, evidence suggests that mercury disrupts the endothelial barrier by altering tight junction proteins, increasing vascular permeability, and promoting leukocyte adhesion and infiltration [22]. These endothelial alterations, in combination with oxidative and inflammatory insults, create a vascular environment conducive to hypertension.

Taken together, these findings demonstrate that endothelial dysfunction is a pivotal mechanism linking mercury exposure to hypertension. The suppression of NO signaling, stimulation of vasoconstrictor pathways, induction of vascular inflammation, and promotion of structural changes in the vascular wall form an integrated framework through which mercury contributes to increased vascular resistance and elevated blood pressure. Mercury exposure—both inorganic and organic (methylmercury)—has emerged as an environmental determinant of endothelial dysfunction that contributes to the development and progression of hypertension.

The vascular effects of mercury are further amplified by systemic interactions: Renal endothelial and tubular injury promote sodium retention and volume expansion, and autonomic/endocrine perturbations (e.g., altered baroreflex sensitivity and sympatho-vagal balance) further raise peripheral resistance. Co-exposure to other cardiotoxic metals (lead, cadmium, arsenic) and common cardiovascular risk modifiers (oxidative stress, inflammation, dyslipidemia) often produces additive or supra-additive effects on endothelial health and blood pressure regulation, explaining heterogeneous findings in low-exposure populations [46, 47]. Together, impaired vasodilator signaling and enhanced vasoconstrictor/inflammatory signaling cause a shift toward a hypertensive vascular state [48]. Experimental models report mercury-induced eNOS uncoupling, increased NADPH oxidase activity, mitochondrial dysfunction in endothelial cells, and upregulation of vasoconstrictors such as ET-1, each of which lowers the threshold for hypertension when combined with other hemodynamic or metabolic stressors. These preclinical data establish temporality and biological plausibility for mercury-driven endothelial injury as a proximate step toward elevated blood pressure [49].

From a translational standpoint, recognizing endothelial dysfunction as a central mechanism linking mercury exposure to hypertension directs both preventive and therapeutic strategies. Mechanistically informed interventions—aimed at restoring redox balance, preserving eNOS coupling, inhibiting NADPH oxidase, or blocking endothelial inflammatory signaling—represent rational adjuncts to traditional antihypertensive therapies in populations with demonstrable mercury burden. Continued research integrating precise exposure assessment, endothelial biomarkers, and interventional trials will be required to move from association to targeted clinical mitigation [50, 51].

Renal Accumulation of Mercury in the Pathogenesis of Hypertension

Since Hg is deemed as a nephrotoxicant, the renal effects are considered to contribute directly and indirectly to the development and maintenance of hypertension. In this regard, the renal accumulation of Hg is known to cause tubular and glomerular injury, activate the pro-hypertensive intrarenal RAS, and aggravate oxidative–inflammatory pathways that promote sodium retention, volume expansion, and long-term structural renal remodeling, which drive the increase in arterial pressure [52]. Mercuric species (inorganic Hg^{2+} and methylmercury) selectively accumulate in the kidney, particularly in proximal tubular epithelial cells, through transport processes that exploit thiol chemistry and organic anion pathways. Once inside tubular cells, mercury binds sulfhydryl groups, depletes intracellular glutathione, impairs mitochondrial function, and inhibits critical enzymes, producing cell stress, apoptosis, and loss of tubular integrity. Clinically, these processes manifest as tubular proteinuria, impaired concentrating ability, and, with prolonged exposure, progressive decline in glomerular filtration [52, 53].

In the kidney, the Hg-induced oxidative stress mediated through mitochondrial ROS generation and NADPH oxidase activation couples tightly with redox-sensitive inflammatory signaling involving the activation of NF-κB and subsequent cytokine release that prolong tubular injury. This sustained injury response drives extracellular matrix deposition and tubular interstitial fibrosis resulting in reduced nephron number and renal functional reserve, both of which predispose to hypertension. Animal and ex vivo studies have demonstrated these aspects and population studies link low-level Hg burden with biomarkers of renal injury and with increased risk of chronic kidney disease [54, 55].

The upregulation of the intrarenal RAS, as evidenced by increased local ACE activity and Ang II signaling, subsequent to exposure to Hg, exerts adverse effects on renal hemodynamics and tubular sodium transport. In this regard, Ang II increases proximal and distal tubular sodium reabsorption and reduces natriuresis; combined with Hg-associated tubular damage, this favors net sodium and water retention, plasma volume expansion, and sustained increases in blood pressure. Indeed, experimental models show that blockade of the RAS attenuates Hg-related pressor responses, supporting a causal role for intrarenal RAS activation in Hg-mediated hypertension [54, 56]. Aside from the direct tubular effects, Hg-driven renal injury disrupts renal autoregulation and baroreflex pathways, increasing the kidney's susceptibility to pressure-mediated damage, and evokes maladaptive cardiorenal feedback loops. Loss of nephron mass and renal microvascular rarefaction increase systemic vascular resistance and amplify hypertensive signals.

Recognition of the kidney as a central target in mercury-related cardiovascular risk underscores two translational priorities. First, early detection of subclinical renal injury (urinary biomarkers, GFR trends, tubular proteinuria) in exposed populations can identify individuals at risk for mercury-driven hypertension. Second, interventions that prevent renal accumulation (reduction of exposure), blunt intrarenal RAS activation (ACE inhibitors/ARBs), or mitigate oxidative–inflammatory cascades (antioxidant and anti-inflammatory strategies) are rational therapeutic approaches to interrupt the trajectory from mercury nephrotoxicity to sustained hypertension.

Clinical and Public Health Implications

The incidence of exposure to Hg represents a global challenge. Public health measures to reduce environmental and dietary mercury intake remain fundamental [9, 57]. The US Environmental Protection Agency (EPA) recommends a reference dose of <0.1 μg/kg/day, a threshold frequently exceeded in high fish consumers and occupationally exposed individuals [58]. Vulnerable populations, including pregnant women and children, require particular protection as fetal neurotoxicity occurs at much lower exposure levels than adult toxicity [7]. The exposure to Hg represents a modifiable but under-recognized risk factor for hypertension. Public health strategies must balance the nutritional benefits of fish consumption with the risks of Hg contamination, while policy-driven reductions in industrial emissions remain essential. In this regard, the Minamata Convention on Mercury, adopted in 2013, represents a major global effort to reduce Hg emissions and human exposure. Continued surveillance, biomonitoring, and public education on safe fish consumption are critical in reducing hypertension burden attributable to Hg.

Conclusion

From the aforementioned, it is evident that through interrelated mechanisms involving oxidative stress, RAS abnormalities, SNS dysregulation, endothelial dysfunction, and renal damage, Hg contributes to the pathogenesis of elevated blood pressure and cardiovascular disease. Future studies should refine dose–response relationships, investigate gene–environment interactions, and evaluate the role of lifestyle interventions in mitigating the adverse cardiovascular effects of Hg.

Acknowledgments We thank the St. Boniface Hospital Albrechtsen Research Centre for providing the infrastructural support for this project.

References

1. Chockalingam A (2007) Impact of world hypertension day. Can J Cardiol 23(7):517–519
2. Carey RM, Calhoun DA, Bakris GL et al (2018) Resistant hypertension: detection, evaluation, and management: a scientific statement from the American Heart Association. Hypertension 72(5):e53–e90
3. Tinkov AA, Ajsuvakova OP, Skalnaya MG et al (2018) Mercury and arterial hypertension: an overview. Environ Res 162:40–52
4. Clarkson TW, Magos L (2006) The toxicology of mercury and its chemical compounds. Crit Rev Toxicol 36(8):609–662
5. Branco V, Aschner M, Carvalho C (2021) Neurotoxicity of mercury: an old issue with contemporary significance. Adv Neurotoxicol 5:239–262
6. Energy Education (2023) Mercury pollution. University of Calgary. https://energyeducation.ca/encyclopedia/Mercury_pollution. Accessed 1 Aug 2025
7. Karagas MR, Choi AL, Oken E et al (2012) Evidence on the human health effects of low-level methylmercury exposure. Environ Health Perspect 120(6):799–806
8. Harada M (1995) Minamata disease: methylmercury poisoning in Japan caused by environmental pollution. Crit Rev Toxicol 25(1):1–24
9. Hu XF, Loan A, Chan HM (2025) Re-thinking the link between exposure to mercury and blood pressure. Arch Toxicol 99(2):481–512
10. Pan Z, Gong T, Liang P (2024) Heavy metal exposure and cardiovascular disease. Circ Res 134(9):1160–1178
11. Kobal AB, Horvat M, Prezelj M et al (2004) The impact of long-term past exposure to elemental mercury on antioxidative capacity and lipid peroxidation in mercury miners. J Trace Elem Med Biol 17(4):261–274
12. Virtanen JK, Rissanen TH, Voutilainen S, Tuomainen TP (2007) Mercury as a risk factor for cardiovascular diseases. J Nutr Biochem 18(2):75–85
13. Vupputuri S, Longnecker MP, Daniels JL et al (2005) Blood mercury level and blood pressure among US women: results from the National Health and Nutrition Examination Survey 1999–2000. Environ Res 97(2):195–200
14. Choi AL, Weihe P, Budtz-Jørgensen E et al (2009) Methylmercury exposure and adverse cardiovascular effects in Faroese whaling men. Environ Health Perspect 117(3):367–372
15. Mozaffarian D, Rimm EB, Willett WC (2011) Fish intake, contaminants, and human health: evaluating the risks and the benefits. JAMA 296(15):1885–1899
16. Rice KM, Walker EM Jr, Wu M et al (2014) Environmental mercury and its toxic effects. J Prev Med Public Health 47(2):74–83
17. da Cunha Martins A Jr, Carneiro MFH, Grotto D et al (2018) Arsenic, cadmium, and mercury-induced hypertension: mechanisms and epidemiological findings. J Toxicol Environ Health B Crit Rev 21(2):61–82
18. Valko M, Jomova K, Rhodes CJ et al (2016) Redox- and non-redox-metal-induced formation of free radicals and their role in human disease. Arch Toxicol 90(1):1–37
19. Houston MC (2011) Role of mercury toxicity in hypertension, cardiovascular disease, and stroke. J Clin Hypertens 13(8):621–627

20. Salonen JT, Seppänen K, Nyyssönen K et al (1995) Intake of mercury from fish, lipid peroxidation, and the risk of myocardial infarction and coronary, cardiovascular, and any death in eastern Finnish men. Circulation 91(3):645–655
21. Sule RO, Rivera GDT, Vaidya T et al (2025) Environmental toxins and oxidative stress: the link to cardiovascular diseases. Antioxidants 14(5):604
22. Farina M, Rocha JBT, Aschner M (2011) Mechanisms of methylmercury-induced neurotoxicity: evidence from experimental studies. Life Sci 89(15–16):555–563
23. Valko M, Leibfritz D, Moncol J et al (2007) Free radicals and antioxidants in normal physiological functions and human disease. Int J Biochem Cell Biol 39(1):44–84
24. Vaziri ND (2008) Mechanisms of lead-induced hypertension and cardiovascular disease. Am J Physiol Heart Circ Physiol 295(2):H454–H465
25. Bridges CC, Zalups RK (2017) Mechanisms involved in the transport of mercuric ions in target tissues. Arch Toxicol 91(1):63–81
26. Wiggers GA, Pecanha FM, Briones AM et al (2008) Low mercury concentrations cause oxidative stress and endothelial dysfunction in conductance and resistance arteries. Am J Physiol Heart Circ Physiol 295(3):H1033–H1043
27. Zalups RK (2000) Molecular interactions with mercury in the kidney. Pharmacol Rev 52(1):113–143
28. Oliveira VA, Pecanha FM, Wiggers GA et al (2015) Mechanisms of mercury-induced hypertension: RAS and oxidative stress crosstalk. Toxicol Lett 232(2):360–368
29. Ferrario CM, Strawn WB (2006) Role of the renin–angiotensin–aldosterone system and proinflammatory mediators in cardiovascular disease. Am J Cardiol 98(1A):121i–128i
30. Zhang Y, Xu Y, Ma J (2018) Role of environmental toxicants in hypertension. J Hypertens 36(8):1493–1501
31. Qiu M, Chen Y, Huang S et al (2025) Angiotensin-(1-7) mitigates progression from acute kidney injury to chronic kidney disease via renin-angiotensin system modulation in a murine model. Inflammation. https://doi.org/10.1007/s10753-025-02341-1
32. Charkiewicz AE, Omeljaniuk WJ, Garley M, Nikliński J (2025) Mercury exposure and health effects: what do we really know? Int J Mol Sci 26(5):2326
33. Nogueira CW, Rocha JBT, Zeni G (2011) Organoselenium compounds and oxidative stress: potential pharmacological targets. Drug Discov Today 16(12):76–85
34. Santos RAS, Sampaio WO, Alzamora AC et al (2018) The ACE2/angiotensin-(1–7)/Mas axis of the renin–angiotensin system: focus on angiotensin-(1–7). Physiol Rev 98(1):505–553
35. Ni W, Yang X, Yang D et al (2020) Role of angiotensin-converting enzyme 2 (ACE2) in COVID-19. Crit Care 24(1):422
36. Bayraktutan U (2025) Angiotensin II and cardiovascular disease: balancing pathogenic and protective pathways. Curr Issues Mol Biol 47(7):501
37. Esler M (2010) The sympathetic system and hypertension. Am J Hypertens 23(11):1160–1168
38. Simões MR, Azevedo BF, Fiorim J et al (2016) Chronic mercury exposure impairs the sympathovagal control of the rat heart. Clin Exp Pharmacol Physiol 43(11):1038–1045
39. Valera B, Dewailly E, Poirier P (2008) Cardiac autonomic activity and blood pressure among Nunavik Inuit adults exposed to environmental mercury: a cross-sectional study. Environ Health 7:29
40. Yilmaz OH, Karakulak UN, Tutkun E et al (2016) Assessment of the cardiac autonomic nervous system in mercury-exposed individuals via post-exercise heart rate recovery. Med Princ Pract 25(4):343–3349
41. Pamphlett R, Kum Jew S, Doble PA, Bishop DP (2021) Mercury in the human adrenal medulla could contribute to increased plasma noradrenaline in aging. Sci Rep 11(1):2961
42. Torres AD, Rai AN, Hardiek ML (2000) Mercury intoxication and arterial hypertension: report of two patients and review of the literature. Pediatrics 105(3):E34
43. Pecanha FM, Wiggers GA, Briones AM et al (2010) Chronic exposure to low concentrations of mercury accelerates hypertension development in spontaneously hypertensive rats: role of oxidative stress and sympathetic overactivity. J Hypertens 28(5):1238–1246

44. Vanhoutte PM, Shimokawa H, Tang EH, Feletou M (2009) Endothelial dysfunction and vascular disease. Acta Physiol 196(2):193–222
45. Abrashev H, Abrasheva D, Nikolov N et al (2025) A systematic review of endothelial dysfunction in chronic venous disease-inflammation, oxidative stress, and shear stress. Int J Mol Sci 26(8):3660
46. Kim DH, Lee S, Jang M, Kim K (2025) Effect of combined exposure to lead, mercury, and cadmium on hypertension: the 2008-2013 Korean National Health and Nutrition Examination Surveys. Int J Occup Med Environ Health 38(3):264–279
47. Akinbode OL, Obeng-Gyasi E (2025) Combined effects of arsenic, cadmium, and mercury with cardiovascular disease risk: insights from the all of us research program. Int J Environ Res Public Health 22(2):239
48. Omanwar S, Fahim M (2015) Mercury exposure and endothelial dysfunction: an interplay between nitric oxide and oxidative stress. Int J Toxicol 34(4):300–307
49. Simões RP, Fardin PBA, Simões MR et al (2020) Long-term mercury exposure accelerates the development of hypertension in prehypertensive spontaneously hypertensive rats inducing endothelial dysfunction: the role of oxidative stress and cyclooxygenase-2. Biol Trace Elem Res 196(2):565–578
50. Blaustein JR, Quisel MJ, Hamburg NM, Wittkopp S (2024) Environmental impacts on cardiovascular health and biology: an overview. Circ Res 134(9):1048–1060
51. Sagheer U, Al-Kindi S, Abohashem S et al (2024) Environmental pollution and cardiovascular disease: part 2 of 2: soil, water, and other forms of pollution. JACC Adv 3(2):100815
52. Pamphlett R, Doble PA, Bishop DP (2021) The prevalence of inorganic mercury in human kidneys suggests a role for toxic metals in essential hypertension. Toxics 9(3):67
53. Onan E, Ulu S, Gungor Ö (2024) Heavy metals and kidney. Turk J Nephrol 33(3):3244–3251
54. Vassallo DV, Simões MR, Giuberti K et al (2019) Effects of chronic exposure to mercury on angiotensin-converting enzyme activity and oxidative stress in normotensive and hypertensive rats. Arq Bras Cardiol 112(4):374–380
55. Arendshorst WJ, Vendrov AE, Kumar N et al (2024) Oxidative stress in kidney injury and hypertension. Antioxidants 13(12):1454
56. Ameer OZ (2022) Hypertension in chronic kidney disease: what lies behind the scene. Front Pharmacol 13:949260
57. World Health Organization. Mercury. https://www.who.int/news-room/fact-sheets/detail/mercury-and-health Accessed 20 Aug 2025
58. Environmental Protection Agency (EPA) (2022) Mercury: Basic information. https://www.epa.gov/mercury/basic-information-about-mercury Accessed 20 Aug 2025

Chapter 14
Occupational Lead Exposure and Its Association with Reproductive Dysfunctions in Both Genders

Phoka C. Rathebe and Mota X. Kholopo

Abstract Occupational lead exposure remains a significant public health concern, particularly due to its detrimental effects on reproductive health in both men and women. Lead, a toxic heavy metal, disrupts multiple physiological pathways, leading to infertility, hormonal imbalances, and adverse pregnancy outcomes. In males, chronic lead exposure is associated with reduced sperm quality, impaired spermatogenesis, and endocrine dysfunction due to its toxic effects on the hypothalamic-pituitary-gonadal (HPG) axis. In females, lead exposure has been linked to menstrual irregularities, spontaneous abortions, and reduced ovarian reserve. Physiological mechanisms involve oxidative stress, DNA damage, endocrine disruption, and impaired calcium metabolism, which collectively contribute to reproductive dysfunction. Clinically, these reproductive impairments translate into increased infertility rates, higher pregnancy complications, and long-term health risks for offspring. Preventive strategies, including stringent occupational safety regulations, regular biomonitoring, chelation therapy, and improved workplace protective measures, are essential to mitigate the risks associated with lead exposure. This review highlights the pathophysiological impact of occupational lead exposure on reproductive health and underscores the need for enhanced occupational health policies to protect workers from its adverse effects.

Keywords Occupational lead exposure · Reproductive dysfunction · Infertility · Endocrine disruption · Oxidative stress

---

P. C. Rathebe (✉) · M. X. Kholopo
Department of Environmental Health, Faculty of Health Sciences, Doornfontein Campus, University of Johannesburg, Johannesburg, South Africa
e-mail: prathebe@uj.ac.za

N. S. Dhalla et al. (eds.), *Functional Biochemistry of Metallic Elements*, Advances in Biochemistry in Health and Disease 36,
https://doi.org/10.1007/978-3-032-14445-4_14

Introduction

Lead (Pb) is a toxic heavy metal that poses significant health risks, particularly in occupational settings where exposure levels are high. Industries such as battery manufacturing, mining, smelting, construction, painting, and automobile repair contribute to occupational lead exposure [1]. Despite stringent regulatory measures in many countries, lead remains a persistent occupational hazard, impacting multiple organ systems, including the reproductive system. Studies have established a strong correlation between lead exposure and reproductive dysfunctions in both men and women, making it a critical issue in occupational health [2, 3]. Lead primarily exerts its toxic effects through oxidative stress, endocrine disruption, and direct damage to reproductive tissues [4, 5]. It has been shown to interfere with the hypothalamic-pituitary-gonadal (HPG) axis, affecting hormone levels essential for reproductive function. Additionally, lead exposure induces the production of reactive oxygen species (ROS), which leads to DNA damage, apoptosis, and impaired gametogenesis [6]. The reproductive effects of lead vary by duration, intensity, and route of exposure, with chronic exposure leading to more severe consequences (León et al. 2020). In men, lead exposure is linked to reduced sperm quality, altered sperm morphology, decreased motility, and lower testosterone levels [7]. Studies have demonstrated that lead accumulates in Sertoli and Leydig cells, impairing their function and leading to testicular atrophy and hormonal imbalances [7]. Moreover, lead disrupts the secretion of luteinizing hormone (LH) and follicle-stimulating hormone (FSH), both of which are essential for spermatogenesis [8]. These effects contribute to male infertility, making lead a major environmental and occupational toxicant.

In women, lead exposure has been associated with irregular menstrual cycles, premature ovarian failure, anovulation, and infertility [8, 9]. Lead disrupts folliculogenesis, resulting in poor oocyte quality and hormonal imbalances. Pregnant women exposed to lead are at a higher risk of miscarriages, stillbirths, preterm labor, and fetal developmental defects [10]. Lead readily crosses the placental barrier, accumulating in fetal tissues and affecting neurodevelopment and overall growth [11]. Postnatal exposure through breastfeeding further exacerbates the risks, potentially leading to cognitive impairment in infants [12]. Given the widespread industrial use of lead, understanding its reproductive effects is essential for developing preventive policies, therapeutic interventions, and occupational safety guidelines. The enforcement of workplace safety measures, regular medical monitoring, and biomarker-based screening is crucial in mitigating the adverse effects of lead exposure [13]. More research is needed to understand the molecular mechanisms behind lead-induced reproductive toxicity, which could lead to novel therapeutic approaches for affected individuals [14]. This review explores the impact of occupational lead exposure on male and female reproductive health, highlighting its physiological mechanisms, clinical implications, and potential preventive strategies. By identifying key areas of risk and proposing intervention methods, this study aims to contribute to the improvement of workplace safety policies and overall

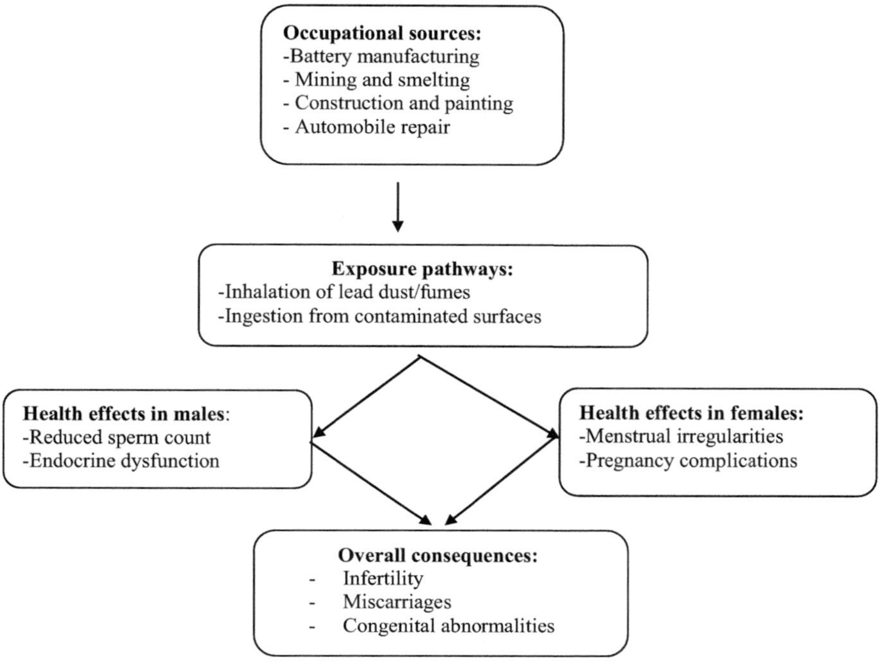

Fig. 14.1 Lead exposure and overall reproductive health

reproductive health in occupational settings. Chronic exposure to lead has been associated with significant reproductive dysfunctions in both men and women (Fig. 14.1), primarily due to its toxic effects on the endocrine system, gametogenesis, and fetal development [1]. Lead interferes with the hypothalamic-pituitary-gonadal (HPG) axis, disrupts hormonal balance, and causes oxidative stress, leading to infertility, miscarriages, and congenital abnormalities [15]. This discussion explores the physiological mechanisms through which lead exposure affects reproductive health, emphasizing its impact on hormonal regulation, gametogenesis, and pregnancy outcomes.

Absorption, Distribution, and Accumulation of Lead in the Body

Lead primarily enters the body through inhalation and ingestion, accumulating in bones, soft tissues, and the bloodstream [16]. Occupational lead exposure occurs via inhalation of lead dust and fumes or ingestion through contaminated hands and surfaces, particularly in high-risk workplaces [17]. Once absorbed, it binds to red blood cells and is distributed to various organs, including the brain, liver, and reproductive system [18]. Lead has a long biological half-life and is predominantly stored in bones, where it can be released into the bloodstream over time, exacerbating chronic exposure effects [19]. Lead exposure significantly alters the function of the HPG axis by disrupting gonadotropin-releasing hormone (GnRH) secretion, reducing luteinizing hormone (LH) and follicle-stimulating hormone (FSH) levels, which are critical for spermatogenesis and testosterone production [15, 20]. A study has demonstrated that exposure decreases testosterone levels by directly affecting Leydig cells in the testes, impairing male fertility [21]. Lead-induced oxidative stress plays a key role in testicular damage. Excessive accumulation in the testes increases the production of reactive oxygen species (ROS), leading to lipid peroxidation, DNA damage, and apoptosis of spermatogenic cells [22]. A study by Gandhi et al. found that occupational lead exposure reduced sperm motility and viability due to elevated oxidative stress markers in seminal plasma [23].

Abnormalities and Infertility

Chronic exposure to lead leads to sperm morphological defects, reduced sperm count, and impaired motility. It interferes with the integrity of the sperm plasma membrane and DNA fragmentation, contributing to male infertility [24]. Furthermore, it accumulates in seminal fluid, negatively affecting fertilization capacity [25]. Lead exposure disrupts ovarian function by interfering with estradiol synthesis and progesterone secretion, leading to menstrual irregularities and anovulation [8, 26]. Studies indicate that exposure of women in occupational settings results in increased risks of oligomenorrhea and early menopause due to its gonadotoxic effects [27, 28]. Lead exposure impairs oocyte quality and reduces ovarian reserve. In animal studies, chronic exposure has been shown to alter follicular development, increase apoptosis in ovarian cells, and disrupt meiotic progression, leading to poor embryo quality [29]. Lead exposure during early pregnancy can result in spontaneous abortions and congenital anomalies due to its teratogenic effects [30]. Lead crosses the placenta and accumulates in fetal tissues, posing risks of developmental defects, low birth weight, and neurotoxicity [31]. Pregnant women exposed to high levels in occupational settings have a greater risk of preterm birth and stillbirths due to lead-induced oxidative damage to placental cells [32].

The reproductive effects of lead exposure extend beyond immediate infertility concerns. Epidemiological studies suggest that prenatal lead exposure increases the risk of childhood neurodevelopmental disorders, cognitive deficits, and behavioral abnormalities [33]. Efforts to minimize occupational lead exposure include workplace safety regulations, personal protective equipment (PPE) use, and routine monitoring of blood lead levels in workers [34]. Chelation therapy and dietary interventions rich in antioxidants, such as vitamin C and zinc, have been suggested to mitigate lead-induced reproductive toxicity [35]. Lead is a widely recognized environmental and occupational toxin that poses significant health risks, particularly to the reproductive system in both men and women. Workers exposed to lead in industries such as battery manufacturing, construction, and smelting often experience infertility, hormonal imbalances, and adverse pregnancy outcomes [1, 36]. The clinical implications of lead exposure extend beyond infertility, affecting fetal development, increasing the risk of miscarriages, and contributing to congenital anomalies [9].

Clinical Manifestations

Chronic lead exposure disrupts the hypothalamic-pituitary-gonadal (HPG) axis, leading to reduced secretion of gonadotropins, testosterone deficiency, and subsequent infertility [20]. Studies indicate that lead inhibits luteinizing hormone (LH) and follicle-stimulating hormone (FSH) secretion, impairing spermatogenesis and reducing sperm production [37]. Clinically, men with occupational lead exposure present with decreased libido, erectile dysfunction, and testicular atrophy due to low testosterone levels [38, 39]. Lead toxicity results in oxidative stress, damaging sperm DNA integrity and plasma membrane structure [40]. Clinical assessments of exposed workers show low sperm concentration and poor motility, affecting fertilization potential [41]; increased sperm morphological abnormalities, including head and tail defects, reducing reproductive success [42]; and elevated sperm DNA fragmentation, leading to failed fertilization and increased miscarriage rates in partners [43]. Men with chronic lead exposure often require assisted reproductive technologies (ART), including in vitro fertilization (IVF) and intracytoplasmic sperm injection (ICSI), due to poor sperm quality [44]. However, high levels of seminal fluid have been associated with lower ICSI success rates due to impaired fertilization capacity [45].

Lead exposure in females is linked to menstrual disturbances, reduced ovarian reserve, and early menopause [12]. Clinically, women working in lead-exposed environments report oligomenorrhea or amenorrhea due to disruption of estrogen and progesterone balance [28, 46]; polycystic ovarian syndrome (PCOS)-like symptoms, including anovulation and irregular cycles, leading to subfertility [47]; and premature ovarian failure (POF), reducing reproductive lifespan and increasing the risk of hormone-dependent disorders [29]. Lead readily crosses the placenta, exposing the fetus to developmental toxicity. Pregnant women exposed to high levels of lead have increased risks of miscarriages and stillbirths due to lead-induced oxidative damage in the placenta [30]; preterm labor and low birth weight, resulting in neonatal complications [31]; and congenital malformations, particularly in the nervous and skeletal systems, affecting long-term child health [32]. Lead accumulates in breast tissue and is excreted into breast milk, exposing infants to lead toxicity [33]. Studies report that infants breastfed by lead-exposed mothers exhibit neurodevelopmental delays, affecting cognitive and motor functions [48], and increased susceptibility to infections, due to weakened immune responses [35]. Children born to lead-exposed parents are at greater risk of cognitive impairment, attention deficits, and behavioral disorders [49]. Lead affects the developing brain, reducing IQ levels and increasing the likelihood of learning disabilities and ADHD [50]. Lead exposure has transgenerational effects, altering epigenetic regulation and predisposing offspring to reproductive disorders in adulthood [51]. Studies suggest that prenatal lead exposure disrupts testicular and ovarian development, leading to subfertility in the next generation [46, 52].

Clinical Management and Prevention Strategies

Regular blood lead level (BLL) screening is essential for workers (Table 14.1) in high-risk industries to detect early signs of lead toxicity [53].

Chelation therapy using dimercaptosuccinic acid (DMSA) or ethylenediaminetetraacetic acid (EDTA) can help reduce lead burden in severely affected individuals [31]. Antioxidants such as vitamins C and E and zinc have been shown to mitigate oxidative damage caused by lead toxicity [30]. Engineering controls (ventilation, enclosed workspaces) to reduce lead dust exposure [54], personal protective equipment (PPE) such as gloves and respirators to minimize inhalation and dermal absorption [55], and regular decontamination protocols to prevent take-home lead exposure, reducing family risks [56], are essential in reducing occupational lead exposure. For individuals experiencing infertility due to chronic lead exposure, fertility preservation options, including sperm and oocyte cryopreservation, should be considered [29]. Clinicians may also recommend assisted reproductive treatments (ART) to improve conception rates in affected couples [57].

Prevention of Occupational Lead Exposure

Lead exposure remains a significant occupational hazard, particularly in industries such as battery manufacturing, mining, construction, and smelting. Chronic exposure can result in reproductive dysfunctions in both men and women, affecting fertility, hormonal balance, and fetal development [1]. Preventive strategies are critical in mitigating these adverse effects, ensuring worker safety, and preserving reproductive health. Government agencies such as the Occupational Safety and Health Administration (OSHA) and the World Health Organization (WHO) have established permissible exposure limits (PEL) for lead to protect workers [58]. It is

Table 14.1 Blood lead level (BLL) clinical management table

BLL (µg/dL)	Clinical significance
>10	Indicates early toxicity, requiring monitoring
>30	30 µg/dL: Associated with reproductive dysfunction, requiring intervention and medical treatment [29]

Table 14.2 Occupational lead exposure limits (OSHA)

Exposure factor	Details
Lead concentration limit	50 μg/m^3 of air
Exposure duration	8-h work shift

recommended to ensure lead exposure does not exceed 50 μg/m^3 of air over an 8-h shift [59] (Table 14.2).

Implementation of engineering controls, including proper ventilation, enclosed workspaces, and air filtration systems, to reduce airborne lead exposure [60] is essential. Routine workplace inspections and exposure assessments should be conducted to identify high-risk areas and implement necessary safety measures [61]. Employers should conduct air and surface monitoring to evaluate lead dust levels; perform biological monitoring, including blood lead level (BLL) screening for workers at risk [62]; and implement mandatory training programs to educate workers on lead hazards, preventive strategies, and safe work practices [63]. BLL testing is crucial for early detection of lead toxicity, preventing long-term reproductive damage. Health guidelines recommend BLL screening every 6 months for workers exposed to lead-based environments [64] and medical removal protection (MRP) if BLL exceeds 30 μg/dL to prevent further exposure-related complications [65]. Given that lead disrupts the hypothalamic-pituitary-gonadal (HPG) axis, routine assessments should include (i) testosterone and estrogen level testing for early signs of endocrine disruption [66]; (ii) semen analysis to monitor sperm count, motility, and DNA fragmentation in male workers [67]; and (iii) menstrual and ovarian function evaluation for early detection of anovulation or ovarian insufficiency in female workers [68].

Chelation therapy is recommended for individuals with significantly elevated BLLs. Dimercaptosuccinic acid (DMSA) and ethylenediaminetetraacetic acid (EDTA) are commonly used to remove lead from circulation [69]. However, chelation therapy should be administered under medical supervision as aggressive treatment may cause mineral imbalances affecting reproductive health [70]. Workers exposed to lead should use appropriate personal protective equipment (PPE) to reduce direct contact and inhalation risks. Recommended PPE includes (i) respirators with HEPA filters to limit inhalation of lead particles [71]; (ii) protective clothing, gloves, and footwear to prevent dermal absorption [30]; and (iii) safety goggles to minimize lead dust contact with mucous membranes [17]. Poor hygiene practices contribute significantly to lead ingestion and secondary exposure. Mandatory handwashing and showering before leaving the workplace [32], regular cleaning of workstations and decontamination of tools to minimize lead accumulation [33], as well as prohibiting eating, drinking, or smoking in contaminated work areas to prevent ingestion of lead particles [48] are crucial measures to reduce occupational lead exposure. Lead dust can be inadvertently transported home, exposing family members, including pregnant women and children.

Occupational lead exposure poses significant risks to reproductive health, necessitating proactive prevention strategies to prevent affected workers. Table 14.3

Table 14.3 Occupational lead exposure prevention strategies

Category	Intervention	Reference
Employer responsibilities	Provision of separate work and personal clothing to prevent cross-contamination	[72]
	Laundering of work uniforms at the workplace rather than at home	[73]
	Regular screening of workers' families for lead exposure risks, especially in households with young children	[74]
Dietary strategies	Calcium-rich foods (dairy, leafy greens) to reduce lead absorption in bones	[75]
	Iron supplementation, particularly for women, as iron deficiency enhances lead uptake	[76]
	Vitamin C and zinc are known for their antioxidant properties, which mitigate oxidative stress induced by lead exposure	[53]
Lifestyle modifications	Avoiding alcohol and smoking as they exacerbate lead toxicity	[77]
	Regular physical exercise, which has been linked to enhanced detoxification and improved sperm parameters in men	[78]
	Stress reduction techniques, such as meditation and mindfulness, to minimize the impact of lead-induced hormonal dysregulation	[79]
Fertility preservation	Sperm banking for male workers with declining semen quality	[80]
	Oocyte cryopreservation for female workers experiencing menstrual irregularities or early ovarian failure	[78]
Assisted reproductive technologies	In vitro fertilization (IVF) to bypass poor sperm quality	[81]
	Intracytoplasmic sperm injection (ICSI) to improve fertilization success in cases of sperm DNA fragmentation	[78]

below summarizes the key prevention strategies, including employer responsibilities, dietary and lifestyle interventions, and reproductive health solutions.

This chapter reviews the physiological, clinical, and preventive aspects of lead-induced reproductive dysfunction, offering a holistic perspective on the topic. It integrates findings from multiple studies to ensure conclusions are based on peer-reviewed data. The discussion on oxidative stress, endocrine disruption, and calcium metabolism highlights key physiological mechanisms. The review also bridges the gap between occupational toxicology and reproductive medicine by outlining the clinical implications of lead exposure, including infertility, hormonal imbalances, and pregnancy complications. The inclusion of preventive strategies, such as workplace regulations, biomonitoring, and protective measures, enhances the practical relevance of the review, guiding occupational health policies. The effects of lead exposure on reproductive health can vary depending on individual susceptibility, duration of exposure, and concentration levels, making it challenging to generalize findings across different populations. While animal studies provide critical insights, long-term human cohort studies are limited, making it difficult to fully assess the chronic reproductive effects of occupational lead exposure. Most studies focus on male reproductive toxicity, while research on female reproductive health outcomes, such as ovarian dysfunction and pregnancy-related complications, remains relatively underexplored. Other workplace hazards, lifestyle factors (e.g., smoking, alcohol consumption), and environmental pollutants may contribute to reproductive dysfunction, complicating the direct attribution of effects to lead exposure alone.

Conclusion

Occupational lead exposure poses a significant risk to reproductive health in both men and women, affecting hormonal balance, gamete quality, and overall fertility. The physiological mechanisms underlying lead-induced reproductive dysfunction include endocrine disruption, oxidative stress, impaired calcium metabolism, and direct cytotoxic effects on reproductive tissues. Clinically, this exposure is associated with reduced sperm quality, menstrual irregularities, pregnancy complications, and an increased risk of adverse reproductive outcomes. Despite existing occupational safety regulations, the persistence of lead exposure in industrial settings highlights the need for stricter enforcement of workplace safety measures, improved biomonitoring, and worker education programs. Preventive strategies such as personal protective equipment (PPE), medical surveillance, and workplace modifications play a critical role in mitigating these risks. However, challenges such as individual variability in lead susceptibility, confounding environmental factors, and gaps in research, particularly regarding female reproductive health, necessitate further investigation.

Addressing occupational lead exposure through interdisciplinary collaboration among policymakers, healthcare professionals, and industry stakeholders is crucial to safeguarding reproductive health. Future research should focus on long-term human studies, more effective preventive measures, and targeted interventions to reduce lead-related reproductive dysfunction, ultimately improving workplace safety and public health outcomes.

References

1. Grandjean P, Landrigan PJ (2014) Neurobehavioural effects of developmental toxicity. Lancet Neurol 13:330–338. https://doi.org/10.1016/S1474-4422(13)70278-3
2. Apostoli P, Bellini A, Porru S, Bisanti L (2000) The effect of lead on male fertility: a time to pregnancy (TTP) study. Am J Ind Med 38:310–315. https://doi.org/10.1002/1097-0274(200009)38:3<310::AID-AJIM10>3.0.CO;2-9
3. Chen C, Wang N, Zhai H, Nie X, Sun H, Han B, Li Q, Chen Y, Cheng J, Xia F et al (2016) Associations of blood lead levels with reproductive hormone levels in men and postmenopausal women: results from the SPECT-China study. Sci Rep 6. https://doi.org/10.1038/srep37809

4. Jiang X, Xing X, Zhang Y, Zhang C, Wu Y, Chen Y, Meng R, Jia H, Cheng Y, Zhang Y et al (2020) Lead exposure activates the Nrf2/Keap1 pathway, aggravates oxidative stress, and induces reproductive damage in female mice. Ecotoxicol Environ Saf 207. https://doi.org/10.1016/j.ecoenv.2020.111231
5. Sorelle D, Ferdinand N, Pascal MF, Bebe N, Laura MNJ, Bertin V, Herman NV, Joseph T (2019) Oxidative stress and reproductive damage induced by lead acetate in female Guinea pig (Cavia Porcellus): curative effects of Hydroethanolic extract of spirulina platensis. Am J Anim Vet Sci. https://doi.org/10.3844/AJAVSP.2019.69.77
6. Jiang N, Wen H, Zhou M, Lei T, Shen J, Zhang D, Wang R, Wu H, Jiang S, Li W (2020) Low-dose combined exposure of carboxylated black carbon and heavy metal Lead induced potentiation of oxidative stress, DNA damage, inflammation, and apoptosis in BEAS-2B cells. Ecotoxicol Environ Saf 206. https://doi.org/10.1016/j.ecoenv.2020.111388
7. He Y, Zou L, Luo W, Yi Z, Yang P, Yu S, Liu N, Ji J, Guo Y, Liu P et al (2019) Heavy metal exposure, oxidative stress and semen quality: exploring associations and mediation effects in reproductive-aged men. Chemosphere 244. https://doi.org/10.1016/j.chemosphere.2019.125498
8. Popovic M, McNeill FE, Chettle DR, Webber CE, Lee CV, Kaye WE (2005) Impact of occupational exposure on lead levels in women. Environ Health Perspect 113(4):478
9. Wu T-N, Chen C-Y (2011) Lead exposure and female infertility. In: Nriagu JO (ed) Encyclopedia of environmental health. Elsevier, Burlington, pp 430–435. ISBN 978-0-444-52272-6
10. Cheng L, Zhang B, Huo W, Cao Z, Liu W, Liao J, Xia W, Xu S, Li Y (2017) Fetal exposure to lead during pregnancy and the risk of preterm and early-term deliveries. Int J Hyg Environ Health 220:984–989. https://doi.org/10.1016/j.ijheh.2017.05.006
11. Singh J, Singh VK, Anand M, Kumar P, Siddiqui MKJ (2010) Placental lead and its interaction with some essential metals among women from Lucknow, India. Asian J Med Sci 1:32–36
12. Lidsky TI, Schneider JS (2003) Lead neurotoxicity in children: basic mechanisms and clinical correlates. Brain 126:5–19. https://doi.org/10.1093/brain/awg014
13. Davari MH, Gharavi M, Salmani M, PiriArdakani M (2022) Health monitoring in employees exposed to lead: review the evidence. TKJ. https://doi.org/10.18502/tkj.v14i1.9860
14. Nemsadze K, Sanikidze T, Ratiani L, Gabunia L, Sharashenidze T (2009) Mechanisms of lead-induced poisoning. Georgian Med News
15. Hamadouche NA, Nesrine S, Abdelkeder A (2013) Lead toxicity and the hypothalamic-pituitary-testicular Axis. Notulae Scientia Biologicae 5:1–6. https://doi.org/10.15835/nsb518038
16. Liu K, Hao J, Zeng Y, Dai F, Gu P (2013) Neurotoxicity and biomarkers of lead exposure: a review. Chin Med Sci J 28:178–188. https://doi.org/10.1016/S1001-9294(13)60045-0
17. Jackson DA, Burr GA, Braun CR, De Perio MA (2018) *Notes from the field:* lead exposures among employees at a bullet manufacturing company—Missouri, 2017. MMWR Morb Mortal Wkly Rep 67:1103. https://doi.org/10.15585/mmwr.mm6739a7
18. B Samuel O (2016) Electrolytes alterations in some organs due to lead exposure. Proc Adv Clin Toxicol 4:1–9
19. Silbergeld EK, Sauk J, Somerman MJ, Todd AC, McNeill FE, Fowler BA, Fontaine A, Buren J (1993) Lead in bone: storage site, exposure source, and target organ. Neurotoxicology 14:225
20. Adebola A, Marvellous LC, Olufunso AB, Phillips AO, Tunde OS, Oluwole OS, David AA, Olawole OO, Ezekiel AA (2024) Chronic Lead exposure impairment of hypothalamic-pituitary-gonadal axis in adult male Wistar rats: biochemical and histomorphological evidence. East Afr Sch J Med Sci. https://doi.org/10.36349/easms.2024.v07i05.006
21. Martynowicz H, Andrzejak R, Mędraś M (2005) The influence of lead on testis function. Med Pr 56:495
22. Barbhuiya SAS, Sengupta M (2015) The toxic effects of Lead on testicular macrophage immunomodulation and sperm cell parameters in mice. Br J Med Med Res 9:1–10. https://doi.org/10.9734/bjmmr/2015/17937
23. Gandhi J, Hernandez RJ, Chen A, Smith NL, Sheynkin Y, Joshi G, Khan S (2017) Impaired hypothalamic-pituitary-testicular axis activity, spermatogenesis, and sperm function pro-

mote infertility in males with lead poisoning. Zygote 25:103–110. https://doi.org/10.1017/S0967199417000028
24. Wijesekara G, Fernando D, Wijeratne S (2020) The effects of Pb on sperm parameters and sperm DNA fragmentation of men investigated for infertility. J Basic Clin Physiol Pharmacol 31. https://doi.org/10.1515/jbcpp-2019-0239
25. He Y, Zou Q, Chen H, Weng S, Luo T, Zeng X (2016) Lead inhibits human sperm functions by reducing the levels of intracellular calcium, CAMP, and tyrosine phosphorylation. Tohoku J Exp Med 238(4):295–303. https://doi.org/10.1620/tjem.238.295
26. He Y, Wang L, Li X, Zhao H (2020) The effects of chronic lead exposure on the ovaries of female juvenile Japanese quails (Coturnix Japonica): developmental delay, histopathological alterations, hormone release disruption and gene expression disorder. Ecotoxicol Environ Saf 205. https://doi.org/10.1016/j.ecoenv.2020.111338
27. Eum K, Weisskopf M, Nie L, Hu H, Korrick S (2014) Cumulative lead exposure and age at menopause in the nurses' health study cohort. Environ Health Perspect 122:229–234. https://doi.org/10.1289/ehp.1206399
28. Ji W (2014) Effects of occupational exposure to lead on female workers' reproductive function: a meta-analysis. Chinese Rural Health Service Administration
29. Abadin H, Taylor J, Buser M, Scinicariello F, Przybyla J, Klotzbach J (2020) Toxicological profile for Lead. Agency for Toxic Substances and Disease Registry, Atlanta. [Citado Marzo 27, 2024]
30. Zhai Q, Narbad A, Chen W (2015) Dietary strategies for the treatment of cadmium and Lead toxicity. Nutrients 7:552–571. https://doi.org/10.3390/nu7010552
31. Flora G, Gupta D, Tiwari A (2012) Toxicity of lead: a review with recent updates. Interdiscip Toxicol 5:47–58. https://doi.org/10.2478/v10102-012-0009-2
32. Sanders T, Liu Y, Buchner V, Tchounwou PB (2009) Neurotoxic effects and biomarkers of lead exposure: a review. Rev Environ Health 24. https://doi.org/10.1515/REVEH.2009.24.1.15
33. Grandjean P, Herz KT (2015) Trace elements as paradigms of developmental neurotoxicants: lead, methylmercury and arsenic. J Trace Elem Med Biol 31:130–134. https://doi.org/10.1016/j.jtemb.2014.07.023
34. Hashtroudi A (2012) Inorganic lead: action required for levels. Occup Med 62:82–85. https://doi.org/10.1093/occmed/kqr185
35. Assi MA, Hezmee MNM, Haron AW, Sabri MYM, Rajion MA (2016) The detrimental effects of lead on human and animal health. Vet World 9:660–671. https://doi.org/10.14202/vetworld.2016.660-671
36. Joffe M, Bisanti L, Apostoli P, Kiss P, Dale A, Roeleveld N, Lindbohm M, Sallmén M, Vanhoorne M, Bonde JP (2003) Time to pregnancy and occupational lead exposure. Occup Environ Med 60:752–758. https://doi.org/10.1136/oem.60.10.752
37. Dursun N, Saraymen R (2001) Chronic occupational lead exposure and testicular function. Trace Elem Electrolytes 18:65
38. Aggarwal T, Sharma V. Pathways and interventions in lead-induced male reproductive toxicity and testosterone dysregulation. | EBSCOhost Available online: https://openurl.ebsco.com/contentitem/doi:10.51470%2Fbca.2024.24.2.1393?sid=ebsco:plink:crawler&id=ebsco:doi:10.51470%2Fbca.2024.24.2.1393. Accessed on 25 Feb 2025
39. Winder C (1993) Lead, reproduction and development. Neurotoxicology 14(2–3):303–317
40. Gupta S, Mukherjee AK, Bhattacharya SK, Roy SK, Chowdhury AR (2007) Lead induced structural and functional alteration of sperm cell among industrial workers. Toxicol Int 14:1–5
41. Naha N, Chowdhury A (2006) Inorganic Lead exposure in battery and paint factory: effect on human sperm structure and functional activity. J UOEH 28(2):157–171. https://doi.org/10.7888/JUOEH.28.157
42. Kleshchev M, Osadchuk A, Osadchuk L (2021) Impaired semen quality, an increase of sperm morphological defects and DNA fragmentation associated with environmental pollution in urban population of young men from Western Siberia, Russia. PLoS One 16:e0258900. https://doi.org/10.1371/journal.pone.0258900

43. Borges E, Zanetti BF, Setti AS, Braga DP d AF, Provenza RR, Iaconelli A (2019) Sperm DNA fragmentation is correlated with poor embryo development, lower implantation rate, and higher miscarriage rate in reproductive cycles of non–male factor infertility. Fertil Steril 112:483–490. https://doi.org/10.1016/j.fertnstert.2019.04.029
44. Mangoli VS, Dandekar S, Desai SK, Mangoli RV (2008) The outcome of ART in males with impaired spermatogenesis. J Hum Reprod Sci 1:73–76. https://doi.org/10.4103/0974-1208.44114
45. Esfandiari N, Burjaq H, Gotlieb L, Casper RF (2008) Seminal hyperviscosity is associated with poor outcome of in vitro fertilization and embryo transfer: a prospective study. Fertil Steril 90:1739–1743. https://doi.org/10.1016/j.fertnstert.2007.09.032
46. Chang S-X, Cheng B, Lee S-L, Chuang H, Yang C-Y, Sung F, Wu T-N (2006) Low blood lead concentration in association with infertility in women. Environ Res 101(3):380–386. https://doi.org/10.1016/J.ENVRES.2005.10.004
47. Ye X, Pan W, Li C, Ma X, Yin S, Zhou J, Liu J (2020) Exposure to polycyclic aromatic hydrocarbons and risk for premature ovarian failure and reproductive hormones imbalance. J Environ Sci 91:1–9. https://doi.org/10.1016/j.jes.2019.12.015
48. Iavicoli I, Fontana L, Bergamaschi A (2009) The effects of metals as endocrine disruptors. J Toxic Environ Health, Part B 12:206–223. https://doi.org/10.1080/10937400902902062
49. Jia Z, Zhang H, Yu L, Qiu F, Lv Y, Guan J, Gang H, Zuo J, Zheng T, Liu H et al (2023) Prenatal lead exposure, genetic factors, and cognitive developmental delay. JAMA Netw Open 6:e2339108. https://doi.org/10.1001/jamanetworkopen.2023.39108
50. Apostoll P (1998) Male reproductive toxicity of lead in animals and humans. Occup Environ Med 55:364–374
51. Nilsson EE, Skinner MK (2015) Environmentally induced epigenetic transgenerational inheritance of reproductive Disease. Biol Reprod 93. https://doi.org/10.1095/biolreprod.115.134817
52. Huang L, Huang S, Luo X, Li L, Chen X, Zan G, Tan Y, Liu C, Hou Q, Ge X et al (2020) Associations of prenatal exposure to multiple metals with testicular volume and anogenital distance in infant boys: a longitudinal cohort study. Environ Int 143:105900. https://doi.org/10.1016/j.envint.2020.105900
53. Thonneau P, Bujan L, Multigner L, Mieusset R (1998) Occupational heat exposure and male fertility: a review. Hum Reprod 13:2122–2125
54. Kokkonen A, Linnainmaa M, Säämänen A, Kanerva T, Sorvari J, Kolehmainen M, Lappalainen V, Pasanen P (2019) Control of dust dispersion from an enclosed renovation site into adjacent areas by using local exhaust ventilation. Ann Work Exposures Health 63(4):468–479. https://doi.org/10.1093/annweh/wxz016
55. Chakr N, Sav A (2024) The role of personal protective equipment (PPE) in reducing firefighter exposure to chemical hazards: a systematic review. J Occup Environ Hyg:1–11. https://doi.org/10.1080/15459624.2024.2400237
56. Ceballos DM, Guerrero M, Kalweit A, Rabin R, Spengler J, Herrick R (2019) One-hour pilot training to prevent workers from taking home workplace contaminants. New Solutions 29:519–529. https://doi.org/10.1177/1048291119887188
57. Bhandari H, Choudhary M, Stewart J (2018) An overview of assisted reproductive technology procedures. Obstet Gynaecol 20:167–176. https://doi.org/10.1111/tog.12509
58. Zhu J, Franko E, Pavelchak N, DePersis R (2012) Worker lead poisoning during renovation of a historic hotel reveals limitations of the OSHA lead in construction standard. J Occup Environ Hyg 9(9):167–171. https://doi.org/10.1080/15459624.2012.700273
59. 1910.1025 – Lead. | Occupational Safety and Health Administration. Available online: https://www.osha.gov/laws-regs/regulations/standardnumber/1910/1910.1025. Accessed on 24 Feb 2025
60. Lange J, Thomulka K (2000) Effectiveness of engineering controls for airborne lead exposure during renovation/demolition of a commercial building. Indoor Built Environ 9:207–215. https://doi.org/10.1177/1420326X0000900310
61. Chao TK, Sun C, Beach J (2017) Developing a tool for identifying high-risk employers for inspection. Occup Med 67(8):609–614. https://doi.org/10.1093/occmed/kqx136

62. Holland M, Cawthon D (2016) Workplace lead exposure. J Occup Environ Med 58(12):371–374. https://doi.org/10.1097/JOM.0000000000000928
63. Hughes J (2012) The critical role of training in protecting workers. New Solutions 22:253–254. https://doi.org/10.2190/NS.22.3.a
64. Guth K, Bourgeois M, Johnson G, Harbison R (2020) Assessment of Lead exposure controls on bridge painting projects using worker blood Lead levels. Regul Toxicol Pharmacol. https://doi.org/10.1016/j.yrtph.2020.104698
65. Tak S, Roscoe R, Alarcon WA, Ju J, Sestito J, Sussell A, Calvert G (2008) Characteristics of US workers whose blood lead levels trigger the medical removal protection provision, and conformity with biological monitoring requirements, 2003–2005. Am J Ind Med 51(9):691–700. https://doi.org/10.1002/ajim.20603
66. Olanrewaju E, Anyaehie B, Ezeh CO, Onyekwelu K, Ezeh R (2017) Effects of endocrine disrupting heavy metals on pituitary and gonadal hormones in normal weight automechanics in Ibadan, Nigeria. Afr J Biomed Res 20:25–35. https://doi.org/10.4314/AJBR.V20I1
67. Hosni H, Selim O, Abbas M, Fathy A (2013) Semen quality and reproductive endocrinal function related to blood lead levels in infertile painters. Andrologia 45. https://doi.org/10.1111/j.1439-0272.2012.01322.x
68. Qu J, Niu H, Wang J, Wang Q, Li Y (2021) A review on the potential mechanism of lead poisoning to the growth and development of ovarian follicle. Toxicology. https://doi.org/10.1016/j.tox.2021.152810
69. Mikirova N, Casciari J, Hunninghake R (2011) Efficacy of oral DMSA and intravenous EDTA in chelation of toxic metals and improvement of the number of stem/progenitor cells in circulation. Transl Biomed 2
70. Joshi P, Patel P (2019) Effect of chelated mineral mixture on blood biochemistry, hormone and mineral status in repeat breeder buffaloes in tribal areas of Dahod District in Gujarat, India. Ind J Vet Sci Biotechnol. https://doi.org/10.21887/IJVSBT.15.1.9
71. Scholz P, Materna B, Harrington D, Uratsu CS (2002) Residential and commercial painters' exposure to Lead during surface preparation. AIHA J 63(1):22–28. https://doi.org/10.1080/15428110208984687
72. Lusk S, Salazar M (1995) Preventing lead poisoning. Workplace Health Saf 43:648–652. https://doi.org/10.1177/216507999504301207
73. Tore G, Green JG, Noguchi J, Echevarría-Ramos M, Peters J, Levy JI, Martinez MPB, Ceballos DM (2024) Development and evaluation of two educational sessions on take-home lead exposure prevention for construction workers and their families. Ann Work Expo Health. https://doi.org/10.1093/annweh/wxae047
74. Hotz M, Kniepmann K, Kohn L (1998) Occupational therapy in pediatric lead exposure prevention. Am J Occup Ther 52(1):53–59. https://doi.org/10.5014/AJOT.52.1.53
75. Kim H-C, Jang T, Chae H, Choi W, Ha M, Ye B, Kim B-G, Jeon M, Kim S-Y, Hong Y-S (2015) Evaluation and management of lead exposure. Ann Occup Environ Med 27. https://doi.org/10.1186/s40557-015-0085-9
76. Williams MK, King E, Walford J (1969) An investigation of lead absorption in an electric accumulator factory with the use of personal samplers. J Occup Environ Med 26(3):202–216
77. Aguilar-Madrid G, Piacitelli G, Juárez-Pérez C, Vázquez-Grameix JH, Hu H, Hernández-Ávila M (1999) Occupational exposure to inorganic lead in a printing plant in Mexico City. Salud publica de Mexico 41(1):42–54
78. Gustafson Å, Hedner P, Schütz A, Skerfving S (1989) Occupational lead exposure and pituitary function. Int Arch Occup Environ Health 61:277–281. https://doi.org/10.1007/BF00381426
79. Gupta A, Tripathi M, Ma BS, Sripad DV, K VR, Mohan DR, Reddy PA, Upparakadiyala R (2024) Effects of the decrease in blood lead levels on renal and neurological functions among occupationally exposed and unexposed populations of South India: a cohort study. Cureus 16. https://doi.org/10.7759/cureus.54432
80. Schwartz B, Hu H (2006) Adult lead exposure: time for change. Environ Health Perspect 115:451–454. https://doi.org/10.1289/ehp.9782
81. Spurgeon A (1994) Review: occupational lead exposure. Indoor Built Environ 3:112–118. https://doi.org/10.1177/1420326X9400300302

Chapter 15
Essential and Toxic: The Dual Nature of Heavy Metals in Human Physiology

Swati Kundu

Abstract Heavy metals present a paradox in biological systems—while elements like zinc, copper, and iron are essential micronutrients, others such as arsenic, mercury, cadmium, and lead are well known for their toxicity. Yet emerging evidence suggests that even these toxic metals can elicit adaptive or beneficial responses when present at ultra-trace levels, underscoring a complex, dose-dependent duality. This chapter synthesizes current understanding of this dichotomy, drawing on both recent literature and my own experimental research. In studies investigating aortic and tracheal hypercontractility, we found that arsenic, mercury, cadmium, and lead impair smooth muscle function via calcium dysregulation, oxidative stress, and altered adrenoceptor sensitivity. Conversely, other reports indicate that minimal exposure to these same metals can activate cytoprotective pathways—including antioxidant defenses and stress response signaling—suggesting a potential hormetic effect. Together, these findings emphasize the importance of nuanced toxicological assessment and open new avenues for exploring therapeutic thresholds in biomedical research.

Keywords Cardiovascular system · Dose-response · Heavy metals · Hormesis · Respiratory system · Therapeutic potential

S. Kundu (✉)
Indian Council of Medical Research Headquarters, Ministry of Health and Family Welfare, New Delhi, India

N. S. Dhalla et al. (eds.), *Functional Biochemistry of Metallic Elements*, Advances in Biochemistry in Health and Disease 36,
https://doi.org/10.1007/978-3-032-14445-4_15

Introduction

The Dual Nature of Heavy Metals

Heavy metals, though naturally occurring in the Earth's crust, have become pervasive environmental contaminants as a result of intensified industrialization, mining activities, fossil fuel combustion, and improper waste disposal [1]. Among these, several metallic elements, such as zinc, copper, selenium, and iron, are considered essential micronutrients that play vital roles in enzymatic catalysis, redox signaling, oxygen transport, and immune function [2–4]. In contrast, heavy metals such as arsenic, mercury, cadmium, and lead have no established biological necessity and are well known for their toxicity at moderate to high levels of exposure. However, recent findings suggest that these traditionally toxic elements can exhibit paradoxical biological effects at ultra-trace concentrations, including growth-promoting, developmental, and reproductive benefits in preclinical models—probably mediated through hormetic or stress-adaptive pathways [5]. These findings challenge the binary classification of metals as purely toxic and point to a dose-dependent duality in their biological actions. However, their narrow margin of safety and cumulative nature warrant caution in interpreting such effects outside tightly controlled experimental settings.

The biological relevance of this dose-dependent duality becomes particularly important when examining organ systems that are sensitive to ionic imbalance and oxidative stress. Cardiovascular and respiratory systems, for example, are highly dependent on regulated calcium signaling, redox homeostasis, and receptor-mediated pathways to maintain vascular tone and patency of the airways [6, 7]. Disruption of these finely tuned systems by toxic metals can result in pathological conditions such as hypertension, endothelial dysfunction, bronchoconstriction, and airway hyperreactivity [8, 9]. At higher concentrations, arsenic, cadmium, mercury, and lead interfere with calcium channels, generate reactive oxygen species (ROS), impair mitochondrial function, and alter gene expression, contributing to cellular injury and tissue dysfunction [10–15]. However, at trace levels, some studies suggest these same metals may transiently activate cellular stress response pathways, such as the upregulation of antioxidant defenses, heat shock proteins, or cytoprotective transcription factors. This phenomenon, known as *hormesis*, describes a biphasic dose-response in which low concentrations of a toxic substance elicit adaptive or even beneficial biological effects while higher concentrations are clearly detrimental [16]. This biphasic response is illustrated in Fig. 15.1.

Building upon these observations, our own investigations have focused on the functional consequences of heavy metal exposure on smooth muscle physiology, particularly within the cardiovascular and respiratory systems. Using isolated rat aortic and tracheal tissues, we demonstrated that arsenic and mercury, even at subtoxic concentrations, significantly enhanced contractile responses to adrenergic agonists—indicating a hyperreactive smooth muscle state [10]. This hypercontractility is likely mediated by disrupted Ca^{2+} homeostasis, increased oxidative stress,

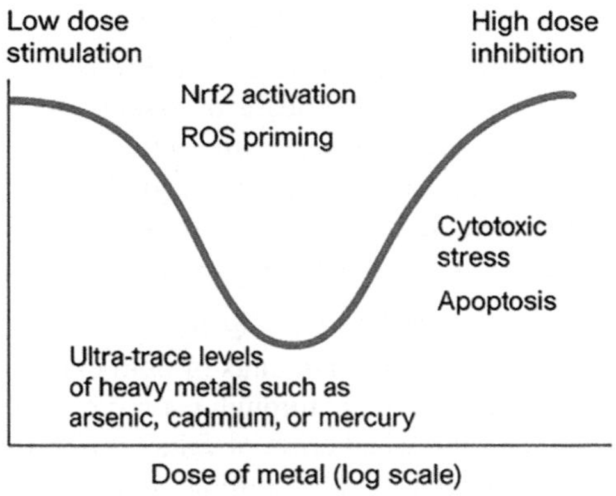

Fig. 15.1 Biphasic dose-response curve of heavy metals: The hormesis model

and modulation of adrenergic receptor sensitivity, particularly through enhanced α-adrenoceptor responsiveness [7, 12, 13]. Heavy metals such as cadmium and lead are known to mimic or displace Ca^{2+} ions, interfering with voltage-gated Ca^{2+} channels and promoting excessive intracellular Ca^{2+} accumulation—a key driver of smooth muscle contraction [10, 17]. Concurrently, metals such as arsenic and mercury impair mitochondrial function and stimulate the production of ROS, which further sensitize contractile machinery and diminish nitric oxide (NO)-mediated vasodilation [13–15].

In preclinical models, ultra-trace exposures to arsenic, mercury, cadmium and lead have been reported to improve certain physiological outcomes, including improved antioxidant enzyme activity, increased body weight, increased fertility, and modulation of stress response genes [5, 16, 18]. These effects are hypothesized to result from mild oxidative or metabolic stress that activates cellular adaptive pathways such as the Nrf2/ARE antioxidant response, mitochondrial biogenesis, or mild inflammatory priming [16]. This biphasic response is known as hormesis, in which low concentrations of a toxic substance elicit adaptive or even beneficial biological effects whereas higher concentrations are clearly detrimental. However, these effects are highly dose and tissue specific, and the window between adaptive and deleterious outcomes remains narrow. Consequently, while such findings offer insight into the nuanced biological interactions of toxic metals, they do not negate the established risks of chronic or cumulative exposure.

Through this chapter, I aim to provide a mechanistic synthesis of how chronic exposure to nonessential heavy metals—specifically arsenic, mercury, cadmium, and lead—impacts the contractile physiology of the cardiovascular and respiratory systems. While these metals are widely recognized for their toxicity at moderate to high levels, emerging evidence also points to subtle biological activity at ultra-trace concentrations, including transient adaptive or stress-mitigating effects in preclinical systems. Drawing on both my experimental findings and the broader toxicological literature, this chapter explores the dose-dependent duality of these metals—where low-level exposures may influence smooth muscle function through calcium signaling and oxidative balance while higher levels induce pathological hyperreactivity.

A. The Toxic Face: Mechanisms and Public Health Concern

(i) *Environmental Presence and Exposure Routes*

The widespread release of arsenic, mercury, cadmium, and lead into the biosphere through anthropogenic activities has significantly increased their concentrations in the air, water, soil and the food chain [1, 8]. These metals are not biodegradable and tend to persist in the environment, resulting in long-term ecological and human health consequences [19].

(ii) *Natural and Anthropogenic Sources*

Although minor amounts of both essential and nonessential metals are released into the environment through natural processes such as volcanic eruptions, soil erosion, and the weathering of mineral-rich rocks, the overwhelming contribution to environmental metal pollution stems from anthropogenic sources. Industrial and agricultural activities, including mining, metal smelting, fossil fuel combustion, battery manufacturing, pesticide application, and improper electronic waste disposal, have drastically increased the environmental burden of toxic metals [20, 21].

For example, arsenic contamination in groundwater, particularly in regions such as West Bengal and Bangladesh, poses a serious public health crisis due to its geogenic origin, exacerbated by overextraction for irrigation [22]. Cadmium contamination is closely related to phosphate fertilizers and electroplating industries [23]. Mercury is commonly released from coal power plants and artisanal gold mining [24], while lead contamination persists due to its legacy use in paints, plumbing, and informal recycling of lead-acid batteries [25]. Their persistence in air, water, and soil contributes to long-term bioaccumulation and biomagnification, contaminating food chains and threatening both ecological and human health [1, 19].

(iii) *Routes of Human Exposure*

Humans are exposed to toxic metals primarily through three major routes: inhalation, ingestion, and dermal contact. These pathways allow toxicants to enter the body and accumulate in target tissues, often by mimicking or displacing essential metals during transport and uptake [8, 19].

Inhalation of heavy metal-contaminated airborne particulate matter (PM) is common in industrial environments, especially near smelting units, coal combustion zones, and urban traffic corridors. Toxic metals such as lead and cadmium are frequently adsorbed onto fine PM2.5, enabling them to penetrate deep into the lungs, reach the alveoli, and subsequently enter systemic circulation [20, 26].

Ingestion is the most prevalent route of chronic exposure. Contaminated drinking water, crops cultivated in polluted soils, and seafood from mercury-laden aquatic systems serve as significant dietary sources [22, 24]. Arsenic contamination in groundwater, particularly in regions such as West Bengal and Bangladesh, affects millions and often exceeds the World Health Organization (WHO) guideline value of 10 μg/L [22]. Unlike essential metals such as iron or zinc, intestinal absorption is tightly regulated, arsenic, mercury, cadmium, and lead are absorbed more erratically and can bypass homeostatic controls [2, 19].

Dermal absorption, although typically less significant, can occur through prolonged contact with contaminated water, soil, or metal-bearing materials—especially in occupational settings such as electroplating, mining, or e-waste recycling [20, 23].

Once inside the body, these toxic metals compete with essential elements for transport proteins and cellular entry points, facilitating their incorporation into biochemical pathways where they exert deleterious effects [1, 7].

(iv) *Bioaccumulation and Biomagnification.*

Once introduced into the body, toxic metals such as arsenic, mercury, cadmium, and lead tend to bioaccumulate in soft tissues, including the liver, kidney, and brain, as well as bone and hair [7, 19]. Their chemical resemblance to essential metals, such as zinc and selenium, allows them to hijack normal transporters and cellular entry pathways. This molecular mimicry enables their uptake by biological systems despite their lack of physiological function [1, 8].

In contrast to essential trace elements, which are tightly regulated by homeostatic mechanisms and are metabolized or excreted as needed, these toxic metals are not converted into less harmful forms and are eliminated very slowly, if at all. As a result, they have prolonged biological half-lives and progressively accumulate with chronic exposure [2, 23].

Within ecosystems, these metals infiltrate aquatic food webs via sediment and plankton. As organisms consume contaminated prey, metals—especially in biologically active forms like methylmercury—undergo biomagnification, increasing in concentration at each successive trophic level [13, 24]. This makes seafood a major source of mercury exposure in human populations, particularly those reliant on fish-based diets. The irony lies in the fact that the very systems evolved to transport and regulate essential micronutrients are exploited by toxic metals, transforming beneficial pathways into conduits of chronic harm.

(v) *Mechanisms of Toxicity*

The toxicological impact of heavy metals such as arsenic, mercury, cadmium, and lead arises not only from their inherent chemical reactivity but also from their ability to masquerade as essential trace elements. In biological systems, metals like calcium, zinc, and selenium are required for enzymatic activity, redox regulation, and signal transduction [2, 3]. However, toxic metals exploit these pathways by mimicking essential metals and displacing them from their physiological binding sites [1, 19]. This competitive interference leads to widespread dysfunction in cellular systems that are finely tuned to trace metal availability [7].

The resulting cellular damage unfolds through multiple interlinked mechanisms, including oxidative stress, dysregulation of calcium homeostasis, mitochondrial dysfunction, and altered gene and protein signaling cascades [7, 13, 15]. These disruptions are particularly detrimental in organs like the heart and lungs, which are heavily dependent on precise ionic and redox balance for smooth muscle contractility and metabolic function [12, 14].

- *Oxidative Stress and Free Radical Generation*
  One of the most studied mechanisms of heavy metal toxicity is ROS generation. Although heavy metals themselves are not redox active, they indirectly stimulate ROS production by inhibiting antioxidant enzymes such as superoxide dismutase (SOD), catalase, and glutathione peroxidase. This leads to oxidative damage to lipids, proteins, and nucleic acids, resulting in cellular dysfunction and apoptosis [6, 7]. For instance, cadmium indirectly generates ROS by displacing zinc in metallothionein, leading to oxidative injury and lipid peroxidation in smooth muscle cells. Arsenic induces oxidative DNA damage through its biotransformation into trivalent and pentavalent species, while methylmercury increases intracellular ROS through inhibition of mitochondrial respiration [27, 28].

- *Disruption of Ca^{2+} Signaling and Ion Homeostasis*
  Several heavy metals, particularly lead and cadmium, interfere with Ca^{2+}-dependent processes. Lead competes with Ca^{2+} ions at voltage-gated calcium channels and intracellular Ca^{2+}-binding proteins, impairing signal transduction in neurons and smooth muscle cells. This mimicry leads to aberrant muscle contractility, neurotransmission disturbances, and vascular hyperresponsiveness [12, 17]. In the cardiovascular system, this dysregulation manifests as sustained vasoconstriction and hypertension. In the respiratory system, it contributes to tracheal hyperreactivity and bronchospasm—effects also observed in our experimental studies using isolated rat tissues [10, 11].

- *Mitochondrial Dysfunction and Apoptosis*
  Heavy metals also target the mitochondria, disrupting oxidative phosphorylation and ATP synthesis. Arsenic impairs the pyruvate dehydrogenase complex and mitochondrial membrane potential, triggering the release of cytochrome c and subsequent apoptosis. Mercury binds to mitochondrial thiol groups, altering membrane permeability and reducing NADH-driven respiration [13, 14]. These disruptions are particularly detrimental to tissues with high energy demands, such as the heart and lungs, and help explain the observed contractile dysfunction in vascular and airway smooth muscles.

- *Alteration of Gene Expression and Cell Signaling*
  Chronic exposure to heavy metals leads to epigenetic modifications, altered transcription factor activity, and inflammatory cytokine production. Arsenic has been shown to inhibit DNA repair enzymes and upregulate pro-inflammatory genes via NF-κB and AP-1 pathways. Cadmium and lead activate MAPK signaling cascades, altering the expression of genes involved in apoptosis and cell cycle regulation [15, 29].

(vi) *System-Specific Toxicity of Heavy Metals*

As summarized in Table 15.1, toxic heavy metals such as arsenic, mercury, cadmium, and lead exert distinct yet overlapping effects on the cardiovascular and respiratory systems. Chronic exposure to arsenic, mercury, cadmium, and lead has been linked to the pathogenesis of hypertension, atherosclerosis, pulmonary constriction, and airway hyperreactivity. This section discusses these

Table 15.1 Cardiopulmonary effects of toxic heavy metals

Heavy metal	Cardiovascular effects	Respiratory effects	Key mechanism
Arsenic	Hypertension, vasospasm	Pulmonary edema	Oxidative stress Ca^{2+} disruption
Mercury	Endothelial damage	Bronchial irritation	Thiol binding Mitochondrial dysfunction
Cadmium	Vasoconstriction	Bronchoconstriction	Calcium channel interference ROS generation
Lead	Vascular hyperreactivity	Airway hyperreactivity	α-adrenergic stimulation Oxidative stress

pathophysiological effects, drawing on experimental and clinical evidence, including our studies using isolated rat trachea and aorta models.

- *Cardiovascular Toxicity*

  The cardiovascular system is highly susceptible to toxic metal exposure due to the direct effect of metals on vascular smooth muscle contractility and endothelial function. Epidemiological studies have linked chronic exposure to arsenic, cadmium, and lead with increased incidence of hypertension, endothelial dysfunction, and vascular stiffness [9, 30, 31]. Our own research has demonstrated that both arsenic and mercury significantly increase contractile responses in isolated rat aortic rings, simulating a hypercontractile vascular state. In a dose-dependent manner, arsenic and mercury induced sustained vasoconstriction, likely mediated by disrupted calcium signaling and oxidative stress [11]. Further, our findings showed that the application of phytochemicals such as eugenol, linalool, and carvone significantly reversed this hypercontractility, indicating their potential as vasoprotective agents. Cadmium and lead have also been shown to induce aortic contraction through both Ca^{2+}-dependent and receptor-mediated mechanisms [32]. These metals mimic Ca^{2+} ions and alter intracellular signaling, leading to increased vascular tone and resistance. Chronic lead exposure, in particular, has been associated with increased peripheral vascular resistance, sympathetic overactivation, and ultimately, sustained hypertension [33].

  The mechanisms underlying these vascular responses include:

- Upregulation of α-adrenergic receptor activity
- Alteration of L-type Ca^{2+} channel permeability
- ROS-mediated impairment of nitric oxide (NO) signaling

These findings collectively suggest that toxic metals disrupt vascular homeostasis at multiple levels—ion channels, receptor targets, and redox balance—contributing to cardiovascular morbidity.

- *Respiratory Toxicity*

  The respiratory system is another major target of metal-induced toxicity due to its extensive surface area and direct exposure to airborne pollutants. Inhalation of cadmium, mercury vapors, or lead-laden particulates can lead to airway inflammation, increased bronchial reactivity, and chronic obstructive airway conditions [34, 35].

In our laboratory studies, we observed that both cadmium and lead significantly increased contractile responses in isolated rat tracheal tissues. These hypercontractile effects were mediated via stimulation of α-adrenergic receptors and voltage-dependent calcium channels, leading to enhanced smooth muscle constriction [10]. These effects were attenuated by natural monoterpenes—linalool and eugenol—supporting their potential to act as bronchodilators through calcium antagonism and antioxidative action.

These findings align with the literature demonstrating that:

- Cadmium can activate pro-inflammatory cytokines in lung tissue and cause emphysema-like changes [36].
- Lead exposure is associated with increased asthma incidence in children and impaired lung function in occupational workers [37].

The mechanistic basis includes:

- Ca^{2+} overload and smooth muscle hyperresponsiveness
- Oxidative damage to airway epithelium
- Increased cholinergic and adrenergic activity

B. Nontoxic and Potentially Beneficial Effects at Ultra-trace Concentrations

Although arsenic, mercury, cadmium, and lead are widely recognized for their toxicity at moderate to high concentrations, an emerging body of evidence suggests that these same metals may exert paradoxical biological effects at ultra-trace levels. This dose-dependent duality is a hallmark of hormesis—a biphasic phenomenon in which low-dose exposures elicit adaptive or beneficial cellular responses whereas higher doses remain toxic as shown in Fig. 15.1.

In controlled preclinical models, ultralow concentrations of arsenic, mercury, and cadmium have been shown to promote a variety of physiological functions (Table 15.2). For example, Bhattacharya [5] reviewed 13 animal studies in which trace-level exposures led to increased litter sizes, enhanced sperm motility, improved body weight, and greater resistance to oxidative stress. These beneficial effects were consistently observed within a narrow concentration window, beyond which the toxic profile of each metal re-emerged. Notably, no similar benefits were reported for lead in these ultra-trace settings.

Mechanistically, these effects appear to stem from mild oxidative or metabolic stress that activates cytoprotective and adaptive signaling pathways without

Table 15.2 Adaptive effects of toxic metals at ultra-trace doses

Metal	Observed adaptive response	Model/ system	Hypothesized mechanism
Arsenic	Enhanced fertility	Rodents	Mild oxidative priming
Cadmium	Improved antioxidant enzyme activity	Zebrafish	Nrf2/ARE pathway activation
Mercury	Increased body weight and growth rate	Mice	Hormetic stress adaptation
Lead	Stimulation of erythropoiesis and mild immunomodulation	Rats	Calcium mimicry and ROS-mediated signaling

triggering cellular injury. At nanomolar concentrations, cadmium has been shown to enhance the activity of antioxidant enzymes such as SOD, catalase, and glutathione peroxidase in rodent tissues [18]. These findings point to a potential "priming" effect—where transient low-level stress prepares cells to better handle future insults.

Further support comes from molecular studies demonstrating that trace metal exposure can activate the Nrf2/ARE pathway, leading to the upregulation of detoxification and stress response genes [15]. Other mechanisms include modulation of inflammatory responses, mitochondrial biogenesis, and even DNA repair processes—all of which are typically suppressed or overwhelmed under high-dose conditions [38–40].

It is important to emphasize that these findings do not imply that toxic metals are beneficial by nature. Rather, they highlight the remarkable sensitivity of biological systems to metal-induced stress and the ability to convert limited exposure into a preparatory, hormetic response. This phenomenon offers intriguing insight into the complexity of dose-response relationships in environmental toxicology and may inform more nuanced regulatory approaches.

Nonetheless, the boundary between beneficial and harmful doses is exceptionally narrow. The observed effects remain context specific, with most evidence confined to tightly regulated laboratory settings. Long-term outcomes, species differences, and potential cumulative toxicity must be rigorously assessed before any translational or therapeutic implications can be drawn.

Public Health Implications and Regulatory Perspective

The global burden of heavy metal contamination presents a growing challenge for public health authorities, particularly in low- and middle-income countries where environmental regulations are often inadequate or poorly enforced. Arsenic, mercury, cadmium, and lead—despite being nonessential and highly toxic—continue to contaminate air, water, soil, and food chains, leading to chronic exposure in both occupational and residential settings.

Environmental Prevalence and Health Burden

In India, arsenic contamination of groundwater is one of the most severe environmental health crises, with over 20 million people exposed to arsenic levels exceeding WHO limits of 10 μg/L, particularly in the states of West Bengal, Bihar, and Uttar Pradesh. Similarly, cadmium and lead are increasingly found in agricultural soils due to phosphate fertilizers and industrial effluents, raising concerns about food safety and long-term dietary exposure [27, 41].

Lead exposure, once thought to be declining post phase out of leaded petrol, persists due to recycling of lead-acid batteries, paint residues, and e-waste.

According to WHO, lead exposure accounts for approximately 900,000 deaths globally each year, primarily through its contribution to hypertension and cardiovascular disease. Mercury, particularly in the form of methylmercury, bioaccumulates in fish and poses a neurotoxic risk to both adults and children in coastal populations [42].

Vulnerable Populations and Long-Term Risks

Children, pregnant women, industrial workers, and communities living near smelters, mines, and informal recycling centers are at heightened risk. Prenatal exposure to these metals has been linked to impaired neurodevelopment, reduced lung function, and increased risk of cardiovascular abnormalities in adulthood [43].

Chronic low-level exposure may not produce immediate symptoms but contributes to "silent toxicity"—a progressive and insidious accumulation of damage in vascular and pulmonary tissues. The additive or synergistic effects of multiple metals further complicate risk assessment and intervention.

Regulatory Frameworks and Gaps

While various national and international bodies—including the Bureau of Indian Standards, Central Pollution Control Board, WHO, and US Environmental Protection Agency (US EPA) —have established permissible exposure limits for toxic metals, enforcement remains inconsistent and often outdated. Notable discrepancies and regulatory gaps include:

- Arsenic: WHO guideline for drinking water = 10 μg/L; Indian standard = 50 μg/L—five times higher than global recommendations, especially concerning arsenic-endemic regions like West Bengal and Bihar.
- Lead: WHO guideline = 10 μg/L in drinking water; however, enforcement and routine monitoring remain inadequate in many Indian states, particularly in rural areas.
- Cadmium: The European Union sets cadmium limits in food between 0.05 and 0.1 mg/kg depending on the food category. India's regulations lag behind in specific food-based cadmium thresholds.
- Mercury: WHO sets a provisional tolerable weekly intake for methylmercury at 1.6 μg/kg body weight. For inorganic mercury, the tolerable weekly intake is 4 μg/kg. The US EPA recommends a reference dose of 0.1 μg/kg/day. India lacks specific mercury exposure guidelines for food, though emissions from thermal plants are loosely regulated.

Despite existing standards, most low- and middle-income countries struggle with weak enforcement, lack of real-time biomonitoring, and outdated risk assessment methods. There is an urgent need to harmonize national guidelines with current toxicological evidence, establish population-based monitoring frameworks, and enforce stricter discharge and product safety regulations—particularly in vulnerable regions affected by industrial pollution, informal e-waste recycling, contaminated groundwater, and inadequate public health infrastructure.

Such reforms must be supported by capacity building in environmental surveillance, increased investment in laboratory infrastructure, and stronger intersectoral coordination between health, industry, and environmental ministries. Without comprehensive and enforceable policies, toxic metal exposure will continue to pose a silent but significant burden on public health, especially in already marginalized communities. There is a pressing need for:

- Re-evaluation of national standards based on recent epidemiological evidence
- Integrated biomonitoring programs to assess metal body burden in at-risk populations
- Stricter industrial discharge norms and soil remediation protocols

Natural Compounds as Preventive Strategies

Our experimental findings, along with a growing body of toxicological evidence, suggest that certain plant-derived compounds—particularly monoterpenes such as eugenol, linalool, carvone, and carvacrol—may offer protective or mitigating effects against heavy metal-induced cardiovascular and respiratory dysfunction. These naturally occurring phytochemicals, found in essential oils of medicinal plants, exhibit a wide range of bioactivities, including antioxidant, anti-inflammatory, calcium channel blocking, and receptor-modulating properties.

In our laboratory studies, we demonstrated that these compounds could significantly attenuate hypercontractile responses in isolated rat aortic and tracheal tissues exposed to subtoxic concentrations of arsenic, mercury, cadmium, and lead. Eugenol, linalool, and carvone effectively reversed arsenic- and mercury-induced aortic hypercontractility, likely through a combination of calcium antagonism and antioxidant activity [11]. In separate studies, eugenol and carvacrol mitigated lead-induced vasoconstriction by modulating α-adrenoceptor responsiveness and reducing oxidative burden [32, 44]. In the airway system, linalool and eugenol significantly reduced cadmium- and lead-induced tracheal hyperreactivity, demonstrating smooth muscle relaxant effects potentially comparable to standard bronchodilators [10].

These results highlight the potential of monoterpenes to restore smooth muscle homeostasis disrupted by heavy metals. Their multimodal mechanisms—ranging from suppression of ROS to inhibition of voltage-gated calcium channels—underscore their versatility as therapeutic agents [45–47].

In light of the growing interest in plant-based therapeutics and functional foods, these phytochemicals present promising candidates for dietary or interventions, particularly in populations with high environmental or occupational metal exposure

[48]. However, the translation of these findings into clinical settings requires careful evaluation. Critical steps include pharmacokinetic and bioavailability profiling, toxicity screening, and human trials to establish safety and efficacy [49]. Additionally, integration into regulatory frameworks and public health policies will be essential to ensure accessibility and standardization [50].

Future Directions

The toxicological impact of heavy metals on the cardiovascular and respiratory systems is now well established through epidemiological, clinical, and experimental studies [1, 7]. However, critical gaps remain in translating mechanistic insights into effective prevention and therapeutic strategies. Given the persistent environmental presence of arsenic, mercury, cadmium, and lead—especially in developing countries like India—there is an urgent need for multidisciplinary approaches that bridge basic science, clinical practice, and public health policy [19].

Mechanistic Research and Molecular Targets

Further investigation into the molecular mechanisms of metal-induced smooth muscle hyperreactivity is essential. While calcium signaling and oxidative stress have emerged as central pathways, other targets such as mitochondrial dynamics, ion channel regulation, epigenetic modifications, and neurohumoral signaling require deeper exploration [12, 15, 17].

The role of adrenoceptors and receptor-coupled G-proteins in mediating metal-induced vasoconstriction and bronchoconstriction, as seen in our studies, warrants further investigation—particularly in human tissue or organ-on-chip models [51]. Transcriptomic and proteomic profiling of metal-exposed tissues could uncover novel biomarkers and drug targets [29, 52].

Phytochemicals: From Bench to Bedside

Our research underscores the therapeutic promise of phytochemicals such as eugenol, linalool, carvone, and carvacrol in reversing metal-induced vascular and airway dysfunction. These compounds act via calcium channel blockade, antioxidative defense, and receptor modulation, offering a natural and potentially safe therapeutic avenue [10, 11].

However, moving from in vitro or ex vivo efficacy to clinical application requires:

- Toxicokinetic studies to assess absorption, distribution, metabolism, and excretion [49]
- Bioavailability enhancement strategies, such as nanoparticle delivery or prodrug formulations [53]
- Randomized clinical trials to validate efficacy in human populations exposed to heavy metals
- Integration of such agents into dietary interventions or nutraceutical products could be an accessible and community-level strategy for mitigating metal toxicity in high-risk regions [48].

Policy and Surveillance Initiatives

Environmental health policy must evolve in step with scientific evidence. Key action areas include:

- Revision of permissible exposure limits based on cumulative and synergistic toxicity of multiple metals [41]
- Mandating biomonitoring in industrial, agricultural, and mining zones [54]
- Incorporating metal toxicity screening in primary health care protocols, especially in endemic regions
- Public awareness campaigns, school-based education, and community surveillance can further empower populations to minimize exposure.

Integrative Approaches and Interdisciplinary Research

Addressing heavy metal toxicity demands interdisciplinary collaboration among toxicologists, clinicians, pharmacologists, environmental scientists, and policymakers. Integrating environmental surveillance data with clinical endpoints through GIS mapping and health informatics can guide targeted interventions [55].

Emerging tools such as organ-on-chip, CRISPR-based gene editing, and machine learning for toxicological prediction can also accelerate the discovery of personalized and precision-based solutions to metal-induced disorders [56].

Concluding Statement

In conclusion, this chapter highlights the complex duality of metallic elements in human physiology. While essential trace metals such as zinc and iron support vital biological functions, toxic metals like arsenic, mercury, cadmium, and lead pose profound health risks—largely due to their chemical mimicry and disruption of fundamental cellular pathways. Despite their environmental persistence and bioaccumulative nature, emerging evidence suggests that the adverse effects of these metals can be counteracted through natural bioactive compounds such as eugenol, linalool, carvone, and carvacrol. These phytochemicals demonstrate promising therapeutic potential by targeting oxidative stress, calcium signaling, and receptor-level dysfunction. By integrating mechanistic insights from experimental models with translational and policy-driven strategies, we can advance both preventive and therapeutic frameworks to mitigate heavy metal toxicity and strengthen global environmental health resilience.

References

1. Tchounwou PB, Yedjou CG, Patlolla AK et al (2012) Heavy metal toxicity and the environment. In: Luch A (ed) Molecular, clinical and environmental toxicology, vol 101. EXS, 3rd edn. Springer, New York, pp 133–164
2. Maret W, Sandstead HH (2006) Zinc requirements and the risks and benefits of zinc supplementation. J Trace Elem Med Biol 20(1):3–18
3. Rayman MP (2000) The importance of selenium to human health. Lancet 356(9225):233–241
4. Prashanth L, Kattapagari KK, Chitturi RT et al (2015) A review on role of essential trace elements in health and disease. J Dr NTR Univ Health Sci 4(2):75–85
5. Bhattacharya S (2024) Can the toxic heavy metals be beneficial at trace levels? Understanding their outranged biological functions. J Environ Pathol Toxicol Oncol 43(1):71–77
6. Florence TM (2001) The role of free radicals in disease. Aust N Z J Ophthalmol 29(6):529–531
7. Valko M, Morris H, Cronin MTD (2005) Metals, toxicity and oxidative stress. Curr Med Chem 12(10):1161–1208
8. Jarup L (2003) Hazards of heavy metal contamination. Br Med Bull 68(1):167–182
9. Navas-Acien A, Silbergeld EK, Pastor-Barriuso R et al (2008) Arsenic exposure and prevalence of type 2 diabetes in US adults. JAMA 300(7):814–822
10. Kundu S, Shabir H, Basir SF et al (2014) Eugenol and carvone as relaxants of arsenic and mercury hypercontracted rat trachea. J Smooth Muscle Res 50:35–43
11. Kundu S, Shabir H, Basir SF et al (2015) Inhibition of As(III) and Hg(II) caused aortic hypercontraction by eugenol, linalool and carvone. J Smooth Muscle Res 50:93–102
12. Garza A, Vega R, Soto E (2006) Cellular mechanisms of lead neurotoxicity. Med Sci Monit 12(3):RA57–RA65
13. Clarkson TW, Magos L (2003) The toxicology of mercury and its chemical compounds. Crit Rev Toxicol 36(8):609–662
14. Hussain S, Rodgers DA, Duhart HM et al (1997) Mercury-induced oxidative stress and neurotoxicity in primary rat brain astrocytes. Toxicol Sci 47(1):93–100
15. Martinez-Finley EJ, Chakraborty S, Slaughter JC et al (2020) Gene-environment inter actions in metal-induced neurotoxicity. Neurotoxicology 79:136–144
16. Calabrese EJ, Baldwin LA (2007) Hormesis: the dose-response revolution. Annu Rev Pharmacol Toxicol 47:297–323
17. Patocka J, Malina L, Stˇredansk'y M (2020) Neurotoxic mechanisms of heavy metals. Neuro Endocrinol Lett 41:155–164
18. Hossain K, Kato M, Yamamoto H et al (2014) Effects of low-level cadmium exposure on male reproductive function in rats. Environ Health Prev Med 19:179–186
19. Jaishankar M, Tseten T, Anbalagan N et al (2014) Toxicity, mechanism and health effects of some heavy metals. Interdiscip Toxicol 7(2):60–72

20. Briffa J, Sinagra E, Blundell R (2020) Heavy metal pollution in the environment and their toxicological effects on humans. Heliyon 6(9):e04691
21. Hernberg S (2000) Lead poisoning in a historical perspective. Am J Ind Med 38(3):244–254
22. Rahman M, Ng JC, Naidu R (2011) Chronic exposure of arsenic via drinking water and its adverse health impacts on humans. Environ Geochem Health 33(2):151–165
23. Genchi G, Sinicropi MS, Lauria G et al (2020) The effects of cadmium toxicity. Int J Environ Res Public Health 17(11):3782
24. Counter SA, Buchanan LH (2004) Mercury exposure in children: a review. Toxicol Appl Pharmacol 198(2):209–230
25. Landrigan PJ, Todd AC (1994) Lead poisoning. West J Med 161(2):153–159
26. Mohmand J, Samas E, Fasola M et al (2015) Human exposure to heavy metals: risk assessment from hair and nail analysis. Biol Trace Elem Res 167:116–130
27. Mukherjee A, Bhattacharya R, Bhattacharjee P (2021) Arsenic-induced toxicity and carcinogenicity: a mechanistic insight. Toxicol Rep 8:916–928
28. Bridgewater L, Sharpe C, Dunlap K (2014) Mercury disrupts neuronal calcium signaling by altering mitochondrial function. Neurotoxicology 45:28–36
29. Chen L, Liu L, Luo Y et al (2008) MAPK and mTOR pathways are involved in cadmium induced neuronal apoptosis. J Neurochem 105(1):251–261
30. Gallagher CM, Meliker JR, Nriagu JO (2008) Arsenic exposure and hypertension: a systematic review. Environ Health Perspect 116(11):1443–1448
31. Chen CJ, Chiou HY, Chiang MH et al (2007) Dose-response relationship between ischemic heart disease mortality and long-term arsenic exposure. Atherosclerosis 192(2):305–313
32. Shabir H, Kundu S, Basir SF et al (2014) Modulation of Pb(II) caused aortal constriction by eugenol and carvacrol. Biol Trace Elem Res 161:116–122
33. Vaziri ND (2008) Mechanisms of lead-induced hypertension and cardiovascular disease. Am J Physiol Heart Circ Physiol 295(2):H454–H465
34. Rahman I, MacNee W (2002) Oxidative stress and regulation of glutathione in lung inflammation. Eur Respir J 16(3):534–554
35. Jomova K, Valko M (2011) Advances in metal-induced oxidative stress and human disease. Toxicology 283(2–3):65–87
36. Richter PA, Bishop EE, Wang J (2009) Trends in smoking and cigarette brand preferences among U.S. adolescents, 1980–2007. Toxicol Sci 111(2):250–256
37. Gilmour MI, Jaques PA, Chen LC (2006) Pulmonary and systemic effects of inhaled lead. Toxicol Sci 89(2):514–522
38. Li J, Qu W, Wang S et al (2007) Cadmium-induced activation of the ERK signaling pathway and NF-B pathway in human bronchial epithelial cells. Environ Toxicol Pharmacol 24(3):274–280
39. Martınez-Reyes I, Cuezva JM (2010) The H^+-ATP synthase: a gate to ROS-mediated cell death or cell survival. Biochim Biophys Acta Bioenerg 1797(6–7):1147–1158
40. Calabrese EJ, Mattson MP (2008) Hormesis provides a generalized quantitative estimate of biological plasticity. J Cell Commun Signal 2(1–2):25–34
41. Rahman A, Vahter M, Ekstr¨om EC et al (2011). Epub 2010 Dec 9) Arsenic exposure in pregnancy increases the risk of lower respiratory tract infection and diarrhea during infancy in Bangladesh. Environ Health Perspect 119(5):719–724
42. World Health Organization (2021) Guidelines for drinking-water quality: fourth edition incorporating the first addendum. Accessed 19 June 2025
43. Geng X, Xu Z, Zhang Y et al (2022) Prenatal exposure to heavy metals and its association with neurodevelopmental and respiratory outcomes in children: a review. Environ Res 204:112306
44. Shabir H, Kundu S, Basir SF et al (2014) Amelioration of lead- and cadmium-induced rat tracheal hypercontraction by linalool and eugenol. Toxicol Environ Chem 96(2):307–317
45. Bakkali F, Averbeck S, Averbeck D et al (2008) Biological effects of essential oils—a review. Food Chem Toxicol 46(2):446–475

46. Liu Y, Song M, Che T et al (2013) Antioxidant and cytoprotective effects of essential oil components against cadmium-induced oxidative damage. Environ Toxicol Pharmacol 36(3):807–814
47. Pettit T, Jackson C, Carrasco-Pozo C (2020) Plant-derived polyphenols and monoterpenes in calcium signaling and oxidative stress modulation. Nutrients 12(9):2747
48. Mishra BB, Tiwari VK (2013) Natural products: an evolving role in future drug discovery. Eur J Med Chem 38(1):232–241
49. Lopes JR, de Sousa DP, de Almeida RN (2017) Pharmacokinetic and bioavailability considerations of monoterpenes in therapeutics. Curr Drug Metab 18(2):103–115
50. Goyal M, Beri K, Jindal N (2022) The global regulatory landscape of herbal products and phytopharmaceuticals: challenges and opportunities. Pharm Policy Pract 15:43
51. Bhatia SN, Ingber DE (2014) Microfluidic organs-on-chips. Nat Biotechnol 32:760–772
52. Cao Z, Yu LR (2022) Mass spectrometry-based proteomics for biomarker discovery. Methods Mol Biol 2486:3–17
53. Patel T, Zhou J, Piepmeier JM et al (2013) Polymeric nanoparticles for drug delivery to the central nervous system. Adv Drug Deliv Rev 64(7):701–705
54. Goyer RA (2004) Issue paper on the human health effects of metals. In: US EPA (ed), pp 1–64
55. Cordeiro CN, Burchiel SW, Andersen ME et al (2021) GIS-integrated toxicology for public health decision-making. Environ Res 199:111254
56. Vulpe C, Sobh A (2019) CRISPR genomic screening informs gene–environment interactions. Curr Opin Toxicol 18:46–53

Chapter 16
Functional Biochemistry of Iron and Its Cardiopulmonary Impact

Keshav Garg, Nitesh Gautam, Prachi Saluja, Allison Shaw Devine, and Jawahar L. Mehta

Abstract Iron is an essential mineral for cardiopulmonary health, playing a crucial role in oxygen transport and cellular metabolism. Iron homeostasis is a complex process, with derangements in iron metabolism playing a role in the pathophysiology of numerous diseases. We explore the clinical impact of dysregulated iron homeostasis, detailing the pathophysiology of both iron deficiency and iron overload. The discussion focuses on the roles of these states in significant cardiopulmonary disorders, including heart failure, pulmonary hypertension, and iron overload cardiomyopathy. Finally, we summarize the evidence from major clinical trials evaluating iron replacement therapies and discuss the therapeutic implications for managing these complex conditions.

Keywords Iron · Heart failure · Pulmonary hypertension · Iron overload · Hemochromatosis

Introduction

The human body's physiological processes depend heavily on essential minerals, which are vital for a wide range of metabolic processes necessary for maintaining cardiopulmonary health [1]. Iron is unique among them since it is required for

K. Garg
Department of Internal Medicine, University of Arkansas for Medical Sciences, Little Rock, AR, USA

N. Gautam · A. S. Devine · J. L. Mehta (✉)
Division of Cardiology, Department of Internal Medicine, University of Arkansas for Medical Sciences, Little Rock, AR, USA
e-mail: mehtajl@uams.edu

P. Saluja
Division of Pulmonary, Critical Care and Sleep Medicine, University of Cincinnati, Cincinnati, OH, USA

N. S. Dhalla et al. (eds.), *Functional Biochemistry of Metallic Elements*, Advances in Biochemistry in Health and Disease 36,
https://doi.org/10.1007/978-3-032-14445-4_16

fundamental biological functions, such as DNA synthesis, cellular energy metabolism, and oxygen delivery [2]. One of the best examples of iron's pharmacokinetic importance is the body's sophisticated system for controlling its absorption, transport, and storage [3]. Thus, disturbances in this delicate equilibrium, resulting in either iron overload or deficiency, have significant clinical consequences for a variety of systemic disorders. In this review, we focus on the functional biochemistry of iron and systemic metabolic processes controlling the cardiac and lung homeostasis. Further, we discuss the implications of derangement of iron metabolism on cardiopulmonary axis, with focus on the impact of iron deficiency in heart failure (HF) and pulmonary hypertension and the impact of iron overload on cardiac function.

Functional Biochemistry of Iron and Its Clinical Interaction with the Cardiopulmonary Axis

The biochemistry of iron is essential in its roles in both health and disease. Its complex metabolism involves cellular uptake and storage, systemic regulation of absorption, and specific interactions in body tissues, particularly the heart and lungs.

Intestinal and Systemic Iron Absorption

The absorption of dietary iron occurs predominantly in the duodenum and proximal jejunum of the small intestine [2]. There are two types of dietary iron: heme iron, which comes from animal sources of hemoglobin and myoglobin, and nonheme iron, which is present in plant-based diets, dairy products, and iron supplements [3]. These two forms are absorbed through distinct mechanisms. Nonheme iron, which is typically in the unabsorbable ferric (Fe^{3+}) state, must first be reduced to the absorbable ferrous (Fe^{2+}) state. Duodenal cytochrome B (Dcytb), a ferric reductase enzyme on the brush border of the enterocytes, can convert the Fe^{3+} ions to Fe^{2+} ions due to the low pH of gastric acid in the proximal duodenum [3]. Following this, the divalent metal transporter 1 (DMT1) carries the resultant Fe^{2+} into the enterocyte [4]. Levels of DMT1 and Dcytb are upregulated in the hypoxic environment of the intestinal mucosa by hypoxia-inducible factor-2 (HIF-2α) [5]. The heme carrier protein 1 (HCP1) carries the more easily absorbed heme iron into the enterocyte, where it is catabolized by heme-oxygenase-1 to release Fe^{2+} into the intracellular environment [6, 7].

Once inside the enterocyte, iron enters a shared labile iron pool and has two potential outcomes. It can either be carried out of the enterocyte to enter the bloodstream or retained intracellularly in a nontoxic form within the ferritin protein shell [8]. Ferroportin is the protein that regulates its exit from the enterocyte [7]. Before entering the circulation, iron must be converted back to its ferric (Fe^{3+}) state by the enzyme hephaestin [8]. The main protein that transports iron, transferrin, attaches to Fe^{3+} in the bloodstream and makes it available to tissues all over the body, especially the bone marrow, where it is used to make red blood cells [9].

The liver-derived peptide hormone hepcidin carefully controls the rate of iron entry into the plasma [8]. Hepcidin acts as the principal negative regulator of systemic iron availability. It functions by binding to ferroportin on the surface of enterocytes and macrophages, inducing its internalization and subsequent degradation. To maintain systemic iron levels within a specific physiological range, hepcidin expression is increased in response to high iron storage and inflammation and decreased in response to iron deficiency and increased erythropoietic demand [10].

Most of this iron is transported to the bone marrow, where it is incorporated into the heme component of hemoglobin in developing red blood cells. Hemoglobin is the essential oxygen-carrying protein in the blood, while myoglobin serves as an oxygen-storage protein within muscle cells, including cardiomyocytes. Both are iron-dependent heme proteins [10]. The most prevalent cause of this supply chain disruption is nutritional insufficiency, which impairs hemoglobin synthesis and causes iron-deficiency anemia.

Iron Homeostasis Within Cardiomyocytes

In the heart, iron is essential for the function of cardiac muscle cells, or cardiomyocytes. These cells have a high energy demand and depend on iron-dependent enzymes and proteins for their metabolic activities [11]. Circulating iron, bound to transferrin, is taken up by cardiomyocytes primarily through the transferrin receptor-1 (TfR1). Once inside the cell, iron is utilized in the mitochondria for the synthesis of heme and iron-sulfur clusters, which are critical components of the electron transport chain that produces ATP, the energy currency of the cell [12]. Additionally, iron plays a crucial role in myoglobin, the protein that stores oxygen in muscle cells, allowing for aerobic respiration to occur. However, because free iron can be toxic, any excess is safely stored within the protein shell of ferritin.

A disruption in myocyte iron homeostasis is central to the pathophysiology of iron overload cardiomyopathy. In conditions such as hereditary hemochromatosis, characterized by deficient hepcidin production, excessive intestinal iron absorption leads to elevated plasma transferrin saturation and the appearance of non-transferrin-bound iron (NTBI). Alternative transporters, such as ZIP14 and the L-type Ca^{2+} channel, enable cardiomyocytes to readily absorb this form of iron [13]. The resulting intracellular iron excess overwhelms the storage capacity of ferritin, leading to an expanded labile iron pool. This free iron catalyzes the production of the harmful reactive oxygen species (ROS) by participating in the Fenton reaction [12]. The resulting oxidative stress leads to lipid peroxidation, mitochondrial dysfunction, and direct damage to cellular proteins and DNA, ultimately resulting in myocyte death, cardiac dysfunction, and the clinical syndrome of iron overload cardiomyopathy [14]. The heart itself expresses proteins of iron metabolism, including ferroportin and hepcidin, indicating a local regulatory axis that is essential for cardiac function [15].

Iron and Lung Homeostasis

The lungs are also critically dependent on iron regulation for their normal function and defense against pathogens and environmental insults [16]. In terms of iron homeostasis, the lung is in a unique situation as it is exposed to both exogenous iron from inhaled particles, such as dust and pollutants, and endogenous iron from the circulation [17]. The lining fluid that surrounds the respiratory tract's epithelial surfaces contains a high amount of iron-binding glycoproteins like lactoferrin and transferrin, which chelate free iron and reduce iron's capacity to cause oxidative damage [17].

Like cardiomyocytes, pulmonary epithelial cells use TfR1 to absorb transferrin-bound iron from the bloodstream for their own metabolic function. These cells can also internalize iron from the airway lumen using transporters such as DMT1, TfR1, and the lactoferrin receptor [18]. Alveolar macrophages play a key role in pulmonary iron defense. These specialized immune cells can sequester significant amounts of iron within ferritin and actively phagocytose pathogens and debris from the alveolar space [18].

Maintaining an optimal iron concentration in the lung tissues is vital. Since iron is essential for the function of immune cells, such as lymphocytes and macrophages, iron deficiency can impact local immunological responses [19]. Conversely, iron excess in the lung can promote oxidative stress, inflammation, and cellular injury, increasing susceptibility to respiratory infections and exacerbating lung diseases, such as chronic obstructive pulmonary disease, asthma, pulmonary fibrosis, and cystic fibrosis [16, 17] For example, an excess of iron in the fluid lining the alveoli worsens oxidant-mediated tissue damage and promotes the proliferation of inhaled infections [20]. Cigarette smoking is also associated with iron dysregulation in the lung [21]. Thus, maintaining lung homeostasis and function requires complex control of iron absorption, storage, and export within the pulmonary environment.

Iron and Cardiac Function: Pathophysiological Links

Given the quintessential role of iron in cardiac metabolism, a deficiency or an overload of iron levels in the cardiac myocytes gives rise to a characteristic disease phenotype. While iron overload acts as a culprit for direct cardiomyocyte damage by oxidative stress and subsequent HF, iron deficiency acts as a catalyst in individuals with preexisting HF, by altering myocardial metabolism, and thereby impacting cardiovascular outcome.

Impact of Iron Deficiency in Heart Failure

Iron deficiency in HF is widespread, with the prevalence being as high as 50% in patients with symptomatic HF [22]. While a ferritin level of <20 ng/mL has been accepted as the traditional cutoff for iron deficiency, the description of iron deficiency in patients with HF is more nuanced. Ferritin is an acute-phase reactant and is a surrogate marker for inflammation. By virtue of it being a pro-inflammatory state, ferritin levels are elevated in HF. This physiology of normal or increased body stores of iron (i.e., normal ferritin values) coupled with decreased iron availability to tissues (i.e., decreased transferrin saturation (Tsat) levels) creates a state of "relative" or "functional" iron deficiency. Hence, to bridge this gap, contemporary guidelines define ferritin <100 μg/mL as at least mild absolute iron deficiency and a ferritin of 100–300 μg/mL and a Tsat<20% as an indicator of functional iron deficiency [23].

Iron deficiency leads to impaired oxygen-carrying capacity of blood, which can exacerbate the symptoms of HF. Moreover, decreased myocardial iron levels lead to altered myocardial metabolism, ultimately resulting in myocardial dysfunction [24]. Sustained neurohormonal activation in HF forms the basis for iron depletion at the myocardial level by decreasing the expression of transferrin receptors and iron regulatory proteins. This in turn leads to decreased myocardial uptake and increased intracellular release of iron, thereby impairing oxidative phosphorylation [25, 26] (Fig. 16.1). In light of the above pathophysiological findings, it is unsurprising that the impact of iron deficiency in HF extends beyond anemia as iron deficiency is associated with a 40% increase in mortality, even after adjusting for other clinical variables, such as NYHA functional class, LVEF, pro-BNP, and the presence of anemia [22]. Given the profound impact of iron deficiency in patients with HF, multiple trials have evaluated the role of iron supplementation in improving outcomes in the aforementioned cohort (Table 16.1) [27–34].

While the wide range of trials conducted have failed to show any cardiovascular mortality benefit, the earlier set of trials did show an improvement in functional capacity. For instance, in the FAIR-HF trial which enrolled 459 patients with HF, the authors discovered a 2.5-fold improvement in the self-reported patient global assessment score at 24 weeks (OR 2.51; 95% CI [1.75 to 3.61]). The study, however, was limited by a relatively short follow-up period, which may have hindered the detection of significant differences in objective clinical outcomes, such as cardiovascular death or hospitalizations for HF [28].

To evaluate the impact of iron repletion on HFH, the AFFIRM-HF study enrolled 1132 patients with iron deficiency and HF, defined by an LVEF ≤50% and recent HF exacerbation. Patients were randomized in a 1:1 fashion to receive either intravenous ferric carboxymaltose or placebo. At 52-week follow-up, a total of 217 HF hospitalizations occurred in the treatment group compared to 294 in the placebo group, with a significant difference (RR 0.74; 95% CI [0.58–0.94], $p = 0.013$) [31].

While the AFFIRM-HF showed promise of iron repletion for reducing HF hospitalizations, subsequent trials failed to demonstrate any difference in the rate of HF hospitalizations following iron repletion therapy in patients with HF and iron

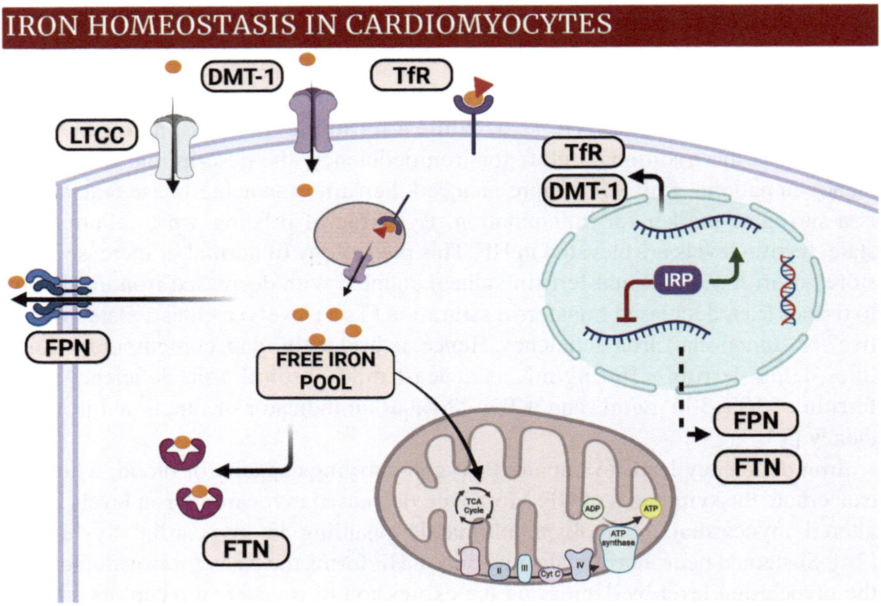

Fig. 16.1 Iron homeostasis in cardiomyocytes. Iron intake into myocytes is governed by the transferrin receptor, in addition to DMT-1 and LTCC. Once internalized, excess iron is coupled to ferritin for iron storage. The labile iron pool is necessary for adequate cellular and mitochondrial function. Iron export occurs through ferroportin, which is, in turn, regulated by hepcidin. Iron regulatory proteins (IRPs) play a crucial role in regulating iron levels within the cell by binding to mRNA, upregulating the production of TfR and DMT-1, and downregulating the production of FPN and FTN. Abbreviations: DMT-1: divalent metal transporter, FPN: ferroportin, FTN: ferritin, LTCC: L-type calcium channel, TfR: transferrin receptor

deficiency [32, 34]. A recent systematic review and meta-analysis demonstrated a reduction in a composite of HF-related hospitalization and cardiovascular mortality at 12 months (RR = 0.72; 95% CI [0.55–0.89]), primarily driven by the decrease in HF hospitalizations at 12 months (RR = 0.69; 95% CI [0.48–0.88]) (Fig. 16.2) [35]. Therefore, while individual trials have yielded conflicting results, pooled evidence suggests a trend toward reducing HF hospitalizations. Ongoing and future trials will be essential to define its hierarchical role in the multifaceted management of patients with HF [23].

Impact of Iron Overload on Cardiomyocytes and Iron Overload Cardiomyopathy

Iron overload can be primary (hereditary hemochromatosis) or secondary, the latter most commonly occurs in individuals with frequent need for blood transfusions (thalassemia, sickle cell disease, etc.). Excessive iron absorption (as seen in

Table 16.1 Summary of randomized controlled trials evaluating the role of intravenous iron repletion in patients with heart failure

Trial	n	Description	Population characteristics	ID definition	Primary outcomes	Results
FERRIC-HF (2008) [27]	35	Open label, randomized trial	NYHA II–IV HF patient with LVEF ≤45% and pVO2 < 18 mL/kg/min	Ferritin <100 ug/L or Tsat <20% with ferritin 100–299 ug/L	Change in peak VO2 from baseline to 18 weeks	Nonstatistically significant change in pVO2 (ΔpVO2 = 96; 95% CI [−12 to 205]; $p = 0.08$) increased pVO2 for Hgb < 12.5 g/dL (ΔpVO2 = 204; 95% CI [31 to 378]; $p = 0.02$)
FAIR-HF (2009) [28]	459	Open label, randomized trial	NYHA II HF patients with LVEF ≤40% NYHA III HF patients with LVEF ≤45%	Ferritin <100 ug/L or Tsat <20% with ferritin 100–299 ug/L	Patient global assessment (PGA) and NYHA class at 24 weeks	Statistically significant increase in PGA (OR 2.51; 95% CI [1.75–3.61]; $p < 0.001$) and improved NYHA class (OR 2.4; 95% CI [1.55–3.71]; $p < 0.001$)
CONFIRM-HF (2015) [29]	304	Double blinded placebo controlled	Symptomatic NYHA II–III HF patients with LVEF ≤45%	Ferritin <100 ug/L or Tsat <20% with ferritin 100–299 ug/L	Change in 6MWT from baseline to 24 weeks	Statistically significant increase in 6MWT (mean treatment effect 33 ± 11 m; $p = 0.002$)
EFFECT-HF (2017) [30]	174	Open label, randomized trial	Mild to moderately symptomatic HF patients with LVEF ≤45%	Ferritin <100 ug/L or Tsat <20% with ferritin 100–299 ug/L	Change in peak VO2 from baseline to 24 weeks	Peak VO2 decreased in the control group (least square means −1.19 ± 0.389 mL/min/kg) and was maintained in the treatment group (−0.16 ± 0.387 mL/min/kg; $P = 0.020$)
AFFIRM-HF (2020) [31]	1132	Double-blinded placebo-controlled trial	LVEF ≤50% and recent discharge for HFH	Ferritin <100 ug/L or Tsat <20% with ferritin 100–299 ug/L)	HFH and CV death	No difference in the CV death (HR 0.96; 95% CI [0.70–1.32]; $p = 0.81$) Statistically significant difference in HFH (RR 0.74; 95% CI [0.58–0.94]; $p = 0.013$)

(continued)

Table 16.1 (continued)

Trial	n	Description	Population characteristics	ID definition	Primary outcomes	Results
IRONMAN (2022) [32]	1137	Open label, randomized, blinded endpoint trial	LVEF ≤45% and symptomatic HF	Ferritin <100 ug/L or Tsat <20% with ferritin 100–299 ug/L)	HFH or CV death	Statistically insignificant difference in primary outcomes (RR 0.82; 95% CI [0.66–1.02]; $p = 0.07$)
HEART-FID (2023) [33]	3065	Double-blinded placebo-controlled trial	LVEF ≤40% with HFH in past 12 months or elevated BNP	Hgb <15 in men and 13.5 in women coupled with relative ID	Primary: Composite of death, HFH, and 6-minute walk test at 6 months from randomization	No difference in primary outcome (overall win ratio 1.10; 99% CI (0.99–1.23)]
FAIR-HF2 (2025) [34]	1105		LVEF ≤45% and symptomatic HF	Ferritin <100 ug/L or Tsat <20% with ferritin 100–299 ug/L)	HFH or CV death; total HFH and time to CV death or first HFH in patients with Tsat <20%	No statistically significant difference in CV death or HFH (HR = 0.79; 95% CI [0.63–0.99]; $P = 0.04$), total HFH (rate ratio = 0.80; 95% CI [0.60–1.06]; $P = 0.12$) or CV death or HFH in patients with Tsat <20% (HR = 0.79; 95% CI [0.61–1.02]; $P = 0.07$)

Abbreviations: *6MWT* 6-min walk test, *CI* confidence interval, *CV* cardiovascular, *HFH* heart failure hospitalization, *Hgb* hemoglobin, *HR* hazard ratio, *LVEF* left ventricular ejection fraction, *NYHA* New York Heart Association, *OR* odds ratio, *RR* risk ratio, *Tsat* transferrin saturation

Study	i.v. iron *n*/*N*	Control *n*/*N*	RR	95% CI
FAIR-HF	13/305	13/154	0.46	(0.18, 1.19)
CONFIRM-HF	22/150	44/151	0.51	(0.28, 0.95)
AFFIRM-AHF	293/558	372/550	0.76	(0.60, 0.96)
IRONMAN	97/527	149/536	0.66	(0.48, 0.91)
HEART-FID	394/1,532	447/1,533	0.89	(0.74, 1.07)
FAIR-HF2	141/558	216/547	0.64	(0.47, 0.86)
Overall (6 studies >200 patients, HN(0.5))			0.72	(0.55, 0.89)
Overall (6 studies, HN(0.1))			0.74	(0.62, 0.86)
Overall (6 studies, KnHa)			0.73	(0.60, 0.89)
Overall (4 studies >1,000 patients, HN(0.5))			0.75	(0.57, 0.97)
Overall (4 studies, HN(0.1))			0.76	(0.65, 0.89)
Overall (4 studies, KnHa)			0.76	(0.59, 0.97)

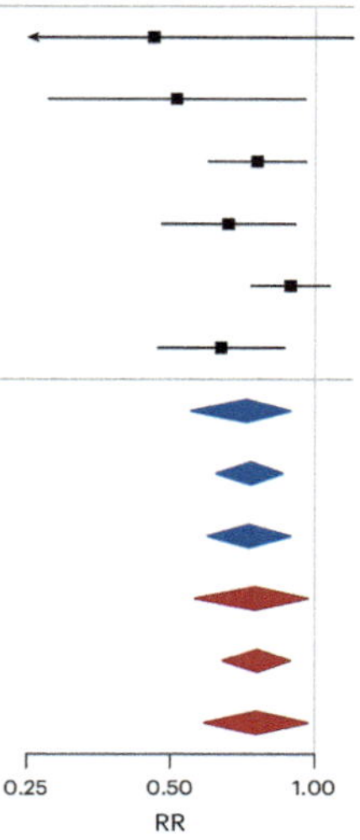

Fig. 16.2 Systematic review evaluating the impact of iron repletion on the composite endpoint of total (first and recurrent) HF hospitalizations and cardiovascular mortality for the first 12 months of follow-up. (Reproduced with permission from Anker et al. [35])

hereditary hemochromatosis) or systemic administration (through blood transfusions) causes an increase in transferrin-bound and non-transferrin-bound iron (NTBI) in blood. Since non-transferrin-bound iron is not regulated by the feedback loops, the excessive cellular iron absorption leads to an increased free iron pool within cells. In the myocytes, the NTBI predominantly enters via the L-type calcium channels (LTCC), which then leads to iron deposition within the myocytes. Moreover, it leads to the production of reactive oxygen species, thereby causing oxidative stress and ultimately resulting in injury to DNA, proteins, and cellular membranes [36]. Moreover, impaired excitation-contraction coupling secondary to iron overload leads to cardiac systolic dysfunction. Apart from cytosolic iron overload, mitochondrial iron overload leads to an increased production of reactive oxygen species and mitochondrial dysfunction, thereby inducing caspase release and subsequent apoptosis [37].

Clinically, iron overload cardiomyopathy is characterized by an early phenotype of restrictive cardiomyopathy followed by a dilated cardiomyopathy, if left untreated. Apart from HF, conduction abnormalities can develop, owing to iron deposition in the cardiac conduction system [38]. While increased serum ferritin levels can be used for screening of the disease, cardiac magnetic resonance can help diagnose iron overload cardiomyopathy. Early diagnosis of disease is important as the disease process is reversible if therapy is started earlier in the disease course [39]. Therapeutic phlebotomy is an option for non-anemic patients, such as individuals with hereditary hemochromatosis [40]. The role of iron chelating agents has been well established for the past four decades in conditions involving secondary iron overload (related to blood transfusions), with chelation therapy having a significant impact on reducing mortality [41, 42]. Apart from iron chelating agents, other therapies such as calcium channel blockers have been shown to reduce cardiac iron content, primarily through LTCC channel blockade, thereby decreasing the uptake of NTBI [39, 43]. Innovative approaches, such as gene therapy and hepcidin modulators, hold promise for the future and are currently undergoing investigative trials [44].

Iron and Pulmonary Function: Pathophysiological Links

Iron in the lungs comes from multiple sources, such as iron adsorption from the serum, red blood cells, and through inhaled particulate matter, such as tobacco smoke, pollution, mining dust, etc., all of which contain iron in varying amounts [45]. Both iron overload and iron deficiency have been implicated in the pathogenesis of lung diseases.

The following section will review the role of dysregulated iron homeostasis in pulmonary diseases such as interstitial lung disease, chronic obstructive pulmonary disease (COPD), asthma, and pulmonary hypertension.

Alveolar macrophages play a critical role in regulating iron and extracellular matrix in the lungs. Transferrin receptor (CD 71) expressed by alveolar macrophages mediates uptake of transferrin complexes and limits free radical production. An increase in alveolar macrophages lacking CD71 has been observed in patients with idiopathic fibrosis when compared to healthy controls [46]. $CD71^-$ alveolar macrophages show phenotypic immaturity, reduced phagocytosis, and profibrotic characteristics and are associated with decreased survival [46]. Despite lacking CD71, alveolar macrophages can accumulate iron through alternative pathways such as NTBI uptake and erythrocyte phagocytosis, along with chronic inflammation-associated downregulation of ferroportin exporter [47, 48]. Murine models and human data have shown that a state of excess iron in the lungs is associated with increased proliferation of fibroblasts and a robust extracellular matrix response through iron-dependent reactive oxygen species generation and mitochondrial dysfunction [49]. To this extent, intranasal iron chelator therapy has been shown to prevent fibrosis and lung function decline in experimental models [49].

Similarly, in COPD, mitochondrial dysfunction in respiratory cells due to iron overload results in impaired ATP production and is associated with disease susceptibility, progression, and impaired host defense. Notably, a lack of iron regulatory protein has been shown to be protective against smoking-related changes in mouse models [50]. In asthma, disrupted iron regulation is associated with heightened airway inflammation, increased bronchial hyperresponsiveness, and structural remodeling. Elevated levels of iron-loaded macrophages and enhanced expression of iron transport proteins have been linked to greater disease severity [51].

On the contrary, iron deficiency in pulmonary arterial hypertension, even in the absence of anemia, is associated with cardiovascular dysfunction [52]. This has been hypothesized to be related to vascular remodeling and increased vasoconstriction [53]. Iron deficiency is related to an increased expression of HIF-1α (hypoxia-inducible factor-1α), HIF-2α, and STAT3 (signal transducer and activation of transcription-3), which subsequently leads to an increased vascular cell proliferation [54].

Whether iron supplementation in pulmonary arterial hypertension improves functional outcomes is a topic of active research. For instance, recent small-scale randomized controlled crossover trials failed to show any functional improvement as evidenced by the 6-minute walk test or cardiopulmonary exercise testing at 12 weeks [55]. A recent systematic review, comprising of 5 studies and 206 patients, demonstrated a positive impact of iron repletion on the 6-min walk test (6MWT) and cardiopulmonary exercise testing [56]. Future large-scale clinical trials, which overcome the challenges of patient recruitment, are needed to determine whether iron repletion leads to any improvement in objective clinical outcomes.

Alongside iron repletion, erythropoiesis-stimulating agents (ESAs) have also been investigated in both heart failure and pulmonary hypertension. In experimental models of pulmonary hypertension, erythropoietin has shown some promise by attenuating vascular remodeling [57]. In heart failure, however, the rationale of correcting anemia to improve cardiac function has not translated into clinical benefit, with large trials showing neutral results and raising safety concerns about thromboembolic events [58, 59]. A detailed discussion on the role of ESAs is beyond the scope of this chapter.

Conclusions

Multiple physiological pathways intricately govern the regulation of iron levels, and disruptions in iron regulation, whether related to iron overload or iron deficiency, are associated with cardiac-pulmonary axis dysfunction. Future studies are necessary to elucidate this role and to determine whether restoration of iron handling leads to improved clinical outcomes.

Acknowledgments None.

Disclosures None.

References

1. Arif L, Hamza M, Iqbal E, Kaleem Z (2024) Role of micronutrients (vitamins & minerals). Int J Multidiscip Sci Arts 3(2):333–337
2. Gulec S, Anderson GJ, Collins JF (2014) Mechanistic and regulatory aspects of intestinal iron absorption. Am J Physiol Gastrointest Liver Physiol 307(4):G397–G409
3. Piskin E, Cianciosi D, Gulec S, Tomas M, Capanoglu E (2022) Iron absorption: factors, limitations, and improvement methods. ACS Omega 7(24):20441–20456

4. Lane DJ, Bae DH, Merlot AM, Sahni S, Richardson DR (2015) Duodenal cytochrome b (DCYTB) in iron metabolism: an update on function and regulation. Nutrients 7(4):2274–2296
5. Xu MM, Wang J, Xie JX (2017) Regulation of iron metabolism by hypoxia-inducible factors. Sheng Li Xue Bao 69(5):598–610
6. Kalman D, Hewlings S, Madelyn-Adjei A, Ebersole B (2025) Dietary heme iron: a review of efficacy, safety and tolerability. Nutrients 17(13)
7. Przybyszewska J, Zekanowska E (2014) The role of hepcidin, ferroportin, HCP1, and DMT1 protein in iron absorption in the human digestive tract. Prz Gastroenterol 9(4):208–213
8. Anderson GJ, Frazer DM (2017) Current understanding of iron homeostasis. Am J Clin Nutr 106(Suppl 6):1559s–1566s
9. Silva AMN, Moniz T, de Castro B, Rangel M (2021) Human transferrin: an inorganic biochemistry perspective. Coord Chem Rev 449:214186
10. Wang CY, Babitt JL (2019) Liver iron sensing and body iron homeostasis. Blood 133(1):18–29
11. Sawicki KT, De Jesus A, Ardehali H (2023) Iron metabolism in cardiovascular disease: physiology, mechanisms, and therapeutic targets. Circ Res 132(3):379–396
12. Paterek A, Mackiewicz U, Mączewski M (2019) Iron and the heart: a paradigm shift from systemic to cardiomyocyte abnormalities. J Cell Physiol 234(12):21613–21629
13. Knutson MD (2019) Non-transferrin-bound iron transporters. Free Radic Biol Med 133:101–111
14. Sumneang N, Siri-Angkul N, Kumfu S, Chattipakorn SC, Chattipakorn N (2020) The effects of iron overload on mitochondrial function, mitochondrial dynamics, and ferroptosis in cardiomyocytes. Arch Biochem Biophys 680:108241
15. Jayakumar D, Narasimhan KKS, Periandavan K (2022) Triad role of hepcidin, ferroportin, and Nrf2 in cardiac iron metabolism: from health to disease. J Trace Elem Med Biol 69:126882
16. Neves J, Haider T, Gassmann M, Muckenthaler MU (2019) Iron homeostasis in the lungs-a balance between health and disease. Pharmaceuticals (Basel) 12(1)
17. Zhang V, Nemeth E, Kim A (2019) Iron in lung pathology. Pharmaceuticals (Basel) 12(1)
18. Philippot Q, Deslée G, Adair-Kirk TL, Woods JC, Byers D, Conradi S et al (2014) Increased iron sequestration in alveolar macrophages in chronic obstructive pulmonary disease. PLoS One 9(5):e96285
19. Ni S, Yuan Y, Kuang Y, Li X (2022) Iron metabolism and immune regulation. Front Immunol 13:816282
20. Khiroya H, Turner AM (2015) The role of iron in pulmonary pathology. Multidiscip Respir Med 10:34
21. Zhang WZ, Butler JJ, Cloonan SM (2019) Smoking-induced iron dysregulation in the lung. Free Radic Biol Med 133:238–247
22. Klip IT, Comin-Colet J, Voors AA, Ponikowski P, Enjuanes C, Banasiak W et al (2013) Iron deficiency in chronic heart failure: an international pooled analysis. Am Heart J 165(4):575–82.e3
23. Beavers CJ, Ambrosy AP, Butler J, Davidson BT, Gale SE, PiÑA IL et al (2023) Iron deficiency in heart failure: a scientific statement from the Heart Failure Society of America. J Card Fail 29(7):1059–1077
24. Alnuwaysir RIS, Hoes MF, van Veldhuisen DJ, van der Meer P, Grote Beverborg N (2021) Iron deficiency in heart failure: mechanisms and pathophysiology. J Clin Med 11(1)
25. Maeder Micha T, Khammy O, dos Remedios C, Kaye David M (2011) Myocardial and systemic iron depletion in heart failure. JACC 58(5):474–480
26. Haddad S, Wang Y, Galy B, Korf-Klingebiel M, Hirsch V, Baru AM et al (2016) Iron-regulatory proteins secure iron availability in cardiomyocytes to prevent heart failure. Eur Heart J 38(5):362–372
27. Okonko DO, Grzeslo A, Witkowski T, Mandal AK, Slater RM, Roughton M et al (2008) Effect of intravenous iron sucrose on exercise tolerance in anemic and nonanemic patients with symptomatic chronic heart failure and iron deficiency FERRIC-HF: a randomized, controlled, observer-blinded trial. J Am Coll Cardiol 51(2):103–112

28. Anker SD, Colet JC, Filippatos G, Willenheimer R, Dickstein K, Drexler H et al (2009) Ferric Carboxymaltose in patients with heart failure and iron deficiency. N Engl J Med 361(25):2436–2448
29. Ponikowski P, van Veldhuisen DJ, Comin-Colet J, Ertl G, Komajda M, Mareev V et al (2015) Beneficial effects of long-term intravenous iron therapy with ferric carboxymaltose in patients with symptomatic heart failure and iron deficiency†. Eur Heart J 36(11):657–668
30. van Veldhuisen DJ, Ponikowski P, van der Meer P, Metra M, Böhm M, Doletsky A et al (2017) Effect of ferric carboxymaltose on exercise capacity in patients with chronic heart failure and iron deficiency. Circulation 136(15):1374–1383
31. Ponikowski P, Kirwan B-A, Anker SD, McDonagh T, Dorobantu M, Drozdz J et al (2020) Ferric carboxymaltose for iron deficiency at discharge after acute heart failure: a multicentre, double-blind, randomised, controlled trial. Lancet 396(10266):1895–1904
32. Kalra PR, Cleland JGF, Petrie MC, Thomson EA, Kalra PA, Squire IB et al (2022) Intravenous ferric derisomaltose in patients with heart failure and iron deficiency in the UK (IRONMAN): an investigator-initiated, prospective, randomised, open-label, blinded-endpoint trial. Lancet 400(10369):2199–2209
33. Mentz RJ, Garg J, Rockhold FW, Butler J, Pasquale CGD, Ezekowitz JA et al (2023) Ferric carboxymaltose in heart failure with iron deficiency. N Engl J Med 389(11):975–986
34. Anker SD, Friede T, Butler J, Talha KM, Placzek M, Diek M et al (2025) Intravenous ferric carboxymaltose in heart failure with iron deficiency: the FAIR-HF2 DZHK05 randomized clinical trial. JAMA 333(22):1965–1976
35. Anker SD, Karakas M, Mentz RJ, Ponikowski P, Butler J, Khan MS et al (2025) Systematic review and meta-analysis of intravenous iron therapy for patients with heart failure and iron deficiency. Nat Med 31(8):2640–2646
36. Buja LM, Roberts WC (1971) Iron in the heart: etiology and clinical significance. Am J Med 51(2):209–221
37. Kumfu S, Chattipakorn SC, Chattipakorn N (2022) Iron overload cardiomyopathy: using the latest evidence to inform future applications. Exp Biol Med 247(7):574–583
38. Ahmed S, Peterson SJ, Parikh MA, Frishman WH (2025) Cardiovascular manifestations of hemochromatosis: a review of pathophysiology, mechanisms, and treatment options. Cardiol Rev 33(4):359–364
39. Gujja P, Rosing DR, Tripodi DJ, Shizukuda Y (2010) Iron overload cardiomyopathy: better understanding of an increasing disorder. J Am Coll Cardiol 56(13):1001–1012
40. Bacon BR, Adams PC, Kowdley KV, Powell LW, Tavill AS (2011) Diagnosis and management of hemochromatosis: 2011 practice guideline by the American Association for the Study of Liver Diseases. Hepatology 54(1):328
41. Olivieri NF, Nathan DG, MacMillan JH, Wayne AS, Liu PP, McGee A et al (1994) Survival in medically treated patients with homozygous beta-thalassemia. N Engl J Med 331(9):574–578
42. Brittenham GM, Griffith PM, Nienhuis AW, McLaren CE, Young NS, Tucker EE et al (1994) Efficacy of deferoxamine in preventing complications of iron overload in patients with thalassemia major. N Engl J Med 331(9):567–573
43. Fernandes JL, Loggetto SR, Veríssimo MPA, Fertrin KY, Baldanzi GR, Fioravante LAB et al (2016) A randomized trial of amlodipine in addition to standard chelation therapy in patients with thalassemia major. Blood 128(12):1555–1561
44. Sandnes M, Reikvam H (2024) Hepcidin as a therapeutic target in iron overload. Expert Opin Ther Targets 28(12):1039–1046
45. Ogger PP, Byrne AJ (2020) Lung fibrosis enters the iron age†. J Pathol 252(1):1–3
46. Allden SJ, Ogger PP, Ghai P, McErlean P, Hewitt R, Toshner R et al (2019) The transferrin receptor CD71 delineates functionally distinct airway macrophage subsets during idiopathic pulmonary fibrosis. Am J Respir Crit Care Med 200(2):209–219
47. Ludwiczek S, Aigner E, Theurl I, Weiss G (2003) Cytokine-mediated regulation of iron transport in human monocytic cells. Blood 101(10):4148–4154

48. Lee J, Arisi I, Puxeddu E, Mramba LK, Amicosante M, Swaisgood CM et al (2018) Bronchoalveolar lavage (BAL) cells in idiopathic pulmonary fibrosis express a complex pro-inflammatory, pro-repair, angiogenic activation pattern, likely associated with macrophage iron accumulation. PLoS One 13(4):e0194803
49. Ali MK, Kim RY, Brown AC, Donovan C, Vanka KS, Mayall JR et al (2020) Critical role for iron accumulation in the pathogenesis of fibrotic lung disease. J Pathol 251(1):49–62
50. Faherty L, Kenny S, Cloonan SM (2023) Iron and mitochondria in the susceptibility, pathogenesis and progression of COPD. Clin Sci (Lond) 137(3):219–237
51. Ali MK, Kim RY, Brown AC, Mayall JR, Karim R, Pinkerton JW et al (2020) Crucial role for lung iron level and regulation in the pathogenesis and severity of asthma. Eur Respir J 55(4):1901340
52. Saluja P, Gautam N, Amisha F, Salah HM, Smith J, Biyani S et al (2022) Influence of iron deficiency on clinical and Haemodynamic parameters in pulmonary arterial hypertension cohorts. Heart Lung Circ 31(12):1594–1603
53. Lakhal-Littleton S, Crosby A, Frise MC, Mohammad G, Carr CA, Loick PAM et al (2019) Intracellular iron deficiency in pulmonary arterial smooth muscle cells induces pulmonary arterial hypertension in mice. Proc Natl Acad Sci USA 116(26):13122–13130
54. Krasinkiewicz JM, Lahm T (2021) Riding the ferrous wheel of iron supplementation in pulmonary arterial hypertension. Ann Am Thorac Soc 18(6):946–948
55. Howard L, He J, Watson GMJ, Huang L, Wharton J, Luo Q et al (2021) Supplementation with iron in pulmonary arterial hypertension. Two randomized crossover trials. Ann Am Thorac Soc 18(6):981–988
56. Baral T, Malakapogu P, Shyma Z, Kurian SJ, Benson R, Manu MK et al (2025) Efficacy of iron replacement in pulmonary hypertension: a systematic review. Nutr Health 31(2):407–418
57. van Loon RLE, Bartelds B, Wagener FADTG, Affara N, Mohaupt S, Wijnberg H et al (2015) Erythropoietin attenuates pulmonary vascular remodeling in experimental pulmonary arterial hypertension through interplay between endothelial progenitor cells and Heme oxygenase. Front Pediatr 3:71
58. Swedberg K, Young JB, Anand IS, Cheng S, Desai AS, Diaz R et al (2013) Treatment of anemia with darbepoetin alfa in systolic heart failure. N Engl J Med 368(13):1210–1219
59. Volpe M, Mastromariona V (2017) Anaemia and heart failure: is there still a role for erythropoiesis-stimulating agents? Arch Turk Soc Cardiol 45(3):217–218

Chapter 17
Iron Metabolism and Its Role in Anemia

Jyoti Upadhyay, Tarnjot Kaur, and Mukesh Nandave

Abstract Iron is an important micronutrient required in various physiological functions, including oxygen transport, energy metabolism, immune function, and DNA synthesis. It can alternate between ferrous and ferric states which enables it to participate in redox reactions necessary for cellular homeostasis. The human body maintains delicate of iron through regulated absorption, transport, storage, and recycling mechanisms. Important proteins like hepcidin, ferroportin, transferrin, and ferritin coordinate for iron regulation. Hepcidin acts as a master regulator by inhibiting iron export during sufficient iron status or inflammation. Imbalance in iron homeostasis can lead to clinical conditions such as iron deficiency anemia (IDA) and anemia of chronic disease (ACD). IDA is generally caused by inadequate dietary intake, impaired absorption, or chronic blood loss, while ACD results from inflammatory-mediated iron sequestration despite adequate stores. Accurate diagnosis depends on a combination of biomarkers such as serum ferritin, transferrin saturation, soluble transferrin receptor (sTfR), and hepcidin levels. Treatment strategies include dietary modifications and oral iron supplements to intravenous iron and erythropoiesis-stimulating agents. Emerging therapies targeting hepcidin and iron-regulatory pathways offer promising alternatives for refractory cases. Furthermore, iron overload disorders such as hereditary hemochromatosis and transfusional hemosiderosis require careful monitoring and management to prevent oxidative damage. Advances in molecular biology and iron-targeted therapeutics are paving the way for personalized treatment approaches. This chapter provides an in-depth overview of iron metabolism, its critical roles in health and disease, and the evolving landscape of diagnosis and treatment of anemia and iron-related disorders.

J. Upadhyay · T. Kaur
Department of Pharmaceutical Sciences, School of Health Sciences and Technology, UPES, Dehradun, Uttarakhand, India

M. Nandave (✉)
Department of Pharmacology, Delhi Pharmaceutical Sciences and Research University (DPSRU), New Delhi, India

N. S. Dhalla et al. (eds.), *Functional Biochemistry of Metallic Elements*,
Advances in Biochemistry in Health and Disease 36,
https://doi.org/10.1007/978-3-032-14445-4_17

Keywords Iron deficiency anemia · Iron metabolism · Hemoglobin · Ferritin · Transferrin · Hepcidin

Introduction

Biological Significance of Iron

Iron is an essential micronutrient that plays a vital role in a various biological process including energy production, oxygen transport, and cellular metabolism. Its chemical versatility, particularly its capacity to exist in multiple oxidation states that is ferric (Fe^{3+}) and ferrous (Fe^{2+}), enables it to perform many of its biological functions. This redox flexibility enables iron to participate in critical cellular activities, including electron transfer, enzymatic catalysis, and regulation of oxidative stress, making it indispensable for maintaining cellular homeostasis and proper physiological function [18, 30].

One of the most critical roles of iron is facilitating the transport and storage of oxygen within the body. It forms the central atom of the heme group in hemoglobin, which is the oxygen-carrying protein in erythrocytes. Each hemoglobin molecule is composed of four heme groups, with each heme containing iron atom that can reversibly bind one molecule of oxygen. This structural arrangement allows hemoglobin to carry up to four oxygen molecules simultaneously, making it highly efficient in transporting oxygen from lungs to the body. The binding and release of oxygen are finely regulated by factors such as pH, carbon dioxide levels, and temperature, ensuring that oxygen delivery matches the metabolic needs of different tissues. This function is crucial for maintaining cellular respiration and overall physiological balance. Similarly, iron is present in myoglobin, a muscle-specific oxygen-binding protein that facilitates oxygen storage and release during muscular activity [20].

Iron is also vital for energy metabolism. It is an essential component of cytochromes in the mitochondrial electron transport chain; it facilitates the transfer of electrons necessary for oxidative phosphorylation and ATP synthesis. This role is important in maintaining cellular energy homeostasis and supporting the metabolic demands of all tissues. Iron also serves as a cofactor in enzymes. It is required for the catalytic activity of enzymes including catalase and peroxidase; these enzymes neutralize reactive oxygen species (ROS) and safeguard the cells from oxidative damage. Iron-containing enzymes such as ribonucleotide reductase are important for DNA synthesis and cytochrome P450 enzymes are involved in the biosynthesis of hormones and the metabolism of xenobiotics [9].

Iron plays an important role in immune system by supporting the proliferation and maturation of lymphocytes, hence contributing to effective immune responses. The body carefully regulates iron availability, especially during infection, because pathogens also rely on iron for their growth. It is known as nutritional immunity; it reflects the dual nature of iron in host-pathogen interactions. Iron is also important for normal neurological development and function. It involves in the production of key neurotransmitters including dopamine, norepinephrine, and serotonin. Iron deficiency, particularly in the crucial time of brain development, has been associated with developmental delays and cognitive impairments [23].

Regulation and Dynamics of Iron Metabolism

Iron metabolism is a complex process; it absorbs, transports, utilizes, stores, and recycles to meet physiological needs while preventing toxicity. It is a trace element but essential for critical biological functions such as DNA synthesis, energy production, neurotransmitter synthesis, immune system function, oxygen transport, and cellular respiration. The body maintains a balance of iron as both deficiency and excess can be harmful [2].

Iron is mainly absorbed in the duodenum and proximal part of the jejunum through tightly regulated mechanisms. Dietary iron is available in two forms: heme iron, which comes from animal-based foods, and nonheme iron, which is found in plant-based foods. Heme iron is absorbed more efficiently via heme carrier proteins, whereas nonheme iron is converted into the ferrous (Fe^{2+}) form through a reduction process and transported into enterocytes by divalent metal transporter 1 (DMT1). Inside the enterocyte, iron may either be stored as ferritin or transported into the bloodstream through ferroportin, the sole known iron exporter at the cellular level. During the export of iron from intestinal cells into the bloodstream, the enzyme hephaestin facilitates the oxidation of iron from its ferrous (Fe^{2+}) to ferric (Fe^{3+}) form. This oxidation is crucial because only the ferric form can effectively bind to transferrin, the plasma protein responsible for transporting iron through the circulation. This step ensures safe and efficient delivery of iron to various tissues and organs where it is needed and enables to bind with transferrin, the primary iron-carrying protein in plasma [7].

Systemic iron homeostasis is regulated by the liver-derived hormone hepcidin; it controls ferroportin activity. When iron stores are adequate or during inflammatory conditions, hepcidin levels rise, causing the breakdown of ferroportin and resulting in reduced iron absorption and limited release of iron from macrophages. Low iron levels or increased erythropoietic demand suppresses hepcidin production, enhancing iron availability. Intracellular iron levels are regulated posttranscriptionally by iron-regulatory proteins (IRPs); they bind to iron-responsive elements (IREs) on mRNAs of iron-related genes to control their translation and stability [3].

The Interplay Between Iron Metabolism and Anemia

Iron homeostasis refers to the precise regulation of iron levels in the body to ensure sufficient red blood cell production (erythropoiesis) while preventing iron deficiency and anemia. This balance is vital for supporting oxygen transport, energy metabolism, and overall physiological function without causing iron overload or toxicity. The body regulates iron absorption, utilization, and storage through tightly controlled mechanisms involving key proteins such as ferroportin, hephaestin, and the regulatory hormone hepcidin. When iron availability is insufficient (due to poor intake, malabsorption, chronic blood loss, or elevated hepcidin levels during inflammation), hemoglobin synthesis is impaired, leading to iron deficiency anemia (IDA) or anemia of chronic disease (ACD). In IDA, depleted iron stores and reduced serum iron limit erythrocytes production, while in ACD, iron is sequestered in macrophages despite normal or increased body stores, resulting in functional iron deficiency [21].

Iron Absorption and Transport

Iron in the Diet: Essential Food Sources

Dietary iron is obtained in two main forms: nonheme iron and heme iron. Source of nonheme iron is plant-based foods including legumes, lentils, leafy green vegetables, fortified cereals, and nuts. It has a lower absorption rate (2–20%) because it is susceptible to dietary inhibitors like phytates, polyphenols, and calcium. Absorption of nonheme iron can be promoted by vitamin C and certain organic acids. In contrast, Heme iron is obtained from animal-based foods including poultry, red meat, and fish. It bounds within hemoglobin and myoglobin and is absorbed more efficiently (typically around 15–35%) via a specific heme carrier protein in the small intestine [30].

Mechanisms of Iron Absorption

Heme iron is taken up directly by intestinal enterocytes through a specific transporter known as heme carrier protein 1 (HCP1), and nonheme iron must first be reduced from the ferric (Fe^{3+}) to ferrous (Fe^{2+}) by ferrireductases. Ferrous (Fe^{2+}) iron is transported inside the cell by DMT1. Within the enterocyte, iron is either retained in the form of ferritin or transported in the blood via ferroportin, the sole identified iron-exporting protein. The exported iron is oxidized back to the ferric form by hephaestin or ceruloplasmin and then bound to transferrin for systemic distribution. The iron absorption is influenced by body iron stores, erythropoietic demand, and the hepatic hormone hepcidin, which downregulates ferroportin activity to reduce iron efflux when stores are sufficient [27].

Ferritin and Transferrin: Central Proteins in Iron Storage and Transport

Ferritin and transferrin are two important proteins involved in the storage and transport of iron. Ferritin is intracellular iron storage protein, capable of sequestering thousands of iron atoms in a safe, bioavailable form in cells. This prevents free iron from catalyzing the formation of harmful reactive oxygen species. Transferrin is a glycoprotein responsible for transporting iron in the plasma. It binds iron in the ferric (Fe^{3+}) state and delivers it to cells via transferrin receptors through receptor-mediated endocytosis. This system maintains iron homeostasis, facilitates iron delivery to tissues such as the bone marrow for erythropoiesis, and limits pathogen access to iron, which is crucial for immune defense [10, 15].

Iron Utilization and Storage

Role of Iron in Hemoglobin Synthesis

Iron is an important component of hemoglobin. Hemoglobin is the protein responsible for carrying oxygen in red blood cells. It contains four heme groups in each of its molecules, and at the core of each heme lies a ferrous iron (Fe^{2+}) atom that reversibly binds oxygen. Ample iron availability is necessary for the synthesis of functional hemoglobin and effective erythropoiesis. In iron deficiency states, hemoglobin production is impaired, leading to microcytic, hypochromic anemia. The regulation of iron uptake by erythroid precursors is tightly coordinated with erythropoietic activity to ensure a continuous supply of iron for hemoglobin assembly [3].

Iron Storage Mechanisms: Roles of Ferritin and Hemosiderin

Ferritin and hemosiderin are iron storage proteins in the body. They act as buffers against iron deficiency and toxicity. Ferritin stores iron in a soluble, bioavailable in spleen, bone marrow, and liver, and releases when needed. Hemosiderin is an insoluble, aggregated form of partially degraded ferritin; it accumulates when iron levels exceed the storage capacity of ferritin. It is often observed in conditions of iron overload. Ferritin and hemosiderin help in maintaining iron homeostasis by sequestering excess iron and protecting cells from oxidative damage induced by free iron [21].

Regulation of Iron Release by Ferroportin and Hepcidin

Ferroportin is the only known iron exporter protein. It transports iron from macrophages, enterocytes, and hepatocytes to the blood. Its function is closely regulated by hepcidin. It is a peptide hormone synthesized by the liver. If systemic iron levels are sufficient or during inflammation, elevated hepcidin levels bind to ferroportin, causing it to be internalized and degraded, which in turn decreases the release of iron into the bloodstream. In iron-deficient or hypoxic states, hepcidin expression decreases, allowing ferroportin to remain active and enhance iron mobilization. This hepcidin-ferroportin axis is central to systemic iron regulation and prevents both iron overload and deficiency [11].

Iron Homeostasis and Regulation

Hepcidin: A Master Regulator of Systemic Iron Balance

Hepcidin is a central hormone in the regulation of systemic iron balance. Hepcidin expression is upregulated in response to elevated body iron stores and inflammation (mediated by IL-6) and downregulated during iron deficiency, anemia, or hypoxia. Dysregulation of hepcidin leads to iron-related disorders. Excess hepcidin leads to anemia of inflammation, while deficiency in hepcidin causes iron overload, as seen in hereditary hemochromatosis. Hence, hepcidin is a master regulator of iron homeostasis by coordinating iron absorption, recycling, and storage [13].

Influence of Hypoxia-Inducible Factors (HIFs) on Iron Metabolic Pathways

HIFs are transcription factors. They help the cells adapt to low oxygen conditions and are key regulators connecting oxygen availability with iron metabolism. In hypoxic conditions or iron deficiency, HIFs become stabilized and translocate to the nucleus, where they activate the transcription of target genes involved in erythropoiesis, iron uptake, and transport. These include genes encoding erythropoietin (EPO), transferrin, transferrin receptor, and DMT1. HIFs also suppress hepcidin expression, thereby enhancing iron availability to meet increased erythropoietic demand. HIFs respond to both oxygen and iron status, thereby ensuring that iron metabolism is regulated to support tissue oxygenation and red blood cell production [16].

The Iron-Regulatory Proteins and Iron-Responsive Elements (IRP-IRE) System in Cellular Iron Regulation

Cellular iron homeostasis is maintained by a posttranscriptional regulatory system made up of iron-responsive elements (IREs) and iron-regulatory proteins (IRP1 and IRP2). IREs are particular RNA stem-loop structures found in the untranslated regions (UTRs) of mRNAs that encode proteins like ferritin, transferrin receptor, and DMT1 that are involved in iron metabolism. Intracellular iron levels cause IRPs to attach to IREs. IRPs stabilize mRNAs with IREs in the 3' UTR (such as transferrin receptor mRNA) to enhance iron uptake in iron-deficient cells while binding to IREs in the 5' UTR of ferritin mRNA to prevent translation and thereby decrease iron storage. IRP binding diminishes in iron-repleted environments, enabling reduced transferrin receptor expression and increased ferritin synthesis. Cells can quickly adjust to changing iron levels, thanks to this precisely calibrated mechanism [31].

Anemia and Iron Deficiency

Definition and Classification of Anemia

Anemia is a condition characterized by hemoglobin levels in the blood falling below the normal reference range, resulting in a decreased ability of red blood cells (RBCs) to carry oxygen. According to the World Health Organization (WHO), anemia is diagnosed when hemoglobin is <12 g/dL in women and hemoglobin is <13 g/dL in men [21]. Anemia may be classified based on its underlying causes and morphological features, as outlined in Table 17.1 for etiology and Table 17.2 for morphology.

Causes of Iron Deficiency Anemia (IDA)

IDA happens when the body's iron supply is less than body's requirements of hemoglobin synthesis and other physiological functions [14]. The various causes of IDA are outlined in detail in Table 17.3.

Table 17.1 Based on etiology, anemia can be classified as follows [4]

S. no.	Type of anemia	Cause
1.	Nutritional deficiency anemias	Iron, vitamin B12, and folate deficiency
2.	Blood loss anemia	Acute or chronic hemorrhage
3.	Hemolytic anemia	Inherited (e.g., thalassemia, sickle cell) or acquired
4.	Bone marrow failure	Aplastic anemia and myelodysplasia
5.	Chronic disease-related	Kidney disease and cancer

Table 17.2 Based on morphology, anemia can be classified as follows [7]

S. no.	Morphology (based on RBC size and color)	Type of anemia
1.	Microcytic, hypochromic anemia	Typically caused by iron deficiency
2.	Normocytic, normochromic anemia	Seen in anemia of chronic disease and acute blood loss
3.	Macrocytic anemia	Usually due to vitamin B12 or folate deficiency

Table 17.3 Causes can be classified into four major categories [6]

S. no.	Cause of iron deficiency anemia	Possible reasons
1.	Decreased iron intake	Poor dietary intake (vegetarian diets without supplementation) Malnutrition or restricted diets Infants on exclusive milk diets without iron fortification
2.	Decreased absorption	Malabsorption syndromes (e.g., inflammatory bowel disease, celiac disease) Postgastrectomy or bariatric surgery Use of proton pump inhibitors or antacids (decreased gastric acidity affects iron absorption)
3.	Increased demand	Pregnancy and lactation Rapid growth during infancy, childhood, and adolescence Menstruating females
4.	Increased loss of iron	Chronic blood loss (e.g., gastrointestinal bleeding, peptic ulcers, colorectal cancer, hemorrhoids) Heavy menstrual bleeding (menorrhagia) Parasitic infections (e.g., hookworm) Genetic factors, environmental conditions (e.g., poor sanitation, infections), and underlying chronic diseases may also cause iron depletion and anemia.

Pathophysiology of IDA: Impaired Hemoglobin Synthesis and Erythropoiesis

Iron plays a central role in the synthesis of hemoglobin, which is essential for oxygen transport. In IDA, a gap between iron demand and supply disrupts this process through the following mechanisms:

Impaired Hemoglobin Synthesis: Inadequate iron reduces the availability of ferrous iron (Fe^{2+}) needed for heme production. As a result, red blood cells become microcytic (small) and hypochromic (pale).

Disrupted Erythropoiesis: Iron deficiency triggers tissue hypoxia and thus activates hypoxia-inducible factors (HIFs), especially HIF-2α, which upregulate erythropoietin (EPO) production by the kidneys to stimulate bone marrow. However, insufficient iron impairs the maturation of erythroid precursors despite EPO stimulation.

Iron Homeostasis Regulation: Hepcidin, a key regulator peptide hormone secreted by hepatocytes, is suppressed in iron deficiency. Low hepcidin increases iron absorption in the intestines (via DMT1 and ferroportin) and release of iron stored in macrophages (spleen, liver). Hepcidin suppression is further regulated by transferrin-bound iron levels and hepatic proteins like TMPRSS6. Ferroportin, the only known iron exporter, is stabilized when hepcidin levels fall, increasing iron availability in the circulation.

Despite these compensatory mechanisms, when the body's iron stores are severely depleted, hemoglobin synthesis remains impaired, causing anemia [6].

Symptoms and Clinical Manifestations of IDA

The clinical presentation of IDA varies depending on the severity and chronicity of iron depletion. Symptoms arise due to both reduced oxygen delivery and iron-dependent enzyme dysfunction. General symptoms include fatigue and weakness; paleness (pallor) of skin, nail beds, and conjunctiva; dizziness or light-headedness; shortness of breath on exertion; cold intolerance; tachycardia; and palpitations. Neurological and cognitive manifestations include difficulty concentrating, irritability, poor academic performance in children, developmental delays, and psychomotor impairment. Epithelial and mucosal changes include brittle nails, spoon-shaped nails (koilonychia), dry skin and hair, sore tongue (glossitis), cracks at the corners of the mouth (angular stomatitis), and dysphagia (due to esophageal web in Plummer-Vinson syndrome).

There are specific risks in vulnerable groups like in pregnancy; there are increased chances of increased maternal and fetal morbidity, preterm labor, low birth weight, and perinatal mortality. In children, IDA can lead to impaired immune function, delayed growth, and increased susceptibility to infections [8].

Iron Overload and Its Implications

Hereditary Hemochromatosis and Secondary Iron Overload

Iron overload can arise from inherited or acquired conditions that lead to excessive accumulation of iron in body tissues. Hereditary hemochromatosis (HH) is the most prevalent inherited disorder leading to iron overload. It is primarily caused by mutations in the HFE gene, especially the C282Y variant, resulting in low hepcidin levels and excessive intestinal absorption of iron. The excess iron deposits in organs such as the liver, skin, heart, joints, and pancreas lead to complications like liver cirrhosis, cardiomyopathy, arthritis, and diabetes. Secondary iron overload is often a result of repeated blood transfusions, chronic hemolytic anemias, or excessive iron intake and commonly affects patients with conditions like thalassemia major or myelodysplastic syndromes. Opposite to HH, where iron accumulates in parenchymal cells, secondary iron overload often leads to iron deposition in macrophages. Timely diagnosis and management, including iron chelation therapy, are necessary to prevent irreversible organ damage [24].

Oxidative Stress and Cellular Damage Due to Excess Iron

Excess iron in the body promotes oxidative stress by participating in the Fenton reaction, which generates harmful hydroxyl free radicals from hydrogen peroxide. These highly reactive radicals cause DNA strand breaks, lipid peroxidation, and protein oxidation, leading to cellular injury and organ dysfunction. Tissues with high metabolic activity like heart, liver, pancreas, and endocrine glands are vulnerable. Over time, this oxidative damage results in conditions like cardiomyopathy, hepatic fibrosis, insulin-dependent diabetes, and endocrine abnormalities. The body cannot actively excrete excess iron, causing cellular toxicity. Hence, to prevent or mitigate oxidative damage and chronic inflammation, it is important to monitor iron levels and use iron chelators when necessary [22].

Role of Iron in Anemia of Chronic Disease (ACD)

Anemia of chronic disease (ACD), also known as anemia of inflammation, happens in autoimmune disorders, chronic infections, malignancies, and chronic kidney disease. It is characterized by impaired iron utilization despite the presence of adequate or elevated total body iron stores. Inflammatory cytokines, especially interleukin-6 (IL-6), stimulate hepatic hepcidin production, which in turn inhibits ferroportin, the iron-exporting protein. This results in iron sequestration in macrophages and enterocytes, reducing iron availability for erythropoiesis. In this condition, patients exhibit low serum iron and transferrin saturation but normal or elevated ferritin levels. In ACD, RBCs are normocytic or mildly microcytic. It is vital to differentiate between ACD and IDA because in ACD, iron supplementation is not always indicated and may even be harmful [26].

Diagnosis and Biomarkers of Iron Status

Accurate diagnosis of iron deficiency and related anemias requires a thorough understanding of iron metabolism biomarkers. These laboratory indicators help in evaluating iron stores, transport, and functional iron availability, thereby guiding differential diagnosis and treatment strategies.

Serum Ferritin, Transferrin Saturation, and Total Iron-Binding Capacity (TIBC)

Serum ferritin is a primary indicator of body iron stores; it is often the first parameter evaluated in suspected iron deficiency. Normal reference range is below 15 ng/mL for assessment of iron deficiency, particularly in the absence of inflammation, but levels can vary between populations. Ferritin is an acute-phase reactant, meaning it can be elevated in infections, chronic diseases, or liver disorders, potentially masking true iron deficiency in these settings [12].

Transferrin saturation (TSAT) is calculated as the ratio of serum iron to TIBC. It reflects the degree of transferrin saturation with iron. A TSAT below 15% suggests iron deficiency, and elevated levels may indicate iron overload conditions such as hemochromatosis. TIBC tends to be elevated in IDA due to compensatory upregulation of transferrin; it is often normal or decreased in ACD due to reduced hepatic production during inflammation [17].

Soluble Transferrin Receptor (sTfR) and Hepcidin Levels

Soluble transferrin receptor (sTfR) levels increase in response to cellular iron deficiency and elevated erythropoietic activity. sTfR is not significantly affected by inflammatory states and hence serves as a valuable biomarker in distinguishing IDA from ACD. An elevated sTfR concentration supports a diagnosis of IDA, and the sTfR/log ferritin index is especially useful in mixed anemia cases, with values greater than 2 typically indicating iron deficiency [28].

Hepcidin, a hormone originating from the liver, plays a crucial role in iron regulation; it plays a central role in iron regulation by controlling ferroportin, the main cellular iron exporter. In IDA, hepcidin levels are low to promote iron absorption and mobilization; in contrast to ACD and inflammatory states, hepcidin levels are elevated, inhibiting iron release and reducing serum iron levels despite adequate stores. Although not yet widely available in routine clinical settings, hepcidin measurement shows promise as a diagnostic marker, particularly when conventional tests yield ambiguous results [29].

Laboratory Tests for Differentiating Iron Deficiency Anemia From Other Anemias

Differentiating IDA from ACD is critical as both may present with similar reductions in serum iron but require distinct treatment approaches. In IDA, complete blood count (CBC) typically reveals microcytic, hypochromic anemia, while ACD is more often normocytic and normochromic. Peripheral blood smears in IDA show

Table 17.4 Diagnostic tests to distinguish iron deficiency anemia (IDA) from anemia of chronic disease (ACD) [12, 17]

S. no.	Parameter	Iron deficiency anemia (IDA)	Anemia of chronic disease (ACD)
1.	Hemoglobin	Decreased	Decreased
2.	MCV	Decreased (microcytic)	Normal or decreased
3.	MCH	Decreased (hypochromic)	Normal or mildly decreased
4.	Serum ferritin	Decreased	Normal or increased
5.	Serum iron	Decreased	Decreased
6.	TIBC	Increased	Decreased or normal
7.	TSAT	Decreased	Decreased
8.	sTfR	Increased	Normal
9.	Hepcidin	Decreased	Increased
10.	Bone marrow iron	Absent	Normal or increased

marked anisopoikilocytosis with microcytic red blood cells, while ACD smears are generally less remarkable [12].

Additional laboratory parameters further support differential diagnosis. Ferritin is typically low in IDA. It is normal or increased in ACD. The TIBC is elevated in IDA and reduced or normal in ACD. TSAT is reduced in both conditions, but a high TIBC in IDA helps differentiate the two. sTfR is elevated in IDA but not in ACD, while hepcidin is suppressed in IDA and elevated in ACD due to inflammatory regulation. Bone marrow iron staining, though rarely performed, shows absent iron in IDA and preserved or increased iron in ACD [17]. Table 17.4 outlines the distinguishing features between IDA and ACD to support precise diagnosis and effective treatment planning.

Therapeutic Approaches

Dietary Interventions and Iron Supplementation

The first approach to treating IDA typically involves dietary modifications and oral iron supplementation. It is advised to increase the intake of iron-rich foods including fish, red meat, poultry, leafy green vegetables, legumes, and fortified cereals. Bioavailability of nonheme iron can be increased by consuming these foods with vitamin C-rich items (e.g., citrus fruits, tomatoes). Oral iron supplements, used mostly in form of ferrous gluconate, ferrous sulfate, or ferrous fumarate, are the most cost-effective and widely used treatment. However, gastrointestinal adverse effects including nausea, constipation, and abdominal discomfort may affect adherence. In cases of mild to moderate deficiency, dietary and oral iron therapy is usually sufficient to replenish iron stores and restore hemoglobin levels [3].

Intravenous Iron Therapy: Indications and Risks

Intravenous (IV) iron therapy is indicated in patients who cannot tolerate oral iron, have poor absorption (e.g., chronic kidney disease, inflammatory bowel disease), require rapid correction of anemia (e.g., preoperative anemia), or are unresponsive to oral supplementation. Frequently used intravenous iron formulations include iron dextran, iron sucrose, and ferric carboxymaltose. While IV iron bypasses gastrointestinal absorption and offers quicker repletion of iron stores, it carries risks such as hypersensitivity reactions, iron overload, and, in rare cases, anaphylaxis. Therefore, its use is generally reserved for selected patients under careful medical supervision [14].

Erythropoiesis-Stimulating Agents (ESAs)

Erythropoiesis-stimulating agents (ESAs) are synthetic forms of erythropoietin. They are used to promote erythrocytes production in anemic patients, particularly in chronic kidney disease, cancer chemotherapy, or certain inflammatory conditions. ESAs are often used along with iron supplements to ensure an adequate supply of iron for hemoglobin synthesis. However, they must be carefully used due to risk of potential adverse effects, including increased risk of thrombosis, hypertension, and cardiovascular events. Appropriate patient selection and careful dose titration are necessary to balance benefits and risks [25].

Emerging Treatments Targeting Hepcidin and Iron Metabolism

Better understanding of iron regulation have led to the development of novel therapies targeting hepcidin and iron metabolism. In conditions like anemia of chronic disease and functional iron deficiency, researchers are exploring ways to lower hepcidin levels. This can be done using treatments like monoclonal antibodies and small interfering RNAs that block hepcidin. These agents restore iron availability by increasing intestinal absorption and promoting iron release from macrophages. Other experimental approaches include ferroportin agonists and modulators of iron-regulatory proteins. Although still under clinical investigation, these emerging therapies hold promise for patients with complex or refractory anemias where traditional iron supplementation is inadequate or ineffective [5].

Future Perspectives and Research Directions

Advances in Iron Metabolism Research

Ongoing research in iron metabolism has enhanced our understanding of how the body regulates iron absorption, transport, storage, and utilization. Inventions related to key regulatory molecules such as hepcidin, ferroportin, and iron-regulatory proteins (IRPs) have provided new insights into the pathophysiology of various types of anemia and iron overload conditions. Advanced genetic and molecular tools have also helped identify mutations associated with inherited iron disorders, such as hemochromatosis and iron-refractory iron deficiency anemia (IRIDA) [1].

Novel Therapeutic Strategies for Anemia and Iron-Related Disorders

Emerging therapies are being developed to address limitations of conventional iron supplementation and to target specific mechanisms involved in iron dysregulation. These include hepcidin antagonists, ferroportin stabilizers, and agents that modulate iron transport and storage proteins. Gene therapy and RNA-based treatments are also being explored for inherited forms of anemia and disorders with chronic inflammation. In addition, there is growing interest in combining iron therapy with anti-inflammatory agents or erythropoiesis-stimulating drugs to improve outcomes in patients with ACD. As research progresses, these novel strategies have the potential to offer safer, more effective, and individualized treatment options for patients with complex iron metabolism disorders [19].

Conclusion

Iron is a fundamental micronutrient involved in oxygen transport, energy metabolism, immune function, and DNA synthesis. Its unique ability to alternate between redox states underlies its biological significance but also necessitates tight regulation to prevent deficiency or toxicity. This chapter has highlighted the mechanisms that affect iron absorption, transport, storage, and utilization, emphasizing the central roles of proteins like hepcidin, ferroportin, ferritin, and transferrin.

Disruptions in iron homeostasis results in range of clinical conditions, IDA and ACD. IDA arises due to inadequate iron availability for hemoglobin synthesis, while ACD results from iron sequestration during inflammation, emphasizing the importance of accurate diagnosis and tailored therapeutic interventions. Advances in biomarkers such as soluble transferrin receptor and hepcidin are enhancing diagnostic precision and helping differentiate between overlapping anemia types.

Therapeutic approaches such as dietary modifications, iron supplementation, and emerging molecular therapies are evolving to address individual patient needs more effectively. Furthermore, cutting-edge research into iron-regulatory pathways and novel treatment modalities is expanding the potential for personalized care in anemia and iron overload disorders.

Understanding iron metabolism in both health and disease is important for developing effective prevention, diagnostic, and management strategies. With ongoing research and innovation, the prospects for addressing iron-related disorders continue to improve, offering hope for better patient outcomes in the years to come.

References

1. Arosio P (2022) New advances in iron metabolism, ferritin and hepcidin research. Int J Mol Sci 23(23):14700. https://doi.org/10.3390/IJMS232314700
2. Bazeley JW, Wish JB (2022) Recent and emerging therapies for iron deficiency in anemia of CKD: a review. Am J Kidney Dis 79:1–9. https://doi.org/10.1053/j.ajkd.2021.09.017
3. Camaschella C (2015) Iron-deficiency anemia. N Engl J Med 372(19):1832–1843. https://doi.org/10.1056/NEJMRA1401038
4. Camaschella C (2019) Iron deficiency. Blood 133(1):30–39. https://doi.org/10.1182/BLOOD-2018-05-815944
5. Camaschella C, Nai A, Silvestri L (2020) Iron metabolism and iron disorders revisited in the hepcidin era. Haematologica 105(2):260. https://doi.org/10.3324/HAEMATOL.2019.232124
6. Chaparro CM, Suchdev PS (2019) Anemia epidemiology, pathophysiology, and etiology in low- and middle-income countries. Ann N Y Acad Sci 1450(1):15–31. https://doi.org/10.1111/NYAS.14092;CTYPE:STRING:JOURNAL
7. Domenica Cappellini M, Motta I (2015) Anemia in clinical practice—definition and classification: does hemoglobin change with aging? Semin Hematol 52(4):261–269. https://doi.org/10.1053/J.SEMINHEMATOL.2015.07.006
8. Duryea EL, Schell RC (2023) Anemia. In: Queenan's management of high-risk pregnancy: an evidence-based approach. Wiley, pp 99–104. https://doi.org/10.1002/9781119636540.ch12
9. Dutt S, Hamza I, Bartnikas TB (2022) Molecular mechanisms of iron and Heme metabolism. Annu Rev Nutr 42:311–335. https://doi.org/10.1146/ANNUREV-NUTR-062320-112625/CITE/REFWORKS
10. Ganz T (2013) Systemic iron homeostasis. Physiol Rev 93(4):1721–1741. https://doi.org/10.1152/PHYSREV.00008.2013
11. Ganz T, Nemeth E (2012) Biochimica et Biophysica Acta Hepcidin and iron homeostasis ☆. BBA - Molecular Cell Res 1823(9):1434–1443. https://doi.org/10.1016/j.bbamcr.2012.01.014
12. Gkamprela E, Deutsch M, Pectasides D (2017) Iron deficiency anemia in chronic liver disease: etiopathogenesis, diagnosis and treatment. Ann Gastroenterol 30(4):405. https://doi.org/10.20524/AOG.2017.0152
13. Gozzelino R, Arosio P (2016) Iron homeostasis in health and disease. Int J Mol Sci 17(1):2–14. https://doi.org/10.3390/ijms17010130
14. Kaur T, Upadhyay J, Nandave M, Alsayari A, Alshehri SA, Pukale S, Wahab S, Ahmad W, Rashid S, Ansari MN (2024) Exploring progress in iron supplement formulation approaches for treating iron deficiency anemia through bibliometric and thematic analysis. Heliyon 10(7). https://doi.org/10.1016/j.heliyon.2024.e29058
15. Kaur T, Upadhyay J, Pukale S, Mathur A, Ansari MN (2022) Investigation of trends in the research on transferrin receptor-mediated drug delivery via a bibliometric and thematic analysis. Pharmaceutics 14(12). https://doi.org/10.3390/PHARMACEUTICS14122574

16. Peyssonnaux C, Nizet V, Johnson RS (2008) Role of the hypoxia inducible factors in iron metabolism. Cell Cycle 7(1):28–32. https://doi.org/10.4161/CC.7.1.5145;REQUESTEDJOURNAL:JOURNAL:KCCY20;WGROUP:STRING:PUBLICATION
17. Rohr M, Brandenburg V, Brunner-La Rocca HP (2023) How to diagnose iron deficiency in chronic disease: a review of current methods and potential marker for the outcome. Eur J Med Res 28(1):1–12. https://doi.org/10.1186/S40001-022-00922-6/METRICS
18. Ru Q, Li Y, Chen L, Wu Y, Min J, Wang F (2024) Iron homeostasis and ferroptosis in human diseases: mechanisms and therapeutic prospects. Signal Transduct Target Ther 9(1):1–64. https://doi.org/10.1038/s41392-024-01969-z
19. Saini M, Trehan K, Thakur S, Modi A, Jain SK (2024) Advances in iron deficiency anaemia management: exploring novel drug delivery systems and future perspectives. Curr Drug Deliv 21. https://doi.org/10.2174/0115672018300804240426070552
20. Sawicki KT, De Jesus A, Ardehali H (2023) Iron metabolism in cardiovascular disease: physiology, mechanisms, and therapeutic targets. Circ Res 132(3):379–396. https://doi.org/10.1161/CIRCRESAHA.122.321667/ASSET/34932E06-A5F2-4EAC-9A46-E69D043F76D1/ASSETS/GRAPHIC/CIRCRESAHA.122.321667.FIG03.JPG
21. Shubham K, Anukiruthika T, Dutta S, Kashyap AV, Moses JA, Anandharamakrishnan C (2020) Iron deficiency anemia: a comprehensive review on iron absorption, bioavailability and emerging food fortification approaches. Trends Food Sci Technol 99:58–75. https://doi.org/10.1016/J.TIFS.2020.02.021
22. Sousa L, Oliveira MM, Pessôa MTC, Barbosa LA (2020) Iron overload: effects on cellular biochemistry. Clin Chim Acta 504:180–189. https://doi.org/10.1016/J.CCA.2019.11.029
23. Surico G, Muggeo P, Muggeo V, Lucarelli A, Martucci T, Daniele RM, Rigillo N (2002) Parenteral iron supplementation for the treatment of iron deficiency anemia in children. Ann Hematol 81(3):154–157. https://doi.org/10.1007/S00277-001-0418-3/METRICS
24. Townsend A, Drakesmith H (2002) Role of HFE in iron metabolism, hereditary haemochromatosis, anaemia of chronic disease, and secondary iron overload. Lancet 359(9308):786–790. https://doi.org/10.1016/S0140-6736(02)07885-6
25. Weir MR (2021) Managing anemia across the stages of kidney disease in those hyporesponsive to erythropoiesis-stimulating agents. Am J Nephrol 52(6):450–466. https://doi.org/10.1159/000516901
26. Weiss G, Goodnough LT (2005) Anemia of chronic disease. N Engl J Med 352(10):1011–1023. https://doi.org/10.1056/NEJMRA041809
27. Wessling-Resnick M (2000) Iron transport. Annu Rev Nutr 20:129–151. https://doi.org/10.1146/ANNUREV.NUTR.20.1.129/CITE/REFWORKS
28. Wians FH, Urban JE, Keffer JH, Kroft SH (2001) Discriminating between iron deficiency anemia and anemia of chronic disease using traditional indices of iron status vs transferrin receptor concentration. Am J Clin Pathol 115(1):112–118. https://doi.org/10.1309/6L34-V3AR-DW39-DH30
29. Wiciński M, Liczner G, Cadelski K, Kołnierzak T, Nowaczewska M, Malinowski B (2020) Anemia of chronic diseases: wider diagnostics—better treatment? Nutrients 12(6):1784. https://doi.org/10.3390/NU12061784
30. Zeidan RS, Martenson M, Tamargo JA, McLaren C, Ezzati A, Lin Y, Yang JJ, Yoon HS, McElroy T, Collins JF, Leeuwenburgh C, Mankowski RT, Anton S (2024) Iron homeostasis in older adults: balancing nutritional requirements and health risks. J Nutr Health Aging 28(5):100212. https://doi.org/10.1016/J.JNHA.2024.100212
31. Zhang DL, Ghosh MC, Rouault TA (2014) The physiological functions of iron regulatory proteins in iron homeostasis - an update. Front Pharmacol 5:92341. https://doi.org/10.3389/FPHAR.2014.00124/REFERENCE

Chapter 18
The Involvement of Copper Deficiency in the Pathogenesis of Nonalcoholic Fatty Liver Disease

Shawn D. George, Elle Garriock, Paramjit S. Tappia, and Bram Ramjiawan

Abstract Nonalcoholic fatty liver disease (NAFLD) is a type of chronic liver disease that occurs due to lipid accumulation in the hepatocytes. The prevalence of NAFLD is increasing in Middle Eastern and Western nations that have been attributed to the increases in obesity, diabetes, and type 2 diabetes. While dietary factors are implicated in the pathogenesis of NAFLD, the deficiency of copper has emerged as an important contributor in the pathophysiology of NAFLD. Accordingly, in this chapter, we discuss the mechanisms involved in the development of NAFLD and how copper status has a role to play. In addition, the complex interplay between copper status, oxidative stress, inflammation, and fibrosis in the development of NAFLD is described. Furthermore, some novel interventions for the treatment of NAFLD are also discussed.

Keywords Copper deficiency · Nonalcoholic fatty liver disease · Metabolic disorders · Oxidative stress · Inflammation

S. D. George · E. Garriock · P. S. Tappia (✉)
Asper Clinical Research Institute and St. Boniface Hospital Albrechtsen Research Centre, Winnipeg, MB, Canada
e-mail: ptappia@sbrc.ca

B. Ramjiawan
Department of Pharmacology and Therapeutics, Max Rady College of Medicine, University of Manitoba, Winnipeg, Canada

N. S. Dhalla et al. (eds.), *Functional Biochemistry of Metallic Elements*, Advances in Biochemistry in Health and Disease 36,
https://doi.org/10.1007/978-3-032-14445-4_18

Introduction

The essential transition metal copper is important in lipid metabolism, redox balance, iron mobilization, and many other critical processes in eukaryotic organisms [1, 2]. The richest sources of copper are found in shellfish, nuts, seeds, legumes, grains, liver, and organ meats. Most fruits and vegetables have intermediate amounts of copper, whereas chicken, many fish, and dairy products contain low amounts. The daily dietary recommended intake is 1.5–3 mg for adults and 0.4–0.6 mg for infants rising to 1.5–2.5 mg for children. Copper is absorbed primarily in the small intestine, with a small amount being absorbed in the stomach. Absorption is thought to occur via a saturable active transport mechanism at low dietary concentrations; however, passive diffusion may occur at high dietary concentrations. Zinc can influence copper absorption. For example, when dietary zinc is high, the amount of copper absorbed is impaired. This is because zinc induces intestinal metallothionein (MT) which traps copper. Following absorption, copper is transported bound primarily to albumin and transcuprein. Most is taken up by the liver and some by the kidney. Once in the liver, copper is incorporated into ceruloplasmin and released into the circulation and delivered to cells. While copper status has been determined in blood and hair, an in vivo bioluminescence visualization imaging technique has been developed to measure cell and tissue copper contents that could have some clinical application [3].

Adult body contains about 50–120 mg of copper, stored in the liver bound to MT-like proteins. The primary route of copper excretion is via bile into the gastrointestinal tract. Healthy individuals excrete only 10–30 μg/day. A low-copper diet increases the efficiency of absorption rapidly. Humans increase endogenous copper excretion when the diet is high in copper and excrete little during copper deficiency or when dietary copper is low. A normal range for serum copper levels is 64–156 μg/dL, and for serum ceruloplasmin, it is 18–40 mg/dL. Functional tests have also been employed to assess copper status that involve measuring the erythrocyte superoxide dismutase (SOD) for which a normal range is recommended at 0.47 ± 0.067 mg/g. Figure 18.1 depicts the absorption, distribution, and excretion of copper [4].

Recent studies suggest that overt copper deficiency is more common than once thought, and further, some have suggested that the copper RDAs for adults may be too low. Additional human balance and interventional studies are necessary and could provide the impetus for reconsidering the copper RDAs in the future [5]. The essential trace mineral copper plays important roles in human physiology and pathophysiology. Disruption of copper homeostasis may underlie the development of ischemic heart disease and connective tissue and neurodegenerative disorders. Copper also likely participates in the host response to bacterial infection and is further implicated more broadly in regulating immunity. A deficiency of copper frequently causes these metabolic disorders such as impaired glucose tolerance and dyslipidemia.

The liver largely regulates most of the metabolism of trace elements, and accordingly, an impairment of liver functions can result in numerous metabolic disorders

Fig. 18.1 Schematic representation of copper metabolism. (Adapted from Turnlund [4])

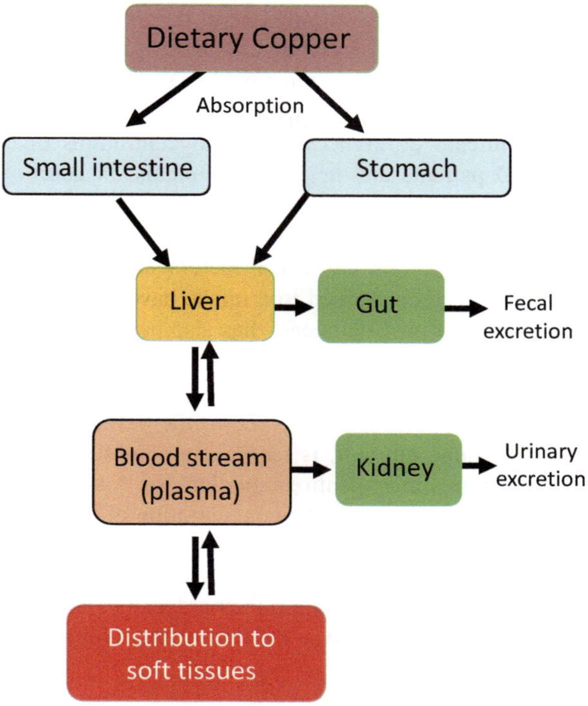

[6, 7]. Furthermore, copper deficiency has also been linked to increases in hepatic iron stores due to decreased expression of ferroportin-1 and decreased ceruloplasmin ferroxidase activity and thus inhibiting liver iron export [8]. Recent studies further associate copper with disruption of lipid homeostasis, as is frequently seen in, for example, nonalcoholic fatty liver disease (NAFLD) [5]. In fact, copper deficiency has been linked to NAFLD [8]. Accordingly, this chapter is intended to describe the pathogenesis of NAFLD and describe some of the mechanisms that may be responsible for the copper deficiency-induced NAFLD. In addition, novel therapeutic approaches for NAFLD related to a copper-deficient state are also discussed.

Copper Deficiency and Nonalcoholic Fatty Liver Disease

NAFLD is a common liver disease that is characterized by an increase in fat in liver cells. The outbreak of NAFLD is estimated to be 32.4% worldwide, with higher rates in Asia and Iran. In fact, a significant portion of the North American population has been suggested to consume insufficient copper [1] with an estimated 30% of the US population affected by NAFLD [9]. Of note, NAFLD in adolescents and children is rapidly becoming one of the most common causes of chronic liver disease worldwide. Globally, metabolic dysfunction-associated steatotic liver disease (MASLD), previously termed NAFLD, is one of the most common liver disorders and is strongly associated with copper deficiency [10]. On the other hand, there are no significant differences in the concentrations of copper and ceruloplasmin in NAFLD patients and healthy individuals; in addition, no association was found between serum copper levels and NAFLD, and no association between copper deficiency and NAFLD risk [11].

Although several nutritional factors such as excessive calorie intake, high fructose intake, and increased iron intake have been reported to play an important role in NAFLD, copper deficiency has also been suggested to have a critical role in the development of NAFLD [12]. Indeed, the development of hepatic steatosis and insulin resistance in response to dietary copper restriction in rats suggests that copper availability may be involved in the development of NAFLD [13]. Copper, an essential micronutrient, is required for lipid metabolism, mitochondrial function, iron metabolism, and antioxidant defense. Although most of the studies assess copper status by measuring the concentration of copper in blood or tissues, lower hair copper concentrations in patients with NAFLD have also been reported and that the risk of NAFLD is elevated in individuals with lower hair copper, suggesting that lower hair copper concentration could also be an important noninvasive test for the diagnosis of NAFLD, even after adjustment for metabolic syndrome factors [14].

Copper Deficiency-Induced Nonalcoholic Fatty Liver Disease and Role of Fructose

Diet is a recognized risk factor in the NAFLD spectrum, which includes nonalcoholic steatohepatitis (NASH) and fibrosis. Low hepatic copper was recently linked to clinical NAFLD/NASH severity. Simple sugar consumption including sucrose and fructose is implicated in NAFLD, while consumption of these macronutrients also decreases liver copper levels [9]. PNPLA3, or patatin-like phospholipase domain-containing 3, is a gene involved in lipid metabolism, including the regulation of fat storage and breakdown in fat and liver cells and has been linked to NAFLD. In this regard, PNPLA3 in NAFLD patients with and without metabolic syndrome has been explored [15]. It was found that the lipid content in hepatocytes was inversely correlated with hepatic copper content; this inverse correlation remained significant only in patients without metabolic syndrome. The presence of minor allele (G) of PNPLA3 was also associated with moderate/severe steatosis both in patients with and without metabolic syndrome but was only associated with NASH and liver fibrosis in patients without metabolic syndrome. It was concluded that copper content in the liver and PNPLA3 mutations are associated with NAFLD patients without metabolic syndrome [15].

It is interesting to note that saturated fatty acid intake is higher, but copper and vitamin E intakes are lower in patients with high BMI [16]. In this regard, in ob/ob mice, it has been reported that there occurs a specific lowering of hepatic liver and thus could provide a useful experimental model to study copper regulation in NAFLD and NASH [17]. It should be mentioned that although experimental studies have shown that increased fructose consumption recapitulates many of the pathophysiological characteristics of the metabolic syndrome in humans, which in turn leads to NAFLD [18], many of the experimental studies involve feeding excessively high levels of fructose (60–70% of total energy intake) which is not reflective of average human intake.

Copper improves lipid metabolism via the adenosine 5′-monophosphate-activated protein kinase (AMPK) pathway. As a copper-binding protein, SCO1 (synthesis of cytochrome c oxidase 1) facilitates AMPK and liver kinase B1 (LKB1) binding, which activates the downstream pathway. Copper, as a signaling molecule, improves hepatic lipid catabolism, and it indicates that targeting copper-SCO1-AMPK signaling pathway ameliorates NAFLD development by modulating AMPK activity [19]. Other studies have shown that a copper-deficient diet and/or fructose feeding in rats induces alterations in hepatic lipid profiles (including content, composition, and saturation), especially ceramide, cardiolipin, phosphatidylcholine, and phosphatidylethanolamine, demonstrating that insufficient dietary copper or high dietary fructose intake induces adverse effects on the hepatic lipid profile and that fructose supplement causes a further hepatic injury in copper deficiency-induced NAFLD [20].

Interestingly, nuclear magnetic resonance metabolomic analyses have provided insights into the mechanisms associated with copper modulation of fructose hepatotoxicity through identification of distinct metabolic phenotypes that appear to be highly correlated with diet and sex [21]. Indeed, NAFLD is a common disease with potential sex-based differences in its prevalence and prognosis. In this regard, sex differences in the relationship between copper concentrations in the blood and NAFLD have been explored [22]. In this study, while blood copper concentrations were lower in NAFLD, it was revealed that low copper concentration is a major contributor to a higher risk of NAFLD in male subjects with metabolic syndrome. However, such association was observed between copper and NAFLD [22]. Furthermore, it was observed that higher copper levels are protective against NAFLD in men, but not in women. Thus, it is plausible that sex-specific intervention with copper is potentially useful for the prevention of development of NAFLD [22].

The association of fructose intake with copper is further demonstrated by the observation that high fructose feeding can impair copper status [23]. In this regard, in marginally copper-deficient rats exposed to fructose, hepatic injury and fat accumulation, as evidenced by an elevated plasma aspartate aminotransferase activity and increased hepatic triglyceride content, have been reported [23]. Furthermore, the endogenous antioxidants in the liver were depressed and lipid peroxidation increased. On the other hand, in this model, the expression levels of hepatic carnitine palmitoyl-CoA transferase I was downregulated, whereas hepatic fatty acid synthase was markedly upregulated. It is also interesting to note that the expression of duodenum Ctr-1 (a copper transporter) was increased in marginal copper deficiency, whereas this increase was prevented by fructose feeding. It was thus suggested that high fructose-induced NAFLD may be partially attributed to inadequate copper in the diet and that the impairment of duodenum Ctr-1 expression due to high fructose feeding may result in a reduction in the absorption of copper and thereby cause copper deficiency [23].

Copper Deficiency-Related Oxidative Stress, Inflammation, and Fibrosis in NAFLD

Oxidative stress is known to have a pathogenic role in NAFLD. In this regard, an important endogenous antioxidant enzyme, Cu/Zn superoxide dismutase (SOD), which depends on the availability of adequate amounts of copper, exhibits reduced activity in patients with NAFLD [24]. Interestingly, an increase in the dietary intake of fructose consumption and inadequate copper intake are two critical risk factors in the development of NAFLD. Mass spectrometry of rat fecal samples has revealed dissimilar fecal metabolome profiles associated with the dietary high fructose and copper level, with some metabolites having the potential to serve as biomarkers of fructose-induced NAFLD [25]. It should be mentioned serum antioxidant capacity owing to ceruloplasmin failure is strongly associated with NAFLD-related damage; however, its precise role in the pathogenesis of NAFLD remains to be established as well as its potential as a diagnostic marker for NAFLD [26].

Copper/zinc SOD1 is a primary antioxidative enzyme that scavenges superoxide anion radicals. Since fatty livers are known to develop in SOD1 knockout mice, the relationship with the development of fibrosis has recently been explored [27]. In this regard, an increase in hepatic collagen accumulation in SOD1 knockout mice has been reported. Furthermore, increases in the hepatic expression levels of HSP47, a chaperone of collagen, and TIMP1, a tissue inhibitor of matrix metalloproteinases (a collagen degrading enzyme), were also observed in SOD1 knockout mice [27]. It was suggested that collagen synthesis is increased in the liver of SOD1 knockout mice; its degradation is inhibited; in addition, altered liver histology of the SOD1 knockout mice attributed to advanced glycation end products (AGEs) was also reported. Thus, hepatic collagen modified with AGEs was seen to be more resistant to the action of collagen degrading enzymes. Taken together, oxidative stress plays an important role in the progression of liver fibrosis [27].

The relationship between fructose-enriched diet and the development of NAFLD has been examined [28]. NAFLD development in fructose-enriched diet in rats is associated with a decrease in hepatic copper concentrations, linked to a decrease in the hepatic superoxide dismutase (SOD) 1 expression; interestingly, treatment with bezafibrate, an antilipemic agent that lowers cholesterol and triglycerides, was shown to increase hepatic copper concentrations and recover hepatic SOD1 expression [28]. SLC25A3 is a gene in humans that encodes a protein called the mitochondrial phosphate carrier. This protein is responsible for transporting inorganic phosphate (Pi) into the mitochondrial matrix, which is crucial for the synthesis of ATP. It should be noted that mitochondrial dysfunction causes oxidative stress, triggering hepatocyte death and inflammation. A reduced expression of SLC25A3 in fatty liver contributes to electron leak from mitochondria by limiting copper availability, rendering hepatocytes more susceptible to oxidative stress. This study provides evidence that SLC25A3 is a novel risk factor for developing NASH [29].

The transcription factor NF-κB plays a critical role in the inflammatory response, and it has been implicated in various diseases, including NAFLD. Although the transient activation of NF-κB is protective of tissues from stress, a prolonged NF-κB activation can exert detrimental effects on tissue homeostasis. Copper Metabolism MURR1 Domain-containing 1 (Commd1), a protein with functions in multiple pathways, has been shown to suppress NF-κB activity. The cell-type-specific contribution of Commd1 to liver inflammation has been explored in hepatocyte and myeloid-specific Commd1-deficient mice [30]. It was demonstrated that although depletion of hepatocyte Commd1 increased NF-κB transactivation subunit p65 (RelA), there was no deterioration of liver inflammation. In difference, deficiency of myeloid Commd1 exacerbated diet-induced liver inflammation. Both hepatic and myeloid Commd1 deficiency in the mice increased hepatic lipid accumulation. While the increase in proinflammatory cytokines in myeloid Commd1-deficient mice may be linked to the increased level of steatosis, this increase was not observed in hepatocyte Commd1-deficient mice despite the increase in hepatic lipid content. It was thus suggested that a cell-type-specific role for Commd1 in suppressing liver inflammation and in the progression of NAFLD may exist [30].

The hepatic expression of genes involved in inflammation and fibrogenesis, including hepatic stellate cell activation, have been examined in a rat model of NAFLD induced by dietary copper deficiency, human-relevant high sucrose intake (30% w/w), or a combination of both. It was revealed that rats fed with either high-sucrose or low-copper diet had increased hepatic expression of genes involved in inflammation and fibrogenesis, including hepatic stellate cell activation, while the combination of diet factors also increased ATP citrate lyase and fatty acid synthase gene transcription [9]. As expected, low dietary copper decreased hepatic and serum copper concentrations and was observed to increase lipid peroxidation and induce NAFLD-like histopathology, while the combination of high sucrose and low copper in the diet resulted in insulin resistance and liver damage. It was thus suggested that low copper and high sucrose in the diet promotes hepatic gene expression and inflammatory/physiological responses linked to NAFLD and NASH, even in the absence of obesity or severe steatosis [9]. Figure 18.2 summarizes the major factors that cause copper deficiency leading to NAFLD and subsequent increased risk for different metabolic diseases.

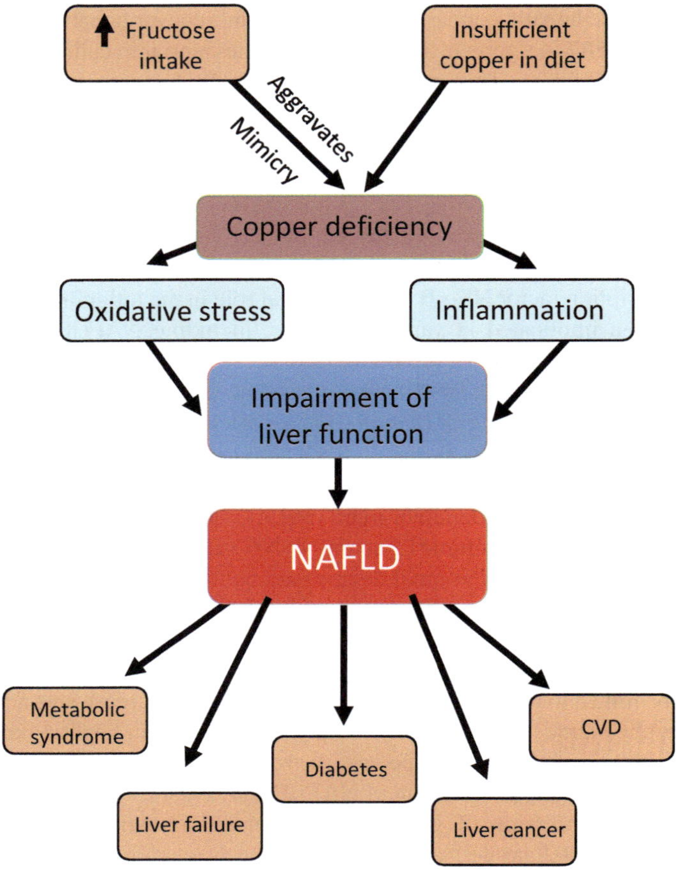

Fig. 18.2 Major aspects of nonalcoholic fatty liver disease and its consequences

Therapeutic Options for NAFLD Due to Copper Deficiency

From the foregoing discussion, it is evident that copper deficiency has been linked to high fructose consumption and thus contribute to the development of NAFLD [20, 23, 31]; thus, restricting fructose intake would be seen to be an appropriate approach. Furthermore, measures undertaken to address copper deficiency through dietary changes or supplementation would also be highly beneficial. In this regard, inclusion of copper-rich foods in the diet including shellfish, organ meats, oysters, nuts, seeds, and beans or copper supplements would be helpful [32]. Interestingly, intermittent fasting (IF) has showed potential positive effects on reducing oxidative stress in NAFLD as well as improved lipid metabolism and liver enzymes [12].

However, while lifestyle modifications (weight management, physical activity, managing other conditions, i.e., diabetes, dyslipidemia) for NAFLD have also been suggested, novel interventions for copper deficiency-induced NAFLD have been explored. For example, biofriendly copper ionophores (HQF) have been developed by constructing pseudonatural flavonols, which exhibit flavonoid bioactivity and enhanced copper transport properties [33]. It was observed that HQF-mediated copper delivery synergistically, safely, and efficiently intervened in the development of fatty liver in experimental models. It is possible that such synergistic copper loading and flavonoid activity for NAFLD may be of benefit in the clinical management of liver disease [33]. *Lactiplantibacillus plantarum* (LPJZ-658) has been shown to be effective in modulating intestinal flora that correlated with differential hepatic lipid species and serum metabolites as well as alleviating copper deficiency combined with sugar-induced middle-aged MASLD and thus may be a possible treatment strategy for improving MASLD [10]. Zinc and antioxidant vitamins have been successfully introduced in the therapy of nonalcoholic liver disease, with positive effects both on insulin resistance and oxidative stress [34, 35].

Conclusion

Metabolic abnormalities due to an increase in the hepatic biosynthesis of fatty acids coupled with an impairment of hepatocyte apolipoprotein secretion or fatty acid β-oxidation result in the accumulation of fat in the liver cells. Indeed, NAFLD is now being termed as a metabolic dysfunction-associated steatotic liver disease (MAFLD). From the aforementioned, it is evident that there is a connection between fructose intake and copper deficiency; indeed, high fructose consumption aggravates copper deficiency-induced NAFLD. This relationship should be an important public health concern given that the growing epidemic of high fructose consumption has been linked to increased prevalence of NAFLD [36]. Moreover, an important factor in the pathophysiology of NAFLD is copper deficiency and that there is a complex interplay between copper status, oxidative stress, inflammation, fibrosis, and development of NAFLD. The progression of NAFLD is augmented in the presence of comorbidities, obesity, type 2 diabetes, dyslipidemia, and insulin resistance. With the observations that consumption of insufficient copper in the diet is paralleled with the increases in NAFLD, it is suggestive of an important role of copper in the pathogenesis of this type of chronic liver disease. Thus, adequate dietary copper intake can be beneficial in managing NAFLD and that maintaining sufficient copper levels can help protect against the disease. Finally, it should be mentioned that a move is currently underway to change the name nonalcoholic fatty liver disease to metabolic dysfunction-associated steatotic liver disease (MASLD), which, given the discussions in this article, would be highly appropriate.

Acknowledgments We thank the St. Boniface Hospital Albrechtsen Research Centre for providing the infrastructural support for this project.

References

1. Morrell A, Tallino S, Yu L, Burkhead JL (2017) The role of insufficient copper in lipid synthesis and fatty-liver disease. IUBMB Life 69(4):263–270
2. Tsang T, Davis CI, Brady DC (2021) Copper biology. Curr Biol 31(9):R421–R427
3. Heffern MC, Park HM, Au-Yeung HY et al (2016) In vivo bioluminescence imaging reveals copper deficiency in a murine model of nonalcoholic fatty liver disease. Proc Natl Acad Sci 113(50):14219–14224
4. Turnlund JR (1999) Copper. In: Shils ME, Olson JA, Shike M, Ross AC (eds) Modern nutrition in health and disease, 9th edn. Williams & Wilkins, Baltimore, pp 241–252
5. Collins JF (2021) Copper nutrition and biochemistry and human (patho)physiology. Adv Food Nutr Res 96:311–364
6. Himoto T, Masaki T (2020) Current trends of essential trace elements in patients with chronic liver diseases. Nutrients 12(7):2084
7. Yu L, Liou IW, Biggins SW et al (2019) Copper deficiency in liver diseases: a case series and pathophysiological considerations. Hepatol Commun 3(8):1159–1165
8. Aigner E, Theurl I, Haufe H et al (2008) Copper availability contributes to iron perturbations in human nonalcoholic fatty liver disease. Gastroenterology 135(2):680–688
9. Tallino S, Duffy M, Ralle M et al (2015) Nutrigenomics analysis reveals that copper deficiency and dietary sucrose up-regulate inflammation, fibrosis and lipogenic pathways in a mature rat model of nonalcoholic fatty liver disease. J Nutr Biochem 26(10):996–1006
10. Li C, Liu Z, Wei W et al (2024) Exploring the regulatory effect of LPJZ-658 on copper deficiency combined with sugar-induced MASLD in middle-aged mice based on multi-omics analysis. Nutrients 16(13):2010
11. Arefhosseini S, Pouretedal Z, Tutunchi H, Ebrahimi-Mameghani M (2022) Serum copper, ceruloplasmin, and their relations to metabolic factors in nonalcoholic fatty liver disease: a cross-sectional study. Eur J Gastroenterol Hepatol 34(4):443–448
12. Karimi M, Nejad MM, Akhgarjand C et al (2023) The effects of intermittent fasting diet in comparison with low-calorie diet on lipid profile, glycemic status, and liver fibrosis in patients with non-alcoholic fatty liver (NAFLD): a study protocol for a randomized controlled clinical trial. BMC Nutr 9(1):145
13. Aigner E, Strasser M, Haufe H et al (2010) A role for low hepatic copper concentrations in nonalcoholic fatty liver disease. Am J Gastroenterol 105(9):1978–1985
14. Lee SH, Kim MJ, Kim YS et al (2018) Low hair copper concentration is related to a high risk of nonalcoholic fatty liver disease in adults. J Trace Elem Med Biol 50:28–33
15. Stättermayer AF, Traussnigg S, Aigner E et al (2017) Low hepatic copper content and PNPLA3 polymorphism in non-alcoholic fatty liver disease in patients without metabolic syndrome. J Trace Elem Med Biol 39:100–107
16. Kamada Y, Takahashi H, Ogawa Y et al (2022) Characterization of nutrient intake in biopsy-confirmed NAFLD patients. Nutrients 14(17):3453
17. Church SJ, Begley P, Kureishy N et al (2015) Deficient copper concentrations in dried-defatted hepatic tissue from ob/ob mice: a potential model for study of defective copper regulation in metabolic liver disease. Biochem Biophys Res Commun 460(3):549–554
18. Yilmaz Y (2012) Review article: fructose in non-alcoholic fatty liver disease. Aliment Pharmacol Ther 35(10):1135–1144
19. Xie L, Yuan Y, Xu S et al (2022) Downregulation of hepatic ceruloplasmin ameliorates NAFLD via SCO1-AMPK-LKB1 complex. Cell Rep 41(3):111498
20. Jiang X, Hu R, Huang Y et al (2023) Fructose aggravates copper-deficiency-induced non-alcoholic fatty liver disease. J Nutr Biochem 119:109402
21. Morrell A, Tripet BP, Eilers BJ et al (2020) Copper modulates sex-specific fructose hepatotoxicity in nonalcoholic fatty liver disease (NALFD) Wistar rat models. J Nutr Biochem 78:108316
22. Lan Y, Wu S, Wang Y et al (2021) Association between blood copper and nonalcoholic fatty liver disease according to sex. Clin Nutr 40(4):2045–2052

23. Song M, Schuschke DA, Zhou Z et al (2012) High fructose feeding induces copper deficiency in Sprague-Dawley rats: a novel mechanism for obesity related fatty liver. J Hepatol 56(2):433–440
24. Nunes VS, Andrade AR, Guedes ALV et al (2020) Distinct phenotype of non-alcoholic fatty liver disease in patients with low levels of free copper and of ceruloplasmin. Arq Gastroenterol 57(3):249–253
25. Wei X, Song M, Yin X et al (2015) Effects of dietary different doses of copper and high fructose feeding on rat fecal metabolome. J Proteome Res 14(9):4050–4058
26. Nobili V, Siotto M, Bedogni G et al (2013) Levels of serum ceruloplasmin associate with pediatric nonalcoholic fatty liver disease. J Pediatr Gastroenterol Nutr 56(4):370–375
27. Sakiyama H, Fujiwara N, Yoneoka Y et al (2016) Cu,Zn-SOD deficiency induces the accumulation of hepatic collagen. Free Radic Res 50(6):666–677
28. Ackerman Z, Skarzinski G, Grozovski M, Oron-Herman M, Sela BA (2014) Effects of antihypertensive and triglyceride-lowering agents on hepatic copper concentrations in rats with fatty liver disease. Basic Clin Pharmacol Toxicol 115(6):545–551
29. Nakanishi T, Kawasaki Y, Nakamura Y et al (2023) An implication of the mitochondrial carrier SLC25A3 as an oxidative stress modulator in NAFLD. Exp Cell Res 431(1):113740
30. Bartuzi P, Wijshake T, Dekker DC et al (2014) A cell-type-specific role for murine Commd1 in liver inflammation. Biochim Biophys Acta 1842(11):2257–2265
31. Song M, Vos MB, McClain CJ (2018) Copper-fructose interactions: a novel mechanism in the pathogenesis of NAFLD. Nutrients 10(11):1815
32. DiNicolantonio J, Mangan D, O'Keefe J (2018) The fructose–copper connection: added sugars induce fatty liver and insulin resistance via copper deficiency. J Met Health 3(1):1–3
33. Jiang J, Chen G, Zhang W et al (2025) Pseudonatural flavonols as novel copper ionophores for NAFLD intervention via synergistic copper delivery and flavonoid activity. J Med Chem 68(6):6450–6461
34. Coni P, Pichiri G, Lachowicz JI et al (2021) Zinc as a drug for Wilson's disease, non-alcoholic liver disease and COVID-19-related liver injury. Molecules 26(21):6614
35. Raza S, Tewari A, Rajak S, Sinha RA (2021) Vitamins and non-alcoholic fatty liver disease: a molecular insight. Liver Res 5(2):62–71
36. Taskinen MR, Söderlund S, Bogl LH et al (2017) Adverse effects of fructose on cardiometabolic risk factors and hepatic lipid metabolism in subjects with abdominal obesity. J Intern Med 282(2):187–201

Chapter 19
Major Minerals: Cardiometabolic Benefits and Their Role in Promoting Healthy Aging

Setor K. Kunutsor and Khushmanjot Kaur

Abstract Major minerals—also known as macrominerals—including calcium, phosphorus, potassium, sodium, magnesium, sulfur, and chloride are essential for physiological homeostasis and disease prevention. Emerging evidence highlights their protective roles against cardiometabolic diseases, yet their contribution to promoting healthy aging and enhancing longevity is not fully understood. This review provides a comprehensive synthesis of observational and interventional evidence on the associations between dietary intake, blood levels, and supplementation of major minerals and cardiometabolic outcomes. Collectively, the findings show that sodium reduction and potassium supplementation yield the most robust effects in lowering blood pressure, while higher intakes of calcium and magnesium are inversely associated with the risks of hypertension, metabolic syndrome, type 2 diabetes, chronic kidney disease, stroke, and all-cause mortality. Elevated phosphorus intake and serum phosphorus concentrations are consistently linked with adverse outcomes, whereas emerging data suggest that higher serum chloride may be protective. Major minerals exert their cardiometabolic benefits through multiple biological pathways, including regulation of vascular tone, insulin sensitivity, inflammation, oxidative stress, lipid metabolism, and mitochondrial function. These effects are closely

S. K. Kunutsor (✉)
Section of Cardiology, Department of Internal Medicine, Max Rady College of Medicine, Rady Faculty of Health Sciences, University of Manitoba, Winnipeg, MB, Canada

Cardiovascular Real-World Evidence (CARDIO-RWE) Research Group, Section of Cardiology, I.H. Asper Institute, St. Boniface Hospital, Winnipeg, MB, Canada

Department of Physiology and Pathophysiology, Rady Faculty of Health Sciences, University of Manitoba, Winnipeg, MB, Canada
e-mail: skk31@cantab.net

K. Kaur
Section of Cardiology, Department of Internal Medicine, Max Rady College of Medicine, Rady Faculty of Health Sciences, University of Manitoba, Winnipeg, MB, Canada

Cardiovascular Real-World Evidence (CARDIO-RWE) Research Group, Section of Cardiology, I.H. Asper Institute, St. Boniface Hospital, Winnipeg, MB, Canada

N. S. Dhalla et al. (eds.), *Functional Biochemistry of Metallic Elements*, Advances in Biochemistry in Health and Disease 36,
https://doi.org/10.1007/978-3-032-14445-4_19

intertwined with mechanisms underpinning healthy aging, suggesting that optimal mineral balance may preserve physiological function, reduce disease burden, and extend health span. Ensuring adequate and balanced intake of major minerals is therefore critical for preventing cardiometabolic diseases and promoting healthy aging. Future research should prioritize prospective studies on understudied minerals, refine dose-response relationships, incorporate biomarker-based assessments, and conduct intervention trials targeting clinically relevant outcomes to fully elucidate the role of major minerals in longevity and health promotion.

Keywords Major minerals · Calcium · Phosphorus · Potassium · Sodium · Magnesium · Sulfur · Chloride · Cardiometabolic · Cardiovascular disease

Introduction

Unhealthy aging has emerged as a major global challenge, marked by a faster-than-normal decline in physiological and metabolic functions, which markedly increases vulnerability to age-related diseases. Among these, cardiometabolic conditions—such as hypertension, metabolic syndrome (MetS), metabolic dysfunction-associated steatotic liver disease (MASLD), type 2 diabetes (T2D), chronic kidney disease (CKD), and cardiovascular diseases (CVDs)—constitute a significant burden on public health systems. Hypertension alone affects over one billion people worldwide and is a key contributor to the development of stroke, heart failure (HF), and renal disease [1, 2]. MetS, characterized by a cluster of abnormalities including abdominal obesity, insulin resistance, dyslipidemia, and elevated blood pressure, greatly increases the risk of T2D and CVDs [3–6]. MASLD, previously termed nonalcoholic fatty liver disease (NAFLD), has become a leading cause of chronic liver disease and is intimately tied to insulin resistance and obesity [7]. T2D, which impacts nearly 500 million adults globally, is defined by persistent hyperglycemia resulting from insulin resistance and/or impaired insulin secretion, leading to serious vascular complications [8]. CKD, characterized by a gradual loss of kidney function, often occurs alongside CVD and MetS, sharing common risk factors such as hypertension and diabetes [9–11]. CVDs, including coronary heart disease (CHD), stroke, and HF, continue to be the primary cause of death globally, accounting for approximately 18 million deaths each year [12, 13]. Together, these interconnected diseases significantly elevate rates of morbidity, mortality, and healthcare expenditure, highlighting the urgent need for preventive interventions annually.

The rising incidence of unhealthy aging and cardiometabolic conditions is strongly influenced by modifiable lifestyle behaviors such as smoking, sedentary behavior, excessive alcohol intake, and poor diet. Intervening on these risk factors presents a vital strategy for curbing the global burden of age-associated diseases and enhancing longevity and health span. In particular, dietary interventions have been extensively studied for their protective role against cardiometabolic risk. Consistent epidemiological and clinical evidence demonstrates that diets abundant in whole grains, fruits, vegetables, nuts, and lean proteins—and limited in processed foods, refined sugars, and trans fats—are associated with reduced inflammation, improved lipid profiles, better insulin sensitivity, and lower blood pressure profiles [14–22]. Additionally, specific dietary components such as fiber, polyphenols, omega-3 fatty acids, and various micronutrients have been implicated in mediating these beneficial effects [23–26].

Within this dietary framework, major minerals—also known as macrominerals—play fundamental roles in physiological homeostasis and disease prevention [27]. These include calcium (Ca), phosphorus (P), potassium (K), sodium (Na), magnesium (Mg), sulfur (S), and chloride (Cl), each contributing to vital metabolic processes such as bone health, electrolyte balance, enzyme function, vascular tone regulation, and cellular signaling [28]. Emerging evidence suggests that these minerals may exert protective effects against cardiometabolic diseases. Although many studies have examined the associations between individual macrominerals and adverse cardiometabolic outcomes, the findings have often been inconsistent [29–35], highlighting the need for a comprehensive review that evaluates all macrominerals collectively to better clarify their potential benefits and implications. Moreover, whether these minerals contribute to healthy aging and improved longevity remains an open question, particularly in the context of growing life expectancy and the rising burden of age-related metabolic disorders. Considering the substantial public health burden posed by adverse cardiometabolic outcomes and the impact of unhealthy aging, this review aims to provide a comprehensive summary of the observational and interventional evidence on the role of the seven major minerals (blood levels, dietary intake, and supplementation) in cardiometabolic disease outcomes. Furthermore, it will explore the underlying biological mechanisms through which these macronutrients may exert their effects, discuss their potential role in promoting healthy aging and longevity, and highlight health, clinical, and policy implications. Finally, this review will identify gaps in the current evidence base and propose future research directions to advance understanding in this critical area of nutrition and public health.

Major Minerals: Roles, Requirements, and Dietary Sources

Major minerals are essential inorganic nutrients required by the human body in relatively large amounts—typically more than 100 mg per day—to maintain physiological homeostasis. They play essential roles in bone health, fluid balance, nerve signaling, muscle function, and metabolic regulation [27, 28]. Unlike trace elements (iron, zinc, copper, iodine, selenium, fluoride, chromium, manganese, molybdenum, and cobalt), which are needed in minute quantities, major minerals serve as key structural components and cofactors in numerous biological processes. The seven major minerals—Ca, P, K, Na, Mg, S, and Cl—are integral to various

Table 19.1 Summary table of major minerals

Mineral	Functions	RDA/AI (mg/day)	Dietary sources
Calcium (Ca)	Bone health, muscle function, nerve signaling, blood clotting	1000–1200	Dairy, leafy greens, fortified foods, fish with bones
Phosphorus (P)	Bone formation, ATP production, acid-base balance	700	Meat, dairy, eggs, nuts, whole grains
Potassium (K)	Fluid balance, blood pressure regulation, nerve function	2600–3400	Bananas, potatoes, leafy greens, dairy, legumes
Sodium (Na)	Fluid balance, nerve transmission, muscle function	<2300	Table salt, processed foods, canned soups
Magnesium (Mg)	Enzyme function, muscle relaxation, heart health	310–420	Nuts, seeds, whole grains, leafy greens
Sulfur (S)	Protein synthesis, detoxification, antioxidant function	No RDA	Meat, fish, eggs, onions, cruciferous vegetables
Chloride (Cl)	Fluid balance, acid-base regulation, digestion	2300	Table salt, seaweed, tomatoes

AI adequate intake, *ATP* adenosine triphosphate, *RDA* recommended daily allowance

biochemical pathways, and their adequate intake is vital for maintaining optimal health and preventing chronic diseases [27, 28]. Their roles, requirements, and dietary sources are described below and summarized in Table 19.1.

Calcium (Ca) Ca is the most abundant mineral in the human body, with over 99% stored in bones and teeth, where it provides structural strength. The remaining 1% is critical for physiological functions, including muscle contraction, nerve transmission, blood clotting, and enzymatic activity. Ca homeostasis is tightly regulated by parathyroid hormone (PTH), vitamin D, and calcitonin, ensuring stable serum levels essential for metabolic function [36]. The recommended dietary allowance (RDA) for Ca varies by age and sex, with adults generally requiring 1000–1200 mg per day. Insufficient intake is associated with reduced bone mineral density, increasing the risk of osteoporosis and fractures, particularly in older adults. Conversely, excessive Ca intake may contribute to hypercalcemia and kidney stone formation. Dietary sources of Ca include dairy products such as milk, cheese, and yogurt, as well as plant-based sources like leafy greens (e.g., kale, bok choy), fortified plant-based milks, almonds, and fish with edible bones (e.g., sardines and salmon) [37].

Phosphorus (P) P, the second most abundant mineral in the body, is a primary component of bones and teeth, forming hydroxyapatite with Ca [38]. Beyond structural roles, P is integral to adenosine triphosphate (ATP) production, cellular energy metabolism, and the regulation of acid-base balance. It is also a component of DNA, RNA, and phospholipids essential for cell membrane integrity.

The RDA for P is approximately 700 mg per day for adults [39]. Although P deficiency is rare due to its widespread presence in food, excessive intake—often from processed foods rich in phosphates—may contribute to an imbalance in Ca metabolism and increase the risk of CVD, particularly in individuals with

CKD. Phosphorus-rich foods include dairy products, meat, fish, poultry, eggs, nuts, seeds, and whole grains [39]. Many processed foods and soft drinks contain added phosphates, which can contribute to excessive intake.

Potassium (K) K is an essential electrolyte that maintains intracellular fluid balance, supports nerve transmission, regulates blood pressure, and aids in muscle contraction [38]. It plays a fundamental role in counteracting the effects of Na, helping to reduce hypertension and the risk of stroke. K also contributes to glucose metabolism and insulin sensitivity, highlighting its importance in the prevention of metabolic diseases. The adequate intake (AI) for K is approximately 2600–3400 mg per day for adults, with higher intakes linked to reduced cardiovascular risk [40]. However, many populations consume suboptimal levels due to low fruit and vegetable intake. K deficiency (hypokalemia) can result in muscle weakness, irregular heart rhythms, and increased blood pressure, while excessive K intake (hyperkalemia) may pose cardiac risks for individuals with impaired kidney function [38, 40]. K is naturally abundant in fruits and vegetables, particularly bananas, oranges, potatoes, spinach, avocados, tomatoes, and legumes. Dairy products, nuts, and fish also contribute to dietary K intake [40].

Sodium (Na) Na is an essential electrolyte that regulates extracellular fluid balance, nerve function, and muscle contractions [38]. It plays a pivotal role in maintaining blood pressure and osmotic equilibrium. However, excessive Na intake is a major public health concern as it is strongly linked to hypertension, CVDs, and kidney dysfunction. The recommended intake for Na is less than 2300 mg per day, with an ideal target of 1500 mg for optimal cardiovascular health [41]. However, the global average sodium intake often exceeds this level, primarily due to the high consumption of processed and fast foods. While Na deficiency (hyponatremia) is uncommon, it can occur in cases of excessive sweating, dehydration, or certain medical conditions. Dietary sources of Na include table salt (sodium chloride), processed foods, canned soups, deli meats, snack foods, and condiments such as soy sauce. Naturally occurring Na is found in small amounts in dairy products, seafood, and vegetables [42].

Magnesium (Mg) Mg is involved in over 300 enzymatic reactions, including those related to energy metabolism, muscle function, nerve signaling, and DNA synthesis [38, 43]. It plays a critical role in bone health, cardiovascular function, and insulin sensitivity. Magnesium also has anti-inflammatory properties and has been implicated in reducing the risk of T2D and MetS. The RDA for magnesium is 310–420 mg per day for adults [43]. Mg deficiency is relatively common due to inadequate dietary intake, excessive alcohol consumption, and chronic diseases such as diabetes. Symptoms of deficiency include muscle cramps, fatigue, irregular heart rhythms, and increased blood pressure. Mg-rich foods include nuts (e.g., almonds, cashews), seeds (e.g., pumpkin seeds, flaxseeds), whole grains, dark leafy greens (e.g., spinach, Swiss chard), legumes, and seafood [43].

Sulfur (S) S is a vital component of amino acids (methionine and cysteine), vitamins (biotin and thiamine), and antioxidants (glutathione) [44]. It is essential for protein synthesis, detoxification, and maintaining connective tissue integrity [45]. S is also involved in cellular respiration and energy metabolism. There is no established RDA for S as it is primarily obtained from dietary protein sources. Deficiency is rare in individuals consuming adequate protein, but inadequate intake may impair detoxification pathways and antioxidant function. Sulfur-rich foods include protein-rich sources such as meat, fish, poultry, eggs, dairy products, legumes, onions, garlic, and cruciferous vegetables (e.g., broccoli, Brussels sprouts) [44, 45].

Chloride (Cl) Cl is a key electrolyte involved in fluid balance, acid-base homeostasis, and digestion as it is a major component of gastric hydrochloric acid [38, 46]. It works in concert with Na and K to regulate osmotic pressure and nerve function [46]. The AI for Cl is 2300 mg per day for adults. Cl deficiency is uncommon, as it is abundant in the diet, but severe depletion can occur due to excessive vomiting, diarrhea, or metabolic disorders. Dietary sources of Cl include table salt (sodium chloride), seaweed, tomatoes, olives, and processed foods.

Methods

This review involved a comprehensive search of MEDLINE and Embase databases up to May 2025, focusing on randomized controlled trials (RCTs), non-RCTs, and observational research such as prospective cohort, nested case-control, case-cohort, and retrospective cohort studies. Emphasis was placed on systematic reviews and meta-analyses, based on the hierarchy of evidence [47]. Keywords used in the search included terms for major minerals ("calcium," "phosphorus," "potassium," "sodium," "magnesium," "sulfur," "chloride") as well as terms related to various cardiometabolic outcomes (e.g., "blood pressure," "hypertension," "metabolic syndrome," "MASLD," "NAFLD," "type 2 diabetes," "chronic kidney disease," "cardiovascular disease," "coronary heart disease," "stroke," "heart failure," "atrial fibrillation," "mortality"). The studies included those evaluating blood level, dietary intakes, or supplements of these major minerals. The review was limited to human studies published in English. Cross-sectional study designs were not included because of inability to examine temporal relationships. We extracted and reported risk estimates (effect estimates (ES), relative risks (RR), odds ratios (ORs), hazard ratios (HRs), mean differences and standardized mean differences (SMDs)) for associations demonstrating significant effects. Fully adjusted risk estimates were reported.

Major Minerals and Cardiometabolic Outcomes

Blood Pressure

The regulation of blood pressure is influenced by several dietary and lifestyle factors, among which the intake of specific major minerals—particularly Na, K, Ca, and Mg—plays a critical role. Extensive research, including observational studies and RCTs, has evaluated the effects of these minerals on systolic blood pressure (SBP) and diastolic blood pressure (DBP), leading to consistent insights as well as areas of uncertainty. In a 2002 meta-analysis of 20 RCTs, Jee and colleagues investigated the effects of Mg supplementation on blood pressure and reported modest reductions in SBP (MD = −0.6 mmHg, 95% CI, −2.2 to 1.0) and DBP (MD = −0.8 mmHg, 95% CI, −1.9 to 0.4). However, a dose-response was evident: Per 10 mmol/day increase in Mg intake, reductions of 4.3 mmHg in SBP (95% CI, 6.3 to 2.2) and 2.3 mmHg in DBP (95% CI, 4.9 to 0.0) were observed [48]. In a 2006 Cochrane review, Dickinson and colleagues assessed Mg supplementation across 12 RCTs and found a significant reduction in DBP (MD = −2.2 mmHg, 95% CI, −3.4 to −0.9) with no significant effect on SBP [49]. In a companion Cochrane review of 13 RCTs, the same team reported that Ca supplementation significantly reduced SBP (MD = −2.5 mmHg, 95% CI, −4.5 to −0.6) but had no significant effect on DBP [50]. Contrasting findings were reported by Beyer et al. (2006) in another Cochrane review, which evaluated the effects of combined K, Ca, and Mg supplementation in primary hypertension. The study found no evidence that such combinations led to significant reductions in blood pressure [51]. Similarly, Dickinson et al. (2006) concluded in a separate review that Mg supplements alone had no important effect on blood pressure [52]. In a 2013 systematic review and meta-analysis, Aburto and colleagues evaluated 22 RCTs and found that increased K intake significantly reduced SBP by 3.49 mmHg (95% CI, 1.82 to 5.15) and DBP by 1.96 mmHg (95% CI, 0.86 to 3.06), particularly in hypertensive individuals. Notably, reductions in SBP of up to 7.16 mmHg were observed at higher K intakes (90–120 mmol/day), though without a clear dose-response trend [53]. In a 2015 Cochrane review, Cormick and colleagues assessed 16 trials and confirmed the anti-hypertensive effect of Ca supplementation, with reductions in SBP (MD = −1.43 mmHg, 95% CI, −2.15 to −0.72) and DBP (MD = −0.98 mmHg, 95% CI, −1.46 to −0.50). The effects were more pronounced in individuals younger than 35 years [54]. Two 2017 meta-analyses further affirmed the blood pressure-lowering effect of K. In one meta-analysis involving 23 RCTs, K supplementation for ≥4 weeks led to significant reductions in SBP (MD = −4.25 mmHg, 95% CI, −5.96 to −2.53) and DBP (MD = −2.53 mmHg, 95% CI, −4.05 to −1.02) [55]. Another meta-analysis by Filippini et al. showed comparable effects, with SBP reduced by 4.48 mmHg (95% CI, 3.07 to 5.90) and DBP by 2.96 mmHg (95% CI, 1.10 to 4.82) [56]. In a 2019 systematic review, McClure and colleagues evaluated 14 intervention and 3 prospective cohort studies to examine the association between dietary P and blood pressure. The analysis found no consistent association, highlighting a lack of robust evidence for P's role in blood pressure regulation [57]. In a 2021 Cochrane review, Cormick and colleagues again confirmed that Ca supplementation is effective in reducing blood pressure in normotensive individuals. The review of 20 trials found significant reductions in SBP (MD = −1.37 mmHg, 95% CI, −2.08 to −0.66) and DBP (MD = −1.45 mmHg, 95% CI, −2.23 to −0.67). The effect was stronger in individuals younger than 35 years and those receiving ≥1500 mg/day of Ca [58]. A 2022 systematic review by Lai et al. evaluated 26 RCTs examining Na reduction strategies. The results showed that dietary Na reduction led to meaningful decreases in SBP (mean reduction of 4.51 mmHg, 95% CI, 3.35 to 5.67) and DBP (2.42 mmHg, 95% CI, 1.61 to 3.23), with the greatest effects observed in interventions involving a low-sodium diet with no added Na [59]. Most recently, in 2025, Amer and colleagues conducted a systematic review of 40 RCTs (24 used in the meta-analysis) to assess the roles of Ca and Mg in hypertension treatment. The findings demonstrated that Ca significantly reduced DBP (MD = −2.04 mmHg, 95% CI, −3.39 to −0.69), while Mg supplementation also significantly reduced DBP (MD = −1.64 mmHg, 95% CI, −3.19 to −0.09). Neither mineral had a significant impact on SBP [60].

Together, these findings highlight that Na reduction and K supplementation yield the most robust and consistent effects in lowering blood pressure, followed by moderate evidence supporting Ca and Mg supplementation. The role of P, S, and Cl in blood pressure regulation remains underexplored and inconclusive.

Hypertension

The role of major minerals in the development and prevention of hypertension has been extensively investigated in prospective cohort studies and meta-analyses. These studies have primarily focused on dietary and circulating levels of Ca, Mg, Na, K, and P, examining their associations with the risk of developing hypertension. In a 2006 prospective cohort study using data from the Women's Health Study (WHS), Song et al. investigated dietary Mg intake and incident hypertension among US women aged ≥45 years. The findings showed a modest inverse association between higher Mg intake and the risk of hypertension, though not statistically significant: (RR = 0.93, 95% CI, 0.86–1.02) comparing the upper with the lower quintile [29]. In a follow-up analysis from the WHS, Wang et al. (2008) examined dietary Ca intake and found a graded inverse association between higher Ca intake and risk of hypertension: (RR = 0.87, 95% CI, 0.81–0.93) comparing the highest vs lowest quintile. Interestingly, supplemental Ca was not significantly associated with hypertension risk [30]. In a 2010 analysis combining data from the Atherosclerosis Risk in Communities (ARIC) and the Multi-Ethnic Study of Atherosclerosis (MESA), Alonso et al. examined the impact of P intake. Although total P intake showed no overall association with hypertension risk, P from dairy sources was inversely associated with incident hypertension (HR = 0.86, 95% CI, 0.76–0.97), an effect consistent across sex, race/ethnicity, and renal function [61]. In a 2013 cohort study from the Prevention of Renal and Vascular End-stage Disease (PREVEND) study, Joosten et al. found that urinary Mg excretion, rather than circulating levels, was inversely associated with hypertension risk in a log-linear fashion (HR = 0.79, 95% CI, 0.71–0.88) per 1-unit increment in log-transformed urinary Mg excretion [62]. In a 2015 cohort study of 4523 normotensive Japanese adults, Takase et al. reported that higher Na intake was associated with increased risk of developing hypertension: (HR = 1.25, 95% CI, 1.04–1.50) for higher vs lower Na intake [63]. In a 2017 systematic review and meta-analysis of 10 cohort studies, Han et al. found an inverse association between dietary Mg intake and the risk of hypertension: (RR = 0.92, 95% CI, 0.86–0.98) comparing the highest with the lowest intake and (RR = 0.95, 95% CI, 0.90–1.00) per 100 mg/day increment. A modest inverse association was also observed between serum Mg concentrations and hypertension (RR = 0.91, 95% CI, 0.80–1.02) [64]. In a 2019 dose-response meta-analysis of eight prospective cohort studies, Jayedi and Zargar demonstrated a significant linear inverse association between dietary Ca intake and risk of hypertension: (RR = 0.89, 95% CI, 0.86–0.93) for the highest vs lowest intake and (RR = 0.93, 95% CI, 0.90–0.97) per 500 mg/day increment. These associations were independent of adiposity and intake of other minerals and were consistent across sex, geographic location, and baseline intake levels [65]. In a 2022 dose-response meta-analysis of 11 cohort studies, Filippini et al. examined dietary Na intake and urinary Na excretion in relation to hypertension. Compared to a reference category of 2 g/day Na, the risk of hypertension increased progressively at higher intakes: (RR = 1.04, 95% CI, 0.96–1.13) for 4 g/day and (RR = 1.21, 95% CI, 1.06–1.37) for 6 g/day, confirming a linear positive association [66]. Most recently, in a 2024 study using the Korean Genome and Epidemiology Study (KoGES), Bae et al. reported that Na intake ≥2.0 g/day was associated with increased hypertension risk (OR = 1.10, 95% CI, 1.03–1.18), while K intake was inversely associated with hypertension risk (OR = 0.96, 95% CI, 0.93–0.99) per 1 g/day increase [67].

Collectively, the evidence underscores the importance of Ca, Mg, K, and Na in the pathogenesis of hypertension. Higher intakes of Ca and Mg, particularly from dietary sources, are generally associated with reduced hypertension risk, while increased Na intake is consistently linked with elevated risk. K, though less studied in long-term cohorts, shows promise as a protective factor. Meanwhile, the role of dietary P remains inconclusive, with possible benefits restricted to specific food sources such as dairy.

Metabolic Syndrome

While a substantial portion of the literature to date has been based on cross-sectional and case-control designs [68–73], a growing number of longitudinal cohort studies and meta-analyses provide more evidence of potential associations. In one of the earliest prospective studies, He and colleagues (2006) examined the association between Mg intake and incident MetS in over 4000 American adults free from MetS and diabetes at baseline. Mg intake was inversely associated with the development of MetS, with a 31% lower risk observed among participants in the highest vs lowest quartile of intake (HR = 0.69, 95% CI, 0.52–0.91). These associations remained consistent across sex and racial groups [31]. In a 2010 prospective cohort study of kidney transplant recipients, Noori et al. investigated whether Mg intake was associated with the onset of MetS. Among 160 patients without MetS or diabetes at baseline, no significant association was found between dietary Mg intake and the risk of MetS [74]. Using data from the French Epidemiological Study on the Insulin Resistance Syndrome (DESIR) cohort, Fumeron and colleagues (2011) assessed the impact of dairy consumption and Ca density on 9-year incidence of MetS and glycemic abnormalities. Higher dietary Ca density was significantly associated with a lower risk of MetS, both when defined using the International Diabetes Federation (IDF) criteria (OR = 0.86, 95% CI, 0.78–0.95) and the National Cholesterol Education Program (NCEP) definition (OR = 0.83, 95% CI, 0.75–0.92), comparing the top and bottom quartiles of intake [75]. In a 2017 retrospective cohort study of more than 12,000 individuals without MetS, Baek et al. found no overall association between serum Ca and incident MetS. However, a subgroup analysis revealed that among individuals with central obesity, those in the highest third of serum Ca had a significantly lower risk of developing MetS (HR = 0.83, 95% CI, 0.74–0.94) [76]. In a 2019 cohort study from Taiwan, Jhuang and colleagues investigated the role of serum P in the development of MetS among older adults (≥60 years). Their findings revealed a positive association: For each unit increase in serum P concentration, the risk of MetS increased by 39% (HR = 1.39, 95% CI, 1.11–1.74), suggesting a potential harmful effect of elevated P levels in this age group [33]. Drawing on data from the Multi-Rural Communities Cohort (MRCohort), Woo et al. (2020) evaluated the associations of total, animal-based, and plant-based Ca intake with MetS risk. Total Ca intake was inversely associated with MetS (IRR = 0.62, 95% CI, 0.48–0.81) comparing the highest to lowest quartile. The inverse association remained significant for both animal (IRR = 0.77, 95% CI, 0.62–0.97) and vegetable Ca (IRR = 0.54, 95% CI, 0.40–0.73), indicating that diverse Ca sources may contribute to reduced MetS risk [77]. Most recently, in a 2022 prospective cohort study using data from the China Health and Nutrition Survey (CHNS), Jiao and colleagues reported a nonlinear inverse relationship between dietary Mg intake and the risk of MetS. Compared to the lowest quintile, participants in the highest quintile of Mg intake had a reduced risk of MetS (HR = 0.84, 95% CI, 0.71–0.99), supporting a protective role of dietary Mg [78].

Taken together, these longitudinal findings provide growing evidence that higher dietary intakes of Ca and Mg are inversely associated with the risk of developing MetS. Notably, Ca from both animal and plant sources appears beneficial. While serum Ca may have context-dependent effects—offering some protection in centrally obese individuals—serum P levels, particularly in older adults, may elevate MetS risk. The current evidence base for other minerals such as Na, K, S, and Cl in relation to MetS remains limited in prospective designs, and further research is warranted to clarify their roles in the etiology of this increasingly prevalent condition.

Metabolic Dysfunction-Associated Steatotic Liver Disease (Formerly NAFLD)

Dietary and circulating levels of major minerals—particularly Mg, Ca, Na, and their ratios—have been increasingly studied for their potential roles in the pathogenesis and prevention of NAFLD/MASLD. While much of the current evidence is derived from cross-sectional and case-control studies [79–85], there is emerging evidence based on prospective data. In a 2019 prospective cohort analysis from the PREVEND study, van den Berg and colleagues examined the association between Na intake—measured via 24-hour urinary sodium excretion—and the risk of NAFLD, assessed using the Fatty Liver Index (FLI) and Hepatic Steatosis Index (HSI). The study found that higher Na intake was associated with a significantly increased risk of NAFLD. Specifically, per 1 standard deviation (SD) increase in 24-hour urinary Na, the odds of having FLI $\geq$60 were 1.30 (95% CI, 1.21–1.41), and for HSI >36, the OR was 1.40 (95% CI, 1.31–1.51). These associations were consistent across sexes [86]. In a 2022 longitudinal analysis using data from the Coronary Artery Risk Development in Young Adults (CARDIA) study with 25 years of follow-up, Lu and colleagues assessed the association between cumulative Mg intake and NAFLD. They reported that higher Mg intake throughout adulthood was associated with a significantly lower odds of NAFLD in midlife: (OR = 0.45, 95% CI, 0.23–0.85) comparing the highest with the lowest quintile of intake [32]. In a 2023 case-control study nested within the Kailuan prospective cohort in China, Yang and colleagues explored the associations between blood levels of Ca, Mg, and the calcium-to-magnesium (Ca/Mg) ratio and NAFLD risk. Inverse linear relationships were found for both serum Ca and the Ca/Mg ratio, with significantly higher odds of NAFLD observed in individuals in the lowest quartiles compared to the highest: OR = 2.12 (95% CI, 1.68–2.67) for Ca and OR = 1.36 (95% CI, 1.08–1.71) for Ca/Mg ratio. Interestingly, the association between serum Mg and NAFLD followed a U-shaped curve, suggesting that both low and high levels may be detrimental [87]. In a 2024 prospective cohort study, Lee and colleagues examined the sex-specific associations of Na intake—again estimated by 24-hour urinary sodium excretion—with NAFLD risk in a community-based population. Elevated Na intake was associated with increased NAFLD risk across all analyses: HR = 1.54 (95% CI, 1.24–1.92) for all participants, HR = 1.50 (95% CI, 1.02–2.19) for men, and HR = 1.51 (95% CI, 1.15–1.98) for women, comparing the highest with the lowest tertiles of Na excretion [88].

Together, these findings suggest that higher dietary Na intake is consistently associated with an increased risk of NAFLD/MASLD, with evidence supported by multiple cohort studies using objective urinary Na measures. Conversely, higher Mg intake—especially when maintained over the life course—appears protective. The relationships involving Ca and Mg balance, particularly as captured by the Ca/Mg ratio, are emerging as important factors in NAFLD pathophysiology. The identification of a U-shaped association for Mg also raises considerations for optimal mineral balance rather than absolute intake alone. Further longitudinal research is needed, particularly involving other macrominerals such as P, S, and Cl, which remain understudied in this context.

Type 2 Diabetes

In a 2010 analysis from the ARIC Study, Chatterjee et al. reported that higher serum K levels (4.5–<5.0 mEq/L) were associated with increased risk of incident T2D compared to levels between 5.0 and 5.5 mEq/L (HR = 1.39, 95% CI, 1.14–1.71), while no association was found for dietary K intake [89]. A 2011 follow-up study by Chatterjee et al. using the same ARIC cohort highlighted a racial disparity, with African Americans experiencing an 18% greater risk of T2D associated with serum K levels compared to Whites [90]. In a 2011 dose-response meta-analysis of 13 prospective cohort studies, Dong et al. demonstrated an inverse association between Mg intake and T2D risk: (RR = 0.78, 95% CI, 0.73–0.84) comparing highest vs lowest intake and (RR = 0.86, 95% CI, 0.82–0.89) per 100 mg/day increment in intake [91]. In 2012, a meta-analysis by Dong and Qin found that higher dietary Ca intake may be modestly protective against T2D: (RR = 0.85, 95% CI, 0.75–0.97). However, this association was attenuated when controlling for Mg (RR = 0.94, 95% CI, 0.85–1.05), suggesting potential confounding or mediation [92]. In a 2015 meta-analysis of 15 cohort studies, Xu et al. reported a nonlinear inverse relationship between Mg intake and T2D: (RR = 0.77, 95% CI, 0.71–0.82) comparing highest vs lowest intake and (RR = 0.84, 95% CI, 0.80–0.88) per 100 mg/day increment [93]. A 2016 meta-analysis by Fang et al. incorporating 25 cohort studies found consistent evidence of a linear inverse association between Mg intake and T2D risk: (RR = 0.83, 95% CI, 0.80–0.86), with a risk reduction of 8–13% per 100 mg/day increase in intake [94]. In 2017, Peng et al. conducted a dose-response meta-analysis of eight cohort studies and showed that higher serum K was associated with lower T2D risk (RR = 0.83, 95% CI, 0.73–0.95 per 1 mmol/L increase), though neither dietary nor urinary K was significantly associated [95]. In a 2018 Korean cohort study, Kim et al. observed that higher dietary Ca intake was associated with reduced T2D risk (HR = 0.88, 95% CI, 0.77–1.00 per unit increase in log-transformed Ca intake), while serum Ca showed no association [96]. Also in 2018, Mancini et al. using the E3N-EPIC cohort reported that high dietary P intake was associated with increased T2D risk: HRs of 1.18, 1.41, and 1.54 across ascending P intake categories compared to <1203 mg/day [97]. In 2019, a meta-analysis by Zhu et al. found that higher circulating Ca levels were associated with increased T2D risk: (RR = 1.14, 95% CI, 1.05–1.24) for albumin-adjusted Ca and (RR = 1.25, 95% CI, 1.10–1.42) for total calcium [98]. Also in 2019, Jhuang et al. demonstrated in a Taiwanese cohort that higher serum P was associated with increased T2D risk in older adults (HR = 1.49, 95% CI, 1.15–1.92 per unit increase) [33]. A 2020 pooled analysis by Zhao et al. of 35 cohort studies further confirmed the protective association of Mg: (RR = 0.77, 95% CI, 0.71–0.82) comparing highest vs lowest intake and (RR = 0.84, 95% CI, 0.80–0.88) per 100 mg/day increment [99]. Also in 2020, Hao et al. using the CHNS showed that higher Na intake and Na-K ratio were associated with greater diabetes risk: (RR = 1.35, 95% CI, 1.02–1.80) comparing top vs bottom Na quartile; K alone was not significantly associated [100]. In a 2022 study using the Framingham Heart Study, Dong et al. reported that higher intake of S amino acids was associated with increased T2D risk: (HR = 1.8, 95% CI, 1.3–2.5) comparing highest vs lowest intake [101]. In a 2023 analysis from the UK Biobank, Wang et al. found that more frequent addition of salt to food—used as a surrogate for Na intake—was associated with increased T2D risk: HRs of 1.11, 1.18, and 1.28 for "sometimes," "usually," and "always" compared to "never/rarely" adding salt [102]. Also in 2023, Wu et al. using the CHNS observed a U-shaped association between dietary P intake and diabetes: P <921.6 mg/day (HR = 0.64, 95% CI, 0.48–0.84 per 1 SD increase) showed reduced risk, while ≥921.6 mg/day was associated with increased risk (HR = 1.33, 95% CI, 1.16–1.53) [103]. In a 2024 cohort study, Li et al. from the Henan Rural Cohort showed that higher K and Mg intakes were associated with decreased T2D risk: ORs = 0.67 (K), 0.76 (Mg), and 0.57 (K-Mg ratio) comparing highest vs lowest quartiles [104]. Most recently, in 2025, Park et al. found no significant association between Na or K intake and T2D in a cohort of approximately 100,000 Korean adults [105].

Collectively, the evidence suggests that higher dietary Mg intake is consistently associated with a reduced risk of T2D, supported by several large-scale meta-analyses and cohort studies. Dietary Ca may also offer modest protective effects, although elevated circulating Ca levels have been linked with increased diabetes risk. Findings on P are more complex, with both high dietary intake and elevated serum levels associated with increased risk, and some evidence suggesting a U-shaped relationship. While Na intake—particularly from added salt—is associated with greater diabetes risk, the role of K appears inconsistent, with protective associations seen for serum but not dietary levels. Emerging evidence also indicates that high intake of S-containing amino acids may elevate the risk of T2D. Research on Cl remains limited and warrants further investigation.

Chronic Kidney Disease

A growing number of longitudinal cohort studies have assessed the associations between intakes of these minerals and CKD risk in diverse populations, providing insight into their preventive or harmful effects. In a 2017 prospective cohort study using data from the KoGES, Yoon and colleagues examined the impact of dietary P density on incident CKD in individuals with normal renal function. The study found that higher P density was associated with a significantly increased risk of developing CKD, but only among participants with diabetes mellitus (HR = 1.68, 95% CI, 1.08–2.63 for top vs bottom quartile) [106]. Using the Ansan-Ansung cohort of the KoGES, Lee and colleagues (2021) explored the effects of dietary Ca and P intake on CKD risk. The results indicated that low intakes of both minerals were independently associated with a higher risk of developing more advanced CKD (Stage 3B and above). Specifically, the risk was substantially elevated for those with P intake <400 mg/day (HR = 6.1, 95% CI, 3.22–11.58) and Ca intake <200 mg/day (HR = 2.0, 95% CI, 1.01–3.62), when compared with individuals consuming recommended amounts (700–1200 mg/day for P and ≥600 mg/day for Ca). No associations were observed for Na or K in this cohort [107]. In a 2022 analysis from the Hispanic Community Health Study/Study of Latinos (HCHS/SOL), Swift and colleagues evaluated the impact of dietary sodium and potassium intakes on incident CKD in a multiethnic urban US population. The study found that each 500 mg decrement in K intake was associated with a higher risk of CKD (IRR = 1.11, 95% CI, 1.00–1.24), while each 500 mg increment in Na intake was modestly associated with increased risk (IRR = 1.05, 95% CI, 0.99–1.12). Notably, a higher Na-to-K ratio was strongly associated with CKD risk (IRR = 1.21, 95% CI, 1.02–1.45 per unit increase) [34]. Most recently, in a 2024 prospective analysis using data from the UK Biobank, Koh and colleagues demonstrated a graded inverse association between dietary Mg intake and incident CKD among adults with clinically normal kidney function at baseline. Individuals in the lowest quintile of Mg intake had a significantly higher risk of developing CKD compared with those in the highest quintile (HR = 1.30, 95% CI, 1.20–1.41), suggesting a protective role of adequate Mg intake in renal function preservation [108].

Together, these studies provide compelling evidence that insufficient intake of certain major minerals—particularly Mg, K, Ca, and P—is associated with elevated risk of developing CKD. Conversely, excessive P intake, especially in vulnerable populations such as those with diabetes, may accelerate CKD onset. A high Na-to-K ratio further increases risk, pointing to the importance of maintaining an optimal dietary mineral balance.

Cardiovascular Disease Outcomes

Evidence from large-scale cohort studies and meta-analyses has evaluated the impact of individual minerals—including Na, K, Ca, Mg, P, and their ratios—on a range of cardiovascular endpoints such as composite CVD, CHD, stroke, myocardial infarction (MI), HF, and CVD mortality. The findings are summarized below:

In a post hoc analysis from the Cholesterol and Recurrent Events (CARE) trial, Tonelli and colleagues (2005) found that elevated serum phosphate concentrations were associated with an increased risk of fatal or nonfatal MI. Compared with participants with serum phosphate levels of 2.5–3.4 mg/dL, those with levels ≥4.0 mg/dL had a 50% higher risk of MI (HR = 1.50, 95% CI, 1.05–2.16) [109]. In a 2011 systematic review and meta-analysis of 11 cohort studies, D'Elia and colleagues showed that a 1.64-g (42 mmol) per day higher K intake was associated with a 21% lower risk of stroke (RR = 0.79, 95% CI, 0.68–0.90), with a trend toward lower risk of CHD and total CVD [110]. In 2013, Xiao and colleagues, using data from the NIH-AARP Diet and Health Study involving approximately 400,000 men and women, reported that high supplemental Ca intake—but not dietary Ca—was associated with increased risk of cardiovascular death in men (RR = 1.20, 95% CI, 1.05–1.36) and death from heart disease (RR = 1.19, 95% CI, 1.03–1.37), comparing 1000 mg/day with no supplementation [35]. That same year, Aburto and colleagues published a systematic review and meta-analysis of 11 prospective cohort studies and showed that higher K intake was inversely associated with risk of stroke (RR = 0.76, 95% CI, 0.66–0.89), although no significant associations were found for total CVD or CHD [53]. In 2014, O'Donnell and colleagues conducted an analysis of over 100,000 participants in 17 countries as part of the PURE study. They found that both low (<3 g/day; OR = 1.27, 95% CI, 1.12–1.44) and high (≥7 g/day; OR = 1.15, 95% CI, 1.02–1.30) estimated Na intakes were associated with increased risk of major cardiovascular events and mortality compared to moderate intake (4.00–5.99 g/day). Conversely, higher K intake (>3.00 g/day) was associated with reduced risk of these outcomes (OR = 0.78, 95% CI, 0.67–0.91) [111]. In 2015, Adebamowo and colleagues examined data from the Nurses' Health Study (NHS) I and II and found that higher intakes of Mg (RR = 0.87, 95% CI, 0.78–0.97) and K (RR = 0.89, 95% CI, 0.80–0.99) were significantly associated with lower risk of stroke in women. A meta-analysis conducted alongside this analysis showed that each 100 mg/day increase in Mg intake reduced stroke risk by 13% (RR = 0.87, 95% CI, 0.83–0.92), and each 1000 mg/day increase in K intake reduced risk by 9% (RR = 0.91, 95% CI, 0.88–0.94) [112]. Also in 2015, Bain and colleagues, using the EPIC-Norfolk cohort, demonstrated that higher dietary Mg intake was associated with reduced risk of stroke in men (HR = 0.59, 95% CI, 0.38–0.93), comparing the top 30% with the bottom 10% of intake [113]. Tian and colleagues (2015) conducted a meta-analysis and found that dairy Ca intake, but not total Ca intake, was inversely associated with stroke risk. The highest vs lowest Ca intake was associated with a 24% lower risk of stroke (RR = 0.76, 95% CI, 0.66–0.86) [114]. In 2016, Mente and colleagues analyzed pooled data from four cohort studies and found that Na excretion levels influenced CVD risk differently based on hypertension status. Among hypertensive individuals, both high (≥7 g/day; HR = 1.23, 95% CI, 1.11–1.37) and low (<3 g/day; HR = 1.34, 95% CI, 1.23–1.47) Na intakes were associated with increased risk compared to 4–5 g/day. Among those without hypertension, only low Na intake (<3 g/day) was significantly associated with increased risk (HR = 1.26, 95% CI, 1.10–1.45) [115]. In the Swedish AMORIS cohort, Rohrmann and colleagues (2016) showed that higher serum Ca concentrations were associated with increased risk of multiple cardiovascular endpoints, including nonfatal CVD (HR = 1.12), MI (HR = 1.19), and stroke (HR = 1.11), as well as fatal CVD (HR = 1.41), fatal MI (HR = 1.41), and fatal stroke (HR = 1.30), comparing the highest vs lowest quintiles. These associations were more pronounced in women [116]. Also in 2016, Fang and colleagues conducted a meta-analysis of 40 cohort studies and found no association between dietary Mg and total CVD or CHD. However, higher Mg intake was associated with a reduced risk of HF (RR = 0.78, 95% CI, 0.69–0.89) and stroke (RR = 0.93, 95% CI, 0.89–0.97), per 100 mg/day increment [117]. That same year, Vinceti and colleagues published a meta-analysis of 16 prospective cohort studies and found an inverse dose-response association between K intake and stroke. Those in the highest intake group had a lower risk compared to the lowest (RR = 0.87, 95% CI, 0.80–0.94), with optimal benefit observed at 90 mmol/day (approximately 3500 mg) [118]. In 2018, Zhu and colleagues conducted a meta-analysis of 16 prospective cohort studies and found that each 100 mmol/day increase in Na intake was associated with higher risks of stroke incidence (RR = 1.10, 95% CI, 1.01–1.19) and stroke mortality (RR = 1.28, 95% CI, 1.07–1.54) [119]. In 2019, Jayedi and colleagues, in a meta-analysis of 14 prospective cohort studies, 1 case-cohort study, and 1 case-control study, found that higher Na intake and a higher dietary Na-to-K ratio were associated with increased risk of stroke. Specifically, stroke risk increased by 6% per 1 g/day increment in Na (RR = 1.06, 95% CI, 1.02–1.10) and by 22% per 1-unit increase in the Na-to-K ratio (RR = 1.22, 95% CI, 1.04–1.41) [120]. In 2020, Goncalves and Abreu conducted a systematic review of 12 studies (including six prospective cohorts) focused on Na and K intake in older adults. Their synthesis found that while the relationship between Na intake and CVD risk was inconsistent, increased K intake and lower Na-to-K ratios were consistently associated with lower CVD risk, particularly for stroke [121]. That same year, Wang and colleagues conducted a meta-analysis of 36 studies and demonstrated a linear dose-response association between Na intake and CVD risk. Compared with the lowest intake group, the highest Na consumers had a 19% higher risk of CVD (RR = 1.19, 95% CI, 1.08–1.30), and each 1 g/day increment in Na was associated with a 6% increase in risk (RR = 1.06, 95% CI, 1.01–1.11) [122]. Zhao and colleagues (2020) evaluated data from 53 cohorts and found a significant inverse association between dietary Mg intake and stroke risk. Compared with the lowest intake, the highest Mg intake was associated with lower risk of total stroke (RR = 0.89, 95% CI, 0.83–0.94) and ischemic stroke (RR = 0.88, 95% CI, 0.81–0.95) [123]. In 2021, Zhu and colleagues analyzed data from the CHNS and found that higher dietary Ca intake was associated with a reduced risk of stroke (HR = 0.53, 95% CI, 0.29–0.96), with a stronger association observed among men. No significant associations were found for dietary Mg or P [124]. Myung and colleagues (2021) pooled a total of 13 double-blind, placebo-controlled RCTs and showed that Ca supplements significantly increased the risk of CVD, specifically in healthy postmenopausal women [125]. Finally, in 2023, Wang and colleagues conducted a meta-analysis of 18 cohort studies and reported a nonlinear inverse association between Ca intake and stroke risk. Stroke risk decreased with Ca intakes of 200–500 mg/day, but no further benefit was seen at intakes ≥700 mg/day. The inverse association was significant in Asian populations but not in non-Asian regions [126].

The evidence indicates that higher K and Mg intakes are consistently associated with reduced risk of stroke and, in the case of Mg, HF. Na intake shows a U- or J-shaped relationship with CVD, where both excessive and insufficient intake may be harmful, particularly in hypertensive individuals. Ca intake demonstrates a complex relationship with CVD; supplemental intake may increase risk, while moderate dietary intake—especially from dairy sources—is protective, particularly in Asian populations. Serum Ca and P levels are positively associated with adverse cardiovascular outcomes, including MI and stroke.

All-Cause Mortality

The potential influence of major minerals on longevity has been extensively studied across large-scale observational cohorts and meta-analyses. Evidence consistently suggests that appropriate intakes and serum levels of Ca, P, K, Na, Mg, S, and Cl are critical in modulating mortality risk, although associations sometimes vary depending on dose, sex, or source (dietary vs supplemental). Some of the relevant findings are summarized below:

In 2014, Graudal and colleagues conducted a meta-analysis of 23 cohort studies and 2 follow-up studies of RCTs, demonstrating a U-shaped association between Na intake and mortality. Both low and high Na intakes were linked with increased mortality risk: The HR for usual vs low Na intake was 0.91 (95% CI, 0.82–0.99), and for high vs usual, Na intake was 1.16 (95% CI, 1.03–1.30) [127]. Also in 2014, Chang and colleagues analyzed NHANES III data and showed that high P intake was associated with elevated mortality risk among individuals consuming over 1400 mg/day of phosphate (HR = 2.23, 95% CI, 1.09–4.5 per 1-unit increase in log-transformed phosphorus intake) [128]. In a 2015 meta-analysis of 22 observational cohort studies, Asemi and colleagues demonstrated that both dietary and supplemental Ca intakes were associated with a reduced risk of all-cause mortality [129]. In 2016, Fang and colleagues conducted a meta-analysis of 40 cohort studies and reported that higher dietary Mg intake was associated with a lower risk of all-cause mortality, with an RR of 0.90 (95% CI, 0.81–0.99) per 100 mg/day increase in Mg intake [117]. That same year, Bai and colleagues published a meta-analysis of six cohort studies, showing that elevated serum P concentrations were independently associated with a 35% higher risk of all-cause mortality (RR = 1.35, 95% CI, 1.15–1.58) among individuals without CKD [130]. Additionally, Cook and colleagues (2016) analyzed data from the Trials of Hypertension Prevention and demonstrated a direct linear association between Na intake and mortality risk: Mortality increased by 12% per 1000 mg/day of Na excretion (HR = 1.12, 95% CI, 1.00–1.26) and similarly with the Na-to-K ratio (HR = 1.13, 95% CI, 1.01–1.27) [131]. Using the US Cancer Prevention Study II Nutrition Cohort, Yang and colleagues (2016) observed sex-specific effects: In women, total and supplemental Ca intakes were associated with lower mortality, whereas in men, supplemental Ca intake $\geq$1000 mg/day was associated with higher all-cause mortality [132]. In a 2021 meta-analysis of 19 studies, Bagheri and colleagues found that higher dietary Mg intake—but not total or supplemental Mg intake—was associated with decreased all-cause mortality (ES = 0.87, 95% CI, 0.79–0.97), with a 6% reduction in mortality risk for each 100 mg/day increment of Mg intake (ES = 0.94, 95% CI, 0.91–0.97) [133]. Naghshi and colleagues (2022) conducted a dose-response meta-analysis of 35 cohort studies and showed that moderate intakes of total Ca (1000–1800 mg/day), dietary Ca (600–1200 mg/day), and supplemental Ca (600–1200 mg/day) were inversely associated with mortality risk. However, higher Ca intakes beyond these ranges were not linked with further reductions in mortality [134]. Using a 10-year community-based cohort, Kwon and colleagues [135] demonstrated that K intake—but not Na intake or the Na-to-K ratio—was inversely associated with all-cause mortality. Participants in the highest quintile of K intake had a 21% lower mortality risk compared to those in the lowest quintile (HR = 0.79, 95% CI, 0.69–0.91) [135]. In 2023, Hou and colleagues analyzed data from the NHANES (1998–2018) cohort and found an inverse L-shaped association between serum Cl concentrations and all-cause mortality. Compared to the lowest quartile (≤101.2 mmol/L), higher quartiles of serum Cl were associated with reduced mortality, with HRs of 0.77 (95% CI, 0.67–0.89), 0.72 (95% CI, 0.63–0.82), and 0.77 (95% CI, 0.65–0.90) from second to fourth quartiles, respectively [136]. In 2024, Yazdanpanah and colleagues used data from the Iranian Golestan Cohort Study to evaluate mineral intake and mortality, finding that higher dietary Ca intake was inversely associated with all-cause mortality (HR = 0.91, 95% CI, 0.85–0.99) when comparing the top vs bottom quintiles, with no significant associations observed for Mg, P, or K [137]. Also in 2024, Liu and colleagues, analyzing NHANES data, showed that daily Na and K intakes were inversely associated with all-cause mortality, while the Na-to-K ratio exhibited a U-shaped association with mortality [138]. In parallel, Gan and colleagues, using the NIH-AARP Diet and Health Study, demonstrated that higher Na intake was associated with increased mortality risk, whereas higher K intake and lower Na-to-K ratios were protective, with stronger associations observed among women [139].

Overall, the evidence suggests that higher dietary Mg, K, and moderate Ca intakes are consistently associated with a reduced risk of all-cause mortality. In contrast, elevated serum K levels and high dietary P intake are linked to an increased mortality risk. The relationship between Na intake and mortality appears U-shaped, with both low and high Na intakes associated with greater mortality risk, while maintaining a lower Na-to-K ratio is generally protective. Emerging data also indicate that higher serum Cl concentrations are inversely associated with mortality.

Potential Pathways Underlying the Beneficial Cardiometabolic Effects of Major Minerals

The beneficial effects of major minerals (Ca, P, K, Na, Mg, S, and Cl) on cardiometabolic outcomes are mediated through a variety of interconnected biological pathways that influence blood pressure regulation, vascular function, glucose metabolism, lipid homeostasis, inflammation, oxidative stress, and overall cellular function (Fig. 19.1). This section discusses the major potential mechanisms through which these minerals exert their cardiometabolic benefits.

Regulation of blood pressure and vascular tone Several minerals directly influence blood pressure regulation by modulating vascular tone, fluid balance, and vascular reactivity. K enhances endothelium-dependent vasodilation by promoting nitric oxide (NO) production and reducing vascular smooth muscle cell contraction [140, 141]. High K intake facilitates natriuresis (Na excretion) and

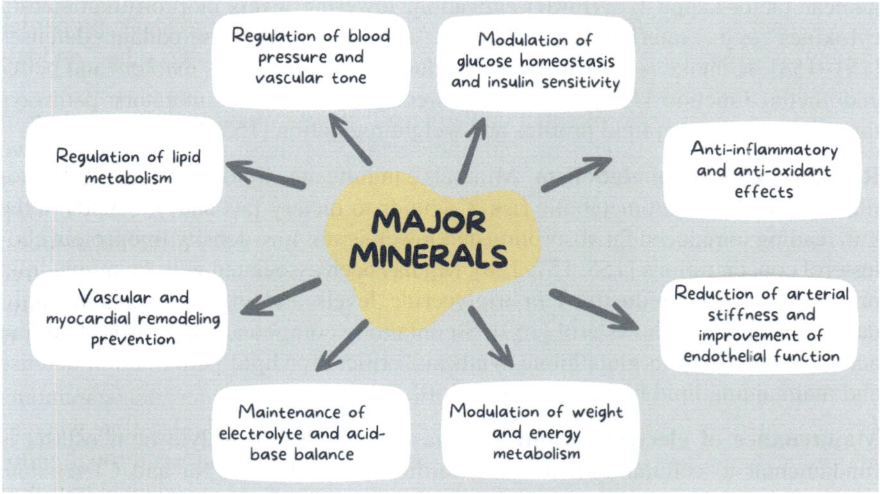

Fig. 19.1 Potential pathways underlying the beneficial cardiometabolic effects of major minerals

counteracts the pressor effects of high Na, leading to reductions in blood pressure [142, 143]. Mg acts as a natural Ca antagonist at vascular smooth muscle cells, leading to vasodilation and decreased peripheral vascular resistance [38]. It also improves endothelial function by enhancing NO bioavailability [144]. Hypomagnesemia has been implicated in the pathogenesis of high blood pressure [145]. Ca regulates vascular smooth muscle contraction. Adequate Ca intake suppresses PTH and 1,25-dihydroxyvitamin D levels, which can otherwise lead to increased intracellular Ca and vasoconstriction [36]. Na is essential for extracellular fluid volume regulation [38]; however, excess Na intake leads to fluid retention, increased blood volume, and elevated blood pressure, emphasizing the importance of balance between Na and K.

Modulation of glucose homeostasis and insulin sensitivity Minerals such as Mg, Ca, and K play critical roles in glucose metabolism and insulin function. Mg serves as a cofactor for various enzymes involved in glucose oxidation, insulin receptor activation, and cellular glucose uptake [38, 43]. It improves insulin sensitivity and reduces systemic inflammation, a key driver of insulin resistance [146, 147]. K helps maintain cellular electrolyte gradients critical for insulin secretion by pancreatic β cells [148]. Hypokalemia impairs insulin secretion and promotes glucose intolerance [149]. Ca is involved in insulin-mediated intracellular processes, including glucose transporter translocation [150]; however, dysregulated intracellular Ca concentrations can impair insulin signaling and β-cell function.

Anti-inflammatory and antioxidant effects Chronic low-grade inflammation and oxidative stress are central in the development of cardiometabolic diseases. Major minerals influence these pathways. Mg reduces systemic inflammation by inhibiting nuclear factor-kappa B (NF-κB) activation, lowering levels of pro-inflammatory cytokines (e.g., interleukin-6, TNF-α), and enhancing antioxidant defenses [151–153]. K intake is associated with reduced oxidative stress markers and better endothelial function [154]. Ca may indirectly modulate inflammatory pathways through its effects on lipid profiles and weight regulation [155, 156].

Regulation of lipid metabolism Minerals can influence lipid metabolism and profiles, impacting cardiometabolic risk. Ca binds to dietary fats and bile acids in the gut, leading to reduced fat absorption and lower serum low-density lipoprotein cholesterol concentrations [155, 157, 158]. Mg has been associated with favorable lipid profiles, including reductions in triglyceride levels and improvements in high-density lipoprotein cholesterol [159]. S-containing compounds (e.g., from S amino acids) are involved in glutathione synthesis, critical for lipid peroxidation defense and maintaining lipid homeostasis [45, 160].

Maintenance of electrolyte and acid-base balance Electrolyte homeostasis is fundamental to cellular function and cardiovascular health. Na and Cl regulate extracellular fluid volume and acid-base balance [38]. Dysregulation can lead to hypertension, endothelial dysfunction, and cardiac remodeling. K counterbalances Na's effects and supports intracellular fluid volume and pH balance [40], protecting against hypertension and arrhythmias.

Vascular and myocardial remodeling prevention Several minerals prevent structural changes in blood vessels and the heart that predispose to cardiometabolic diseases. Mg inhibits vascular calcification by preventing the transformation of vascular smooth muscle cells into osteoblast-like cells [161, 162]. P, particularly when elevated, promotes vascular calcification and myocardial fibrosis [163, 164]; maintaining normal P balance is therefore essential for vascular integrity. Ca and Mg together influence myocardial contractility and excitability [165, 166], helping to maintain normal cardiac rhythm and function.

Modulation of weight and energy metabolism Obesity is a major driver of cardiometabolic diseases. Some minerals are implicated in body weight regulation. Ca may promote fat oxidation and thermogenesis, while reducing lipid absorption, collectively contributing to weight regulation [155, 156]. Mg deficiency is associated with increased risk of MetS and obesity due to impaired energy metabolism and heightened inflammation [167, 168].

Reduction of arterial stiffness and improvement of endothelial function K and Mg intake have been shown to reduce arterial stiffness [169–171], a major independent predictor of cardiovascular risk [172]. Cl, although less studied, plays a role in vascular smooth muscle function and endothelial barrier integrity [173], indirectly influencing arterial stiffness and cardiovascular outcomes.

Impact of Major Minerals on Longevity and Healthy Aging

Longevity refers to an extended lifespan, typically characterized by survival beyond the average life expectancy [174], while healthy aging is defined by the World Health Organization as "the process of developing and maintaining the functional ability that enables well-being in older age" [175]. Unlike mere lifespan extension, healthy aging emphasizes quality of life, preserving physical and cognitive function, and minimizing the burden of chronic disease and disability. Strategies to enhance longevity and promote healthy aging encompass a range of modifiable lifestyle factors, including regular physical activity, avoidance of tobacco and harmful alcohol use, psychological well-being, and above all, adherence to a balanced and nutrient-rich diet [176–178]. In this context, the intake of major minerals—Ca, P, K, Na, Mg, S, and Cl—is increasingly recognized as essential, not only for preventing nutritional deficiencies but also for promoting systemic health, preventing chronic diseases, and possibly extending both lifespan and health span.

Cardiometabolic Benefits and Mortality Risk Reduction: Pathways to Longevity

Accumulating evidence from prospective cohort studies and meta-analyses suggests that adequate and balanced intake of major minerals confers significant protection against a broad spectrum of cardiometabolic diseases, including hypertension, T2D, CKD, stroke, and CVD—conditions that are leading causes of morbidity and mortality in aging populations. For instance, higher intakes of Mg and K and moderate levels of Ca have been consistently associated with lower risks of CVD events and all-cause mortality [53, 111, 112, 114, 121]. Conversely, dysregulated levels of P, Na, and Ca and low serum Cl have been linked to increased risk of vascular calcification, arterial stiffness, CVD, and early mortality [109, 115, 116, 122, 127, 128, 130]. By modulating these cardiometabolic risk factors and reducing mortality, major minerals may exert a longevity-enhancing effect, not only by extending lifespan but also by preserving physiological resilience, functionality, and independence into older age.

Biological Processes Linking Major Minerals to Healthy Aging

The beneficial effects of major minerals on healthy aging are mediated through several fundamental biological processes that govern cellular health, systemic homeostasis, and age-related physiological decline (Fig. 19.2). These include the following:

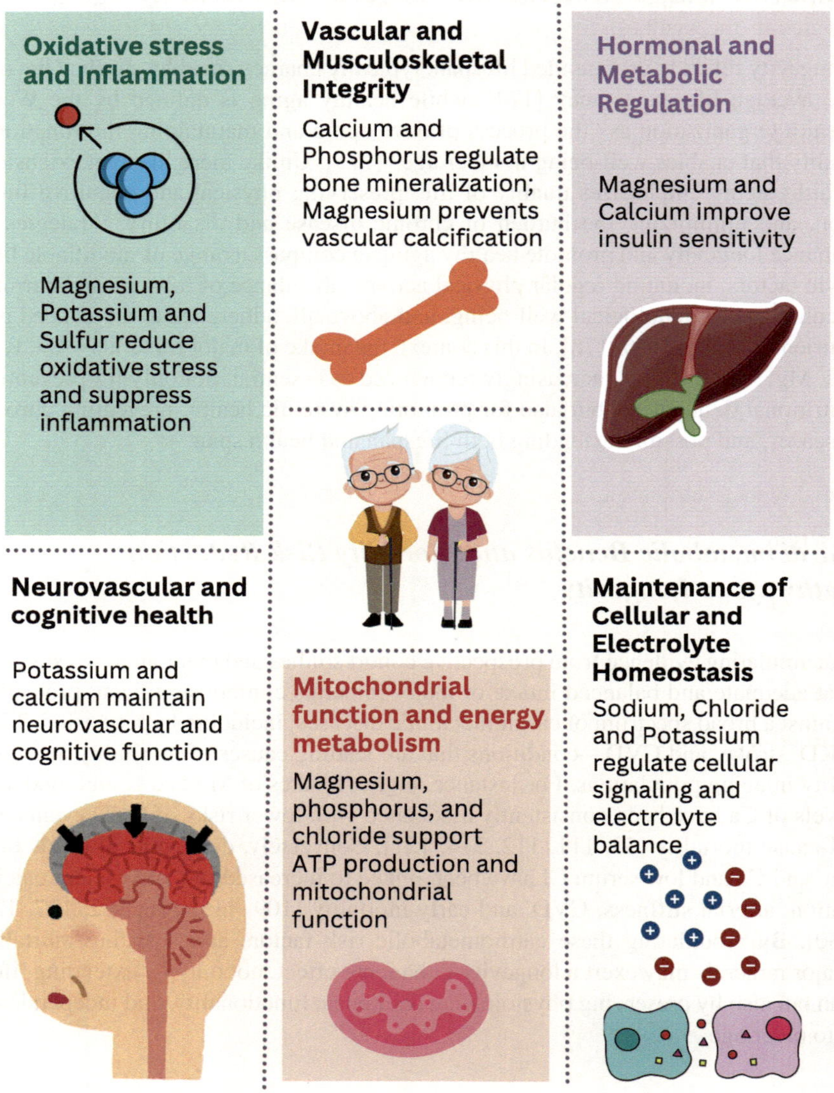

Fig. 19.2 Potential pathways to healthy aging

Oxidative stress and inflammation Aging is accompanied by increased oxidative stress and chronic low-grade inflammation ("inflammaging"), which accelerate tissue damage and the development of age-related diseases [179–185]. Mg and K have anti-inflammatory properties, attenuating cytokine production and oxidative damage through modulation of NF-κB and antioxidant enzyme systems [151–153, 186].

S, primarily through S-containing amino acids like methionine and cysteine, contributes to the synthesis of glutathione, a master antioxidant critical for cellular detoxification and redox balance.

Mitochondrial function and energy metabolism Mitochondrial dysfunction is a hallmark of aging and contributes to reduced cellular energy availability, increased reactive oxygen species (ROS) generation, and apoptosis [187–190]. Mg plays a pivotal role in mitochondrial ATP production, stabilizing mitochondrial membranes, and supporting the function of enzymes in the tricarboxylic acid (TCA) cycle [38, 43]. P is critical for ATP synthesis and energy storage [38, 40], while Cl maintains ionic gradients necessary for mitochondrial membrane potential [38, 46].

Neurovascular and cognitive health Age-related cognitive decline is closely linked with vascular aging and neuroinflammation. K, through its vasodilatory effects, enhances cerebral blood flow and supports neurovascular coupling [140, 141]. Adequate Ca intake is essential for synaptic transmission and memory function [191], although excessive Ca—especially in supplement form—has been associated with cerebrovascular risk [192], underlining the need for moderation.

Vascular and musculoskeletal integrity The integrity of vascular and musculoskeletal systems is crucial for maintaining mobility, independence, and quality of life in older age. Ca and P are essential for bone mineralization [36, 38], while Mg prevents vascular calcification by inhibiting osteogenic transformation of vascular smooth muscle cells [161, 162]. These minerals also contribute to muscle function, reducing the risk of sarcopenia and frailty.

Maintenance of cellular and electrolyte homeostasis Major minerals are key regulators of electrolyte balance, acid-base homeostasis, and cellular signaling. Dysregulation in Na, Cl, or K levels can precipitate fluid imbalances, metabolic acidosis, and impaired neuromuscular function [38, 46]—common challenges in aging. Maintaining these mineral levels within physiological ranges supports organ function and cellular vitality.

Hormonal and metabolic regulation Minerals such as Mg and Ca interact with key hormones, including insulin, PTH, and vitamin D [36, 193, 194], which are intimately linked with metabolic aging. For example, Mg enhances insulin signaling and reduces insulin resistance [146, 147], while Ca modulates PTH activity and skeletal homeostasis [36], both contributing to metabolic resilience.

Clinical and Public Health Implications

The collective evidence on the associations between major mineral intake and a spectrum of cardiometabolic outcomes has important clinical and public health implications. It underscores the critical need to recalibrate dietary guidelines and population-level strategies toward optimizing mineral intake—not merely to prevent deficiencies but to promote long-term cardiometabolic health, enhance longevity, and support healthy aging.

Among all the major minerals, Na and K emerge as the most influential in blood pressure regulation and overall cardiovascular risk, with consistently strong and reproducible effects. The evidence supports a reduction in dietary Na intake, particularly in hypertensive and salt-sensitive individuals, and a simultaneous increase in K consumption as a central strategy in the clinical management of hypertension and stroke prevention. Public health programs should emphasize Na reduction—especially in processed foods—and promote whole-food sources of K, such as fruits, vegetables, and legumes, which may also confer additional fiber, antioxidant, and micronutrient benefits. Maintaining an optimal Na-to-K ratio is particularly important as imbalances have been consistently linked to higher risks of CKD, CVD, and premature mortality [34, 120, 138].

Mg intake, largely from dietary sources, also shows strong protective associations across a wide range of cardiometabolic endpoints—including MetS, T2D, CKD, stroke, HF, and all-cause mortality [31, 91, 108, 112, 117]. Its role as a cofactor in enzymatic, anti-inflammatory, and vascular regulatory processes [38, 43] makes it a promising target for both clinical nutrition interventions and public health campaigns. Increasing dietary Mg through nuts, leafy greens, and whole grains should be prioritized, especially in older adults who are at increased risk of Mg insufficiency due to dietary inadequacies and decreased absorption with age.

Ca shows a more complex pattern. While moderate dietary Ca intake, particularly from dairy and plant-based sources, appears beneficial for reducing the risk of hypertension, MetS, stroke, CVD, and mortality [30, 65, 114, 129], the evidence regarding supplemental Ca is less favorable. In several studies, supplemental Ca—especially in high doses—was associated with adverse cardiovascular outcomes in men and possible harm when serum levels become elevated [35, 116, 125]. These findings support a food-first approach to Ca intake, with supplementation considered only when dietary intake is clearly inadequate and individualized clinical indications exist.

The role of P is similarly complex. While P from dairy foods may exert protective effects, high total intake—especially from phosphate additives and processed foods—and elevated serum P levels are associated with increased risks of MetS, T2D, CKD, cardiovascular events, and mortality [33, 97, 106, 109, 128]. These findings warrant closer regulatory scrutiny of inorganic phosphate additives and support efforts to improve labeling transparency and public education regarding hidden dietary sources of P.

S and Cl, though less extensively studied, are beginning to emerge as relevant contributors to cardiometabolic and aging-related outcomes. Elevated intake of sulfur-containing amino acids may increase diabetes risk [101], while higher serum Cl concentrations are associated with reduced all-cause and cardiovascular mortality [136]. These early signals call for expanded research to determine the physiological thresholds, optimal ranges, and potential interactions of these minerals within the broader dietary matrix.

Taken together, these findings reinforce the concept that a balanced mineral profile—not just the avoidance of excess or deficiency—plays a central role in delaying disease onset, preserving physiological function, and enhancing life expectancy. This is especially critical given the rising burden of noncommunicable diseases in aging populations. Interventions focused on dietary mineral optimization offer a scalable, low-cost strategy for promoting healthy aging across diverse populations and settings.

Clinically, this evidence supports greater integration of mineral intake assessments into routine dietary counselling, particularly in cardiometabolic risk management. At the public health level, these findings strengthen the case for policies that promote dietary diversity, reduce reliance on ultra-processed foods, and ensure equitable access to nutrient-rich whole foods. As global populations age, the dual pursuit of longevity and quality of life—health span and lifespan—will depend in part on our ability to leverage foundational elements like major minerals to support metabolic resilience and cardiovascular integrity well into older adulthood.

Gaps in the Evidence and Future Directions

Despite growing interest and an expanding evidence base on the role of major minerals in cardiometabolic health, healthy aging, and longevity, several important gaps remain that limit the ability to draw firm conclusions or develop targeted guidelines. These limitations highlight the need for more robust, comprehensive, and mechanistically informed research to better inform clinical and public health practice.

A key gap is the limited prospective evidence for several of the major minerals beyond Ca, Mg, K, and Na. While these four minerals have been extensively investigated—particularly in relation to blood pressure, cardiovascular outcomes, and mortality—the roles of P, S, and Cl remain comparatively underexplored in longitudinal studies. Most of the available data on these minerals come from cross-sectional analyses or studies involving patients with preexisting diseases such as CKD, limiting generalizability to broader populations. For example, the potential harmful effects of excessive P—particularly from inorganic additives—and the emerging protective associations of serum Cl require validation in well-designed, population-based prospective cohorts. Moreover, the dose-response relationships and optimal intake ranges for several minerals remain inadequately defined. U- or J-shaped associations have been observed for Na, Mg, Ca, and P in relation to mortality and cardiometabolic outcomes [87, 103, 127], suggesting that both deficiency and excess may confer risk. However, the thresholds at which these risks manifest, and whether they differ across age, sex, ethnicity, or disease states, are poorly understood. More detailed research is needed to establish safe and effective intake windows, particularly in older adults, who are more susceptible to both deficiency and iatrogenic excess through supplementation.

Another notable gap is the limited integration of biomarkers of mineral status in many observational studies. Reliance solely on dietary intake estimates from food frequency questionnaires (FFQs) or 24-hour recalls may not fully capture individual absorption, bioavailability, or endogenous regulation. Incorporating serum, urinary, and tissue-level biomarkers, and relating them to both dietary intake and clinical outcomes, could significantly strengthen causal inference and improve risk stratification.

There is also a scarcity of RCTs evaluating mineral supplementation as a preventive or therapeutic strategy in general populations. While some RCTs exist for Ca, Mg, and K—primarily in the context of blood pressure and bone health [48, 49, 53, 54]—few have assessed long-term cardiometabolic or mortality outcomes. Moreover, very few trials have been adequately powered or long enough in duration to assess the impact of mineral balance on disease progression or healthy aging trajectories. Pragmatic, long-term RCTs and factorial trials investigating mineral combinations and interactions (e.g., Ca:Mg or Na:K ratios) are urgently needed.

The interactions between minerals and other dietary components or medications also represent an underexplored frontier. For example, the cardiometabolic effects of Ca may differ depending on concurrent vitamin D status, while Mg's benefits may be attenuated or enhanced by fiber intake or diuretic use [195, 196]. Future studies should systematically examine these synergistic and antagonistic relationships within the broader dietary and pharmacological milieu.

Additionally, the life course perspective is missing from much of the existing literature. Most studies assess mineral intake and outcomes at a single time point, failing to capture the cumulative effects of long-term intake patterns. Emerging evidence suggests that sustained mineral sufficiency over decades, particularly beginning in early adulthood, may be critical for preventing late-onset conditions such as NAFLD, T2D, and cognitive decline. Longitudinal cohort studies with repeated dietary assessments and longer follow-up durations are needed to address this issue.

Finally, there is a pressing need for population-specific research, especially in underrepresented groups such as ethnic minorities, low- and middle-income countries, and older adults with multimorbidity. Differences in dietary patterns, baseline nutritional status, genetic predispositions, and healthcare access may all modulate the effects of mineral intake on health outcomes, necessitating context-specific guidelines and interventions.

Conclusions

In summary, current evidence underscores the important role of major minerals—particularly Ca, Mg, K, and Na—in cardiometabolic health, mortality risk reduction, and the promotion of longevity and healthy aging. Optimizing mineral intake, notably by reducing Na while increasing K, Mg, and moderate Ca consumption from dietary sources, holds substantial promise as a clinical and public health strategy for mitigating the burden of chronic diseases and supporting functional aging. Despite these advances, significant gaps remain, particularly regarding the long-term effects of P, S, and Cl, optimal intake ranges, biomarker validation, and life course impacts. Future research should focus on addressing these uncertainties through well-designed prospective studies and intervention trials. Collectively, the findings advocate for a shift toward balanced mineral nutrition as a cornerstone of preventive healthcare and aging policy, with the potential to meaningfully extend both lifespan and health span in diverse global populations.

Authors' contributions Setor K. Kunutsor: Conceptualization, methodology, data curation, formal analysis, investigation, original draft, writing—review and editing. Khushmanjot Kaur: Conceptualization, methodology, investigation, writing—review and editing.

Funding statement This work was supported by a Research Start-up Grant awarded to SKK. SKK is the holder of the Evelyn Wyrzykowski Research Chair in Cardiology, Section of Cardiology, Department of Internal Medicine, University of Manitoba.

Conflict of interest The authors declare that they have no conflicts of interest.

Ethics approval and consent to participate Not applicable.

Consent for publication Not applicable

Data availability This is a narrative review; no new scientific data was generated, and all data are within the paper.

References

1. Mills KT, Stefanescu A, He J (2020) The global epidemiology of hypertension. Nat Rev Nephrol 16(4):223–237
2. World Health Organization (2023) Hypertension: key facts. https://www.who.int/news-room/fact-sheets/detail/hypertension. Accessed on 22 Dec 2023
3. Sattar N, Gaw A, Scherbakova O, Ford I, O'Reilly DS, Haffner SM et al (2003) Metabolic syndrome with and without C-reactive protein as a predictor of coronary heart disease and diabetes in the West of Scotland Coronary Prevention Study. Circulation 108(4):414–419
4. Laaksonen DE, Lakka HM, Niskanen LK, Kaplan GA, Salonen JT, Lakka TA (2002) Metabolic syndrome and development of diabetes mellitus: application and validation of recently suggested definitions of the metabolic syndrome in a prospective cohort study. Am J Epidemiol 156(11):1070–1077
5. Isomaa B, Almgren P, Tuomi T, Forsen B, Lahti K, Nissen M et al (2001) Cardiovascular morbidity and mortality associated with the metabolic syndrome. Diabetes Care 24(4):683–689
6. Lakka HM, Laaksonen DE, Lakka TA, Niskanen LK, Kumpusalo E, Tuomilehto J et al (2002) The metabolic syndrome and total and cardiovascular disease mortality in middle-aged men. JAMA 288(21):2709–2716
7. Lekakis V, Papatheodoridis GV (2024) Natural history of metabolic dysfunction-associated steatotic liver disease. Eur J Intern Med 122:3–10
8. Zheng Y, Ley SH, Hu FB (2018) Global aetiology and epidemiology of type 2 diabetes mellitus and its complications. Nat Rev Endocrinol 14(2):88–98
9. Burnier M, Damianaki A (2023) Hypertension as cardiovascular risk factor in chronic kidney disease. Circ Res 132(8):1050–1063

10. World Health Organization. Fact sheets. The top 10 causes of death. https://www.who.int/news-room/fact-sheets/detail/the-top-10-causes-of-death. Retrieved on 10 Sep 2021
11. Couser WG, Remuzzi G, Mendis S, Tonelli M (2011) The contribution of chronic kidney disease to the global burden of major noncommunicable diseases. Kidney Int 80(12):1258–1270
12. World Health Organization. Cardiovascular diseases. https://www.who.int/health-topics/cardiovascular-diseases#tab=tab_1. Accessed on 4 Dec 2024
13. Roth GA, Mensah GA, Johnson CO, Addolorato G, Ammirati E, Baddour LM et al (2020) Global burden of cardiovascular diseases and risk factors, 1990-2019: update from the GBD 2019 study. J Am Coll Cardiol 76(25):2982–3021
14. Akbar Z, Fituri S, Ouagueni A, Alalwani J, Sukik A, Al-Jayyousi GF et al (2023) Associations of the MIND diet with cardiometabolic diseases and their risk factors: a systematic review. Diabetes Metab Syndr Obes 16:3353–3371
15. Petersen KS, Flock MR, Richter CK, Mukherjea R, Slavin JL, Kris-Etherton PM (2017) Healthy dietary patterns for preventing Cardiometabolic disease: the role of plant-based foods and animal products. Curr Dev Nutr 1(12):cdn.117.001289
16. Widmer RJ, Flammer AJ, Lerman LO, Lerman A (2015) The Mediterranean diet, its components, and cardiovascular disease. Am J Med 128(3):229–238
17. Filippou CD, Tsioufis CP, Thomopoulos CG, Mihas CC, Dimitriadis KS, Sotiropoulou LI et al (2020) Dietary Approaches to Stop Hypertension (DASH) diet and blood pressure reduction in adults with and without hypertension: a systematic review and meta-analysis of randomized controlled trials. Adv Nutr 11(5):1150–1160
18. Ungvari Z, Kunutsor SK (2024) Coffee consumption and cardiometabolic health: a comprehensive review of the evidence. Geroscience 46(6):6473–6510
19. Paterson KE, Myint PK, Jennings A, Bain LKM, Lentjes MAH, Khaw KT et al (2018) Mediterranean diet reduces risk of incident stroke in a population with varying cardiovascular disease risk profiles. Stroke 49(10):2415–2420
20. Feart C, Samieri C, Rondeau V, Amieva H, Portet F, Dartigues JF et al (2009) Adherence to a Mediterranean diet, cognitive decline, and risk of dementia. JAMA 302(6):638–648
21. Cherian L, Wang Y, Holland T, Agarwal P, Aggarwal N, Morris MC (2021) DASH and mediterranean-dash intervention for neurodegenerative delay (MIND) diets are associated with fewer depressive symptoms over time. J Gerontol A Biol Sci Med Sci 76(1):151–156
22. de Lorgeril M, Salen P, Martin JL, Monjaud I, Delaye J, Mamelle N (1999) Mediterranean diet, traditional risk factors, and the rate of cardiovascular complications after myocardial infarction: final report of the Lyon Diet Heart Study. Circulation 99(6):779–785
23. Parikh M, Netticadan T, Pierce GN (2018) Flaxseed: its bioactive components and their cardiovascular benefits. Am J Physiol Heart Circ Physiol 314(2):H146–HH59
24. Dinu M, Sofi F, Lotti S, Colombini B, Mattioli AV, Catapano AL et al (2024) Effects of omega-3 fatty acids on coronary revascularization and cardiovascular events: a meta-analysis. Eur J Prev Cardiol 31(15):1863–1875
25. Qi X, Zhu H, Ya R, Huang H (2023) Omega-3 polyunsaturated fatty acids supplements and cardiovascular disease outcome: a systematic review and meta-analysis on randomized controlled trials. Rev Cardiovasc Med 24(1):24
26. Godos J, Micek A, Currenti W, Franchi C, Poli A, Battino M et al (2024) Fish consumption, cognitive impairment and dementia: an updated dose-response meta-analysis of observational studies. Aging Clin Exp Res 36(1):171
27. Farag MA, Abib B, Qin Z, Ze X, Ali SE (2023) Dietary macrominerals: updated review of their role and orchestration in human nutrition throughout the life cycle with sex differences. Curr Res Food Sci 6:100450
28. Ali AAH (2023) Overview of the vital roles of macro minerals in the human body. J Trace Elem Miner 4:100076
29. Song Y, Sesso HD, Manson JE, Cook NR, Buring JE, Liu S (2006) Dietary magnesium intake and risk of incident hypertension among middle-aged and older US women in a 10-year follow-up study. Am J Cardiol 98(12):1616–1621

30. Wang L, Manson JE, Buring JE, Lee IM, Sesso HD (2008) Dietary intake of dairy products, calcium, and vitamin D and the risk of hypertension in middle-aged and older women. Hypertension 51(4):1073–1079
31. He K, Liu K, Daviglus ML, Morris SJ, Loria CM, Van Horn L et al (2006) Magnesium intake and incidence of metabolic syndrome among young adults. Circulation 113(13):1675–1682
32. Lu L, Chen C, Li Y, Guo W, Zhang S, Brockman J et al (2022) Magnesium intake is inversely associated with risk of non-alcoholic fatty liver disease among American adults. Eur J Nutr 61(3):1245–1254
33. Jhuang YH, Kao TW, Peng TC, Chen WL, Chang PK, Wu LW (2019) Serum phosphorus as a risk factor of metabolic syndrome in the elderly in Taiwan: a large-population cohort study. Nutrients 11(10)
34. Swift SL, Drexler Y, Sotres-Alvarez D, Raij L, Llabre MM, Schneiderman N et al (2022) Associations of sodium and potassium intake with chronic kidney disease in a prospective cohort study: findings from the Hispanic Community Health Study/Study of Latinos, 2008-2017. BMC Nephrol 23(1):133
35. Xiao Q, Murphy RA, Houston DK, Harris TB, Chow WH, Park Y (2013) Dietary and supplemental calcium intake and cardiovascular disease mortality: the National Institutes of Health-AARP diet and health study. JAMA Intern Med 173(8):639–646
36. Yu E, Sharma S. Physiology, calcium. [Updated 2023 Aug 14]. In: StatPearls [Internet]. Treasure Island: StatPearls Publishing; 2025. Available from: https://www.ncbi.nlm.nih.gov/books/NBK482128/
37. Institute of Medicine (US) Committee to Review Dietary Reference Intakes for Vitamin D and Calcium; Ross AC, Taylor CL, Yaktine AL, et al., editors. Dietary reference intakes for calcium and vitamin D. Washington (DC): National Academies Press (US); 2011. 2, Overview of Calcium. Available from: https://www.ncbi.nlm.nih.gov/books/NBK56060/
38. Shrimanker I, Bhattarai S (2025) Electrolytes. [Updated 2023 Jul 24]. In: StatPearls [Internet]. StatPearls Publishing, Treasure Island. Available from: https://www.ncbi.nlm.nih.gov/books/NBK541123/
39. National Institute of Health Office of Dietary Supplements. Phosphorus - Fact Sheet for Health Professionals. https://ods.od.nih.gov/factsheets/Phosphorus-HealthProfessional/. Accessed on 29 April 2025
40. National Institute of Health Office of Dietary Supplements. Potassium - Fact Sheet for Health Professionals. https://ods.od.nih.gov/factsheets/Potassium-HealthProfessional/. Accessed on 29 April 2025
41. MacLeod SM, Cairns JA (2015) Controversial sodium guidelines: scientific solution or perpetual debate? CMAJ 187(2):95–96
42. Guideline: Sodium Intake for Adults and Children. Geneva: World Health Organization; 2012. Annex 2, Examples of sodium content in various foods and food groups. Available from: https://www.ncbi.nlm.nih.gov/books/NBK133307/
43. National Institute of Health Office of Dietary Supplements. Magnesium - Fact Sheet for Health Professionals. https://ods.od.nih.gov/factsheets/Magnesium-HealthProfessional/. Accessed on 29 April 2025
44. Nimni ME, Han B, Cordoba F (2007) Are we getting enough sulfur in our diet? Nutr Metab (Lond) 4:24
45. Parcell S (2002) Sulfur in human nutrition and applications in medicine. Altern Med Rev 7(1):22–44
46. Berend K, van Hulsteijn LH, Gans ROB (2012) Chloride: the queen of electrolytes? Eur J Intern Med 23(3):203–211
47. OCEBM Levels of Evidence Working Group. The Oxford 2011 levels of evidence. Oxford Centre for Evidence-Based Medicine. https://www.cebm.ox.ac.uk/resources/levels-of-evidence/ocebm-levels-of-evidence. Accessed on 08 Oct 2020

48. Jee SH, Miller ER 3rd, Guallar E, Singh VK, Appel LJ, Klag MJ (2002) The effect of magnesium supplementation on blood pressure: a meta-analysis of randomized clinical trials. Am J Hypertens 15(8):691–696
49. Dickinson HO, Nicolson DJ, Campbell F, Cook JV, Beyer FR, Ford GA et al (2006) Magnesium supplementation for the management of essential hypertension in adults. Cochrane Database Syst Rev 19(3):CD004640
50. Dickinson HO, Nicolson DJ, Cook JV, Campbell F, Beyer FR, Ford GA et al (2006) Calcium supplementation for the management of primary hypertension in adults. Cochrane Database Syst Rev 19(2):CD004639
51. Beyer FR, Dickinson HO, Nicolson DJ, Ford GA, Mason J (2006) Combined calcium, magnesium and potassium supplementation for the management of primary hypertension in adults. Cochrane Database Syst Rev 19(3):CD004805
52. Dickinson HO, Nicolson DJ, Campbell F, Beyer FR, Mason J (2006) Potassium supplementation for the management of primary hypertension in adults. Cochrane Database Syst Rev 19(3):CD004641
53. Aburto NJ, Hanson S, Gutierrez H, Hooper L, Elliott P, Cappuccio FP (2013) Effect of increased potassium intake on cardiovascular risk factors and disease: systematic review and meta-analyses. BMJ 346:f1378
54. Cormick G, Ciapponi A, Cafferata ML, Belizan JM (2015) Calcium supplementation for prevention of primary hypertension. Cochrane Database Syst Rev 2015(6):CD010037
55. Poorolajal J, Zeraati F, Soltanian AR, Sheikh V, Hooshmand E, Maleki A (2017) Oral potassium supplementation for management of essential hypertension: a meta-analysis of randomized controlled trials. PLoS One 12(4):e0174967
56. Filippini T, Violi F, D'Amico R, Vinceti M (2017) The effect of potassium supplementation on blood pressure in hypertensive subjects: a systematic review and meta-analysis. Int J Cardiol 230:127–135
57. McClure ST, Rebholz CM, Medabalimi S, Hu EA, Xu Z, Selvin E et al (2019) Dietary phosphorus intake and blood pressure in adults: a systematic review of randomized trials and prospective observational studies. Am J Clin Nutr 109(5):1264–1272
58. Cormick G, Ciapponi A, Cafferata ML, Cormick MS, Belizan JM (2021) Calcium supplementation for prevention of primary hypertension. Cochrane Database Syst Rev 8(8):CD010037
59. Lai JS, Aung YN, Khalid Y, Cheah SC (2022) Impact of different dietary sodium reduction strategies on blood pressure: a systematic review. Hypertens Res 45(11):1701–1712
60. Amer SA, Abo-Elnour DE, Abbas A, Abdelrahman AS, Hamdy HM, Kenawy S et al (2025) Calcium, magnesium, and vitamin D supplementations as complementary therapy for hypertensive patients: a systematic review and meta-analysis. BMC Complement Med Ther 25(1):89
61. Alonso A, Nettleton JA, Ix JH, de Boer IH, Folsom AR, Bidulescu A et al (2010) Dietary phosphorus, blood pressure, and incidence of hypertension in the atherosclerosis risk in communities study and the multi-ethnic study of atherosclerosis. Hypertension 55(3):776–784
62. Joosten MM, Gansevoort RT, Mukamal KJ, Kootstra-Ros JE, Feskens EJ, Geleijnse JM et al (2013) Urinary magnesium excretion and risk of hypertension: the prevention of renal and vascular end-stage disease study. Hypertension 61(6):1161–1167
63. Takase H, Sugiura T, Kimura G, Ohte N, Dohi Y (2015) Dietary sodium consumption predicts future blood pressure and incident hypertension in the Japanese normotensive general population. J Am Heart Assoc 4(8):e001959
64. Han H, Fang X, Wei X, Liu Y, Jin Z, Chen Q et al (2017) Dose-response relationship between dietary magnesium intake, serum magnesium concentration and risk of hypertension: a systematic review and meta-analysis of prospective cohort studies. Nutr J 16(1):26
65. Jayedi A, Zargar MS (2019) Dietary calcium intake and hypertension risk: a dose-response meta-analysis of prospective cohort studies. Eur J Clin Nutr 73(7):969–978

66. Filippini T, Malavolti M, Whelton PK, Vinceti M (2022) Sodium intake and risk of hypertension: a systematic review and dose-response meta-analysis of observational cohort studies. Curr Hypertens Rep 24(5):133–144
67. Bae E, Ji Y, Jo J, Kim Y, Lee JP, Won S et al (2024) Effects of polygenic risk score and sodium and potassium intake on hypertension in Asians: a nationwide prospective cohort study. Hypertens Res 47(11):3045–3055
68. Han D, Fang X, Su D, Huang L, He M, Zhao D et al (2019) Dietary calcium intake and the risk of metabolic syndrome: a systematic review and meta-analysis. Sci Rep 9(1):19046
69. Cheng L, Hu D, Jiang W (2019) Dietary calcium intake and the risk of metabolic syndrome: evidence from observational studies. Public Health Nutr 22(11):2055–2062
70. Cai X, Li X, Fan W, Yu W, Wang S, Li Z et al (2016) Potassium and obesity/metabolic syndrome: a systematic review and meta-analysis of the epidemiological evidence. Nutrients 8(4):183
71. Moradmand Z, Rouhani P, Saneei P (2025) Association between circulating levels of magnesium and metabolic syndrome: a systematic review and meta-analysis of observational studies with GRADE assessment. Nutr Rev 83:1631
72. Soltani S, Kolahdouz Mohammadi R, Shab-Bidar S, Vafa M, Salehi-Abargouei A (2019) Sodium status and the metabolic syndrome: a systematic review and meta-analysis of observational studies. Crit Rev Food Sci Nutr 59(2):196–206
73. Saltevo J, Niskanen L, Kautiainen H, Teittinen J, Oksa H, Korpi-Hyovalti E et al (2011) Serum calcium level is associated with metabolic syndrome in the general population: FIN-D2D study. Eur J Endocrinol 165(3):429–434
74. Noori N, Nafar M, Poorrezagholi F, Ahmadpoor P, Samadian F, Firouzan A et al (2010) Dietary intakes of fiber and magnesium and incidence of metabolic syndrome in first year after renal transplantation. J Ren Nutr 20(2):101–111
75. Fumeron F, Lamri A, Abi Khalil C, Jaziri R, Porchay-Balderelli I, Lantieri O et al (2011) Dairy consumption and the incidence of hyperglycemia and the metabolic syndrome: results from a french prospective study, Data from the Epidemiological Study on the Insulin Resistance Syndrome (DESIR). Diabetes Care 34(4):813–817
76. Baek JH, Jin SM, Bae JC, Jee JH, Yu TY, Kim SK et al (2017) Serum calcium and the risk of incident metabolic syndrome: a 4.3-year retrospective longitudinal study. Diabetes Metab J 41(1):60–68
77. Woo HW, Lim YH, Kim MK, Shin J, Lee YH, Shin DH et al (2020) Prospective associations between total, animal, and vegetable calcium intake and metabolic syndrome in adults aged 40 years and older. Clin Nutr 39(7):2282–2291
78. Jiao Y, Li W, Wang L, Jiang H, Wang S, Jia X et al (2022) Relationship between dietary magnesium intake and metabolic syndrome. Nutrients 14(10)
79. Emamat H, Ghalandari H, Totmaj AS, Tangestani H, Hekmatdoost A (2021) Calcium to magnesium intake ratio and non-alcoholic fatty liver disease development: a case-control study. BMC Endocr Disord 21(1):51
80. Salehi-Sahlabadi A, Mirfazli E, Teymoori F, Roosta S, Mokari A, Azadi M et al (2021) The association between dietary intake of sodium, potassium, and Na:K ratio with the risk of NAFLD: a case-control study among Iranian Adults. Int J Prev Med 12:179
81. Chen HK, Lan QW, Li YJ, Xin Q, Luo RQ, Wang JJ (2024) Association between dietary potassium intake and nonalcoholic fatty liver disease and advanced hepatic fibrosis in U.S. Adults. Int J Endocrinol 2024:5588104
82. Choi Y, Lee JE, Chang Y, Kim MK, Sung E, Shin H et al (2016) Dietary sodium and potassium intake in relation to non-alcoholic fatty liver disease. Br J Nutr 116(8):1447–1456
83. Zhou L, Yang Y, Feng Y, Zhao X, Fan Y, Rong J et al (2021) Association between dietary sodium intake and non-alcoholic fatty liver disease in the US population. Public Health Nutr 24(5):993–1000

84. Han E, Kim MK, Im SS, Kim HS, Kwon TK, Jang BK (2023) High sodium intake, as assessed by urinary sodium excretion, is associated with nonalcoholic fatty liver disease or sarcopenia. Gut Liver 17(3):456–465
85. Huh JH, Lee KJ, Lim JS, Lee MY, Park HJ, Kim MY et al (2015) High dietary sodium intake assessed by estimated 24-h urinary sodium excretion is associated with NAFLD and hepatic fibrosis. PLoS One 10(11):e0143222
86. van den Berg EH, Gruppen EG, Blokzijl H, Bakker SJL, Dullaart RPF (2019) Higher sodium intake assessed by 24 hour urinary sodium excretion is associated with non-alcoholic fatty liver disease: the PREVEND cohort study. J Clin Med 8(12)
87. Yang C, Wu S, Lan Y, Chen S, Zhang D, Wang Y et al (2023) Association between blood calcium, magnesium, and non-alcoholic fatty liver disease in adults: a cohort-based case-control study. Biol Trace Elem Res 201(10):4625–4636
88. Lee J, Lee JY, Yang YJ (2024) Sex-specific association between sodium intake estimated by 24-hour urinary sodium excretion and nonalcoholic fatty liver disease: the community-based prospective cohort study. Nutrients 16(4)
89. Chatterjee R, Yeh HC, Shafi T, Selvin E, Anderson C, Pankow JS et al (2010) Serum and dietary potassium and risk of incident type 2 diabetes mellitus: the Atherosclerosis Risk in Communities (ARIC) study. Arch Intern Med 170(19):1745–1751
90. Chatterjee R, Yeh HC, Shafi T, Anderson C, Pankow JS, Miller ER et al (2011) Serum potassium and the racial disparity in diabetes risk: the Atherosclerosis Risk in Communities (ARIC) Study. Am J Clin Nutr 93(5):1087–1091
91. Dong JY, Xun P, He K, Qin LQ (2011) Magnesium intake and risk of type 2 diabetes: meta-analysis of prospective cohort studies. Diabetes Care 34(9):2116–2122
92. Dong JY, Qin LQ (2012) Dietary calcium intake and risk of type 2 diabetes: possible confounding by magnesium. Eur J Clin Nutr 66(3):408–410
93. Xu T, Chen GC, Zhai L, Ke KF (2015) Nonlinear reduction in risk for type 2 diabetes by magnesium intake: an updated meta-analysis of prospective cohort studies. Biomed Environ Sci 28(7):527–534
94. Fang X, Han H, Li M, Liang C, Fan Z, Aaseth J et al (2016) Dose-response relationship between dietary magnesium intake and risk of type 2 diabetes mellitus: a systematic review and meta-regression analysis of prospective cohort studies. Nutrients 8(11)
95. Peng Y, Zhong GC, Mi Q, Li K, Wang A, Li L et al (2017) Potassium measurements and risk of type 2 diabetes: a dose-response meta-analysis of prospective cohort studies. Oncotarget 8(59):100603–100613
96. Kim KN, Oh SY, Hong YC (2018) Associations of serum calcium levels and dietary calcium intake with incident type 2 diabetes over 10 years: the Korean Genome and Epidemiology Study (KoGES). Diabetol Metab Syndr 10:50
97. Mancini FR, Affret A, Dow C, Balkau B, Clavel-Chapelon F, Bonnet F et al (2018) High dietary phosphorus intake is associated with an increased risk of type 2 diabetes in the large prospective E3N cohort study. Clin Nutr 37(5):1625–1630
98. Zhu J, Xun P, Bae JC, Kim JH, Kim DJ, Yang K et al (2019) Circulating calcium levels and the risk of type 2 diabetes: a systematic review and meta-analysis. Br J Nutr 122(4):376–387
99. Zhao B, Deng H, Li B, Chen L, Zou F, Hu L et al (2020) Association of magnesium consumption with type 2 diabetes and glucose metabolism: a systematic review and pooled study with trial sequential analysis. Diabetes Metab Res Rev 36(3):e3243
100. Hao G, Liu K, Halbert JD, Chen H, Wu J, Jing C (2020) Dietary sodium and potassium and risk of diabetes: a prospective study using data from the China health and nutrition survey. Diabetes Metab 46(5):377–383
101. Dong Z, Richie JP Jr, Gao X, Al-Shaar L, Nichenametla SN, Shen B et al (2022) Cumulative consumption of sulfur amino acids and risk of diabetes: a prospective cohort study. J Nutr 152(11):2419–2428
102. Wang X, Ma H, Kou M, Tang R, Xue Q, Li X et al (2023) Dietary sodium intake and risk of incident type 2 diabetes. Mayo Clin Proc 98:1641

103. Wu Q, Ye Z, Zhang Y, Yang S, Zhou C, Liu M et al (2023) A U-shaped association between dietary phosphorus intake and new-onset diabetes: a nationwide cohort study in China. Nutr Metab Cardiovasc Dis 33(10):1932–1940
104. Li J, Li Y, Wang C, Mao Z, Yang T, Li Y et al (2024) Dietary potassium and magnesium intake with risk of type 2 diabetes mellitus among rural China: the Henan rural cohort study. Biol Trace Elem Res 202(9):3932–3944
105. Park SK, Oh CM, Ryoo JH, Jung JY (2025) Intake of sodium and potassium, sodium-potassium intake ratio, and their relation to the risk of diabetes mellitus. Sci Rep 15(1):4411
106. Yoon CY, Park JT, Jhee JH, Noh J, Kee YK, Seo C et al (2017) High dietary phosphorus density is a risk factor for incident chronic kidney disease development in diabetic subjects: a community-based prospective cohort study. Am J Clin Nutr 106(1):311–321
107. Lee J, Oh KH, Park SK (2021) Dietary micronutrients and risk of chronic kidney disease: a cohort study with 12 year follow-up. Nutrients 13(5)
108. Koh HB, Kim HJ, Heo GY, Kim HW, Jung CY, Han SH et al (2024) Association between dietary magnesium intake and incident chronic kidney disease: a prospective observational cohort study. Am J Clin Nutr 120(4):964–972
109. Tonelli M, Sacks F, Pfeffer M, Gao Z, Curhan G, Cholesterol and Recurrent Events Trial Investigators et al (2005) Relation between serum phosphate level and cardiovascular event rate in people with coronary disease. Circulation 112(17):2627–2633
110. D'Elia L, Barba G, Cappuccio FP, Strazzullo P (2011) Potassium intake, stroke, and cardiovascular disease a meta-analysis of prospective studies. J Am Coll Cardiol 57(10):1210–1219
111. O'Donnell M, Mente A, Rangarajan S, McQueen MJ, Wang X, Liu L et al (2014) Urinary sodium and potassium excretion, mortality, and cardiovascular events. N Engl J Med 371(7):612–623
112. Adebamowo SN, Spiegelman D, Willett WC, Rexrode KM (2015) Association between intakes of magnesium, potassium, and calcium and risk of stroke: 2 cohorts of US women and updated meta-analyses. Am J Clin Nutr 101(6):1269–1277
113. Bain LK, Myint PK, Jennings A, Lentjes MA, Luben RN, Khaw KT et al (2015) The relationship between dietary magnesium intake, stroke and its major risk factors, blood pressure and cholesterol, in the EPIC-Norfolk cohort. Int J Cardiol 196:108–114
114. Tian DY, Tian J, Shi CH, Song B, Wu J, Ji Y et al (2015) Calcium intake and the risk of stroke: an up-dated meta-analysis of prospective studies. Asia Pac J Clin Nutr 24(2):245–252
115. Mente A, O'Donnell M, Rangarajan S, Dagenais G, Lear S, McQueen M et al (2016) Associations of urinary sodium excretion with cardiovascular events in individuals with and without hypertension: a pooled analysis of data from four studies. Lancet 388(10043):465–475
116. Rohrmann S, Garmo H, Malmstrom H, Hammar N, Jungner I, Walldius G et al (2016) Association between serum calcium concentration and risk of incident and fatal cardiovascular disease in the prospective AMORIS study. Atherosclerosis 251:85–93
117. Fang X, Wang K, Han D, He X, Wei J, Zhao L et al (2016) Dietary magnesium intake and the risk of cardiovascular disease, type 2 diabetes, and all-cause mortality: a dose-response meta-analysis of prospective cohort studies. BMC Med 14(1):210
118. Vinceti M, Filippini T, Crippa A, de Sesmaisons A, Wise LA, Orsini N (2016) Meta-analysis of potassium intake and the risk of stroke. J Am Heart Assoc 5(10)
119. Zhu Y, Zhang J, Li Z, Liu Y, Fan X, Zhang Y et al (2018) Association of sodium intake and major cardiovascular outcomes: a dose-response meta-analysis of prospective cohort studies. BMC Cardiovasc Disord 18(1):192
120. Jayedi A, Ghomashi F, Zargar MS, Shab-Bidar S (2019) Dietary sodium, sodium-to-potassium ratio, and risk of stroke: a systematic review and nonlinear dose-response meta-analysis. Clin Nutr 38(3):1092–1100
121. Goncalves C, Abreu S (2020) Sodium and potassium intake and cardiovascular disease in older people: a systematic review. Nutrients 12(11)
122. Wang YJ, Yeh TL, Shih MC, Tu YK, Chien KL (2020) Dietary sodium intake and risk of cardiovascular disease: a systematic review and dose-response meta-analysis. Nutrients 12(10)

123. Zhao B, Zeng L, Zhao J, Wu Q, Dong Y, Zou F et al (2020) Association of magnesium intake with type 2 diabetes and total stroke: an updated systematic review and meta-analysis. BMJ Open 10(3):e032240
124. Zhu HL, Liu Y, Zhang J, Wang MX, Jiang H, Guo F et al (2021) Dietary calcium, magnesium, and phosphorus intakes and risk of stroke in Chinese adults. Sci Rep 11(1):11270
125. Myung SK, Kim HB, Lee YJ, Choi YJ, Oh SW (2021) Calcium supplements and risk of cardiovascular disease: a meta-analysis of clinical trials. Nutrients 13(2)
126. Wang ZM, Bu XX, Zhou B, Li YF, Nie ZL (2023) Dietary calcium intake and the risk of stroke: meta-analysis of cohort studies. Nutr Metab Cardiovasc Dis 33(5):934–946
127. Graudal N, Jurgens G, Baslund B, Alderman MH (2014) Compared with usual sodium intake, low- and excessive-sodium diets are associated with increased mortality: a meta-analysis. Am J Hypertens 27(9):1129–1137
128. Chang AR, Lazo M, Appel LJ, Gutierrez OM, Grams ME (2014) High dietary phosphorus intake is associated with all-cause mortality: results from NHANES III. Am J Clin Nutr 99(2):320–327
129. Asemi Z, Saneei P, Sabihi SS, Feizi A, Esmaillzadeh A (2015) Total, dietary, and supplemental calcium intake and mortality from all-causes, cardiovascular disease, and cancer: a meta-analysis of observational studies. Nutr Metab Cardiovasc Dis 25(7):623–634
130. Bai W, Li J, Liu J (2016) Serum phosphorus, cardiovascular and all-cause mortality in the general population: a meta-analysis. Clin Chim Acta 461:76–82
131. Cook NR, Appel LJ, Whelton PK (2016) Sodium intake and all-cause mortality over 20 years in the trials of hypertension prevention. J Am Coll Cardiol 68(15):1609–1617
132. Yang B, Campbell PT, Gapstur SM, Jacobs EJ, Bostick RM, Fedirko V et al (2016) Calcium intake and mortality from all causes, cancer, and cardiovascular disease: the cancer prevention study II nutrition cohort. Am J Clin Nutr 103(3):886–894
133. Bagheri A, Naghshi S, Sadeghi O, Larijani B, Esmaillzadeh A (2021) Total, dietary, and supplemental magnesium intakes and risk of all-cause, cardiovascular, and cancer mortality: a systematic review and dose-response meta-analysis of prospective cohort studies. Adv Nutr 12(4):1196–1210
134. Naghshi S, Naemi M, Sadeghi O, Darooghegi Mofrad M, Moezrad M, Azadbakht L (2022) Total, dietary, and supplemental calcium intake and risk of all-cause cardiovascular, and cancer mortality: a systematic review and dose-response meta-analysis of prospective cohort studies. Crit Rev Food Sci Nutr 62(21):5733–5743
135. Kwon YJ, Lee HS, Park G, Lee JW (2022) Association between dietary sodium, potassium, and the sodium-to-potassium ratio and mortality: a 10-year analysis. Front Nutr 9:1053585
136. Hou X, Xu W, Zhang C, Song Z, Zhu M, Guo Q et al (2023) L-shaped association of serum chloride level with all-cause and cause-specific mortality in American adults: population-based prospective cohort study. JMIR Public Health Surveill 9:e49291
137. Yazdanpanah MH, Sharafkhah M, Poustchi H, Etemadi A, Sheikh M, Kamangar F et al (2024) Mineral intake and cardiovascular disease, cancer, and all-cause mortality: findings from the Golestan cohort study. Nutrients 16(3)
138. Liu D, Tian Y, Wang R, Zhang T, Shen S, Zeng P et al (2024) Sodium, potassium intake, and all-cause mortality: confusion and new findings. BMC Public Health 24(1):180
139. Gan L, Zhao B, Inoue-Choi M, Liao LM, Graubard BI, Weinstein SJ et al (2024) Sex-specific associations between sodium and potassium intake and overall and cause-specific mortality: a large prospective U.S. cohort study, systematic review, and updated meta-analysis of cohort studies. BMC Med 22(1):132
140. Jackson WF (2017) Potassium channels in regulation of vascular smooth muscle contraction and growth. Adv Pharmacol 78:89–144
141. Haddy FJ, Vanhoutte PM, Feletou M (2006) Role of potassium in regulating blood flow and blood pressure. Am J Physiol Regul Integr Comp Physiol 290(3):R546–R552

142. Huang L, Li Q, Wu JH, Tian M, Yin X, Yu J et al (2024) The contribution of sodium reduction and potassium increase to the blood pressure lowering observed in the salt substitute and stroke study. J Hum Hypertens 38(4):298–306
143. Gritter M, Rotmans JI, Hoorn EJ (2019) Role of dietary K(+) in natriuresis, blood pressure reduction, cardiovascular protection, and renoprotection. Hypertension 73(1):15–23
144. Darooghegi Mofrad M, Djafarian K, Mozaffari H, Shab-Bidar S (2018) Effect of magnesium supplementation on endothelial function: a systematic review and meta-analysis of randomized controlled trials. Atherosclerosis 273:98–105
145. Touyz RM (2008) Transient receptor potential melastatin 6 and 7 channels, magnesium transport, and vascular biology: implications in hypertension. Am J Physiol Heart Circ Physiol 294(3):H1103–H1118
146. Rodriguez-Moran M, Guerrero-Romero F (2003) Oral magnesium supplementation improves insulin sensitivity and metabolic control in type 2 diabetic subjects: a randomized double-blind controlled trial. Diabetes Care 26(4):1147–1152
147. Kim DJ, Xun P, Liu K, Loria C, Yokota K, Jacobs DR Jr et al (2010) Magnesium intake in relation to systemic inflammation, insulin resistance, and the incidence of diabetes. Diabetes Care 33(12):2604–2610
148. Chanut F (2006) Potassium channels rule over insulin release with an ion fist. PLoS Biol 4(2):e53
149. Plavinik FL, Rodrigues CI, Zanella MT, Ribeiro AB (1992) Hypokalemia, glucose intolerance, and hyperinsulinemia during diuretic therapy. Hypertension 19(2 Suppl):II26–II29
150. Lanner JT, Katz A, Tavi P, Sandstrom ME, Zhang SJ, Wretman C et al (2006) The role of Ca2+ influx for insulin-mediated glucose uptake in skeletal muscle. Diabetes 55(7):2077–2083
151. Sugimoto J, Romani AM, Valentin-Torres AM, Luciano AA, Ramirez Kitchen CM, Funderburg N et al (2012) Magnesium decreases inflammatory cytokine production: a novel innate immunomodulatory mechanism. J Immunol 188(12):6338–6346
152. Maier JA, Castiglioni S, Locatelli L, Zocchi M, Mazur A (2021) Magnesium and inflammation: advances and perspectives. Semin Cell Dev Biol 115:37–44
153. Ashique S, Kumar S, Hussain A, Mishra N, Garg A, Gowda BHJ et al (2023) A narrative review on the role of magnesium in immune regulation, inflammation, infectious diseases, and cancer. J Health Popul Nutr 42(1):74
154. D'Elia L, Cappuccio FP, Masulli M, La Fata E, Rendina D, Galletti F (2023) Effect of potassium supplementation on endothelial function: a systematic review and meta-analysis of intervention studies. Nutrients 15(4)
155. Morvaridzadeh M, Agah S, Alibakhshi P, Heydari H, Hoseini AS, Palmowski A et al (2021) Effects of calcium and vitamin D co-supplementation on the lipid profile: a systematic review and meta-analysis. Clin Ther 43(9):274–296
156. Li P, Fan C, Lu Y, Qi K (2016) Effects of calcium supplementation on body weight: a meta-analysis12. Am J Clin Nutr 104(5):1263–1273
157. Major GC, Alarie F, Dore J, Phouttama S, Tremblay A (2007) Supplementation with calcium + vitamin D enhances the beneficial effect of weight loss on plasma lipid and lipoprotein concentrations. Am J Clin Nutr 85(1):54–59
158. Kjolbaek L, Lorenzen JK, Larsen LH, Astrup A (2017) Calcium intake and the associations with faecal fat and energy excretion, and lipid profile in a free-living population. J Nutr Sci 6:e50
159. Salehidoost R, Taghipour Boroujeni G, Feizi A, Aminorroaya A, Amini M (2022) Effect of oral magnesium supplement on cardiometabolic markers in people with prediabetes: a double blind randomized controlled clinical trial. Sci Rep 12(1):18209
160. Oda H (2006) Functions of sulfur-containing amino acids in lipid metabolism. J Nutr 136(6 Suppl):1666S–1669S
161. Ter Braake AD, Tinnemans PT, Shanahan CM, Hoenderop JGJ, de Baaij JHF (2018) Magnesium prevents vascular calcification in vitro by inhibition of hydroxyapatite crystal formation. Sci Rep 8(1):2069

162. Louvet L, Buchel J, Steppan S, Passlick-Deetjen J, Massy ZA (2013) Magnesium prevents phosphate-induced calcification in human aortic vascular smooth muscle cells. Nephrol Dial Transplant 28(4):869–878
163. Cozzolino M, Ciceri P, Galassi A, Mangano M, Carugo S, Capelli I et al (2019) The key role of phosphate on vascular calcification. Toxins (Basel) 11(4)
164. Amann K, Törnig J, Kugel B, Gross M-L, Tyralla K, El-Shakmak A et al (2003) Hyperphosphatemia aggravates cardiac fibrosis and microvascular disease in experimental uremia. Kidney Int 63(4):1296–1301
165. Rafaqat S, Rafaqat S, Khurshid H, Rafaqat S (2022) Electrolyte's imbalance role in atrial fibrillation: pharmacological management. Int J Arrhythm 23(1):15
166. Noordam R, Young WJ, Salman R, Kanters JK, van den Berg ME, van Heemst D et al (2019) Effects of calcium, magnesium, and potassium concentrations on ventricular repolarization in unselected individuals. J Am Coll Cardiol 73(24):3118–3131
167. Cazzola R, Della Porta M, Piuri G, Maier JA (2024) Magnesium: a defense line to mitigate inflammation and oxidative stress in adipose tissue. Antioxidants (Basel) 13(8)
168. Pelczynska M, Moszak M, Bogdanski P (2022) The role of magnesium in the pathogenesis of metabolic disorders. Nutrients 14(9)
169. Schutten JC, Joris PJ, Mensink RP, Danel RM, Goorman F, Heiner-Fokkema MR et al (2019) Effects of magnesium citrate, magnesium oxide and magnesium sulfate supplementation on arterial stiffness in healthy overweight individuals: a study protocol for a randomized controlled trial. Trials 20(1):295
170. Joris PJ, Plat J, Bakker SJ, Mensink RP (2016) Long-term magnesium supplementation improves arterial stiffness in overweight and obese adults: results of a randomized, double-blind, placebo-controlled intervention trial. Am J Clin Nutr 103(5):1260–1266
171. Sun Y, Byon CH, Yang Y, Bradley WE, Dell'Italia LJ, Sanders PW et al (2017) Dietary potassium regulates vascular calcification and arterial stiffness. JCI Insight 2(19)
172. Vlachopoulos C, Aznaouridis K, Stefanadis C (2010) Prediction of cardiovascular events and all-cause mortality with arterial stiffness: a systematic review and meta-analysis. J Am Coll Cardiol 55(13):1318–1327
173. Goto K, Kitazono T (2022) Chloride ions, vascular function and hypertension. Biomedicine 10(9)
174. Newman AB, Murabito JM (2013) The epidemiology of longevity and exceptional survival. Epidemiol Rev 35(1):181–197
175. Browning CJ, Thomas SA, Kendig H, Ory MG (2017) Healthy aging. In: Pachana NA (ed) Encyclopedia of geropsychology. Springer Singapore, Singapore, pp 1033–1041
176. Fekete M, Szarvas Z, Fazekas-Pongor V, Feher A, Csipo T, Forrai J et al (2022) Nutrition strategies promoting healthy aging: from improvement of cardiovascular and brain health to prevention of age-associated diseases. Nutrients 15(1)
177. Lo WC, Hu TH, Shih CY, Lin HH, Hwang JS (2024) Impact of healthy lifestyle factors on life expectancy and lifetime health care expenditure: Nationwide cohort study. JMIR Public Health Surveill 10:e57045
178. Li Y, Pan A, Wang DD, Liu X, Dhana K, Franco OH et al (2018) Impact of healthy lifestyle factors on life expectancies in the US population. Circulation 138(4):345–355
179. Fekete M, Major D, Feher A, Fazekas-Pongor V, Lehoczki A (2024) Geroscience and pathology: a new frontier in understanding age-related diseases. Pathol Oncol Res. https://doi.org/10.3389/pore.2024.1611623
180. Csiszar A, Wang M, Lakatta EG, Ungvari ZI (2008) Inflammation and endothelial dysfunction during aging: role of NF-{kappa}B. J Appl Physiol 105(4):1333–1341
181. Rebelo-Marques A, De Sousa Lages A, Andrade R, Ribeiro CF, Mota-Pinto A, Carrilho F et al (2018) Aging hallmarks: the benefits of physical exercise. Front Endocrinol (Lausanne) 9:258
182. Garatachea N, Pareja-Galeano H, Sanchis-Gomar F, Santos-Lozano A, Fiuza-Luces C, Moran M et al (2015) Exercise attenuates the major hallmarks of aging. Rejuvenation Res 18(1):57–89

183. Wilson DM 3rd, Cookson MR, Van Den Bosch L, Zetterberg H, Holtzman DM, Dewachter I (2023) Hallmarks of neurodegenerative diseases. Cell 186(4):693–714
184. Lopez-Otin C, Blasco MA, Partridge L, Serrano M, Kroemer G (2013) The hallmarks of aging. Cell 153(6):1194–1217
185. Ungvari Z, Tarantini S, Donato AJ, Galvan V, Csiszar A (2018) Mechanisms of vascular aging. Circ Res 123(7):849–867
186. Wang W, Soltero L, Zhang P, Huang XR, Lan HY, Adrogue HJ (2007) Renal inflammation is modulated by potassium in chronic kidney disease: possible role of Smad7. Am J Physiol Renal Physiol 293(4):F1123–F1130
187. Gonzalez-Freire M, de Cabo R, Bernier M, Sollott SJ, Fabbri E, Navas P et al (2015) Reconsidering the role of mitochondria in aging. J Gerontol A Biol Sci Med Sci 70(11):1334–1342
188. Canto C, Menzies KJ, Auwerx J (2015) NAD(+) metabolism and the control of energy homeostasis: a balancing act between mitochondria and the nucleus. Cell Metab 22(1):31–53
189. Marzetti E, Csiszar A, Dutta D, Balagopal G, Calvani R, Leeuwenburgh C (2013) Role of mitochondrial dysfunction and altered autophagy in cardiovascular aging and disease: from mechanisms to therapeutics. Am J Physiol Heart Circ Physiol 305:H459
190. Silaghi CN, Farcas M, Craciun AM (2021) Sirtuin 3 (SIRT3) pathways in age-related cardiovascular and neurodegenerative diseases. Biomedicine 9(11)
191. Pikor D, Hurla M, Slowikowski B, Szymanowicz O, Poszwa J, Banaszek N et al (2024) Calcium ions in the physiology and pathology of the central nervous system. Int J Mol Sci 25(23)
192. Barreira-Hernandez D, Rodriguez-Martin S, Gil M, Mazzucchelli R, Izquierdo-Esteban L, Garcia-Lledo A et al (2023) Risk of ischemic stroke associated with calcium supplements and interaction with oral bisphosphonates: a nested case-control study. J Clin Med 12(16)
193. Rodriguez-Ortiz ME, Canalejo A, Herencia C, Martinez-Moreno JM, Peralta-Ramirez A, Perez-Martinez P et al (2014) Magnesium modulates parathyroid hormone secretion and upregulates parathyroid receptor expression at moderately low calcium concentration. Nephrol Dial Transplant 29(2):282–289
194. Pittas AG, Harris SS, Stark PC, Dawson-Hughes B (2007) The effects of calcium and vitamin D supplementation on blood glucose and markers of inflammation in nondiabetic adults. Diabetes Care 30(4):980–986
195. Coudray C, Demigne C, Rayssiguier Y (2003) Effects of dietary fibers on magnesium absorption in animals and humans. J Nutr 133(1):1–4
196. Grober U (2019) Magnesium and drugs. Int J Mol Sci 20(9)

Chapter 20
Major Minerals (Na, K, Ca, Mg, P) in Physiological Function and Disease

Ambika Binesh, Kaliyamurthi Venkatachalam, Amit Ranjan, Sisira Joy, and Anusha Prasannan

Abstract This chapter examines the five essential macro-minerals, sodium, potassium, calcium, magnesium, and phosphorus, based on their physiological significance, dietary sources, absorption mechanisms, and systemic functions. These minerals are needed in moderate levels; however, they are crucial to bone metabolism, fluid management, acid-base balance, homeostasis, metabolic and neuromuscular activity, and the prevention of chronic diseases. The influence of modern diets on mineral status, along with the medical implications of their excesses and deficiencies, and a focus on WHO/FAO dietary recommendations are discussed. The chapter summarizes the increasing importance of mineral supplementation in immune system regulation, critical illness, and the prospective use of precision medicine in combination with personalized nutrition to maximize mineral intake for better health outcomes.

Keywords Dietary intake · Disease prevention · Macro-minerals · Macro-mineral interactions · Personalized medicine · Recommended Dietary Allowance (RDA)

Introduction

Definition and Classification of Major Minerals (vs. Trace Elements)

Minerals are chemical elements, in the context of human nutrition, that are essential for normal human physiology and maintaining good health. They serve as vital nutrients, derived from the diet, to support the biochemical reactions of metabolism

A. Binesh (✉) · K. Venkatachalam · A. Ranjan · S. Joy · A. Prasannan
Department of Basic Sciences, Institute of Fisheries Post Graduate Studies, Tamil Nadu Dr. J. Jayalalithaa Fisheries University (TNJFU), Chennai, Tamil Nadu, India
e-mail: ambikabinesh@tnfu.ac.in

N. S. Dhalla et al. (eds.), *Functional Biochemistry of Metallic Elements*, Advances in Biochemistry in Health and Disease 36,
https://doi.org/10.1007/978-3-032-14445-4_20

in living organisms. They are broadly classified into two: macro-minerals (calcium (Ca^{2+}), chlorine (Cl^-), phosphorus(PO_4^{3+}), potassium(K^+), sodium (Na^+), and magnesium (Mg^{2+})) and micro-minerals or trace elements (sulfur (S), iron ($Fe^{2+/3+}$), cobalt ($Co^{2+/3+}$), copper ($Cu^{1+/2+}$), zinc (Zn^{2+}), manganese (Mn), molybdenum (Mo), iodine (I^-), and selenium (Se^{2+})). This classification is based solely on the quantity needed, but both categories play a crucial role in ensuring the proper functioning of the body. It is reported that daily, more than 100 mg/dl of macro-minerals and less than 100 mg/dl of micro-minerals are required. Globally, significant health issues are encountered due to the deficiency of essential minerals in the body that can be caused by genetics, diet, and lifestyle habits [44].

Overview of the Five Minerals (Sodium (Na), Potassium (K), Calcium (Ca), Magnesium (Mg), Phosphorus (P)) and Their Importance in Human Physiology

The human body has 4–6% minerals, including the most prevalent macro-minerals (calcium, magnesium, sodium, potassium, and phosphorus). Calcium maintains bone integrity, hemostasis, neurotransmission, muscular contraction, and biochemical activity, and its deficiency results in conditions such as osteoporosis, rickets, and hypertension [37, 44]. Magnesium is involved in energy production, protein synthesis, neuromuscular coordination, and cardiovascular well-being [7, 28]. Through the sodium-potassium pump, sodium and potassium serve as the main extracellular and intracellular electrolytes that regulate neuronal transmission, fluid equilibrium, and muscle contractions [77, 78]. Phosphorus forms an essential component of adenosine triphosphate (ATP), nucleic acids, and phospholipids, making it essential for energy metabolism, bone growth, and the maintenance of acid-base equilibrium [77, 78]. Figure 20.1 gives an overview of the functions and conditions caused by their deficiency.

To ensure the body's metabolic, structural, and regulatory functions run effectively, macro-minerals must be found in the right amounts or the right ratio. They neutralize the toxic substances, aiding their elimination, and also by hindering the buildup of toxins in the body. These macro-minerals are needed in comparatively large quantities, and their dietary intake has to be balanced to avoid deficiencies that might end up in serious health problems such as metabolic disorders, chronic illnesses, and weakness in muscles.

Brief Note on Dietary Sources and Global Health Relevance

Milk and dairy products are the main source of calcium. Some green leafy vegetables, fish, meat, and grains are other sources. It is necessary for bone strength, neurotransmission, and blood coagulation [79]. Inadequate calcium intake is associated

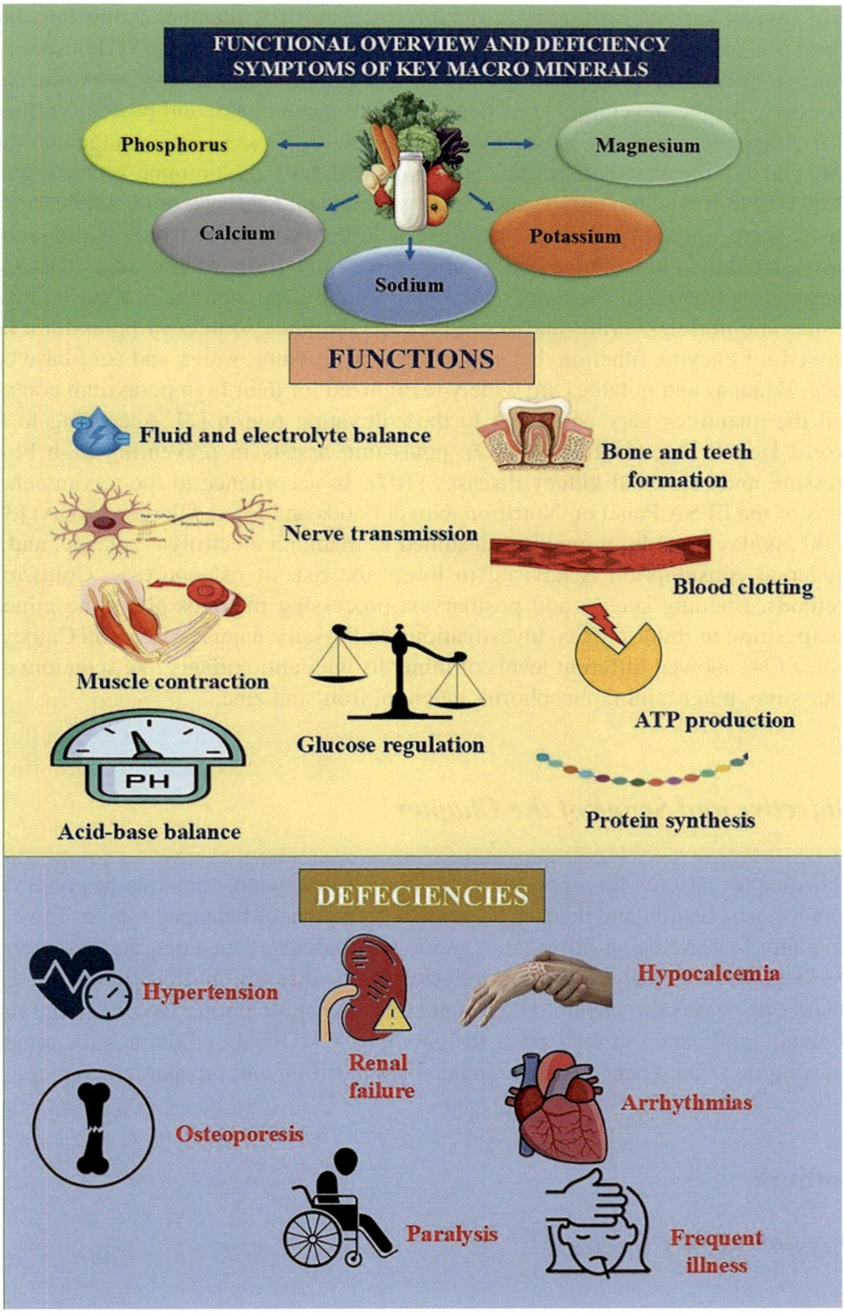

Fig. 20.1 Functional roles and deficiency symptoms of key macromolecules in human health

with several serious health risks, notably osteoporosis, metabolic diseases, high blood pressure, and cancer of the gastrointestinal tract and breast [33]. Phosphorus, which is present in meat, fish, grains, nuts, legumes, milk, and dairy products, is necessary for the development of bones, energy metabolism, and protein synthesis [79]. Adequate consumption of phosphorus promotes bone health, collagen production, and cellular metabolism. The recommended daily requirement is 700 mg/day [33]. Whole foods, including green leafy vegetables, legumes, nuts, whole grains, seeds, dairy, animal products, and fruits, include magnesium, which is the second most abundant intracellular cation after potassium [64]. Magnesium makes up approximately 60% of the body's total magnesium level, and the rest exists in soft tissues and muscles. Fruits and vegetables are abundant sources of potassium. It is needed for enzyme function, balancing osmotic pressure, water, and acid-base balance. Bananas and potatoes are widely recognized for their high potassium content, and the quantities vary according to the cultivation region [2]. According to the World Health Organization (WHO), potassium assists in preventing high blood pressure and heart and kidney diseases [107]. In accordance to the recommendations of the EFSA Panel on Nutrition, Novel Foods and Food Allergen (NDA) [36], 2300 mg/day of sodium may be consumed to maintain electrolyte balance, and its moderate consumption is advised to lower the risk of osteoporosis. Cultivation methods, ripening stages, and postharvest processing methods alters the mineral composition of food sources. Investigations on Brassica napus L. [99] and Calocybe indica [94] showed different levels of minerals with antioxidants like selenium and potassium, magnesium, phosphorus, calcium, iron, and zinc.

Objective and Scope of the Chapter

This chapter ensures the importance of the five key macro-minerals in preserving homeostasis, health, and disease prevention by means of balanced intake. The contents aim to serve as an informative guide for students, educators, and researchers working in the fields of nutrition, physiology, healthcare, and biochemistry. This chapter discusses the physiological functions, crosstalk among minerals, and their clinical significance. It also gives insights into WHO/FAO guidelines for mineral consumption and recent advances in personal nutrition and precision medicine.

Sodium

Physiological Functions

Sodium (Na^{+}), a monovalent cation, is a major ion in extracellular fluid (ECF) and is responsible for maintaining fluid homeostasis, excitability of the neuronal elements, and muscle activity. It contributes to nearly 88% of the osmolarity of ECF in combination with anions like chloride and bicarbonate. Homeostasis of sodium is maintained by the equilibrium between intake through diet and excretion by urine, feces, and sweating [9]. Sodium also involved in regulation of extracellular fluid (ECF) volume and osmolarity. It is responsible for 90–95% ECF osmolarity by binding with chloride and bicarbonate. Alterations in sodium levels have a direct impact on osmotic gradients, affecting water distribution between intracellular and extracellular spaces [19]. Sodium is essential for neuron excitability in facilitating and conducting action potentials by voltage-gated Na^+ channels, mainly at nodes of Ranvier on myelinated neurons. In muscle, neuromuscular junction Na^+ influx leads membrane depolarization, release of Ca^{2+} via DHPR-RyR coupling, and muscle contraction. The Na^+/K^+-ATPase is the reason for resting potential in neurons and muscle. Also, Na^+ maintains acid-base balance as a component of the bicarbonate buffer ($NaHCO_3$), allowing for H^+ buffering and CO_2 elimination during acidosis, and provides plasma electroneutrality [49].

Absorption, Transport, and Regulation

Absorption of sodium is important for fluid balance, regulation of pH, and nutrient absorption. In the small intestine, Na^+ enters by passive diffusion, Na^+/glucose symporters (such as SGLT1), and Na^+/H^+ exchangers (NHE3), and basolateral Na^+/K^+-ATPase is essential for regulating gradients. Electroneutral Na^+–Cl^- absorption in the colon is through NHE3 and Cl^-/HCO_3^- exchangers. Paracellular pathways aids with absorption, particularly in the jejunum. In the kidney, approximately 99% reabsorption of filtered sodium occurs: 65–70% in the proximal tubule, 25% in the thick ascending limb, 5% in the distal tubule, and the remainder in the collecting duct through ENaC, which is aldosterone-regulated [32, 62].

Deficiency and Excess

Hyponatremia (serum Na^+ <135 mEq/L) is the most frequent electrolyte abnormality, resulting from water retention caused by inappropriately secreted ADH. It is divided into volume status: hypovolemic (loss of Na^+ > water loss), euvolemic (SIADH, hypothyroidism), and hypovolemic (heart failure, cirrhosis). The presentation varies from mild cognitive symptoms to seizure and coma. Risk and more significant sodium disturbances are frequently observed in susceptible populations. It leads to neural dehydration, resulting in lethargy, seizures, or coma. Cerebral edema can be precipitated by acute correction of chronic cases. It indicates fluid imbalance and high fatality. Salt-sensitivity hypertension is a condition of amplified BP response to Na^+ resulting in the inability to excrete Na^+ by kidneys, RAAS/SNS overactivity, and mechanisms associated with obesity. It increases cardiovascular risk and is more prevalent in older, obese, and Black individuals. It is treated with sodium restriction and directed therapy [53, 76, 80, 88, 91].

Disease

Excessive sodium consumption (>5 g/day) increases blood pressure (BP), which elevates risks of hypertension, cardiovascular disease, and stroke. Decreasing sodium decreases BP and cardiovascular morbidity, with even modest decreases being effective in all populations. Elevated sodium promotes water retention, expanded blood volume, and pressure natriuresis, which cause vascular remodelling and endothelial injury, even in normotensive individuals. Renal absorption under aldosterone, sympathetic activity, and natriuretic peptides primarily regulate sodium balance. Whereas healthy kidneys compensate for sodium changes, salt-sensitive subjects exhibit BP elevations with excess sodium (~1–2 g/day). In CKD, disrupted sodium excretion results in volume expansion and worsening blood pressure. In nephrotic syndrome, stimulation by plasmin (through proteinuria) of ENaC causes sodium retention regardless of aldosterone; amiloride has potential therapeutic utility. Excessive dietary sodium also enhances urinary calcium loss, leading to parathyroid hormone-mediated bone reabsorption and increased fracture risk. Low sodium triggers RAAS and oxidative stress, raises osteoclasts, and elevates fall risk, particularly in older individuals and sex hormone deficiency patients. In zebrafish, gill ionocytes are specialized for active regulation of Na^+ uptake and excretion and are of interest in osmoregulation research. Rodent models show sodium intake by diet affects blood pressure and renal sodium regulation, often through Na^+/K^+ATPase and epithelial sodium channels (ENaC). Evidence from mice reveals the genetic changes and the pathophysiological processes underlying hypertension and sodium-sensitive cardiovascular disease. Sodium balance is a determinant of neuronal signaling and behavior in Drosophila, with mutants of sodium transporters having altered locomotor activity and synaptic transmission [12, 50, 54, 80].

Potassium (K)

Physiological Functions

Potassium (K^+), the major intracellular cation, is required to sustain membrane potential, nerve conduction, muscle contraction, and cell homeostasis. Approximately 98% of body K^+ is intracellular (120–150 mM), whereas extracellular concentration (3.5–5.0 mM) is maintained low by the Na^+/K^+-ATPase pump that utilizes ATP to exchange intracellular K^+ for Na^+ to sustain the resting membrane potential and permit electrical excitability. In muscle cells and neurons, K^+ efflux via leak and voltage-gated channels facilitates after-hyperpolarization and repolarization during action potentials. Any disruption leads to impaired muscle function and nerve conduction and to arrhythmias or paralysis. In muscle contraction, K^+ repolarizes membranes to facilitate calcium reuptake and termination of contraction. Beyond excitability, K^+ supports protein synthesis, enzyme activity, osmotic balance, intracellular pH, and epithelial transport in the kidney and the gut. Regulation of K^+ is crucial because slight disturbances may cause severe physiological dysfunction [26, 49, 95].

Absorption, Transport, and Regulation

Potassium (K^+) is essential for cardiac, neuromuscular, and cellular processes. Approximately 98% is found intracellularly, with strict control over the extracellular concentration (3.5–5.0 mEq/L). Although the kidneys are the main controllers, the GI tract is the major organ for K^+ absorption and secretion. Daily intake (~70–100 mEq) is balanced by GI secretions (90–100 mEq), predominantly reabsorbed in the small intestine. Loss via feces is only ~5–10 mEq/day. Potassium absorption is passive in the small intestine (through solvent drag and electrochemical gradients) and active in the colon through Na^+/K^+-ATPase, H^+/K^+-ATPase, and K^+ channels. Regulatory factors include hormones such as aldosterone, which increases potassium secretion, as well as luminal pH, where an acidic environment promotes potassium uptake. Nutrients are another regulatory factor; this includes glucose and insulin which stimulate absorption. Transit time is also an important part; fast transit compromises absorption, with augmented fecal loss of K^+ [95].

Deficiency and Excess

Potassium (K^+) is required for the maintenance of membrane potential, conduction of nerve impulses, and muscle function. Both low and elevated serum potassium levels may cause severe, frequently lethal complications. Hypokalemia (K^+ <3.5 mEq/L) is due to inadequate intake, loss with diarrhea, vomiting, diuretics, or transcellular shifts. It leads to muscle cramps and weakness, constipation, paralysis, and life-threatening arrhythmias. Treatment is with oral or IV potassium replacement with ECG monitoring [65]. Hyperkalemia (K^+ >5.0 mEq/l) typically results from renal failure to excrete, increased intake, or drugs such as ACE inhibitors. It will compromise cardiac and neuromuscular function, manifesting clinically with weakness, paresthesias, and ECG findings of peaked T waves and widened QRS complexes. Treatment in an emergency involves IV calcium to stabilize the heart, agents to move K^+ into cells, and measures to remove it from the body. Impaired potassium balance changes excitability of the cardiac cells, hypokalemia causes hyperpolarization and torsades de pointes, whereas hyperkalemia diminishes action potential formation and results in sustained depolarization and weakness. Prompt diagnosis and treatment are essential to avoid arrhythmias, paralysis, and fatalities [55].

Disease

Potassium is a dietary mineral that has profound protective actions, especially on cardiovascular and renal function. Excessive dietary potassium conception reduces blood pressure through natriuresis, extracellular fluid, volume reduction, and vasodilation by hyperpolarization of vascular smooth muscle. Potassium also increases baroreceptor sensitivity, decreases oxidative stress through NADPH oxidase downregulation, and maintains endothelial function. Clinical trials like DASH and large-scale epidemiological studies, like the Nurses' Health study, all repeatedly demonstrate that increased potassium consumption is correlated with decreased pressure and stroke risk. Potassium also prevents kidney stones particularly as potassium citrate by raising urinary citrate, acidifying urine, and accelerating urine flow, with up to a 56% decrease in stone risk being reported. Furthermore, potassium regulates the sympathetic nervous system, enhances insulin sensitivity, inhibits vascular fibrosis, and stabilizes cardiac electrophysiology, thus reducing the incidence of arrhythmias in hypertension and heart failure. Global health organizations such as the WHO and AHA promote potassium-dense diets composed of fruits, vegetables, and legumes backed by large-population-based evidence like INTERSALT and TOHP that attests to its extensive benefits in disease prevention and cardiovascular health. Potassium is essential to maintain cellular membrane potential and cardiac function. To mimic the human system, zebrafish embryos with disrupted K^+ channel expression displaying cardiac arrhythmias and developmental defects were investigated. For renal excretion studies, the rodent models with low K^+ diets associated with hypertension and metabolic disease are used to investigate potassium homeostasis. Potassium channels associated with synaptic transmission and neuron excitability have been revealed by Drosophila models, illuminating epilepsy and neurodegenerative disease [1, 13, 38, 56, 66].

Calcium (Ca)

Physiological Functions

Calcium (Ca^{2+}) is a critical mineral that makes up 1–2% of body weight, of which more than 99% is deposited in bones and teeth, mainly as hydroxyapatite. It contributes to the structural integrity and mechanical strength of the skeleton, which is also a dynamic reservoir that maintains blood calcium homeostasis through ongoing remodelling. During development, particularly adolescence, calcium is needed for maximum bone mass acquisition, and lack of it raises the risk of osteoporosis in adulthood. In the teeth, do not remodel, and after development, calcium deficiency has no effect on their mineral content. Aside from structural functions, calcium is at the heart of many physiological processes. In muscle contraction, released calcium from the sarcoplasmic reticulum binds to troponin, inducing conformational changes enabling actin-myosin interaction, allowing contraction. Relaxation results from the re-sequestration of calcium. In neuron, influx of calcium through voltage-gated channels on depolarization causes release of neurotransmitters by exciting synaptic vesicle replenishment and exocytosis. Calcium is also essential in blood clotting, serving as a cofactor in several stages of the clotting cascade. It helps to promote assembly of clotting complexes on phospholipid surfaces, such as conversion of prothrombin to thrombin and activity of vitamin K-dependent clotting factors [24, 63, 97].

Absorption, Transport, and Regulation

Calcium does not synthesize by body; sufficient dietary supply and effective absorption are essential. Intestinal absorption involves two major pathways: passive intracellular diffusion and active transcellular transport. The latter is vitamin D-dependent and includes apical uptake through TRPV5/6 channels, intracellular transport by calbidin-D9K, and basolateral extrusion by PMCA1b and Na^+/Ca^{2+} exchanger. These processes are controlled by the active vitamin D form, $1,25D(OH)_2D_3$, by binding through the vitamin D receptor (VDR). VDR knockout models demonstrate diminished calcium uptake and decreased expression of the transporter, reinforcing implication of vitamin. Calcium homeostasis is stringently regulated by parathyroid hormone (PTH), calcitriol, and calcitonin. Decreased plasma calcium stimulates PTH, which increases bone reabsorption, renal calcium reabsorption, and calcitriol production. Calcitriol also stimulates intestinal and renal calcium absorption and mobilizes bone calcium in cooperation with PTH. Calcitonin, on the other hand, is released during hypercalcemia to suppress bone reabsorption, but it has little effect on kidney and intestines. The kidneys have a significant role to play by reabsorbing more than 98% of filtered calcium, especially in the distal nephron under hormonal control. Bones are a dynamic reservoir of calcium, constantly remodelled to match physiological needs [18, 25, 40, 41].

Deficiency and Excess

Hypocalcemia is low serum calcium and mainly influences neuromuscular function. Calcium stabilizes neuronal membrane; decreasing levels cause an increase in membrane excitability. This results in findings such as tetany, a severe sustained muscle spasm, commonly occurring in hands, face, and seizures. Clinical presentations such as Chvostek's and Trousseau's suggest increased neuromuscular irritability. At the cellular level, a low calcium decreases then threshold for generation of the action potential by enhancing sodium influx into the neurons. The hyperexcitability results in reparative firing of nerves and, in the brain, induces generalized or focal seizures that are usually unresponsive to therapy until the calcium has been repleted. Hypercalcemia, or increased serum calcium, has symptoms that vary with severity and speed of onset. It can be asymptomatic in mild cases or cause nonspecific symptoms such as weakness, fatigue, constipation, and mood disturbances. More severe hypercalcemia causes gastrointestinal, dehydration, confusion, coma, and cardiovascular manifestations [15]. Precipitation of calcium salts can lead to formation of kidney stones, particularly in chronic hypercalcemia as in hyperparathyroidism. In the heart, too much calcium changes the electrical conduction, decreased qt interval, and can cause life-threatening arrhythmias or heart block. Hypercalcemia associated with malignancy is usually very rapid and needs to be corrected urgently. Early diagnosis and treatment are needed to avoid life-threatening complications [16, 97, 101].

Disease

Osteoporosis is characterized by lower bone mineral density and compromised bone structure, raising the risk of fracture especially in postmenopausal women and older men. Bone strength is reliant on calcium in hydroxyapatite form, deposited during years of maximum bone mass. Calcium absorption is diminished, and bone reabsorption is facilitated by aging and estrogen deficiency. Vitamin D with calcium preserves bone integrity and lowers the fracture risk, but not adequate as monotherapy. Rickets, an illness of childhood, and osteomalacia, its adult equivalent, both occur due to abnormal bone mineralization. Nutritional types are treatable with correction of diet, whereas genetic types can be treated with specific therapy. Calcium regulates vascular muscle contraction, neuronal transmission, and hormonal signaling (e.g., PTH, vitamin D, RAAS). Both calcium deficiency and hypercalcemia affect blood pressure. Calcium channel blockers lower vascular tone and BP, and calcium supplementation in some populations may be useful in controlling hypertension. In colon cancer, increased calcium conception can lower colorectal cancer risk by sequestering toxic bile/fatty acids and stimulating the calcium-sensing receptor (CaR), which inhibits cell growth and induces apoptosis in colonic cells. The presence of CaR in normal and malignant colon tissue validates the protective action of calcium. Kidney stones occur when urine becomes supersaturated with solute such as calcium oxalate or phosphate. Contrary to expectations, low calcium diets enhance oxalate absorption and stone formation. Hypercalciuria may be genetic in origin and is one of the causes of kidney stone formation. Preventive measures intend to control dietary calcium and fluid and decrease urinary crystal saturation. Calcium plays significant roles in bone growth, muscle contraction, and intracellular signaling. Zebrafish are utilized to model bone mineralization and calcification under calcium dependence where calciotropic hormones and Ca^{2+} transporters influence skeletal morphogenesis. The role of Ca^{2+} in bone mineral density, cardiovascular physiology, and parathyroid hormone regulation was illustrated by using rodent models. The regulation of muscle function and neural activity by calcium signaling in Drosophila and mutations of calcium channels have been implicated in learning and memory defects [14, 21, 60, 84, 108].

Magnesium (Mg)

Physiological Functions

Magnesium (Mg^{2+}) is an important intracellular mineral that is a cofactor in more than 300 enzymatic reactions that are critical to energy metabolism, nucleic acid synthesis, and protein synthesis. It stabilizes ATP by the creation of Mg-ATP, which is the biologically active form needed for cell function. In the mitochondria, magnesium regulates major enzymes of the TCA cycle and ATP synthesis, controlling cellular energy equilibrium. Neurologically, magnesium stabilizes neurons by inhibiting NMDA receptors and regulating ion channels, stopping excitotoxicity, and enhancing neurotransmitter release. Deficiency results in muscle cramps and neuromuscular diseases. In the cardiovascular system, magnesium preserves normal heart rhythm by controlling Na^+/K^+-ATPase and calcium channels, stabilizing action potentials, and arresting arrythmias. Magnesium also has a structural function in bones, where 50–60% of the magnesium in the body is reserved. It facilitates hydroxyapatite formation, enhances bone elasticity, and modulates osteoblast and osteoclast activity. Bone serves as a buffer and releases magnesium in deficiency. Imbalances with fluoride or calcium can compromise bone strength and mineralization [5, 106].

Absorption and Regulation

Magnesium (Mg^{2+}) is taken in from the diet and absorbed in the small intestine through two mechanisms: passive paracellular diffusion, prevalent during high intake, and active transcellular transport by TRPM6 and TRPM7 channels, of particular importance during low intake. TRPM6 expression is increased when intracellular Mg^{2+} is low, increasing efficiency of absorption. Absorption is magnesium status dependent—increasing to 80% in deficiency and decreasing to ~20% in excess—and is affected by health of gut, transit time, and medications. Uptake is disrupted by diseases such as IBD. Renal control is important for magnesium homeostasis. Of the ~2400 mg filtered per day by the kidneys, 95–99% are reabsorbed. The proximal tubule passively recovers 10–30%, whereas the thick ascending limb (TAL) reabsorbs 50–70% through a paracellular mechanism generated by NKCC2- and ROMK-governed voltage. This involves the claudin-16 and claudin-19 proteins, which are critical for their loss due to mutations (e.g., in FHHNC). The distal convoluted tubule (DCT) actively reabsorbs the balance of 5–10% through TRPM6 channels under hormonal regulation, which can be increased by estrogen. Urinary excretion of magnesium is very variable (0.5–70% of filtered load), providing for a tight systemic regulation [29, 81].

Deficiency and Excess

Hypomagnesemia, serum Mg^{2+} <1.46 mg/dL, is usually not symptomatic until it falls below 1.2 mg/dL. It results from chronic diseases, alcoholism, GI or renal losses, or inherited syndromes such as Gitelman or EAST. Magnesium depletion impairs cardiac and neuromuscular stability, causing tremors, cramps, seizures, and fatal arrhythmias such as torsade de pointes. It inhibits parathyroid hormone release, resulting in hypocalcemia, and enhances renal loss of potassium, resulting in hypokalemia. ECG abnormalities may consist of QT prolongation and ST depression. Treatment consists of magnesium supplementation, particularly in cardiac or critically ill patients [74]. Hypermagnesemia (serum Mg^{2+} >1.1 mmol/L) is uncommon and typically iatrogenic, commonly caused by the inappropriate use of magnesium-containing antacids, laxatives, or IV therapy, especially in renal failure. It inhibits neuromuscular transmission and may result in weakness of the muscles, hyporeflexia, flaccid paralysis, bradycardia, hypotension, and cardiac arrest. It also exacerbates hypocalcemia by parathyroid hormone inhibition. Symptoms are frequently nonspecific, and an early diagnosis must be made. Treatment involves discontinuation of magnesium intake, IV calcium for symptom management, and dialysis in advanced or renal-compromised patients [100].

Disease

Magnesium is crucial for the etiopathogenesis and treatment of manty chronic illness, such as type 2 diabetes (T2DM), hypertension, metabolic syndrome, cardiovascular disease (CVD), and migraine. In T2DM, magnesium supports insulin signaling and glucose uptake. Deficiency disrupts these processes, enhancing insulin resistance and oxidative stress and contributing to nephropathy and cardiac dysfunction [84]. Observational data associate increased magnesium intake with lowered diabetes risk, but results of supplementation trials remain conflicting. Magnesium regulates blood pressure by countering calcium-induced vasoconstriction and increasing nitric oxide production. Deficiency may elevate blood pressure through calcium accumulation and RAAS stimulation, as evidenced in studies such as ARIC. In metabolic syndrome, reduced magnesium exacerbates insulin resistance, lipid disorder, and inflammation, further elevating disease risk. Cardiovascular effects of magnesium involve stabilizing heart rhythm, decreasing oxidative stress, and enhancing vascular function, thereby reducing risk of arrhythmias, stroke, and atherosclerosis. In migraine, magnesium deficiency alters neurotransmission and mitochondrial function, initiating attacks. Supplementation is found to be particularly useful in menstrual and migraines symptoms. Magnesium is crucial for enzymatic activity, ATP stability, and neuromuscular transmission. Mg^{2+} deficiency leads to muscle dysfunction and abnormal development in zebrafish. Zebrafish mutants lacking TRPM7, an important Mg^{2+} transporter, show phenotypes similar to human hypomagnesemia. Mg^{2+} deficiency in the rodent models aggravates inflammation, cardiovascular disease, and insulin resistance. Drosophila models have also been employed to uncover genes that control Mg^{2+} uptake and its metabolic activity within the cell, which may result in high throughput screening of Mg^{2+} modulators [10, 34, 83, 96].

Phosphorus (P)

Physiological Functions

Phosphorus is an essential mineral that plays a crucial role in structural and metabolic processes within the human body. Approximately 85% of body phosphorous is contained within bones and teeth in the form of hydroxyapatite, a calcium phosphate crystal that contributes to skeletal rigidity. It helps in bone development by facilitating osteoblast function, mineralization, and osteoclast-mediated resorption regulation, with enzymes such as alkaline phosphate and transporters such as P1T1 and PHOSPHO1 being central. At the molecular level, phosphorus, like phosphoramidite, is employed for synthetic DNA/RNA synthesis. Phosphorus plays an important role in cellular energy, as well, in the formation of ATP, which aids metabolism. Phosphate groups of ATPS engage in protein phosphorylation to control signaling pathways, such as MAPK/ERK, through kinase enzymes. Additionally, phosphate buffers the acid-base balance, particularly in metabolic acidosis, to maintain intracellular and urinary pH. Nevertheless, over-mobilization of phosphate from bones results in bone demineralization [75, 90, 92].

Absorption and Regulation

Phosphorus is taken up in the intestine predominantly as an inorganic phosphate (PO_4^{3-}). About 70% of it is taken up passively in the small intestine, and active transport through the NPT2b (SLCC34A2) sodium-phosphate cotransporter becomes important with low phosphate intake. 1,25-dihydroxyvitamin D_3 [$1,25(OH)_2D$] stimulates this active uptake by enhancing NPT2b expression. Three hormones such as PTH, fibroblast growth factor 23 (FGF23), and $1,25(OH)_2D$ are responsible for regulating phosphate homeostasis.

- PTH increases blood calcium but lowers phosphate by facilitating bone reabsorption and lowering renal phosphate reabsorption.
- FGF23, osteocyte-released, reduces serum phosphate by inhibiting renal phosphate reabsorption and vitamin D activation.
- $1,25(OH)_2D$ increases absorption of phosphate and calcium in the intestine and inhibits both PTH and FGF23.

All these mechanisms cooperate to produce systemic phosphate balance necessary for bone health and function [8, 26].

Deficiency and Excess

Hypophosphatemia, serum phosphate <2.5 mg/dL, disturbs energy metabolism, bone, and neurological function. Even mild deficiency is often asymptomatic, but moderate to severe deficiency causes muscle weakness, bone pain, and confusion. Muscle weakness results from impaired energy metabolism, while defective bone mineralization leads to fractures and manifests as osteomalacia in adults or rickets in children. Neurological symptoms are due to decreased ATP-dependent neuron function. The usual causes are malabsorption, renal phosphate wasting, and intracellular redistribution during refeeding, sepsis, or alkalosis. Early diagnosis and phosphate replacement oral or IV depending on severity are crucial. Interprofessional management enhances outcomes and avoids complications [60]. Hyperphosphatemia, which is prevalent in chronic kidney disease (CKD), is a major factor in vascular calcification and cardiovascular risk. Abnormal phosphate excretion results in high serum phosphate, which promotes vascular smooth muscle cells (VSMCs) to take on bone-like features, expressing proteins like Runx2 and osteocalcin. This remodelling enhances arterial calcification, leading to decreased elasticity and the potential for increased risk of hypertension, cardiac hypertrophy, and vascular events. Disruption of hormone balance with FGF23, PTH, and vitamin D further disturbs phosphate balance and potentiates calcification. Careful management—dietary control of phosphate, phosphate binders, and optimal dialysis—is crucial to minimize vascular damage and enhance outcomes in CKD [48].

Disease

Chronic kidney disease (CKD) is widespread among older cats and occurs in more than 30% of cats over 15 years of age. Among dogs, CKD is not as common but varies according to breed. Nephron loss in CKD compromises the excretion of phosphorus, producing hyperphosphatemia, which is critical determinant of disease progression. High phosphorus causes secondary renal hyperparathyroidism (SRHP), and elevated parathyroid hormone (PTH) most affected cats and dogs. High phosphorus is also associated with decreased survival, so its management is crucial. The IRIS staging system employs creatinine as a measure of CKD severity and provides stage-specific targets for phosphorus levels. Phosphorus management—diet and phosphate binders—is critical for the diminishment of CKD progression and to enhance outcomes in both species [45]. Renal osteodystrophy (ROD) is a skeletal disorder seen with chronic kidney disease (CKD), caused by deranged mineral metabolism and compromised skeletal remodelling. The central factor in ROD is phosphate retention that occurs early in CKD because of reduced renal excretion. Increased serum phosphate provokes secondary hyperparathyroidism (sHPT), a compensatory but ultimately maladaptive process. In excess phosphate, parathyroid hormone (PTH) secretion is stimulated, which enhances bone turnover and causes osteitis fibrosa cystica, with fibrosis and structural weakening of the bones. Chronic sHPT is a cause of high turnover ROD, but excessive treatment to reduce PTH can cause adynamic bone disease (ABD), an abnormal low-turnover condition with impaired bone regeneration. Both high and low turnover types of ROD enhance the risk of fractures, osteopenia, and vascular calcification. Hence, control of phosphate by dietary restriction, phosphate binders, vitamin D analogues, and calcimimetics is pivotal to ensure balanced bone remodelling and reduce skeletal complications in CKD and end-stage kidney disease (ESKD) [57]. Phosphorus is a principal component in cardiovascular calcification, particularly in CKD and ESRD. Inefficient phosphate excretion results in hyperphosphatemia, which is responsible for vascular smooth muscle cell differentiation to bone-like cells. This creates calcium phosphate deposits in coronary arteries and heart valves, leading to arterial stiffness, left ventricular hypertrophy, and compromised coronary perfusion. Phosphate-induced calcification is responsible for intimal and medical vascular calcification, aortic valve stenosis, and major complications such as calcific uremic arteriolopathy (CUA). Control of serum phosphate by diet, binders, and dialysis is necessary to decrease cardiovascular morbidity in CKD patients [11, 30].

Interaction and Balance Among Major Minerals

Sodium and potassium are the two main monovalent cations that regulate electrolyte balance and electrochemical gradient in the human body. The tightly controlled distribution across cell membranes is essential for supporting osmotic equilibrium, nerve and muscle excitability, acid-base homeostasis, and nutrient transport. Although Na^+ is the predominate ion in the extracellular fluid (ECF), K^+ is mostly intracellular, with about 98% being found inside cells. This asymmetric ionic distribution is actively maintained by the Na^+/K^+-ATPase, a critical membrane pump that expels three Na^+ ions from the cell in return for two K^+ ions at the expense of ATP. This process not only maintains the resting membrane potential (~−70 to −90 mV) but also forces secondary active transport of glucose, amino acids, and other solutes by creating steep sodium gradients. Sodium role in the ECF is crucial for regulating osmolarity, fluid volume, and blood pressure. In contrast, potassium's intracellular dominance contributes to cell volume regulation, enzyme activity, and electrical signaling. The exchange between Na^+ and K^+ directly affects action potential generation and propagation in excitable tissues like neurons and muscle fibers, where Na^+ influx initiates depolarization and K^+ efflux restores the resting state. Hormonal mechanisms like the renin-angiotensin-aldosterone system (RAAS) regulate sodium reabsorption and potassium secretion in the kidney in a complex manner. Aldosterone increases sodium entry through epithelial Na^+ channels (ENaC)

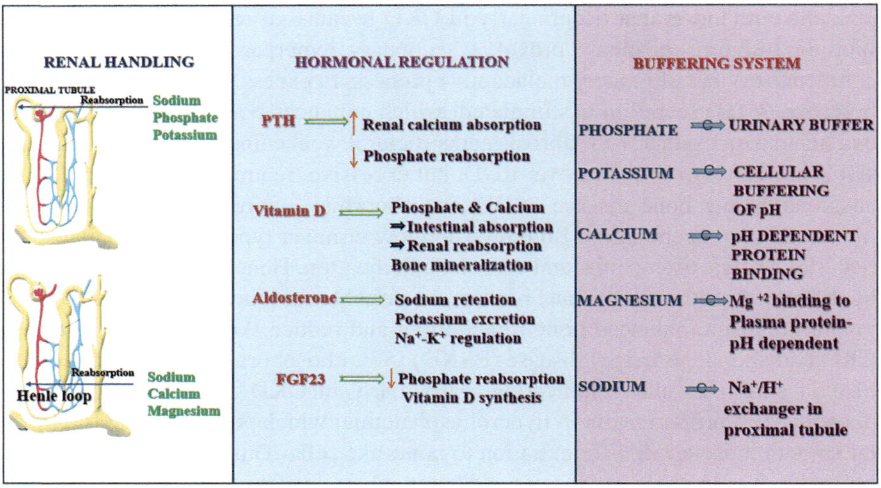

Fig. 20.2 Integrated control of electrolytes via renal transport, hormonal signaling, and acid-base buffering mechanism

while stimulating K^+ loss through apical K^+ channels (e.g., ROMK), demonstrating their mutual regulation. In addition, potassium and sodium are also implicated in acid-base balance. Sodium is structurally a part of the bicarbonate buffer system as $NaHCO_3$, whereas potassium affects intracellular pH through its action on H^+/K^+ exchangers. Abnormality of Na^+–K^+ balance can result in life-threatening states like hyponatremia, hyperkalemia, or arrhythmias, which attest to their pivotal position in maintaining homeostasis. Therefore, the interaction of Na^+ and K^+ is critical to the preservation of physiological balance and the coordinate function of organ systems [49, 51]. Figure 20.2 gives an overview of integrated control of electrolytes via renal transport, hormonal signaling, and acid-base buffering mechanisms.

Bone metabolism depends on the coordinated efforts of calcium, magnesium, and phosphate. These three minerals not only coexist at a structural level in bone but also control each other's functions in the development, remodelling, and mineral homeostasis of the skeleton. Calcium and phosphate compose the crystalline hydroxyapatite $[Ca_{10}(PO_4)_6(OH)_2]$ structure, the main mineral responsible for the strength and rigidity of bone. Their conjoint deposition onto the collagen matrix is critical to bone mineralization. Calcium offers rigidity in structure, whereas phosphate adds crystal growth, hardness, as well as intracellular signaling in bone development. The minerals should be in specific ratios, whereby in excess, phosphate affects hydroxyapatite formation even in the presence of ample calcium. Magnesium has a twofold function: Structurally, it is integrated into bone mineral by replacing some of the calcium in hydroxyapatite crystals to form a more soluble, metabolically active bioapatite. Functionally, magnesium also affects the efficiency of calcium by controlling calcium channels and transporters at the cellular levels, competing with calcium for sites on bone crystals; an imbalance, specifically high calcium with low magnesium, can cause brittle bone mineral and defective remodelling. In addition, magnesium influences phosphate metabolism. Through its influence on the activity of enzymes like alkaline phosphate and PHOSPHO1, magnesium regulates local phosphate availability during mineralization. Magnesium deficiency perturbs these enzymatic processes, decreasing the effectiveness of phosphate incorporation into bone compromising mineral strength. The dynamic interaction is also observed in bone remodelling: Phosphate and calcium are mobilized to resorption to maintain extracellular concentrations, and magnesium regulates the sensitivity of bone cells to hormonal stimuli like PTH and vitamin D, both of which control phosphate and calcium [59, 111]. The human body achieves mineral and electrolyte homeostasis by complex and highly integrated mechanisms involving renal handling, hormonal regulation, and buffering systems. Five key ions, phosphorus (P), sodium (Na^+), potassium K^+), calcium (Ca^+), and magnesium (Mg^{2+}), are managed through overlapping routes that both reflect their distinct physiological roles and their interdependence in sustaining fluid balance, membrane potential, acid-base status, and bone mineralization. The kidney has a key role in maintaining these ions by selective reabsorption, secretion, and mechanisms. The main extracellular cation is sodium, and its active reabsorption by the Na^+/K^+-ATPase pump in the proximal tubule and loop of Henle is crucial not just for sodium balance but also as a driving force for other ion reabsorption. As an example, phosphate reabsorption in the proximal tubule is mediated by Na^+-dependent cotransporters (NaPi-IIa and IIc), showing how sodium gradients promote phosphate preservation. Potassium, mainly intracellular, is absorbed in the proximal tubule and secreted in the distal nephron in response to aldosterone. Its management is closely associated with sodium reabsorption in the collecting duct increases K^+ secretion through ROMK channel. In contrast, calcium and magnesium are largely absorbed by in the loop of Henle's thick ascending limb via paracellular pathways, under control of electrochemical gradients due to sodium reabsorption. Magnesium is also processed in the distal tubule through the TRPM6 channel and calcium through TRPV5/6, both of which are under hormonal regulation. Several hormones regulate the management of these ions, commonly acting on more than one system at a time. PTH is an important factor in enhancing renal reabsorption of calcium and reducing phosphate reabsorption, thus raising serum calcium at the cost of phosphate. Vitamin D (1,2-dihydroxyvitamin D_3) has synergic action with PTH to enhance intestinal absorption and renal reabsorption of calcium and phosphate, which is important for bone metabolism. Aldosterone, a mineralocorticoid, stimulates the retention of sodium and the excretion of potassium in the distal nephron in direct association between Na^+ and K^+ regulation. In contrast, fibroblast growth factor (FGF23), which is secreted in the presence of increased phosphate, suppresses phosphate reabsorption and vitamin D production and hence creates a negative feedback loop. These hormones exhibit a tightly coupled hormonal axis wherein regulation of one ion have direct effects on the status of others. Aside from electrolyte and volume control, several ions are involved in buffering systems to ensure acid-base homeostasis. Phosphate functions as a significant urinary buffer, especially in the distal nephron, when it sequesters hydrogen ions (H^+) as $H_2PO_4^-$. Sodium mediates exchange for H^+

in the Na^+/H^+ exchanger (NHE3), favoring acid extraction and bicarbonate reabsorption. Potassium is closely involved in acid–base buffering through transcellular ion shifts; during acidosis, hydrogen ions move into cells in exchange for potassium, resulting in hyperkalemia. Calcium and magnesium buffering are also pH-dependent; their binding to plasma proteins is pH-dependent and increases ionized Ca^{2+} in acidosis, with reduction by alkalosis. The physiological regulation of phosphorus, sodium, potassium, calcium, and magnesium is a model of systemic coordination; the kidney, responding to hormonal cues and buffering needs, directs the tightly regulated equilibrium of the ions by cooperative transporters, crosstalk between hormones, and acid-base interactions. Dysfunction on one pathway tends to influence others, emphasizing the interdependence necessary for homeostasis. Appreciation of these cooperative mechanisms offers insights into pathophysiological states such as electrolyte disturbances, metabolic acidosis, renal osteodystrophy, and hypertension [17, 52, 70, 101].

Dietary Recommendations and Public Health Implications

RDA/AI for Adults

RDA (Recommended Dietary Allowance) refers to the average daily nutrient requirements for a healthy individual, based on gender and specific life stages, while AI (Adequate Intake) indicates the average daily nutrient amount derived from observed or experimental data (Table 20.1).

Dietary Sources and Fortification

Food fortification refers to the addition of vitamins and minerals to foods consumed with the vision of increasing their nutritional value, thereby preventing deficiencies. Table 20.2 provides the information regarding the major dietary sources and fortification practices of major minerals.

Table 20.1 RDA/AI values of minerals for adults ([a]AI)

Mineral	RDA/AI (mg/day)	Reference
Sodium	2300	Farag et al. [37]
Potassium	2600 (female), 3400 (male)[a]	
Calcium	1000	
Magnesium	310 (female), 420 (male)[a]	
Phosphorus	700	

Table 20.2 Dietary sources and fortification practices of major minerals

Mineral	Primary natural source	Fortification practices	Reference
Sodium (Na)	Processed foods (canned soups, snacks, cured meat) and table salt	Iodized salt	Calvo and Uribarri [20]
Potassium	Fruits (bananas, oranges, apricots, avocado, melons), vegetables (potatoes, spinach, sweet potato), legumes	Limited fortification. Sometimes, potassium chloride is used as a salt substitute.	McLean and Wang [67]
Calcium (Ca)	Dairy products (milk, cheese, yogurt), leafy greens, tofu	Fortified plant-based milk (soya, almond), orange juice	Calvo and Uribarri [20]
Magnesium (Mg)	Nuts (almonds, cashews), seeds (pumpkin, chia), whole grain, legumes, dark chocolate	Occasional fortification	Rosanoff et al. [89]
Phosphorus (P)	Animal products (meat, poultry, fish, dairy), legumes, nuts, beans	Widely added as phosphate additives in processed meats, soft drinks, and baked goods	National Institute of Health [72]

Impact of Western Diets vs. Traditional Diets

Changes in the dietary and physical activity habits of the population in the present world have been associated with cardiovascular diseases, obesity, type 2 diabetes, depression and anxiety, and inflammation. Living organisms survive best in their environment and diet they are evolutionarily adapted which has regular sun exposure, sleeping patterns aligning with the circadian rhythm, regular physical activity, social interaction, lack of exposure to pollutants, and consumption of fresh food [22], while the Western diets include red meat, high-sugar drinks, fried foods, high-fat dairy products, prepackaged foods, refined grains, and a lack of intake of fruits, nuts, whole grains, and vegetables, which are the important sources of vitamins, minerals, fiber, and antioxidants essential for health. The Western diet is low in antioxidants and high in omega-6 fatty acids and advanced glycation end products (AGEs), which are prooxidant substances. Higher intakes of saturated and trans fats, salt, and added sugars have been linked to an increase in the risk of cardiovascular diseases (CVD), with the leading cause of death worldwide. The Western diet also influences gene expression and epigenetics [17, 27, 93].

WHO/FAO Guidance on Mineral Intake and Disease Prevention

With the goal to promote optimal health and aid in avoidance of chronic diseases, the Food and Agriculture Organization (FAO) and the World Health Organization (WHO) have been working together to develop complete dietary recommendations. In many societies, particularly those with limited resources, deficiencies in calcium and magnesium remain concerning. However, excessive consumption of sodium

Table 20.3 WHO/FAO recommendations on mineral intake and associated disease prevention

Minerals	WHO/FAO recommendation for adults (mg/day)	Disease prevention	Reference
Calcium (Ca)	<2000	Prevents osteoporosis, osteomalacia, rickets, and colon cancer and improves heart function	Drewnowski et al. [35]
Sodium (Na)	<2000 (equivalent to 5 g salt)	Reduced Na usage leads to reduced risk of CVD, and excess leads to high blood pressure, stroke, and heart disease.	Food and agriculture Organization of the United Nations [42]
Potassium (K)	3510	Reduces blood pressure, stroke, and CVD risk	Drewnowski et al. [35]
Phosphorus (P)	700	Supports bone strength when taken in combination with (P); excess intake leading to CVD	NIH [73]
Magnesium (Mg)	300–600	Support heart health, insulin sensitivity (reduce type 2 diabetes), and bone density	WHO/FAO [105]

and phosphorus, particularly from processed foods, has grown into a significant public health concern. The guidance strongly recommends the intake of potassium and magnesium through natural food sources like fruits, legumes, whole grains, and vegetables, thereby reducing sodium consumption.

These recommendations (Table 20.3) are laid down on the basis of randomized controlled trials, meta-analyses, epidemiological studies, and population-level research. The objective is not only to identify the optimum intake amounts for preventing disease but also to evaluate dietary biomarkers, enhance guidance for particular age groups, and monitor long-term health outcomes. Importantly, the FAO and WHO encourage nations to integrate these recommendations into population-level policies that are effective.

Emerging Research and Therapeutic Applications

Mineral Supplementation in Critical Illness and Metabolic Disorders

Assessing mineral status during illness is quite challenging as underlying inflammatory states can mask actual mineral deficiencies. Over-supplementation of minerals also leads to toxicity risks and worsens the condition for which personalized supplementation and precise dosing are essential [46]. Factors like genetics, diet, lifestyle, and bioavailability often limit the effectiveness of traditional supplements, emphasizing the need for more bioavailable and tolerable formulations [44]. Magnesium is directly involved in glucose metabolism and insulin action; chronic stress and inflammation, which are common in critical illness, can actively deplete magnesium levels and worsen the condition. People with type 2 diabetes also have compromised magnesium levels. Similarly, in postmenopausal women, low magnesium levels have been associated with osteoporosis [58]. For adults over 50, a daily intake of 1000 mg calcium and 8000 IU of vitamin D is recommended to maintain bone health [82].

Therefore, mineral supplementation holds significant therapeutic promise in managing critical illness and metabolic disorders. The available clinical approaches have limitations due to unreliable diagnostic markers, variation in bioavailability, and the risks of over-supplementation. Prioritizing the development of functional biomarkers that will reflect the exact cellular mineral status and enabling precise individualized supplementation strategies will help overcome the existing challenges. There is a need for well-designed large-scale randomized controlled trials in order to determine optimal dosing, duration, and combinations of minerals. Instead of following the traditional single-nutrient approach, experimenting with the combined effects of multi-mineral supplementation may also offer better clinical outcomes.

Role in Immune Modulation, Inflammation, and Neurodegeneration

Studies have proven that magnesium influences acute phase response, inflammation, T-cell development and proliferation, and macrophage function. Its insufficiency exposes cells to oxidative stress and causes potential health complications like diabetes, lymphoma, hypereosinophilia, histaminosis, and thymus atrophy. Its deficiency contributes to low-grade inflammation, which is a risk factor for pathological conditions like hypertension and CVD. Its deficiency is associated with the production of proinflammatory cytokines (IL-1, IL-6, and tumor necrosis factor (TNF-α)), which increases the production of reactive oxygen species (ROS) [4]. It also serves as a cofactor for C3 convertase and immunoglobulin synthesis, supporting its role in immune modulation [71].

Calcium and phosphorus also influence regulating inflammation, proliferation, differentiation, and release of cytokines from T cells, B cells, and macrophages [87, 110]. For the normal neuronal function, mitochondrial Ca^{2+}buffering is essential, and any dysfunction in this serves as the cause for many neurological diseases [103]. High salt (Na^{2+}) content in the diet was found to play a crucial role in mediating immune-mediated inflammatory diseases (IMIDs). It favored the differentiation of $CD4^{+}$ cells into pathogenic Th17 cells, which results in several inflammatory diseases. An increase in salt causes chronic inflammation and increased vascular endothelial growth factor levels in cancer, and it also reduces nitric oxide production in endothelial cells, leading to hypertension [6, 104].

Innovations in Personalized Nutrition and Precision Medicine

Genomic, proteomic, and metabolomic studies provide a deeper understanding of individual characteristics and contribute to the development of more effective pharmaceutical and nutritional therapies. This can be achieved not only by the questionnaire-based method but also by the use of different omics approaches to develop optimal and personalized dietary support, which promotes health and disease maintenance [69].

A study by Celis-Morales et al. [23] stands out to be the largest and most comprehensive personalized nutrition study where five genotypes (fate mass and obesity associated gene (*FTO*), fatty acid desaturase 1 (*FADS1*), transcription factor 7-like 2 (*TCF7L2*), apolipoprotein E (APOE), and methylenetetrahydrofolate reductase (*MTHFR*)) known to impact nutrient metabolism was studied. Genome-wide association studies (GWAS) have shown their contribution to identifying several variables that influence genes associated with appetite, inflammation, insulin resistance, energy intake, and lipid metabolism [47]. A holistic algorithm approach was designed that helped to create genetic risk scores based on 95 single-nucleotide polymorphisms (SNPs) related to energy homeostasis, which helped in enabling personalized prescription of diets in a group of Spanish subjects with overweight/obesity [86]. Similarly, *APOA1*(rs670) gene polymorphism study by de Luis et al. [31] showed effects on insulin resistance, adiposity, and cholesterol levels when subjected to a hypocaloric diet.

Zeevi et al. [108] reported using a machine learning algorithm that incorporated lifestyle information, gut microbiota, dietary habits, and anthropometric data to accurately predict personalized diet rather than reducing the carbohydrate intake suggested in traditional methods. The obtained predictions were validated and the algorithm-based intervention resulted in significantly lower postprandial responses, which kept blood sugar levels under control. Several microRNA studies revealed that miRNAs can be modified by dietary intake and they influence various cellular processes like inflammation or apoptosis [68] and potential biomarkers [43]. Plasma miR-23a-3p expression revealed that it was positively influenced by sodium intake [39]. The health challenges faced by perimenopausal women were studied by Andrade et al. [3], which focused on SNPs in the COMT-, FUT2, and MTHFR genes. This approach enabled precision medical aspects and personalized nutrition to mitigate the metabolic and psychological effects of perimenopause in women.

Conclusion

Even though macro-minerals (sodium, potassium, calcium, magnesium, and phosphorus) are required in small quantities, they have a vast influence over various physiological systems and human health. Emerging insights from personalized nutrition and precision medicine highlight genetic, epigenetic, and microbiome variability in mineral requirements and metabolism. The elderly, postmenopausal women, and those with chronic illnesses are among the most vulnerable populations who need a paradigm modification beyond a broad generalized dietary recommendation. These minerals do not function in isolation; their biological availability and activity are interconnected. Recent advances in precision medicine and personalized nutrition focus on the variation in mineral demands and metabolism, which depends on genetics, microbiota, and epigenetics, of the recommendations, and toward personalized consumption regimes. These minerals additionally function in regulating oxidative stress, cell-cell communication cascades, inflammatory conditions, and immunological responses, which offer novel possibilities for therapeutic strategies. Understanding the molecular and systemic interactions between these macronutrients will be crucial as scientific research advances to develop strategies that prevent chronic diseases and restore physiological equilibrium. In summary, a balanced and nutritious mineral intake, guided by relevant clinical studies and public health policies, is crucial for maintaining systemic well-being.

Funding Declaration This study was supported by the grant from the Indian Council of Medical Research, Govt of India, ICMR – IIRP – 0508/2023.

Competing Interests The authors declare that they have no competing interests.

References

1. Aaron KJ, Sanders PW (2013) Role of dietary salt and potassium intake in cardiovascular health and disease: a review of the evidence. Mayo Clin Proc 88(9):987–995. Elsevier
2. Afzal MF, Khalid W, Akram S, Khalid MA, Zubair M, Kauser S et al (2022) Bioactive profile and functional food applications of banana in food sectors and health: a review. Int J Food Prop 25(1):2286–2300
3. Andrade P, Santamarina AB, de Freitas JA, Marum ABRF, Pessoa AFM (2025) Personalized nutrition and precision medicine in perimenopausal women: a minireview of genetic polymorphisms COMT, FUT2, and MTHFR. Clinics 80:100549
4. Arancibia-Hernández YL, Hernández-Cruz EY, Pedraza-Chaverri J (2023) Magnesium (Mg2+) deficiency, not well-recognized non-infectious pandemic: origin and consequence of chronic inflammatory and oxidative stress-associated diseases. Cell Physiol Biochem 57(S1):1–23
5. Baker WL (2017) Treating arrhythmias with adjunctive magnesium: identifying future research directions. Eur Heart J Cardiovasc Pharmacother 3(2):108–117
6. Balan Y, Packirisamy RM, Mohanraj PS (2020) High dietary salt intake activates inflammatory cascades via Th17 immune cells: impact on health and diseases. Arch Med Sci 18(2):459
7. Bawiec P, Jaworowska A, Sawicki J, Czop M, Szalak R, Koch W (2025) In vitro evaluation of bioavailability of Mg from daily food rations, dietary supplements and medicinal products from the polish market. Nutrients 17(5):748
8. Bergwitz C, Jüppner H (2010) Regulation of phosphate homeostasis by PTH, vitamin D, and FGF23. Annu Rev Med 61(1):91-104
9. Bie P (2018) Mechanisms of sodium balance: total body sodium, surrogate variables, and renal sodium excretion. Am J Phys Regul Integr Comp Phys 315(5):R945–R962

10. Binesh A (2021) Decades-long involvement of signalling pathways in cardiovascular research using zebrafish model and its global trends. Rev Aquac 13(1):556–566
11. Binesh A, Venkatachalam K (2024) Copper in human health and disease: a comprehensive review. J Biochem Mol Toxicol 38(11):e70052
12. Binesh A, Devaraj SN, Halagowder D (2018a) Atherogenic diet induced lipid accumulation induced NFκB level in heart, liver and brain of Wistar rat and diosgenin as an anti-inflammatory agent. Life Sci 196:28–37
13. Binesh A, Devaraj SN, Devaraj H (2018b) Inhibition of nuclear translocation of notch intracellular domain (NICD) by diosgenin prevented atherosclerotic disease progression. Biochimie 148:63–71
14. Binesh A, Devaraj SN, Devaraj H (2019) Molecular interaction of NFκB and NICD in monocyte-macrophage differentiation is a target for intervention in atherosclerosis. J Cell Physiol 234(5):7040–7050
15. Binesh A, Devaraj SN, Devaraj H (2020a) Expression of chemokines in macrophage polarization and downregulation of NFκB in aorta allow macrophage polarization by diosgenin in atherosclerosis. J Biochem Mol Toxicol 34(2):e22422
16. Binesh A, Devaraj Sivasitambaram N, Halagowder D (2020b) Monocytes treated with ciprofloxacin and oxyLDL express myristate, priming atherosclerosis. J Biochem Mol Toxicol 34(3):e22442
17. Binesh A, Venkatachalam K, Saravanan R (2025) Investigating the therapeutic potential of Gracilaria edulis extract in copper-exposed zebrafish embryos: an anti-inflammatory perspective. Nat Prod J 15(5):e22103155294834
18. Bouillon R, Van Cromphaut S, Carmeliet G (2003) Intestinal calcium absorption: molecular vitamin D mediated mechanisms. J Cell Biochem 88(2):332–339
19. Bourque CW (2008) Central mechanisms of osmosensation and systemic osmoregulation. Nat Rev Neurosci 9(7):519–531
20. Calvo MS, Uribarri J (2013) Public health impact of dietary phosphorus excess on bone and cardiovascular health in the general population. Am J Clin Nutr 98(1):6–15
21. Carpenter TO, Shaw NJ, Portale AA, Ward LM, Abrams SA, Pettifor JM (2017) Rickets (Primer). Nat Rev Dis Primers 3(1)
22. Carrera-Bastos P, Fontes-Villalba M, O'Keefe JH, Lindeberg S, Cordain L (2011) The western diet and lifestyle and diseases of civilization. Res Rep Clin Cardiol:15–35
23. Celis-Morales C, Livingstone KM, Marsaux CF, Macready AL, Fallaize R, O'Donovan CB et al (2017) Effect of personalized nutrition on health-related behaviour change: evidence from the Food4Me European randomized controlled trial. Int J Epidemiol 46(2):578–588
24. Chen Y, Naik SG, Krzystek J, Shin S, Nelson WH, Xue S et al (2012) Role of calcium in metalloenzymes: effects of calcium removal on the axial ligation geometry and magnetic properties of the catalytic diheme center in MauG. Biochemistry 51(8):1586–1597
25. Christakos S, Dhawan P, Porta A, Mady LJ, Seth T (2011) Vitamin D and intestinal calcium absorption. Mol Cell Endocrinol 347(1–2):25–29.0
26. Civan MM (1978) Intracellular activities of sodium and potassium. Am J Physiol 234(4):F261–F269
27. Clemente-Suárez VJ, Beltrán-Velasco AI, Redondo-Flórez L, Martín-Rodríguez A, Tornero-Aguilera JF (2023) Global impacts of western diet and its effects on metabolism and health: a narrative review. Nutrients 15(12):2749
28. Costello R, Wallace TC, Rosanoff A (2016) Magnesium. Adv Nutr 7(1):199-201
29. Cozzolino M, Ciceri P, Galassi A, Mangano M, Carugo S, Capelli I, Cianciolo G (2019) The key role of phosphate on vascular calcification. Toxins 11(4):213
30. Curry JN, Alan SL (2018) Magnesium handling in the kidney. Adv Chronic Kidney Dis 25(3):236–243
31. de Luis DA, Izaola O, Primo D, Aller R (2018) Role of rs670 variant of APOA1 gene on lipid profile, insulin resistance and adipokine levels in obese subjects after weight loss with a dietary intervention. Diabetes Res Clin Pract 142:139–145

32. De Wardener HE (1978) The control of sodium excretion. Am J Physiol 235(3):F163–F173
33. Divya BJ, Suman B, Venkataswamy M, Thyagaraju K (2017) A study on phytochemicals, functional groups and mineral composition of Allium sativum (garlic) cloves. Int J Curr Pharm Res 9(3):42–45
34. Dolati S, Rikhtegar R, Mehdizadeh A, Yousefi M (2020) The role of magnesium in pathophysiology and migraine treatment. Biol Trace Elem Res 196(2):375–383
35. Drewnowski A, Rehm CD, Maillot M, Mendoza A, Monsivais P (2015) The feasibility of meeting the WHO guidelines for sodium and potassium: a cross-national comparison study. BMJ Open 5(3):e006625
36. EFSA Panel on Nutrition, Novel Foods and Food Allergens (NDA), Turck D, Castenmiller J, de Henauw S, Hirsch-Ernst KI, Kearney J et al (2019) Dietary reference values for sodium. EFSA J 17(9):e05778
37. Farag MA, Abib B, Qin Z, Ze X, Ali SE (2023) Dietary macrominerals: updated review of their role and orchestration in human nutrition throughout the life cycle with sex differences. Curr Res Food Sci 6:100450
38. Ferraro PM, Mandel EI, Curhan GC, Gambaro G, Taylor EN (2016) Dietary protein and potassium, diet–dependent net acid load, and risk of incident kidney stones. Clin J Am Soc Nephrol 11(10):1834–1844
39. Ferrero G, Carpi S, Polini B, Pardini B, Nieri P, Impeduglia A et al (2021) Intake of natural compounds and circulating microRNA expression levels: their relationship investigated in healthy subjects with different dietary habits. Front Pharmacol 11:619200
40. Fleet JC (2022) Vitamin D-mediated regulation of intestinal calcium absorption. Nutrients 14(16):3351
41. Fleet JC, Schoch RD (2010) Molecular mechanisms for regulation of intestinal calcium absorption by vitamin D and other factors. Crit Rev Clin Lab Sci 47(4):181–195
42. Food and Agriculture Organization of the United Nations (n.d.) Human vitamin and mineral requirements—minerals. https://www.fao.org/nutrition/requirements/minerals/en/. Accessed on 17 July 2025
43. Garcia MC, Diaz AF, Godinez R, Sanchez JA (1992) Effect of sodium deprivation on contraction and charge movement in frog skeletal muscle fibres. J Muscle Res Cell Motil 13:354–365
44. Ge MX, Chen RP, Zhang L, Wang YM, Chi CF, Wang B (2023) Novel Ca-chelating peptides from protein hydrolysate of Antarctic krill (Euphausia superba): preparation, characterization, and calcium absorption efficiency in Caco-2 cell monolayer model. Mar Drugs 21(11):579
45. Geddes RF, Finch NC, Syme HM, Elliott J (2013) The role of phosphorus in the pathophysiology of chronic kidney disease. J Vet Emerg Crit Care 23(2):122–133
46. Gombart AF, Pierre A, Maggini S (2020) A review of micronutrients and the immune system–working in harmony to reduce the risk of infection. Nutrients 12(1):236
47. Goni L, Cuervo M, Milagro FI, Martínez JA (2016) Future perspectives of personalized weight loss interventions based on nutrigenetic, epigenetic, and metagenomic data. J Nutr 146(4):905S–912S
48. Goyal R, Jialal I (2019) Hyperphosphatemia
49. Grider MH, Jessu R, Kabir R (2019) Physiology, action potential
50. Grillo A, Salvi L, Coruzzi P, Salvi P, Parati G (2019) Sodium intake and hypertension. Nutrients 11(9):1970
51. Hanck DA, Fozzard HA (2007) Voltage-gated sodium channels. In: Biological membrane ion channels: dynamics, structure, and applications. Springer, New York, pp 219–239
52. Hanna RM, Ahdoot RS, Kalantar-Zadeh K, Ghobry L, Kurtz I (2022) Calcium transport in the kidney and disease processes. Front Endocrinol 12:762130
53. Hirohama D, Fujita T (2019) Evaluation of the pathophysiological mechanisms of salt-sensitive hypertension. Hypertens Res 42(12):1848–1857
54. Hong S, Choi JW, Park JS, Lee CH (2022) The association between dietary sodium intake and osteoporosis. Sci Rep 12(1):14594

55. Hoppe LK, Muhlack DC, Koenig W, Carr PR, Brenner H, Schöttker B (2018) Association of abnormal serum potassium levels with arrhythmias and cardiovascular mortality: a systematic review and meta-analysis of observational studies. Cardiovasc Drugs Ther 32:197–212
56. Houston M (2011) The role of magnesium in hypertension and cardiovascular disease. J Clin Hypertens 13(11):843–847
57. Hruska KA, Saab G, Mathew S, Lund R (2007) Phosphorus metabolism and management in chronic kidney disease: renal osteodystrophy, phosphate homeostasis, and vascular calcification. Semin Dial 20(4):309–315
58. Hsu SH, Chen LR, Chen KH (2024) Primary osteoporosis induced by androgen and estrogen deficiency: the molecular and cellular perspective on pathophysiological mechanisms and treatments. Int J Mol Sci 25(22):12139
59. Jahnen-Dechent W, Ketteler M (2012) Magnesium basics. Clin Kidney J 5(Suppl_1):i3–i14
60. Kaur J, Castro D (2024) Hypophosphatemia. In: StatPearls [Internet]. StatPearls Publishing
61. Khan SR, Pearle MS, Robertson WG, Gambaro G, Canales BK, Doizi S et al (2016) Kidney stones. Nat Rev Dis Prim 2(1):1–23
62. Lee J, deWardener HE (1974) Neurosecretion and sodium excretion. Kidney Int 6(5):323–330
63. Lee S, Park HI, Sang QXA (2007) Calcium regulates tertiary structure and enzymatic activity of human endometase/matrilysin-2 and its role in promoting human breast cancer cell invasion. Biochem J 403(1):31–42
64. Mahadik AB, Giri AB, Bhambre AS, Mundhe SA, Shinde RD (2021) Osteoporosis and lifestyle medicine. Immunodeficiency 4(6):7
65. Mandal AK (1997) Hypokalemia and hyperkalemia. Med Clin North Am 81(3):611–639
66. Matsui H, Shimosawa T, Uetake Y, Wang H, Ogura S, Kaneko T et al (2006) Protective effect of potassium against the hypertensive cardiac dysfunction: association with reactive oxygen species reduction. Hypertension 48(2):225–231
67. McLean RM, Wang NX (2021) Potassium. In: Advances in food and nutrition research, vol 96. Academic, pp 89–121
68. Milenkovic D, Jude B, Morand C (2013) miRNA as molecular target of polyphenols underlying their biological effects. Free Radic Biol Med 64:40–51
69. Moore JB (2020) From personalised nutrition to precision medicine: the rise of consumer genomics and digital health. Proc Nutr Soc 79(3):300–310
70. Murray SL, Wolf M (2024) Calcium and phosphate disorders: Core curriculum 2024. Am J Kidney Dis 83(2):241–256
71. Napiórkowska-Baran K, Treichel P, Czarnowska M, Drozd M, Koperska K, Węglarz A et al (2024) Immunomodulation through nutrition should be a key trend in type 2 diabetes treatment. Int J Mol Sci 25(7):3769
72. National Institutes of Health (2024) Phosphorus fact sheet for health professionals; 2023. Accessed 9 July
73. National Institutes of Health, Office of Dietary Supplements (2023 – Updated) Phosphorus—fact sheet for health professionals. U.S. Department of Health and Human Services. https://ods.od.nih.gov/factsheets/Phosphorus-HealthProfessional/
74. Negru AG, Pastorcici A, Crisan S, Cismaru G, Popescu FG, Luca CT (2022) The role of hypomagnesemia in cardiac arrhythmias: a clinical perspective. Biomedicine 10(10):2356
75. Nicholls JW, Chin JP, Williams TA, Lenton TM, O'Flaherty V, McGrath JW (2023) On the potential roles of phosphorus in the early evolution of energy metabolism. Front Microbiol 14:1239189
76. Nishimoto M, Griffin KA, Wynne BM, Fujita T (2024) Salt-sensitive hypertension and the kidney. Hypertension 81(6):1206–1217
77. Paul SH, Usman AA, Adenyi OD, Olutoye MA (2018a) Formulation of a multifunctional flora composite for the prevention and management of cancer, diabetes, blood pressure and stroke. Curr Trends Biomed Eng Biosci 11(1):1–11

78. Paul SH, Usman A, Gana I, Manase A, Adeniyi O, Olutoye M (2018b) Comparative study of mineral and nutritional composition of a multifunctional flora composite formulated from seven medicinal plants and their applications to human health
79. Peloza OC, Žeželj SP, Jovanović GK, Pavičić I, Radetić ATJ, Cvek SZ et al (2021) Osteoporosis and dietary inflammatory index. In: Osteoporosis-recent advances, new perspectives and applications. IntechOpen
80. Peri A (2019) Management of hyponatremia: causes, clinical aspects, differential diagnosis and treatment. Expert Rev Endocrinol Metab 14(1):13–21
81. Philipp Schuchardt J, Hahn A (2017) Intestinal absorption and factors influencing bioavailability of magnesium-an update. Curr Nutr Food Sci 13(4):260–278
82. Pitaraki EE (2019) The role of the calcium intake in the development of cardiovascular calcification. J Res Pract Musculoskeletal Syst 3(1):26–31
83. Piyushbhai MK, Binesh A, Shanmugam SA, Venkatachalam K (2023) Exposure to low-dose arsenic caused teratogenicity and upregulation of proinflammatory cytokines in zebrafish embryos. Biol Trace Elem Res 201(7):3487–3496
84. Piyushbhai MK, Binesh A, Shanmugam SA, Venkatachalam K (2024) Effects of mercury chloride on antioxidant and inflammatory cytokines in zebrafish embryos. J Biochem Mol Toxicol 38(1):e23589
85. Porter JL, Varacallo M, Castano M (2023) Osteoporosis (nursing). In: StatPearls. StatPearls Publishing
86. Ramos-Lopez O, Cuervo M, Goni L, Milagro FI, Riezu-Boj JI, Martinez JA (2020) Modeling of an integrative prototype based on genetic, phenotypic, and environmental information for personalized prescription of energy-restricted diets in overweight/obese subjects. Am J Clin Nutr 111(2):459–470
87. Rocha I (2024) The influence of calcium and phosphorus on immune cell activity. Int J Immunol Res 6:31–34
88. Rondon H, Badireddy M (2017) Hyponatremia
89. Rosanoff A, Weaver CM, Rude RK (2012) Suboptimal magnesium status in the United States: are the health consequences underestimated? Nutr Rev 70(3):153–164
90. Salcedo-Betancourt JD, Moe OW (2024) The effects of acid on calcium and phosphate metabolism. Int J Mol Sci 25(4):2081.3
91. Sanada H, Jones JE, Jose PA (2011) Genetics of salt-sensitive hypertension. Curr Hypertens Rep 13:55–66
92. Serna J, Bergwitz C (2020) Importance of dietary phosphorus for bone metabolism and healthy aging. Nutrients 12(10):3001
93. Severino A, Tohumcu E, Tamai L, Dargenio P, Porcari S, Rondinella D et al (2024) The microbiome-driven impact of western diet in the development of noncommunicable chronic disorders. Best Pract Res Clin Gastroenterol 72:101923
94. Shashikant M, Bains A, Chawla P, Fogarasi M, Fogarasi S (2022) The current status, bioactivity, food, and pharmaceutical approaches of Calocybe indica: a review. Antioxidants 11(6):1145
95. Spencer RP (1959) Potassium metabolism and gastrointestinal function; a review: a semiquantitative approach. Am J Dig Dis 4:145–158
96. Tangvoraphonkchai K, Davenport A (2018) Magnesium and cardiovascular disease. Adv Chronic Kidney Dis 25(3):251–260
97. Terrell K, Choi S, Choi S (2023) Calcium's role and signaling in aging muscle, cellular senescence, and mineral interactions. Int J Mol Sci 24(23):17034
98. Thirunavukarasu S, Devi V, Usheva S, Arumugam D, Ramanujam A (2018) A rare case of acute symptomatic seizure, tetany and poor sensorium secondary to autosomal dominant hypocalcemia and hypercalciuria. J Rare Disord Diagn Ther 4:2
99. Tileuberdi N, Turgumbayeva A, Yeskaliyeva B, Sarsenova L, Issayeva R (2022) Extraction, isolation of bioactive compounds and therapeutic potential of rapeseed (Brassica napus L.). Molecules 27(24):8824

100. Van Laecke S (2023) An update on hypomagnesemia and hypermagnesemia. Kidney Dial 4(1):1–14
101. Wagner CA (2024) The basics of phosphate metabolism. Nephrol Dial Transplant 39(2):190–201
102. Walker MD, Shane E (2022) Hypercalcemia: a review. JAMA 328(16):1624–1636
103. Walters GC, Usachev YM (2023) Mitochondrial calcium cycling in neuronal function and neurodegeneration. Front Cell Dev Biol 11:1094356
104. Willebrand R, Kleinewietfeld M (2018) The role of salt for immune cell function and disease. Immunology 154(3):346–353
105. World Health Organization, Food and Agriculture Organization (2004) Vitamin and mineral requirements in human nutrition (2nd ed.). https://www.who.int/publications/i/item/9241546123. Accessed on 17 July 2025
106. Yamanaka R, Tabata S, Shindo Y, Hotta K, Suzuki K, Soga T, Oka K (2016) Mitochondrial Mg2+ homeostasis decides cellular energy metabolism and vulnerability to stress. Sci Rep 6(1):30027
107. Yawson DO, Adu MO, Ason B, Armah FA, Boateng E, Quansah R (2018) Ghanaians might be at risk of excess dietary intake of potassium based on food supply data. J Nutr Metabol 2018(1):5989307
108. Zeevi D, Korem T, Zmora N, Israeli D, Rothschild D, Weinberger A et al (2015) Personalized nutrition by prediction of glycemic responses. Cell 163(5):1079–1094
109. Zemel MB (2001) Calcium modulation of hypertension and obesity: mechanisms and implications. J Am Coll Nutr 20(sup5):428S–435S
110. Zhou H, Hu YY, Tang ZX, Jiang ZB, Huang J, Zhang T et al (2024) Calcium transport and enrichment in microorganisms: a review. Foods 13(22):3612
111. Zhu K, Prince RL (2012) Calcium and bone. Clin Biochem 45(12):936–942

Chapter 21
Sodium Homeostasis and Roles of Sodium-Dependent Transport in Health

Takeshi Y. Hiyama, Kotaro Yamamoto, Mana Hatakeyama, and Robby Alkhusairy

Abstract The sodium (Na^+) gradient across the plasma membrane, generated by the Na^+/K^+-ATPase, constitutes the fundamental driving force for a diverse array of secondary active transport systems essential to numerous human tissues. This electrochemical gradient facilitates the function of Na^+-dependent cotransporters and exchangers, enabling the translocation of nutrients and ions against their respective concentration gradients, thereby ensuring efficient absorption and systemic distribution. These transport processes are critical for the maintenance of mineral homeostasis and overall physiological equilibrium.

This section delineates key Na^+-dependent transport mechanisms involved in the cellular uptake and systemic regulation of major minerals such as calcium (Ca^{2+}) and magnesium (Mg^{2+}), essential vitamins including vitamin C and biotin, as well as fundamental nutrients like glucose, amino acids, and phosphate. These substrates utilize Na^+ gradients to traverse cellular membranes efficiently, facilitating their participation in metabolic pathways, enzymatic functions, and cellular signaling networks.

A comprehensive understanding of these Na^+-coupled transport pathways is imperative as their dysregulation is implicated in a spectrum of pathophysiological conditions, including mineral imbalances, metabolic disorders, and nutrient deficiencies. Moreover, this review addresses the hormonal and molecular regulatory systems that sustain sodium homeostasis, notably the renin–angiotensin–aldosterone system (RAAS) and natriuretic peptides, which collectively modulate Na^+ levels to preserve fluid balance and physiological function.

T. Y. Hiyama (✉)
Department of Integrative Physiology, Tottori University Graduate School and Faculty of Medicine, Tottori University, Yonago, Japan

International Platform for Dryland Research and Education, Tottori University, Tottori, Japan
e-mail: hiyama@tottori-u.ac.jp

K. Yamamoto · M. Hatakeyama · R. Alkhusairy
Department of Integrative Physiology, Tottori University Graduate School and Faculty of Medicine, Tottori University, Yonago, Japan

N. S. Dhalla et al. (eds.), *Functional Biochemistry of Metallic Elements*, Advances in Biochemistry in Health and Disease 36,
https://doi.org/10.1007/978-3-032-14445-4_21

Elucidating the intricate relationship between Na^+ gradients and nutrient transport advances our understanding of mineral and vitamin metabolism and informs the development of targeted therapeutic strategies for related diseases.

Keywords Sodium homeostasis · Na^+-dependent transport · Mineral metabolism · Vitamin transport · Nutrient absorption · Renin–angiotensin–aldosterone system (RAAS) · Natriuretic peptides · Na_x · Antidiuretic hormone (ADH) · Body fluids

Introduction

Sodium is one of the principal constituents of body fluids, including blood, interstitial fluid, and cerebrospinal fluid, and the maintenance of sodium homeostasis is critical for sustaining osmotic balance. In addition, sodium ion (Na^+) plays an essential role in the generation of neuronal action potentials, synaptic transmission, muscle contraction, and active transport of biologically essential substances (Fig. 21.1). Because the Na^+ concentration in body fluids is greatly influenced by the volume of body water, the maintenance of Na^+ concentration homeostasis must be considered in conjunction with the regulation of total body water. The Na^+ gradient generated by the Na^+/K^+-ATPase constitutes the fundamental driving force for numerous secondary active transport mechanisms essential across diverse human tissues. Na^+-dependent cotransporters and exchangers exploit this Na^+ gradient to facilitate the movement of target nutrients and ions against their own concentration gradients, thereby playing a pivotal role in maintaining mineral homeostasis [1]. This article focuses on the mechanisms by which major minerals (calcium, magnesium), vitamins (vitamin C, biotin), and nutrients (glucose, amino acids, phosphate) are transported via Na^+-dependent pathways and discusses their implications for human health and disease. In addition, we provide an overview of the regulatory processes that sustain Na^+ homeostasis.

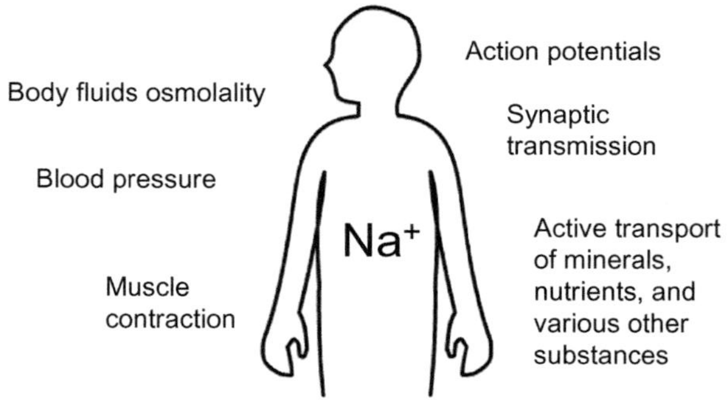

Fig. 21.1 Physiological significance of sodium homeostasis. Body Na^+ plays an essential role in the generation of neuronal action potentials, synaptic transmission, muscle contraction, and active transport of biologically essential substances

Mechanisms for Maintaining Sodium Homeostasis

Sodium homeostasis in the body is maintained through the regulation of oral intake and excretion via sweat and urine. The control of oral intake is conducted primarily through regulation of salt appetite. The kidney plays a central role in sodium balance by filtering plasma Na^+ and reabsorbing the necessary amount to maintain systemic homeostasis. Approximately 99% of the filtered Na^+ is reabsorbed at various nephron sites, including the proximal tubule, thick ascending limb of the loop of Henle, distal tubule, and collecting duct. Additionally, body fluid volume significantly influences Na^+ levels.

A critical hormonal cascade involved in sodium homeostasis is the renin–angiotensin–aldosterone system (RAAS). This system functions to retain Na^+ and water in the body, thereby maintaining fluid volume. Juxtaglomerular cells in the kidney secrete renin in response to decreased renal perfusion pressure, sympathetic nervous system activation, or reduced Na^+ delivery to the distal tubule [2]. Renin cleaves liver-derived angiotensinogen to angiotensin I, which is then converted by angiotensin-converting enzyme (ACE) to angiotensin II (Ang II). Ang II is produced when fluid volume maintenance or blood pressure is compromised, such as during hemorrhage, dehydration, or Na^+ deficiency [3]. Ang II acts on water- and salt-sensitive neurons in the subfornical organ of the brain, inducing thirst and salt appetite [4] (Fig. 21.2). During dehydration, plasma Na^+ levels increase, activating the Na^+ sensor Na_x expressed in glial cells of the subfornical organ (SFO). This activation suppresses salt neuron activity via inhibitory neurons mediated by

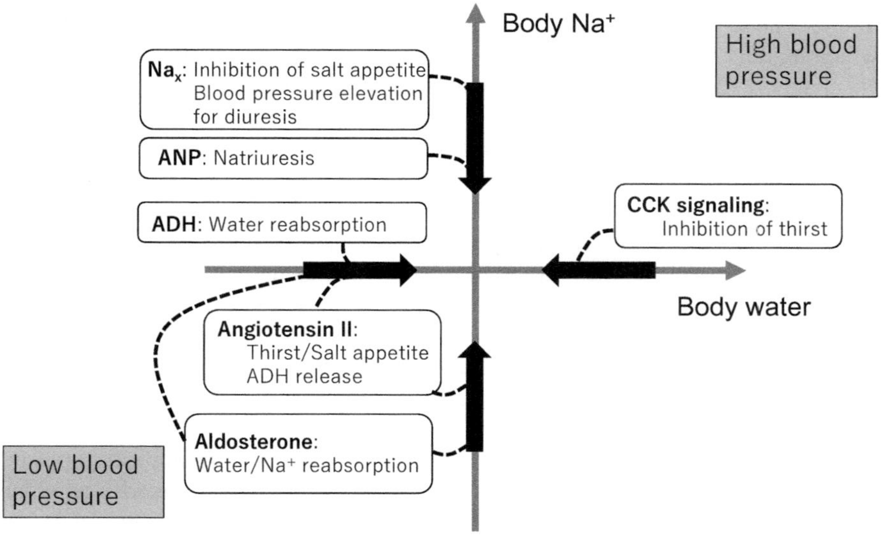

Fig. 21.2 Mechanisms for maintaining Na^+ homeostasis. Mechanisms controlling the oral intake and renal excretion of Na^+ and water to maintain body fluid volume and blood pressure

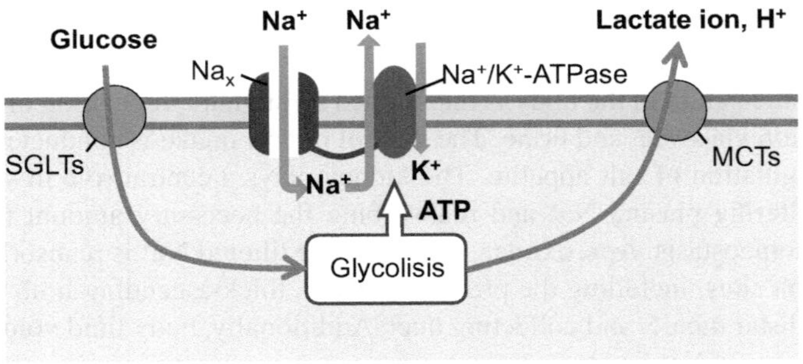

Fig. 21.3 Mechanisms by which activation of Na_x leads to the release of lactate ions and protons. The Na^+ sensor Na_x is a sodium channel, with Na^+/K^+-ATPase and the glycolytic enzyme triosephosphate isomerase (TPI) directly bound to its C-terminus. Upon sensing an increase in extracellular Na^+ concentration, Na_x opens, allowing a sustained influx of Na^+. To remove the accumulated intracellular Na^+, the ATP consumption of Na^+/K^+-ATPase increases, which in turn activates glycolysis leading to lactate production. The produced lactate is subsequently exported via monocarboxylate transporters (MCTs)

lactate, thus reducing salt appetite (Fig. 21.3). Conversely, under salt deficiency, cholecystokinin (CCK)-producing neurons are activated, inhibiting water neuron activity [5] (Fig. 21.2).

Conversely, atrial natriuretic peptide (ANP) promotes Na^+ excretion and reduces blood volume and pressure, thereby preventing volume overload conditions such as heart failure. ANP is secreted from atrial myocytes in response to atrial stretch due to increased blood volume. Acting antagonistically to the renin–angiotensin system and aldosterone, ANP induces natriuresis and vasodilation [6]. ANP binds to natriuretic peptide receptor A (NPR-A) in the kidney, activating the cGMP pathway, which inhibits Na^+ reabsorption in the collecting duct by downregulating epithelial Na channels (ENaC), promotes afferent arteriole vasodilation to increase glomerular filtration rate (GFR), and suppresses renin and aldosterone secretion [7].

There is a hypothesis that salt-sensitive hypertension is a result of a physiological response to promote Na^+ excretion via pressure natriuresis. In fact, elevations in NaCl concentrations in plasma and cerebrospinal fluid have been reported to enhance sympathetic nerve activity, leading to increases in blood pressure [8]. Animal studies have shown that blockade of sympathetic nerve activity markedly suppresses blood pressure elevations induced by high salt intake or increases in Na^+ levels in body fluids [9, 10]. Furthermore, the Na_x channel expressed in glial cells of the organum vasculosum lamina terminalis (OVLT) senses increases in Na^+ levels in blood and cerebrospinal fluid and activates neurons that regulate sympathetic nerve activity via acid-sensing ion channel 1a (ASIC1a) [11].

Body fluid volume alters plasma Na^+ dilution, significantly impacting plasma Na^+ concentration. Antidiuretic hormone (ADH, vasopressin) primarily regulates water reabsorption. In response to increased plasma osmolality or hypovolemia, ADH is secreted from the posterior pituitary and binds to V2 receptors on collecting duct cells, promoting insertion of aquaporin-2 channels into the apical membrane and increasing water permeability [12].

Fluctuations in body fluid Na^+ concentration are typically no more than about 10%, yet they can cause physiological changes such as those observed in dehydration. Moreover, when considering variations in local intracellular Na^+ concentrations, the Na^+ gradient across the plasma membrane can be substantially greater. The following section provides an overview of substance transport processes influenced by Na^+ concentration.

Na^+-Dependent Metal Ion Transport

Na^+-Dependent Magnesium Transporter: NME

Magnesium (Mg^{2+}) is the second most abundant intracellular cation after potassium and serves as an essential cofactor for numerous enzymatic reactions, including ATP metabolism, DNA/RNA synthesis, and protein synthesis. Moreover, magnesium is indispensable for the regulation of muscle contraction, blood pressure, insulin metabolism, myocardial excitability, vascular tone, neurotransmission, and neuromuscular conduction [13]. Mg^{2+} also plays a critical role in blocking N-methyl-D-aspartate (NMDA) receptor channels, which are involved in memory, learning, and the development of neural circuits.

NME (encoded by *SLC41A1*) protein functions as a Na^+/Mg^{2+} exchanger, mediating the efflux of Mg^{2+} to maintain intracellular magnesium homeostasis [14]. This activity is strictly dependent on extracellular Na^+, and its transport function is regulated via phosphorylation by protein kinase A, suggesting hormonal modulation of its activity [14]. Mutations in this gene have been associated with nephronophthisis-like phenotypes [15]. Furthermore, NME has been linked to a range of severe disorders, including neurodegenerative, neurological, and psychiatric conditions such as Alzheimer's disease (AD) [16], stroke [17], aggressive behavior [18], increased stress susceptibility [19], and hyperactivity [20]. Notably, NME is located within the PARK16 locus, which has been implicated in the onset and progression of Parkinson's disease and other neurodegenerative disorders [21].

Magnesium also serves as a cofactor in enzymatic pathways of the tricarboxylic acid (TCA) cycle and the electron transport chain (ETC). Alterations in NME activity or dietary Mg^{2+} intake can influence myocardial mitochondrial enzyme activity. In mice lacking *Slc41a1* gene under magnesium-deficient dietary conditions, compensatory changes in mitochondrial energy enzymes, such as aconitase and isocitrate dehydrogenase, have been observed, indicating that NME function may affect cardiac energy metabolism [14].

Na^+-Coupled Calcium Transporters: NCX and NCKX

Two families of Na^+/Ca^{2+} exchangers have been identified: the K^+-independent NCX (*SLC8* gene family) and the K^+-dependent NCKX (*SLC24* gene family) [22]. NCX serves as a major mechanism for Ca^{2+} efflux, operating through an exchange transport mode in which three Na^+ are taken up into the cell in exchange for the extrusion of one Ca^{2+}. Due to its high transport capacity, NCX is well suited for rapidly removing large amounts of Ca^{2+} from the cell [23]. There are three subtypes of NCX (NCX1–3), each with distinct tissue distributions, with NCX1 playing a central role in maintaining intracellular ionic homeostasis. All three NCX subtypes are expressed in the brain [24], where under physiological conditions they are crucial for Ca^{2+} clearance following neuronal action potentials. However, during cerebral ischemia, the disruption of Na^+ and Ca^{2+} ion gradients can cause NCX to operate in reverse mode, importing Ca^{2+} into the cell rather than extruding it, leading to elevated intracellular Ca^{2+} concentrations and excitotoxic neuronal injury [25]. NCKX transporters are important for maintaining Na^+, Ca^{2+}, and K^+ homeostasis in the central nervous system, and evidence suggests that NCKX2 plays a protective role against focal ischemic brain injury [26].

In the heart, NCX1 is abundantly expressed in the sarcolemma of cardiomyocytes and functions as the primary Ca^{2+} extrusion system, supporting the calcium cycle essential for excitation–contraction (E–C) coupling and myocardial relaxation. NCX is an electrogenic transporter and, under normal membrane potentials, primarily operates in the forward mode (Ca^{2+} efflux), though it can operate in reverse mode depending on the membrane potential [27]. Increased NCX activity can contribute to heart failure, ischemic injury, and arrhythmogenesis [28]. During myocardial ischemia, reduced Na^+/K^+-ATPase activity leads to elevated intracellular Na^+ concentration, which is followed by an increase in intracellular Ca^{2+}. This Ca^{2+} accumulation is dependent on the rise in cytosolic Na^+ and is thought to be mediated via reverse-mode NCX activity [29].

NCX1 activity is also modulated by a process known as Na^+-dependent inactivation, whereby elevated intracellular Na^+ suppresses NCX function. Introduction of the K229Q mutation, which abolishes Na^+-dependent inactivation, results in enhanced NCX1 activity, prolonged action potential duration, increased frequency of afterdepolarizations, faster decay of Ca^{2+} transients, and reduced cell shortening [30]. These findings suggest that Na^+-dependent inactivation plays a role in modulating arrhythmia risk.

Na^+-Dependent Manganese (Mn^{2+}) Transport and ZIP14

Manganese ions (Mn^{2+}) are essential trace elements involved in numerous physiological processes, including antioxidant defense, metabolism, and enzymatic reactions required for neurotransmitter synthesis. However, excess Mn^{2+} exhibits neurotoxicity; therefore, the uptake, distribution, and excretion of Mn^{2+} must be tightly regulated to maintain homeostasis [31].

The *SLC39A14* gene encodes ZIP14, a member of the Zrt- and Irt-like protein (ZIP) family of metal ion transporters. ZIP14 is a transmembrane protein that transports divalent metal ions such as Cd^{2+}, Mn^{2+}, Zn^{2+}, and Fe^{2+}. Through Na^+ dependence, it can actively transport Mn^{2+} against its concentration gradient, enabling efficient absorption and intracellular accumulation. ZIP14 has been identified as a major transporter mediating Na^+-dependent cellular Mn^{2+} uptake [32, 33]. Patients carrying mutations in ZIP14 develop manganese toxicity and early-onset dystonia.

ZIP14 is highly expressed in hepatocytes, where it plays a central role in systemic Mn^{2+} clearance via biliary excretion. In intestinal epithelial cells, it contributes to dietary Mn absorption, while in pancreatic cells, it influences Mn metabolism [34]. It is also expressed in brain tissue, participating in Mn regulation in the nervous system. Loss of ZIP14 function impairs Mn clearance, promoting Mn accumulation in the brain and exacerbating neurotoxicity [33].

Excessive Mn accumulation in the basal ganglia results in manganism, a neurotoxic syndrome characterized by tremor, dystonia, and cognitive impairment, presenting with Parkinson's disease-like symptoms. Consistently, mutations in *SLC39A14* can lead to hypermanganesemia accompanied by dystonia and parkinsonism [35].

Autosomal recessive *SLC39A14* mutations are associated with HMNDYT2 (hypermanganesemia with dystonia type 2), a disorder characterized by markedly elevated blood Mn levels, progressive dystonia, developmental delay, and neurodegeneration. Patients typically present with motor symptoms in early childhood, with the underlying pathology being impaired Mn transport due to loss of ZIP14 function [36].

Na^+-Dependent Vitamin Transport

Na^+-Dependent Multivitamin Transporter (SMVT) and Its Role in the Transport of Biotin (B7), Pantothenic Acid (B5), and Lipoic Acid

The Na^+-dependent multivitamin transporter (SMVT), encoded by the *SLC5A6* gene, is an essential membrane protein expressed in various tissues including the placenta, intestine, brain, liver, lung, kidney, cornea, retina, and heart [37]. SMVT mediates the active, Na^+-dependent uptake of multiple water-soluble vitamins and micronutrients such as biotin (vitamin B7), pantothenic acid (vitamin B5), and lipoic acid [38, 39]. These micronutrients serve as critical cofactors in numerous enzymatic reactions indispensable for cellular metabolism, antioxidant defense, and energy production.

For example, biotin functions as a coenzyme for carboxylases involved in fatty acid synthesis, gluconeogenesis, and amino acid catabolism. Moreover, biotinylation of histones epigenetically regulates gene expression [40]. Biotin is widely present in many foods, and its deficiency causes severe clinical manifestations such as neurological disorders and dermatological symptoms; in animals, biotin deficiency can result in embryonic growth retardation, congenital malformations, and lethality. Pantothenic acid is required for the biosynthesis of coenzyme A (CoA) and acyl carrier proteins, playing an indispensable role in the tricarboxylic acid (TCA) cycle, fatty acid metabolism, and acyl group transfer reactions involved in neurotransmitter and steroid hormone synthesis [41]. Lipoic acid acts as a cofactor for mitochondrial enzyme complexes such as pyruvate dehydrogenase and α-ketoglutarate dehydrogenase. It also exhibits potent antioxidant properties through scavenging reactive oxygen species and regenerating other antioxidants [42].

SMVT expression in intestinal epithelial cells is essential for the absorption of dietary biotin, pantothenic acid, and lipoic acid. The intestine is exposed to two sources of biotin: dietary and microbiota derived. Protein-bound biotin is first released by digestion into free biotin, which is then absorbed via SMVT [43]. Similarly, pantothenic acid is available from both dietary and microbial sources; dietary protein-bound pantothenic acid is hydrolyzed in the intestinal lumen to free pantothenic acid and absorbed through SMVT [43]. Furthermore, biotin and pantothenic acid transport across the luminal side of the blood–brain barrier (BBB) is Na^+-dependent, with SMVT playing a significant role [38, 44].

SMVT has been found to be significantly overexpressed in lung adenocarcinoma [45]. Overexpression of *SLC5A6* promotes cell proliferation and migration, whereas knockout of *SLC5A6* inhibits these processes and induces apoptosis through mitochondrial dysfunction [46]. These findings suggest that the Na^+-dependent transport activity of SMVT may play an important role in cancer growth and could represent a potential therapeutic target for cancer treatment.

Na^+-Dependent Vitamin C Transporters SVCT1 and SVCT2

Vitamin C (ascorbic acid) is a water-soluble essential micronutrient that plays a crucial role in numerous biological functions, including antioxidant defense, collagen synthesis, immune regulation, and neurotransmitter production. Humans lack L-gulonolactone oxidase, the enzyme necessary for endogenous vitamin C synthesis, making dietary intake indispensable. Cellular uptake and systemic distribution of vitamin C rely primarily on two specific Na^+-dependent vitamin C transporters, SVCT1 and SVCT2, encoded by the *SLC23A1* and *SLC23A2* genes, respectively [47]. These transporters utilize the Na^+ gradient to actively transport reduced ascorbate into cells, thereby maintaining intracellular vitamin C homeostasis and enabling normal physiological function and protection against oxidative stress.

SVCT1 is predominantly expressed in epithelial tissues such as the intestinal epithelium and renal proximal tubules. It exhibits relatively low affinity but a high capacity for vitamin C transport, making it suitable for bulk absorption and reabsorption of vitamin C. In contrast, SVCT2 displays high affinity and is widely expressed in metabolically active tissues including the brain, eyes, adrenal glands, and placenta. SVCT2 ensures efficient cellular vitamin C uptake and is critical in tissues requiring strong antioxidant capacity and metabolic support [48, 49].

Vitamin C deficiency leads to scurvy, characterized by impaired collagen synthesis, bleeding, poor wound healing, and connective tissue abnormalities. While dietary insufficiency is the primary cause, defects in SVCT1 function that impair absorption or renal reabsorption can exacerbate deficiency. SVCT2 is essential for maintaining brain vitamin C levels and has been implicated in neurodegenerative diseases such as Alzheimer's and Parkinson's disease [50–52]. Given that oxidative stress contributes to neuronal damage, reduced expression or dysfunction of SVCT2 may weaken antioxidant defenses and worsen disease progression. Furthermore, vitamin C exhibits both antioxidant and prooxidant properties that influence cancer biology. Expression levels of SVCT1 and SVCT2 vary across different cancers, affecting intracellular ascorbate concentrations, tumor growth, and chemotherapy sensitivity. Notably, high SVCT2 expression in certain tumors enhances vitamin C uptake, impacting redox status and therapeutic response [53].

Na^+-Dependent Nutrient Transport

Na^+-Dependent Glucose Transporters SGLTs

Glucose absorption in the human body is essential for cellular energy supply, and Na^+-dependent transporters play a central role in facilitating the uptake of these nutrients across cell membranes. The primary transporter systems include Na^+-glucose cotransporters 1 through 5 (SGLT1–5). Among these, SGLT1, encoded by the *SLC5A1* gene, is a high-affinity Na^+-glucose cotransporter primarily responsible for glucose absorption in the small intestine, while low-affinity SGLT2 (*SLC5A2*) functions predominantly in the kidney, mediating bulk reabsorption of filtered glucose [54].

SGLT couples the transport of one glucose molecule with two Na^+, utilizing the Na^+ gradient maintained by Na^+/K^+-ATPase to actively uptake glucose against its concentration gradient. SGLT1 is indispensable for efficient absorption of dietary glucose, directly influencing postprandial blood glucose levels and energy availability. In the kidney, SGLTs contribute to glucose reabsorption, preventing glycosuria under physiological conditions.

Mutations in *SLC5A1* cause glucose-galactose malabsorption, a rare autosomal recessive disorder characterized by severe diarrhea and dehydration due to impaired intestinal absorption of glucose and galactose [55]. Moreover, altered expression or function of SGLT2 has been implicated in diabetes mellitus, where enhanced intestinal glucose uptake may exacerbate hyperglycemia [56]. Consequently, SGLT1 inhibitors are under development as antidiabetic agents aimed at reducing glucose absorption and lowering blood glucose levels [57].

Na$^+$-Dependent Amino Acid Transporters: SLC1, SLC6, and SLC38 Family

The *SLC1* and *SLC6* gene family comprises Na$^+$- and/or chloride-dependent amino acid transporters, including glycine transporters (GlyT1, GlyT2, encoded by *SLC6A5* and *SLC6A9*, respectively), gamma-aminobutyric acid transporters (GAT1-3 and BGT-1, encoded by *SLC6A1*, *SLC6A13*, *SLC6A11*, and *SLC6A12*, respectively), and various neutral and charged amino acid transporters, including EAAT1-3 (encoded by *SLC1A3*, *SLC1A2*, and *SLC1A1*, respectively) and SNAT1-7 (encoded by *SLC38A1-7*) for glutamate, and ASCT1/2 (encoded by *SLC1A4* and *SLC1A5*, respectively) for alanine, serine, and cysteine [58]. These transporters utilize the electrochemical gradient of Na$^+$ to facilitate the uptake of amino acids into cells, playing essential roles in neurotransmission, metabolism, and cellular homeostasis [59].

SLC6 amino acid transporters are highly expressed in the brain, intestine, and kidney. In the intestine, they mediate absorption of dietary amino acids, while in the brain, they regulate the clearance and recycling of neurotransmitters in the synaptic cleft [60–62]. Mutations or dysregulation of SLC6 transporters have been linked to neurological disorders such as epilepsy, autism spectrum disorders, and glycine encephalopathy, due to altered neurotransmitter homeostasis [63]. In the kidney and intestine, impaired amino acid transport can cause malabsorption syndromes and metabolic disturbances [64].

Na$^+$-Dependent Phosphate Transporters NaPi-IIa, NaPi-IIb, and NaPi-IIc

In the kidney, phosphate (Pi) reabsorption occurs primarily in the proximal tubule, where Na$^+$-dependent phosphate transporters play a crucial role. Notably, type II Na$^+$-phosphate cotransporters NaPi-IIa (*SLC34A1*) and NaPi-IIc (*SLC34A3*) are localized to the brush border membrane and mediate phosphate reabsorption from the tubular lumen [65]. The activity of these transporters is tightly regulated by parathyroid hormone (PTH), fibroblast growth factor 23 (FGF23), and vitamin D, thereby maintaining systemic phosphate homeostasis. Both PTH and FGF23 downregulate the expression of NaPi transporters, promoting phosphate excretion, whereas active vitamin D (1,25-dihydroxyvitamin D_3) enhances phosphate reabsorption [66].

In the intestine, dietary phosphate absorption occurs predominantly in the small intestine via both passive diffusion and active transport mechanisms. Na^+-dependent phosphate uptake is primarily mediated by the NaPi-IIb (*SLC34A2*) transporter expressed on the brush border membrane of intestinal epithelial cells, whose activity is regulated according to dietary phosphate levels and vitamin D status. This pathway is especially important during growth and periods of high phosphate demand [67].

Intracellular phosphate is indispensable for numerous metabolic pathways, including ATP synthesis, nucleic acid formation, and phospholipid production. Na^+-dependent phosphate transport is also observed in mitochondria and other organelles, directly contributing to energy metabolism [68].

Dysfunction of Na^+- and phosphate-transport mechanisms can lead to diseases such as osteomalacia, hypophosphatemic rickets, and chronic kidney disease (CKD) [69]. In CKD, impaired phosphate excretion causes hyperphosphatemia and secondary hyperparathyroidism, which promote cardiovascular calcification and bone metabolism abnormalities. Understanding these Na^+-dependent transport systems is essential for developing therapeutic strategies, including phosphate binders and hormonal modulation therapies.

Conclusions

Na^+ is a major extracellular cation essential for maintaining fluid balance, osmolarity, nerve impulse conduction, and muscle function. Sodium homeostasis is maintained through a complex interplay of renal function, dietary intake, and hormonal regulation (Fig. 21.4). Hormonal systems such as the renin–angiotensin–aldosterone system (RAAS), atrial natriuretic peptide (ANP), and antidiuretic hormone (ADH, vasopressin), along with the brain-expressed Na^+ sensor Na_x that monitors systemic Na^+ concentration, collectively preserve Na^+ balance and thereby support these critical physiological functions. Disruptions of sodium homeostasis due to dehydration, intravenous fluid administration, or diabetes insipidus can significantly impact the various physiological processes discussed in this article.

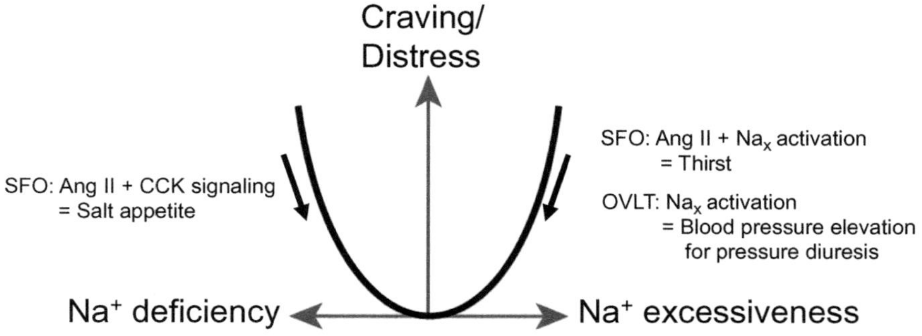

Fig. 21.4 Thirst and salt appetite controls for sodium homeostasis. The subfornical organ (SFO) contains multiple types of neurons expressing the angiotensin II receptor AT1a. Among these are water neurons that encode thirst and salt neurons that encode salt appetite. Angiotensin II (Ang II) is a hormone that helps maintain blood pressure by regulating body fluid volume and inducing vasoconstriction. When body fluid sodium levels are low, circulating Ang II increases and activates AT1a receptors. Concurrently, cholecystokinin (CCK)-producing neurons in the SFO are activated, which suppress water neuron activity via activation of inhibitory neurons, resulting in the exclusive expression of salt appetite. During dehydration, plasma sodium concentration rises due to decreased body fluid volume. Although circulating Ang II similarly increases as in sodium deficiency, activation of Na_x leads to inhibitions of salt neuron activity, leaving only thirst. Na_x is also involved in the induction of thirst in the SFO. Furthermore, activation of Na_x in the organum vasculosum lamina terminalis (OVLT) leads to sympathetic nerve activation and an increase in blood pressure

References

1. Gagnon K, Delpire E (2021) Sodium transporters in human health and disease. Front Physiol 11:588664
2. Friis UG, Madsen K, Stubbe J, Hansen PBL, Svenningsen P et al (2013) Regulation of renin secretion by renal juxtaglomerular cells. Pflugers Arch 465:25–37
3. Yang T, Xu C (2017) Physiology and pathophysiology of the intrarenal renin-angiotensin system: an update. J Am Soc Nephrol 28:1040–1049
4. Matsuda T, Hiyama TY, Niimura F, Matsusaka T, Fukamizu A et al (2017) Distinct neural mechanisms for the control of thirst and salt appetite in the subfornical organ. Nat Neurosci 20:230–241
5. Hiyama TY (2024) Brain sodium sensing for regulation of thirst, salt appetite, and blood pressure. Physiol Rep 12:e15970
6. Sarzani R, Allevi M, Di Pentima C, Schiavi P, Spannella F et al (2022) Role of cardiac natriuretic peptides in heart structure and function. Int J Mol Sci 23:14415
7. Wong PCY, Guo J, Zhang A (2017) The renal and cardiovascular effects of natriuretic peptides. Adv Physiol Educ 41:179–185
8. Stocker SD, Monahan KD, Browning KN (2013) Neurogenic and sympathoexcitatory actions of NaCl in hypertension. Curr Hypertens Rep 15:538–546
9. Kawano Y, Ferrario CM (1984) Neurohormonal characteristics of cardiovascular response due to intraventricular hypertonic NaCl. Am J Phys 247:H422–H428

10. Ribeiro N, Panizza Hdo N, Santos KM, Ferreira-Neto HC, Antunes VR (2015) Salt-induced sympathoexcitation involves vasopressin V1a receptor activation in the paraventricular nucleus of the hypothalamus. Am J Physiol Regul Integr Comp Physiol 309:R1369–R1379
11. Nomura K, Hiyama TY, Sakuta H, Matsuda T, Lin CH (2019) [Na^+] increases in body fluids sensed by central Nax induce sympathetically mediated blood pressure elevations via H+-dependent activation of ASIC1a neuron. Neuron 101:60–75.e6
12. Kortenoeven MLA, Pedersen NB, Rosenbaek LL, Fenton RA (2015) Vasopressin regulation of sodium transport in the distal nephron and collecting duct. Am J Physiol Renal Physiol 309:F280–F299
13. Gröber U (2015) Magnesium in prevention and therapy. Nutrients 7:8199–8226
14. Kolisek M, Nestler A, Vormann J, Schweigel-Röntgen M (2012) Human gene SLC41A1 encodes for the Na^+/Mg^{2+} exchanger. Am J Physiol Cell Physiol 302:C318–C326
15. Hurd TW, Otto EA, Mishima E, Gee HY, Inoue H et al (2013) Mutation of the Mg^{2+} transporter SLC41A1 results in a nephronophthisis-like phenotype. J Am Soc Nephrol 24:967–977
16. Barbagallo M, Belvedere M, Di Bella G, Dominguez LJ (2011) Altered ionized magnesium levels in mild-to-moderate Alzheimer's disease. Magnes Res 24:S115–S121
17. Nie ZL, Wang ZM, Zhou B, Tang ZP, Wang SK (2012) Magnesium intake and incidence of stroke: meta-analysis of cohort studies. Nutr Metab Cardiovasc Dis 23:169–176
18. Johnson S (2001) The multifaceted and widespread pathology of magnesium deficiency. Med Hypotheses 56:163–170
19. Sartori SB, Whittle N, Hetzenauer A, Singewald N (2012) Magnesium deficiency induces anxiety and HPA axis dysregulation: modulation by therapeutic drug treatment. Neuropharmacology 62:304–312
20. Mousain-Bosc M, Roche M, Rapin J, Bali JP (2004) Magnesium VitB6 intake reduces central nervous system hyperexcitability in children. J Am Coll Nutr 23:545S–548S
21. Kolisek M, Sponder G, Mastrototaro L, Smorodchenko A, Launay P et al (2013) Substitution p.A350V in Na^+/Mg^{2+} exchanger SLC41A1, potentially associated with Parkinson's disease, is a gain of function mutation. PLoS One 8:e71096
22. Kimura J, Watanabe Y, Li L, Watano T (2002) Pharmacology of Na^+/Ca^{2+} exchanger. Ann N Y Acad Sci 976:513–519
23. Carafoli E, Santella L, Branca D, Brini M (2001) Generation, control, and processing of cellular calcium signals. Crit Rev Biochem Mol Biol 36:107–260
24. Canitano A, Papa M, Boscia F, Castaldo P, Sellitti S et al (2002) Brain distribution of the Na^+/Ca^{2+} exchanger-encoding genes NCX1, NCX2, and NCX3 and their related proteins in the central nervous system. Ann N Y Acad Sci 976:394–404
25. Shenoda B (2015) The role of Na^+/Ca^{2+} exchanger subtypes in neuronal ischemic injury. Transl Stroke Res 6(3):181–190
26. Cuomo O, Gala R, Pignataro G, Boscia F, Secondo A et al (2008) A critical role for the potassium-dependent sodium-calcium exchanger NCKX2 in protection against focal ischemic brain damage. J Neurosci 28:2053–2063
27. Bers DM (2002) Cardiac excitation-contraction coupling. Nature 415:198–205
28. Pott C, Eckardt L, Goldhaber JI (2011) Triple threat: the Na^+/Ca^{2+} exchanger in the pathophysiology of cardiac arrhythmia, ischemia and heart failure. Curr Drug Targets 12:737–747
29. Imahashi K, Kusuoka H, Hashimoto K, Yoshioka J, Yamaguchi H et al (1999) Intracellular Na+ accumulation during ischemia as the substrate for reperfusion injury. Circ Res 84:1401–1406
30. Scranton K, John S, Angelini M, Steccanella F, Umar S et al (2024) Cardiac function is regulated by the sodium-dependent inhibition of the sodium-calcium exchanger NCX1. Nat Commun 15:3831
31. Aschner M, Erikson KM, Dorman DC (2005) Manganese dosimetry and neurotoxicity. Toxicol Appl Pharmacol 204:278–293
32. Liuzzi JP, Aydemir F, Nam H, Knutson MD, Cousins RJ (2006) ZIP14 mediates non-transferrin-bound iron uptake into cells. Proc Natl Acad Sci USA 103:13612–13617

33. Jenkitkasemwong S, Akinyode A, Paulus E, Weiskirchen R, Hojyo S et al (2018) SLC39A14 deficiency alters manganese homeostasis and excretion resulting in brain manganese accumulation and motor deficits in mice. Proc Natl Acad Sci USA 115:E1769–E1778
34. Winslow JWW, Limesand KH, Zhao N (2020) The functions of ZIP8, ZIP14, and ZnT10 in the regulation of systemic manganese homeostasis. Int J Mol Sci 21:3304
35. Zeglam A, Abugrara A, Kabuka M (2018) Autosomal-recessive iron deficiency anemia, dystonia and hypermanganesemia caused by new variant mutation of the manganese transporter gene slc39a14. Acta Neurol Belg 119:379–384
36. Tuschl K, Meyer E, Valdivia LE, Zhao N, Dadswell C et al (2016) Mutations in SLC39A14 disrupt manganese homeostasis and cause childhood-onset parkinsonism-dystonia. Nat Commun 7:11601
37. Vadlapudi AD, Vadlapatla RK, Mitra AK (2012) Sodium dependent multivitamin transporter (SMVT): a potential target for drug delivery. Curr Drug Targets 13:994–1003
38. Uchida Y, Ito K, Ohtsuki S, Kubo Y, Suzuki T et al (2015) Major involvement of Na^+-dependent multivitamin transporter (SLC5A6/SMVT) in uptake of biotin and pantothenic acid by human brain capillary endothelial cells. J Neurochem 134(1):97–112
39. Quick M, Shi L (2015) The sodium/multivitamin transporter: a multipotent system with therapeutic implications. Vitam Horm 98:63–100
40. Mock DM (2017) Biotin: from nutrition to therapeutics. J Nutr 147(8):1487–1492
41. Leonardi R, Jackowski S (2007) Biosynthesis of pantothenic acid and coenzyme A. EcoSal Plus 2. https://doi.org/10.1128/ecosalplus.3.6.3.4
42. Packer L, Witt EH, Tritschler HJ (1995) alpha-Lipoic acid as a biological antioxidant. Free Radic Biol Med 19:227–250
43. Said HM, Mohammed ZM (2006) Intestinal absorption of water-soluble vitamins: an update. Curr Opin Gastroenterol 22:140–146
44. Park S, Sinko PJ (2005) The blood-brain barrier sodium-dependent multivitamin transporter: a molecular functional in vitro-in situ correlation. Drug Metab Dispos 33:1547–1554
45. Leithner K, Hirschmugl B, Li Y, Tang B, Papp R et al (2016) TASK-1 regulates apoptosis and proliferation in a subset of non-small cell lung cancers. PLoS One 11:e0157453
46. Zheng C, Hu W, Wu D, Chen R, Xu C et al (2025) Foxd3/SLC5A6 axis regulates apoptosis in LUAD cells by controlling mitochondrial biotin uptake. Cell Signal 125:111473
47. Savini I, Rossi A, Pierro C, Avigliano L, Catani MV (2008) SVCT1 and SVCT2: key proteins for vitamin C uptake. Amino Acids 34:347–355
48. Bürzle M, Hediger MA (2012) Functional and physiological role of vitamin C transporters. Curr Top Membr 70:357–375
49. Takanaga H, Mackenzie B, Hediger MA (2004) Sodium-dependent ascorbic acid transporter family SLC23. Pflugers Arch 447:677–682
50. Qiu S, Li L, Weeber EJ, May JM (2007) Ascorbate transport by primary cultured neurons and its role in neuronal function and protection against excitotoxicity. Neurosci Res 85:1046–1056
51. Moretti M, Fraga DB, Rodrigues ALS (2017) Preventive and therapeutic potential of ascorbic acid in neurodegenerative diseases. CNS Neurosci Ther 23:921–929
52. Salazar K, Martínez F, Pérez-Martín M (2018) SVCT2 expression and function in reactive astrocytes is a common event in different brain pathologies. Mol Neurobiol 55:5439–5452
53. Muñoz-Montesino C, Peña E, Roa FJ et al (2021) Transport of vitamin C in cancer. Antioxid Redox Signal 35:61–74
54. Koepsell H (2017) The Na^+-D-glucose cotransporters SGLT1 and SGLT2 are targets for the treatment of diabetes and cancer. Pharmacol Ther 170:148–165
55. Gorboulev V, Schürmann A, Vallon V, Kipp H, Jaschke A et al (2012) Na^+-D-glucose cotransporter SGLT1 is pivotal for intestinal glucose absorption and glucose-dependent incretin secretion. Diabetes 61:187–196
56. Cannizzaro M, Jarošová J, De Paepe B (2019) Relevance of solute carrier family 5 transporter defects to inherited and acquired human disease. J Appl Genet 60:305–317

57. Maccari R, Ottanà R (2022) Sodium-glucose cotransporter inhibitors as antidiabetic drugs: current development and future perspectives. J Med Chem 65:10848–10881
58. Kristensen AS, Andersen J, Jørgensen TN et al (2011) SLC6 neurotransmitter transporters: structure, function, and regulation. Pharmacol Rev 63:585–640
59. Pochini L, Scalise M, Galluccio M, Indiveri C (2014) Membrane transporters for the special amino acid glutamine: structure/function relationships and relevance to human health. Front Chem 2:61
60. Bröer S (2008) Amino acid transport across mammalian intestinal and renal epithelia. Physiol Rev 88:249–286
61. Mizzi N, Blundell R (2025) Glycine receptors: structure, function, and therapeutic implications. Mol Asp Med 103:101360
62. Andersen JV (2025) The glutamate/GABA-glutamine cycle: insights, updates, and advances. J Neurochem 169:e70029
63. Wang T, Shi L, Luo Q, Wang Y, Zhao H (2025) SLC6 transporters as pharmacological targets in depression: molecular mechanisms and therapeutic strategies. Biochem Pharmacol 242:117210
64. Bröer S (2023) Intestinal amino acid transport and metabolic health. Annu Rev Nutr 43:73–99
65. Werner A, Patti M, Zinad HS, Fearn A, Laude A et al (2016) Molecular determinants of transport function in zebrafish Slc34a Na-phosphate transporters. Am J Physiol Regul Integr Comp Physiol 311:R1213–R1222
66. Wilson R, Mukherjee-Roy N, Gattineni J (2024) The role of fibroblast growth factor 23 in regulation of phosphate balance. Pediatr Nephrol 39:3439–3451
67. Hernando N, Gagnon K, Lederer E (2021) Phosphate transport in epithelial and nonepithelial tissue. Physiol Rev 101:1–35
68. Kritmetapak K, Kumar R (2021) Phosphate as a signaling molecule. Calcif Tissue Int 108:16–31
69. Perumal NL, Padidela R (2024) Phosphate homeostasis and disorders of phosphate metabolism. Curr Pediatr Rev 20:412–425

Index

N. S. Dhalla et al. (eds.), *Functional Biochemistry of Metallic Elements*, Advances in Biochemistry in Health and Disease 36,
https://doi.org/10.1007/978-3-032-14445-4

GPSR Compliance

The European Union's (EU) General Product Safety Regulation (GPSR) is a set of rules that requires consumer products to be safe and our obligations to ensure this.

If you have any concerns about our products, you can contact us on ProductSafety@springernature.com

In case Publisher is established outside the EU, the EU authorized representative is:

Springer Nature Customer Service Center GmbH
Europaplatz 3
69115 Heidelberg, Germany

Batch number: 10755222

Printed by Printforce, the Netherlands